中　外　物　理　学　精　品　书　系

本书出版得到“国家出版基金”资助

中 外 物 理 学 精 品 书 系

高 瞻 系 列 · 5

Applied Symbolic Dynamics and Chaos

实用符号动力学与混沌

郝柏林
郑伟谋 著

图书在版编目(CIP)数据

实用符号动力学与混沌=Applied symbolic dynamics and chaos: 英文/郝柏林，郑伟谋著. —北京: 北京大学出版社, 2014.10
(中外物理学精品书系)
ISBN 978-7-301-24756-3

Ⅰ. ①实…　Ⅱ. ①郝…　②郑…　Ⅲ. ①动力系统(数学)—理论—英文 ②混沌理论—英文　Ⅳ. ①O19　②O415.5

中国版本图书馆 CIP 数据核字（2014）第 205742 号

书　　　名: **Applied Symbolic Dynamics and Chaos(实用符号动力学与混沌)**
著作责任者: 郝柏林　郑伟谋　著
责 任 编 辑: 尹照原　曾琬婷
标 准 书 号: ISBN 978-7-301-24756-3/O・1003
出 版 发 行: 北京大学出版社
地　　　址: 北京市海淀区成府路 205 号　100871
网　　　址: http://www.pup.cn
新 浪 微 博: @北京大学出版社
电 子 信 箱: zpup@pup.pku.edu.cn
电　　　话: 邮购部 62752015　发行部 62750672　编辑部 62752021
　　　　　　出版部 62754962
印　刷　者: 北京中科印刷有限公司
经　销　者: 新华书店
　　　　　　730 毫米×980 毫米　16 开本　34 印张　653 千字
　　　　　　2014 年 10 月第 1 版　2014 年 10 月第 1 次印刷
定　　　价: 95.00 元

序　言

物理学是研究物质、能量以及它们之间相互作用的科学。她不仅是化学、生命、材料、信息、能源和环境等相关学科的基础，同时还是许多新兴学科和交叉学科的前沿。在科技发展日新月异和国际竞争日趋激烈的今天，物理学不仅囿于基础科学和技术应用研究的范畴，而且在社会发展与人类进步的历史进程中发挥着越来越关键的作用。

我们欣喜地看到，改革开放三十多年来，随着中国政治、经济、教育、文化等领域各项事业的持续稳定发展，我国物理学取得了跨越式的进步，做出了很多为世界瞩目的研究成果。今日的中国物理正在经历一个历史上少有的黄金时代。

在我国物理学科快速发展的背景下，近年来物理学相关书籍也呈现百花齐放的良好态势，在知识传承、学术交流、人才培养等方面发挥着无可替代的作用。从另一方面看，尽管国内各出版社相继推出了一些质量很高的物理教材和图书，但系统总结物理学各门类知识和发展，深入浅出地介绍其与现代科学技术之间的渊源，并针对不同层次的读者提供有价值的教材和研究参考，仍是我国科学传播与出版界面临的一个极富挑战性的课题。

为有力推动我国物理学研究、加快相关学科的建设与发展，特别是展现近年来中国物理学者的研究水平和成果，北京大学出版社在国家出版基金的支持下推出了“中外物理学精品书系”，试图对以上难题进行大胆的尝试和探索。该书系编委会集结了数十位来自内地和香港顶尖高校及科研院所的知名专家学者。他们都是目前该领域十分活跃的专家，确保了整套丛书的权威性和前瞻性。

这套书系内容丰富，涵盖面广，可读性强，其中既有对我国传统物理学发展的梳理和总结，也有对正在蓬勃发展的物理学前沿的全面展示；既引进和介绍了世界物理学研究的发展动态，也面向国际主流领域传播中国物理的优秀专著。可以说，“中外物理学精品书系”力图完整呈现近现代世界和中国物理科学发展的全貌，是一部目前国内为数不多的兼具学术价值和阅读乐趣的经典物理丛书。

“中外物理学精品书系”另一个突出特点是，在把西方物理的精华要义“请进来”的同时，也将我国近现代物理的优秀成果“送出去”。物理学科在世界范围内的重要性不言而喻，引进和翻译世界物理的经典著作和前沿动态，可以满足当前国内物理教学和科研工作的迫切需求。另一方面，改革开放几十年来，我国的物理学研究取得了长足发展，一大批具有较高学术价值的著作相继问世。这套丛书首次将一些中国物理学者的优秀论著以英文版的形式直接推向国际相关研究的主流领域，使世界对中国物理学的过去和现状有更多的深入了解，不仅充分展示出中国物理学研究和积累的“硬实力”，也向世界主动传播我国科技文化领域不断创新的“软实力”，对全面提升中国科学、教育和文化领域的国际形象起到重要的促进作用。

值得一提的是，“中外物理学精品书系”还对中国近现代物理学科的经典著作进行了全面收录。20 世纪以来，中国物理界诞生了很多经典作品，但当时大都分散出版，如今很多代表性的作品已经淹没在浩瀚的图书海洋中，读者们对这些论著也都是“只闻其声，未见其真”。该书系的编者们在这方面下了很大工夫，对中国物理学科不同时期、不同分支的经典著作进行了系统的整理和收录。这项工作具有非常重要的学术意义和社会价值，不仅可以很好地保护和传承我国物理学的经典文献，充分发挥其应有的传世育人的作用，更能使广大物理学人和青年学子切身体会我国物理学研究的发展脉络和优良传统，真正领悟到老一辈科学家严谨求实、追求卓越、博大精深的治学之美。

温家宝总理在 2006 年中国科学技术大会上指出，“加强基础研究是提升国家创新能力、积累智力资本的重要途径，是我国跻身世界科技强国的必要条件”。中国的发展在于创新，而基础研究正是一切创新的根本和源泉。我相信，这套“中外物理学精品书系”的出版，不仅可以使所有热爱和研究物理学的人们从中获取思维的启迪、智力的挑战和阅读的乐趣，也将进一步推动其他相关基础科学更好更快地发展，为我国今后的科技创新和社会进步做出应有的贡献。

“中外物理学精品书系”编委会　主任

中国科学院院士，北京大学教授

王恩哥

2010 年 5 月于燕园

Preface for the Second Edition

The authors are very happy to see the second revised edition of this monograph appearing in a joint effort of the Peking University Press and the World Scientific Publishing Co., Inc., Singapore. The printing of this book in Beijing greatly increases the availability of the book to readers within China.

The hard work of revising the text and figures was mainly done by Dr. Wei-mou Zheng. The revisions concern mainly the application of symbolic dynamics to ordinary differential equations via constructing two-dimensional symbolic dynamics of the corresponding Poincare maps for the ODEs. I would like to emphasize once more that the way of getting into two-dimensional maps and ODEs was paved by Dr. Zheng almost single-handed since the early 1990s. This approach significantly extends the qualitative study of ODEs by numerical means under the guidance of topology, as symbolic dynamics is topological in nature. However, many difficult problems remain unsolved regarding the relation of symbolic dynamics to knot theory and formal language theory. These problems are only touched briefly in the last chapters with the hope to inspire further studies. Criticism and feedback from the readers are mostly welcome.

Special thanks go to Ms. Xiao-hong Chen and Mr. Zhao-yuan Yin from the Peking University Press who have been very patient and helpful during the yearly long process of preparing the second edition.

Bai-lin Hao
1 August 2014, Beijing

Preface

Symbolic dynamics is a coarse-grained description of dynamics. It has been a long-studied chapter of the mathematical theory of dynamical systems, but its abstract formulation has kept away many practitioners of physical sciences and engineering from appreciating its simplicity, beauty, and power. At the same time, symbolic dynamics provides almost the only rigorous way to understand global systematics of periodic and, especially, chaotic motion in dynamical systems. In a sense, everyone who enters the field of chaotic dynamics should begin with the study of symbolic dynamics. However, this has not been an easy job for non-mathematicians to accomplish. On one hand, the method of symbolic dynamics has been developed to such an extend that it may well become a practical tool in studying chaotic dynamics, both on computers and in laboratories. On the other hand, most of the existing literature on symbolic dynamics is mathematics-oriented. This book is an attempt at partially filling up this apparent gap by emphasizing the applied aspects of symbolic dynamics without pretending to mathematical rigor.

No previous knowledge of dynamical systems theory is required in order to read this book. The mathematics used does not exceed basic calculus taught at engineering schools. Starting from simple one-dimensional maps, we go through circle maps and two-dimensional maps to arrive at numerical study of some ordinary differential equations under the guidance of symbolic dynamics. Instead of numbered formal definitions and proofs, the reader will find many examples and figures which embody the idea and method of symbolic dynamics. We have also included two kinds of computer programs in the book. A few short BASIC programs, implementing one or another procedure just described in the text, may help the reader to understand the method thoroughly. These programs may be considered part of the text or be skipped at first reading. Some more sophisticated programs written in C language are listed in the Appendix. These may be easily modified to treat systems not studied in the book and are aimed at the research need of some readers.

The book is organized as follows.

Chapter 1 is a brief introduction to the general idea of symbolic dynamics.

Chapter 2 studies symbolic dynamics of unimodal, one-hump map of the interval, the simplest yet very rich dynamical system. Recent development of the applied direction of symbolic dynamics, what we call *Applied Symbolic Dynamics*[1], has drawn much inspiration from the unimodal map.

Chapter 3 studies one-dimensional maps of the interval with multiple critical points or points of discontinuity. These maps occur naturally in many applications and they are needed in understanding ordinary differential equations with dissipation, i.e., those equations that allow the existence of strange attractors.

Chapter 4 is devoted to symbolic dynamics of circle maps as the simplest model of physical systems with competing frequencies. The symbolic dynamics of circle maps possesses some specific features absent in interval maps. The knowledge is of much help in the study of periodically forced systems described by ordinary differential equations.

Chapter 5 develops symbolic dynamics of two-dimensional maps. A mostly analytical study will be carried out on some piecewise linear 2D maps, providing hints and clues to deal with more general maps. Being a major progress in the last decade, the symbolic dynamics of 2D maps are essential for the symbolic dynamics study of differential equations as the usual Poincaré maps are two-dimensional.

Chapter 6 focuses on numerical study of ordinary differential equations under the guidance of symbolic dynamics. This mostly "experimental", at present time, approach is capable to provide some global understanding of the system that cannot be reached neither by purely analytical means nor by numerical methods alone. For example, we are able to list and locate all short periodic orbits, stable as well as unstable ones, in fairly large region of the parameter space for the Lorenz model and for the periodically forced Brusselator.

Chapter 7 provides the complete solution of a counting problem which is related to but goes beyond symbolic dynamics, namely, the number of periods

[1] The name was suggested by Ian Percival during his visit to Beijing in 1988.

in continuous maps of the interval with multiple critical points. It also contains partial result for maps with discontinuity.

Symbolic sequences fit naturally into the framework of formal languages. The well-established machinery of formal language theory is of great help in the classification of orbital types and their complexity in 1D maps. Many new results have been obtained in the last few years. Since there has just appeared a nice book on this topic (Xie [B1996]), we confine ourselves to a brief summary of this line of study in Chapter 8.

Chapter 9 discusses the relation of symbolic dynamics with another topological approach to dynamical systems, namely, the study of periodic orbits as knots and links. Although knots and links are objects in three-dimensional space, many problems may be posed using 1D maps. Instead of presenting finished results, this Chapter, we hope, may inspire some new research interest.

There is a quite detailed *Table of Contents* and an *Index*, which may help the reader to see the scope of the book and to look for interested topics.

This book is not a mathematical treatise. However, we hope mathematicians may also find a few new tricks or some interesting applications of their abstract theory, including some contributions of Chinese scientists that are not readily available elsewhere.

Some years ago one of the authors published a book entitled *Elementary Symbolic Dynamics and Chaos in Dissipative Systems* (Hao [B1989]). So many new results and a much deeper understanding have been achieved since then that the present monograph can hardly be considered as an update of the 1989 book. We refer to that book only in a few occasions when something not directly related to symbolic dynamics is touched.

A few words about the reference convention in this book. References to the list at the end of the book are given as, e.g., Poincaré [B1899], the capital B indicating the first part of the References on "Books', or Xie [1996], addressing a paper in the second part "Papers" of the References. A few citations to sources not included in the References are given in footnotes. No efforts have been made to clarify the chronology of one or another statement. In a rapidly expanding and interdisciplinary field like *Chaos* there have been many rediscoveries of important facts. It is better to leave these to the historians of science.

Our own work on chaos has been partially supported by the Division of Mathematics and Physics, Academia Sinica (1983–1985), the Chinese Natural Science Foundation (1986–1988), and the Project on Nonlinear Science (1990–1995). In 1989 Sun Microsystems, Inc., donated a Sun Workstation to the Nonlinear Dynamics Group at the Institute of Theoretical Physics, Academia Sinica, Beijing. Wolfram Research, Inc., donated the Sun version of Mathematica 1.0 software. These were great help to our earlier research. Later on the Local Network of ITP and the State Key Laboratory on Scientific and Engineering Computation of Academia Sinica provided computing facilities. We thank all these organizations for their support.

We would also like to acknowledge the inspiring discussion and interaction with many colleagues and former students over the years. An incomplete list includes Shi-gang Chen, Ming-zhou Ding, Hai-ping Fang, Jun-xian Liu, Li-sha Lu, Shou-li Peng, Zuo-bin Wu, Fa-gen Xie, Hui-ming Xie, and Wan-zhen Zeng.

The text was typeset by the authors using LaTeX of Leslie Lamport with indispensable help from the staff of World Scientific Publishing Co. Pte. Ltd. In particular, we would like to thank Dr. K. K. Phua, the Editor-in-Chief, and Dr. Lock-Yee Wong, the editor, for their patience and advice.

Bai-lin Hao and Wei-mou Zheng

Contents

1
Introduction

Human's observation and measurement of natural phenomena are always carried out with finite precision. This precision improves with the advance of science and technology, but it can never reach a state of absolute exactness. On the other hand, the ultimate goal of our observation and measurement is to draw rigorous conclusions on the law of nature and on the basic properties of the process under study. Can we achieve the goal in spite of the restriction of finite precision data?

Precise measurements usually bring about a huge amount of data. However, the characteristics of a natural phenomenon usually consist of a small set of quantities. Is it always necessary to proceed from the huge amount of data in order to extract a few characteristics?

In the development of physical sciences it is inevitable to work at various levels of "coarse-grained" or "reduced" description. The balance between finite precision and rigorous conclusion, between the huge amount of data and the small set of characteristics, is reached by way of coarse-graining. So far this program has been best realized in the study of dynamical systems. For a practitioner in physical sciences or in engineering, symbolic dynamics is nothing but a coarse-grained description of dynamics. Symbolic dynamics provides a rigorous description of dynamics based on finite precision observation.

This book is devoted to *Applied Symbolic Dynamics* as a practical tool in the study of nonlinear physical systems. Of course, it is based on the mathematical theory of dynamical systems, including the *Abstract Symbolic Dynamics*. However, we will highlight the connection only very briefly in this introductory chapter. Accordingly, this chapter is organized as follows. Section 1.1 introduces the notion of dynamical systems. The notion of phase space and parameter space will be explained briefly. Several examples, whose symbolic dynamics will be studied in this book, are listed. Section 1.1.3 discusses the basic idea of symbolic dynamics as a coarse-grained description of the dynamics. In fact, symbolic dynamics is the simplest dynamics one can ever imagine. The usefulness of symbolic dynamics roots in the comparison of more realistic dynamics with the simplest dynamics by way of coarse-graining. Section 1.3 explains the relation between the applied symbolic dynamics and the abstract symbolic dynamics. Section 1.4 indicates a few important works that have paved the road to applied symbolic dynamics.

1.1 Dynamical Systems

Generally speaking, any physical system, whose state evolves with time is a *dynamical system*. To be more specific, one describes the physical state by a number of time-dependent dynamical variables. These variables span a *phase space*.

1.1.1 Phase Space and Orbits

The phase space may be a space of finite or infinite dimensions, depending on the number of variables being finite or infinite. We note that the term "phase space" is used here mainly in a broad sense, in contrast to that in mechanics, where a phase space is spanned by pairs of "conjugate variables", e.g., coordinate and momentum, hence always of even dimensions.

There is a more essential distinction of such a phase space from that in mechanics. In mechanics of conservative systems a phase space is spanned by generalized coordinates and momenta. The Liouville Theorem guarantees the conservation of phase volume in the process of time evolution. However, dissipation is inevitable in most physical and engineering systems of practical significance. In the presence of dissipation an initial phase volume shrinks with time. Eventually the long-time dynamical evolution may take place on some "attractor" whose dimension may be smaller than the dimension of the original phase volume. Dissipation reduces the description in a natural way and enables many low-dimensional systems to simulate the long-time dynamics of higher-dimensional systems. This is the ultimate reason why symbolic dynamics of one- and two-dimensional maps happens to be quite useful in the study of higher-dimensional systems. Most models studied in this book are dissipative systems. In most cases we will use the term "phase space" in a loose sense, and do not care about its dimension being even or being related to coordinate and momentum.

A point in the phase space represents a particular state of the system. One can visualize a given "dynamics" by watching the motion of a representative point in the phase space. The locus of the representative point in the phase space, called a trajectory or an orbit.

1.1.2 Parameters and Bifurcation of Dynamical Behavior

The environment, in which the system evolves, may be characterized by a number of *parameters*. Generally speaking, there is no clearcut distinction between dynamical variables and parameters. Those variables which are under our control and may be kept constant during the period of observation are considered as parameters.

In general, we are interested not only in the dynamical behavior of a system at a fixed parameter set but also concerned with how the dynamics changes qualitatively when the parameters vary. For example, in some parameter range the system may exhibit periodic motion, while in another region of the parameter space chaotic behavior shows off. The qualitative change of dynamical behavior usually takes place abruptly at some well-defined parameter value. This is called a *bifurcation* in mathematics and may be analyzed thoroughly using, for example, the Implicit Function Theorem. However, this is beyond the scope of this book and the reader may consult a text on nonlinear dynamics, e.g., the book by Thompson and Stewart [B1986].

1.1.3 Examples of Dynamical Systems

We give a few examples of dynamical systems. Take the Sun, Earth and Moon to be mass points. The system of equations of motion, based on Newtonian mechanics, is a dynamical system. It is a conservative system, and not treated in this book. However, we will study a much-simplified conservative system, the stadium billiard problem by using the method of symbolic dynamics.

The Quadratic Map

Under some over-simplified assumptions the equation governing the population change of insects without generation overlap may be reduced to the following difference equation:

$$x_{n+1} = \mu x_n(1 - x_n), \tag{1.1}$$

where x_n takes value from a real interval $I = [0, 1]$ and describes the normalized population; $\mu \in (2, 4]$ is a control parameter. This is a quadratic map, known also as the logistic map. The mapping function has a parabolic shape with one maximum and two monotone branches, hence belongs to the unimodal map. The maximum point of the map is a turning point between monotone increasing and decreasing behaviors. It is a *critical point* of the map. Being a one-dimensional dynamical system, the subscript n represents discrete time. In fact, the unimodal map is a starting point for the entire applied symbolic dynamics. We will study the symbolic dynamics of unimodal maps in great detail in Chapter 2, using the quadratic map for demonstration.

The Circle Map

Another important class of one-dimensional dynamical systems, used widely in modeling systems with two competing frequencies, is a map of the circumference of a circle to itself, called a circle map. The circle may be obtained by identifying the two end points of an interval. The length of the circle is usually take to be 1. A general circle map is given by

$$x_{n+1} = x_n + A + Bg(x_n) \pmod{1}. \tag{1.2}$$

Here A and B are parameters. The meaning of taking modulus (mod 1) is to keep the fraction part of the result, throwing away its integer part. In (1.2) $g(x)$ is a function of period 1:

$$g(x + 1) = g(x).$$

The symbolic dynamics of circle maps, whose phase space is a closed circle, has some distinctive new features. Chapter 4 will be devoted to its study.

The Lorenz Model

The thermal convection above a flat earth surface is described by a set of partial differential equations. After much simplification it may be reduced to a set of three ordinary differential equations (Saltzman [1962], Lorenz [1963]):

$$\begin{aligned} \dot{x} &= \sigma(y - x), \\ \dot{y} &= rx - xz - y, \\ \dot{z} &= xy - bz. \end{aligned} \tag{1.3}$$

It is known as the Lorenz model. The three-dimensional phase space is spanned by the three coordinates x, y, and z. There are three parameters σ, b, and r. Usually two of the three parameters are kept constant, e.g., $\sigma = 10$, $b = 8/3$. Many periodic and chaotic orbits are observed when r is varied over a wide range, say, from 1 to 350. The aperiodic orbit observed by Lorenz at $r = 28$ was one of the earliest examples of strange attractors.

The Lorenz model (1.3) has a discrete symmetry: it remains unchanged when the signs of x and y are reversed while z is kept unchanged. In other words, it is invariant under the transformation $x \to -x$, $y \to -y$, and $z \to z$. This anti-symmetry makes the Lorenz model close to the following one-dimensional anti-symmetric cubic map

$$x_{n+1} = Ax_n^3 + (1-A)x_n, \quad x_n \in (-1,1), \quad A \in (0,4], \tag{1.4}$$

and a few other 1D maps with the same symmetry.

The anti-symmetric and more general cubic maps provide a bridge to maps with many monotone branches. These important maps call for extension of the symbolic dynamics of unimodal case to maps with multiple critical points and discontinuities. This will be studied in Chapter 3.

Periodically Forced Systems

Many ordinary differential equations with two variables are nonlinear oscillators. The most complex behavior in such planar systems is periodic motion. Chaotic motion cannot appear. However, if a planar system is driven by periodic external force, chaotic behavior may come into play. There are many systems incorporating the interplay between the internal and external frequencies. For example, the periodically forced Brusselator

$$\begin{aligned} \dot{x} &= A - (B+1)x + x^2y + \alpha\cos(\omega t), \\ \dot{y} &= Bx - x^2y. \end{aligned} \tag{1.5}$$

Another example is the forced Duffing equation, which describes the nonlinear oscillation of a magnetic beam:

$$\begin{aligned} \dot{x} &= y, \\ \dot{y} &= -x^3 - \delta y + \alpha\cos(\omega t). \end{aligned} \tag{1.6}$$

In the parameter space of these periodically driven nonlinear oscillators there are regions of harmonic, subharmonic, quasi-periodic, and chaotic oscillations.

A harmonic oscillates at integer multiples of the fundamental frequency; a subharmonic oscillates at fractions of the fundamental frequency or, equivalently, its period is a multiple of the fundamental period. In Chapter 6 we will see that the symbolic dynamics of one-dimensional interval maps may be compared with these systems in some regions of the parameter space, while in other regions the symbolic dynamics of circle maps is more appropriate.

Differential Equations with Time-Delay

Differential equations with time delay appear in many applications, e.g., in models of optical bistable devices or in physiological models. Formally, a time-delayed equation may contain only one variable, e.g.,

$$\varepsilon \frac{dx(t)}{dt} + x(t) = \mu f\left(x(t-T)\right), \tag{1.7}$$

where $f(x(t))$ is a nonlinear function of $x(t)$, μ is a parameter. However, the simple fact that the time in $f(\mu, x(t))$ is taken at $t - T$ with a constant delay T has made it a system of infinite order. Time-delayed systems are notorious in numerical practice for their instabilities. In fact, they have very rich bifurcation and chaos "spectra", see, e.g., Li and Hao [1989], Losson, Mackey and Longtin [1993]. When $\varepsilon \ll 1$, away from rapid transition regions, we have

$$x \approx \mu f\left(x(t-T)\right).$$

This is simply a one-dimensional map. One map obtained in this way, the sine-square map, will be studied in Chapter 3.

Partial differential equations are manifestly infinite-dimensional systems. Unless one truncates them to finite-order ordinary differential equations like the Lorenz model, they are beyond the reach of applied symbolic dynamics for the time being. Therefore, we will not touch on any example of such systems.

1.2 Symbolic Dynamics as Coarse-Grained Description of Dynamics

We have emphasized repeatedly that symbolic dynamics is nothing but a coarse-grained description of dynamics. There is deep reason to introduce coarse-graining. We can never grasp a natural process in all its details and connections. One has to focus on a certain level of observation, smearing out finer pictures. Any measuring instrument has a limited resolution power. However, the limited resolution does not mean that we cannot reach any rigorous conclusion. For instance, if the answer to some question may be reduced to a "yes" or "no" type phrase, then even crude measurement may provide precise answer. The success depends on the nature of the problem and on how we approach the data. What just said is the essence of symbolic dynamics as a tool for qualitative analysis.

In fact, we are doing symbolic description in our everyday laboratory and industry measurements. Take, for example, an analog-digital converter (ADC) used for input to any digital apparatus. Readings on a 10-bit ADC yield numbers from 0 to 1023. Any minor change smaller than 1/1024 of the maximal measuring range cannot be reported. These readings are in essence 1024 symbols. Yet with this limited number of symbols we wish to describe our observation as precise as we can. So far, coarse-graining has been best realized in the study of dynamics. We hope experience and inspiration obtained by applying symbolic dynamics to nonlinear systems may contribute to the improvement of the general methodology of science and technology.

1.2.1 Fine-Grained and Coarse-Grained Descriptions

Speaking about dynamics, among the enormous variety of description levels there are two extremes. One can perform a fine-grained description by recording the coordinates of the representative point at each and every time instant. This is what people do in carrying out molecular dynamics. In principle, any characteristics of the system may be recovered from the huge amount of data. Usually, there is also a big amount of unused information.

The other extreme is to carry out a very crude description, say, by dividing the whole phase space into a small number of subregions. Assign a unique symbol to each subregion and record a symbol at each observation. The alternation of symbols provides a symbolic orbit. In so doing one loses a great amount of information, but some essential property may be kept. The first such property is periodicity. Clearly, periodic motion shows off not only in the fine-grained description, but also in a rather crude description. Other properties that may be kept include chaoticity of the dynamics, as we will see in the sequel. If the partition of phase space is carried out in accordance with the geometric and physical specification of the system, e.g., taking into account the intrinsic symmetry of the system, we can achieve even more with a crude partition. In a sense, most of the partitions to be used in this book are the crudest possible ones with a minimal number of symbols.

1.2.2 Symbolic Dynamics as the Simplest Dynamics

Symbolic dynamics is perhaps the simplest dynamics one can ever imagine. In order to define it as a dynamical system, we have to indicate the phase space and the dynamics.

First, the **phase space**. Take, for example, an *alphabet* of two letters, say, R and L, and form all possible sequences or "words" made of these symbols. For simplicity we only consider semi-infinite sequences for the time being. We will encounter bi-infinite symbolic sequences in Chapters 5 and 6 when studying two- and higher-dimensional dynamical systems. Take each symbolic sequence to be a *point* and form a space of all possible points, i.e., a space of all possible symbolic sequences. One point in this space, e.g., may be the following sequence

$$x_1 = RLRRLLLRLRLRRL\cdots .$$

We have given this point a name or label x_1 for convenience. Another point x_2 may be

$$x_2 = LRRLLLRLRLRRL\cdots .$$

One can even define distance between points and devise other structures in the space of symbolic sequences. We will see some of them later in this book, e.g., "ordering" and "metric representation" of symbolic sequences (see, for example, Section 2.3.2 and Section 2.8.6).

Second, the **dynamics**. In fact, the two points, given above as examples, are related by a very simple "dynamics", namely, if one discards the first symbol in the first sequence or, equivalently, shift the origin by one symbol to the right, one gets the second sequence. This is called a *shift* of the sequence. We define a *shift operator* $\mathcal{S}$:

$$\mathcal{S}s_0s_1s_2\cdots s_ks_{k+1}\cdots s_ns_{n+1}\cdots = s_1s_2\cdots s_ks_{k+1}\cdots s_ns_{n+1}\cdots \tag{1.8}$$

or, applied to a symbol in the sequence:

$$\mathcal{S}s_k = s_{k+1}, \quad k = 0, 1, 2, \cdots , \tag{1.9}$$

where s_k is either R or L which has appeared as the k-th symbol in the first sequence. The shift operator may be used repeatedly to get, for instance,

$$\mathcal{S}^k s_0 s_1 s_2 \cdots s_k s_{k+1} \cdots s_n s_{n+1} \cdots = s_k s_{k+1} \cdots s_n s_{n+1} \cdots ,$$

where $\mathcal{S}^k$ means to apply $\mathcal{S}$ for k times. In terms of the shift operator $\mathcal{S}$ the relation between the two points is

$$x_2 = \mathcal{S} x_1 .$$

Using this simple phase space and dynamics we can demonstrate many basic notions in dynamical system theory.

Taking an initial point x_0 and applying the shift operator repeatedly, we get a series of points. All sequences, connected by shifts, make a *trajectory* or an *orbit.* A sequence like

$$LLLLLLLLLLLLLL\cdots$$

does not change under any number of shifts and thus represents a *fixed point* under the shift dynamics. The sequence

$$RLRRLRRLRRLRRLR\cdots$$

returns to itself after every three shifts, so it is a *periodic orbit* of period 3 or a 3-cycle. The above two sequences have clear regularity and can be written concisely as L^∞ and $(RLR)^\infty$. It can be seen that in a symbolic dynamics based on an alphabet of two symbols, there are only two possible fixed points R^∞ and L^∞, one period 2 orbit $(RL)^\infty$, two period 3 orbits $(RLR)^\infty$ and $(RLL)^\infty$, etc. The study and extension of this combinatorial problem will be continued in Chapter 7.

One can define an *eventually periodic sequence* as an aperiodic sequence formed by appending a periodic sequence to a finite string of symbols. One example is $RLL(RL)^\infty$. Those sequences, for which there is no regularity to simplify their writing, or, equivalently, the only way to present them is to copy them letter by letter, are *random* sequences. In fact, this may lead to a general definition of chaotic orbits. We will say more in Chapter 8 on grammatical complexity of symbolic sequences.

That is all for the simplest symbolic dynamics. However, one cannot get very far in this manner without establishing a correspondence between more realistic dynamics in the phase space of physical systems and the simple symbolic dynamics. In making this connection, there are, so to speak, two opposite directions of pursuing. One may either widen the extension of the concept by weakening the conditions imposed on the systems under study or enrich its intention by confining oneself to specific classes of dynamical systems. The former goes in line of *abstract* symbolic dynamics of mathematicians; the latter develops into *applied* symbolic dynamics that is closer to the approach of practitioners in physical sciences and engineering.

1.3 Abstract versus Applied Symbolic Dynamics

Both abstract and applied symbolic dynamics may be abbreviated as ASD. This is a book on ASD the *Applied*. However, in order to give the reader a certain flavor, we imitate a few lines from ASD the *Abstract*.

1.3.1 Abstract Symbolic Dynamics

Suppose our phase space X is a compact space and there is a deterministic dynamics f defined on X which maps X into itself:

$$f : X \to X.$$

Since X is a compact space, there exists a finite open covering of X. We label each covering by a letter from a finite alphabet. Using all possible symbolic sequences made of these letters as points, we form a space Σ of symbolic sequences. The shift operator $\mathcal{S}$ defines a map of Σ to itself:

$$\mathcal{S} : \Sigma \to \Sigma,$$

called a *shift automorphism* in mathematics.

Ignoring the precise location of a point $x \in X$ and watching only under what covering it falls, we replace the point x by the label of the covering. Thus we have performed the coarse-graining and we establish a correspondence

$$\Psi : x \to \text{sequence in } \Sigma.$$

This is a quite general correspondence. For example, we have not required that different coverings do not overlap. In general, the correspondence between trajectories in X and trajectories in Σ may be many-to-one. This correspondence must be quite loose, as the nature of the two spaces may be very different. For instance, usually X is a continuous, differentiable manifold, but Σ consists of discrete points. Of course, one can introduce metric and topology in Σ to create richer mathematical structures. One can try to construct the covering in such a way as to realize a "Markovian partition" of X. This would establish a correspondence of the deterministic dynamics f on X with a topological Markovian chain on Σ. This discussion will be resumed in Section 2.8.5.

Symbolic dynamics developed along the above abstract line applies both to dynamics evolving in a continuous space and to discrete dynamical systems. On one hand, symbolic dynamics based on coarse-graining is capable to cope with situations when differential calculus based on traditional infinitesimal analysis can hardly work. On the other hand, for problems unreachable by numerical means symbolic dynamics can often yield qualitative results. Therefore, symbolic dynamics has naturally become a powerful tool for theorem-proving in the theory of dynamical systems. We stop here with abstract symbolic dynamics. To readers interested in this direction we recommend the review by Alekseev and Yakobson [1981].

1.3.2 Applied Symbolic Dynamics

Practitioners in physical sciences and engineering pay more attention to the intention of concepts. In carrying out coarse-graining one must take into account the geometric and physical peculiarity of the system under study. A theory established in this way applies only to a selected class of dynamical systems, but it has much richer content.

In this book we are primarily interested in physical systems with dissipation. The presence of dissipation reduces significantly the effective number of degrees of freedom. Therefore, low-dimensional symbolic dynamics may capture the essence of many higher-dimensional dynamical systems. In particular, one-dimensional systems will have a big share in this book. Due to the correspondence of points on a line with the natural order of real numbers one can obtain a great amount of useful results in symbolic dynamics of one-dimensional maps. We will go from simple systems to more complicated ones, from one-hump maps to maps with multiple maxima and minima, from continuous maps to maps with discontinuities, from one-parameter systems to those with many parameters.

Mappings from a circle to circle is one-dimensional in nature. However, the identification of the two end points of an interval brings about new topological property into the dynamics. Owing to their importance in understanding higher-dimensional systems with frequency competition circle maps will be studied in a separate chapter.

The symbolic dynamics of circle maps and maps of interval enjoys some unifying concepts and methods. They share the following common features:

1. The partition of the phase space is based on monotone branch of the mapping function. Each monotone branch corresponds to a symbol. Although the *turning points* between monotone branches, including maxima, minima, and discontinuities, are also labeled by symbols for convenience, they are only limiting cases of the symbols that represent subintervals. This will become more clear when we show how to replace these symbols by combinations of the others in Section 2.8.6.

2. An *ordering* is established for all symbolic sequences. This ordering has a natural correspondence to the ordering of points in the 1D phase space. Symbolic sequences which start from the first iterate of turning points have received a special name—kneading sequences. In fact, kneading sequences provide the most natural parametrization of a map. The ordering of kneading sequences orders the maps in the parameter space.

3. Obviously, an arbitrarily given symbolic sequence may not correspond to any actual orbit in the dynamics. We must formulate the conditions for a symbolic sequence to be allowed by the dynamics. These *admissibility conditions* are based on the ordering of sequences with respect to kneading sequences. Kneading sequences obey the same admissibility conditions.

4. Symbolic dynamics must provide a way to generate admissible words, including all admissible words of a given length and median words between two given sequences.

5. Rules to generate kneading sequences from known ones by replacing letters with strings are called composition rules. We will introduce the $*$-*composition* and *generalized composition* rules for unimodal and cubic maps as well as the *Farey transformation* for circle maps.

6. Symbolic dynamics should be capable to characterize chaotic orbits. In particular, it provides method to calculate topological entropy and to com-

pare complexity of orbits. Symbolic dynamics also outlines the borderlines of topological chaos, i.e., regions of chaotic motion with a positive topological entropy, in the parameter space.

7. For a given map it is desirable to be able to determine the parameter for a particular type of orbits to occur. To this end we will introduce the *word-lifting technique* and the *bisection method.*

8. A dynamical system with a certain symmetry may undergo *symmetry breakings* and *restorations* when the parameter is varied. Symbolic dynamics helps to analyze these phenomena.

9. Symbolic dynamics may help to answer some global questions on the system, e.g., the number of different periodic orbits of a given length. This problem has been solved completely for continuous maps with multiple critical points and will be elucidated in Chapter 7.

Symbolic dynamics approach has been applied to experimental data, mainly for identification of periodic regimes (see, e.g., Simoyi, Wolf and Swinney [1982]; Coffman, McCormick and Swinney [1986]). We will not touch this theme in the book.

Applied symbolic dynamics complies with the basic principles of abstract symbolic dynamics and develops within the framework of the latter. The great amount of concrete results obtained in applying symbolic dynamics to various classes of physical systems has enriched the general theory. Having benefited from applied symbolic dynamics, a practitioner in nonlinear science may wish to learn more about the mathematical theory. We give a brief summary of the literature in the next Section.

1.4 Literature on Symbolic Dynamics

According to a historical note in Procaccia, Thomae and Tresser [1987] the use of symbolic sequences dated back to the middle of 19th century. The idea of symbolic dynamics originated from the work of J. Hadamard [1898] and M. Morse [1921]. Starting from the 1920s, it has become a powerful tool for theorem-proving in the hand of mathematicians.

A detailed account of symbolic dynamics in the topological theory of dynamical systems was given by Morse and Hedlund [1938]. Further contribution to symbolic dynamics has been made by R. Bowen, D. Ruelle, Ya. Sinai and many others in connection with ergodic theory and the theory of differentiable dynamical systems.

V. M. Alekseev edited a collection (Bowen [B1979]) of Bowen's papers (Bowen [1970, 1973a, 1973b, 1975a, 1975b]; Bowen and Ruelle [1975]) in Russian translation and, with M. V. Yakobson, wrote an Appendix to this book. The English translation of this Appendix later appeared in *Physics Reports* (Alekseev and Yakobson [1981]). In his Foreword to that issue J. Ford emphasized the far-reaching implications of symbolic dynamics for the fundamentals of physics, praising the Editor "for exhibiting the daring foresight required to publish this highly abstract, mathematical review article in a physical journal. A comparing daring could have been achieved some decades ago, had *The Physical Review* published an abstract mathematical review of Hilbert Space Operator Algebra in, say, 1922." As we know it was only in the 1930s the mathematical structure of quantum mechanics was finally understood as theory of linear operators in the Hilbert space. Further development of symbolic dynamics may show that these words of J. Ford do not compose an overstatement.

In the history of science it has happened many times that even before the inception of an important development many scientists had prepared for it. Two years before T. Y. Li and J. Yorke used the term *chaos* in the title of their 1975 paper (Li and Yorke [1975]), Metropolis, Stein and Stein [1973] published a paper in a journal on combinatorial theory. Apparently the authors did not emphasize its meaning for dynamics. They discovered some universal feature in the systematics of periodic orbits in unimodal maps by using symbolic dynamics. The physicists Derrida, Gervois and Pomeau [1978] made further contribution to symbolic dynamics by introducing so-called $*$-composition for symbolic sequences of unimodal maps. This algebraic operation turns out to be a convenient tool to describe the self-similar structure seen in the bifurcation diagram (see Sections 2.2.2 and 2.5.1). These two papers marked the beginning of *Applied Symbolic Dynamics.*

The contribution of Milnor and Thurston to symbolic dynamics of one-dimensional maps in a long-circulated 1977 Princeton preprint has been known under the name of *kneading theory.* An update of this once hardly available paper finally appeared later in the open literature (Milnor and Thurston [1977, 1988]). Guckenheimer's lecture in Guckenheimer, Moser and Newhouse

[B1980] contains a detailed exposition of the kneading theory. The book of Collet and Eckmann [B1980], though written in a fairly mathematical style, adds some concrete results on symbolic dynamics of one-dimensional maps. Since then papers along the line of applied symbolic dynamics has been increasing slowly but steadily. We have put the titles of those papers that have been referred to in this book at the end of the book.

The nonlinear dynamics group at the Institute of Theoretical Physics, the Chinese Academy of Sciences, has been using symbolic dynamics to study one- and two-dimensional maps, and some ordinary differential equations as well since early 1980s. Some of our early results were collected in a book (Hao [B1989]), in a review (Zheng and Hao [1990]), and in a Chinese booklet (Zheng and Hao [1994]). This book is partially based on our own work and includes some new results.

2
Symbolic Dynamics of Unimodal Maps

Historically, Applied Symbolic Dynamics was first developed for maps with two monotone branches, so-called unimodal or one-hump maps. It requires only two symbols to realize the coarse-graining or "partition" of the one-dimensional phase space—the interval. Simple as it is, the symbolic dynamics possesses all the important ingredients of more general cases.

A unimodal map

$$x_{n+1} = f(\mu, x_n) \tag{2.1}$$

is defined by a real nonlinear continuous function $f(\mu, x)$ of a real variable x, which depends on a parameter μ. The function $f(\mu, x)$ for a fixed parameter μ is often simply written as $f(x)$. It has the following general property:

1. The function $f(\mu, x)$ is defined on an interval I of real numbers and transforms a point $x \in I$ to another point $x' \in I$.

2. It reaches a maximal value at a *Central* or *Critical* point $C \in I$, hence the name *unimodal* or *one-hump* map. The point C is located somewhere within the interval I, not necessarily at the exact middle point of I.

3. The function $f(x)$ is monotonically increasing for $x < C$ and monotonically decreasing for $x > C$.

4. It depends on a parameter μ whose essential role is to determine the height $f(\mu, C)$ of the maximum. Depending on the parameter, the height of the map may change. We note that formally μ may denote a set of parameters, however, in the unimodal case only one parameter matters.

Examples of function $f(\mu, x)$, satisfying the above conditions, are shown in Fig. 2.1 (a) $x_{n+1} = \mu x_n(1 - x_n^2)$ with $I = [0, 1]$ and (b) $x_{n+1} = 1 - \mu x_n^2$ with $I = [-1, 1]$. In fact, a linear transformation brings one to the other. We emphasize that for the purpose of symbolic dynamics the precise form of the function is not important. In particular, the precise values of the function at the two end points of the interval I are not essential. It is crucial, however, that the function must be monotonically increasing on the left subinterval of $x < C$, and at the same time be monotonically decreasing on the right subinterval of $x > C$. The left and right subintervals are labeled by the letters L and R, respectively. These letters are examples of symbols which we shall play with throughout this book.

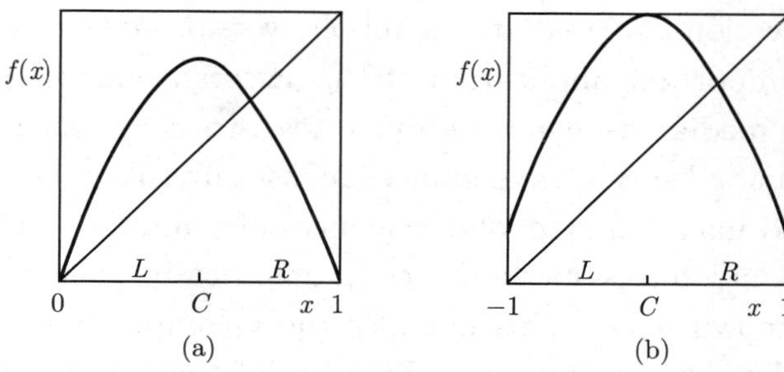

Figure 2.1 Examples of unimodal map

Moreover, in Fig. 2.1 the function $f(\mu, x)$ is shown to change continuously from increasing to decreasing behavior around C, thus it may have a zero derivative $\partial f/\partial x = 0$ at $x = C$. However, we do not impose this requirement in the discussion. The derivative may have a discontinuity at $x = C$, as will be seen, e.g., in the *tent map* to be discussed in Sections 2.6.1 and 2.8.1.

The symbolic sequence which starts from the first iterate of C, i.e., from $f(C)$, plays an important role in applied symbolic dynamics of one-dimensional maps. It carries a special name *kneading sequence* (Milnor and Thurston [1977, 1988]). We have indicated in Chapter 1 that kneading theory is just another name for symbolic dynamics of 1D maps.

This Chapter is organized as follows. Section 2.1 is a general discussion on symbolic sequences in unimodal maps. It is centered on the correspondence of a numerical orbit with a symbolic sequence and a nested, composite functional relationship embodied in the same set of symbols. This correspondence leads directly to the word-lifting technique, a trick to determine the parameter of any given superstable or eventually periodic orbit for an explicitly known unimodal map. The notion of fixed points and periodic orbits and their stability criterion is also introduced in this Section for later use.

Section 2.2 describes a specific model of unimodal maps—the quadratic map and its modifications. We will frequently use this map to demonstrate one or another aspects of the general theory of symbolic dynamics. In particular, we will show a bifurcation diagram of this map and discuss a few distinguished features of the bifurcation diagram, including the dark lines and periodic windows in the diagram.

Section 2.3 touches a central theme of applied symbolic dynamics—the admissibility condition for symbolic sequences, which is based on the ordering of sequences. Section 2.4 treats the symbolic version of the Periodic Window Theorem. An important application of this theorem yields a simple method to generate the median sequence between any two given kneading sequences, hence to construct a list of all kneading sequences up to a given length between the minimal and maximal symbolic sequences of a unimodal map.

Section 2.5 introduces methods to systematically generate kneading sequences from known ones. This includes the $*$-composition and generalized composition rules. An important application of the generalized composition rules is the description of coarse-grained chaos by comparison with the well-understood chaotic orbits in the surjective quadratic map. This is done in Section 2.6.

Section 2.7 discusses a weak characteristic of chaos—topological entropy. In a few important circumstances it may be calculated from the kneading sequence alone without knowing the metric property of the map. We will return to the problem of topological entropy several times in this book in different context.

The last Section 2.8 describes a physical realization of symbolic dynamics of two symbols, namely, the piecewise linear tent and shift maps. We will introduce a metric representation of symbolic sequences, which will be extensively used in the construction of symbolic dynamics for two-dimensional maps later in Chapter 5. Transfer matrix of symbolic sequences and topological entropy of maps will also be discussed in this Section.

2.1 Symbolic Sequences in Unimodal Maps

In this Section we discuss the correspondence among three objects:

1. A numerical orbit (2.2).

2. A symbolic sequence (2.16).
3. A functional composition (2.20).

This correspondence is a key concept in applied symbolic dynamics. We will conclude this Section by giving a first application of this correspondence—the *word-lifting technique* to determine the parameter of a particular superstable orbit for an explicitly given unimodal map.

2.1.1 Numerical Orbit and Symbolic Sequence

Now consider a general unimodal map (2.1) at a fixed parameter μ, as the one shown in Fig. 2.1. Starting from an initial point $x_0 \in I$, we get a *numerical orbit*

$$x_0, x_1, x_2, \cdots, x_n, \cdots \tag{2.2}$$

by iterating the map:

$$\begin{aligned} x_1 &= f(x_0), \\ x_2 &= f(x_1) = f^2(x_0), \\ &\cdots \\ x_n &= f(x_{n-1}) = f^n(x_0), \\ &\cdots \end{aligned} \tag{2.3}$$

where we have use a shorthand for composition of functions:

$$f^n(x) \equiv \underbrace{f \circ f \circ \cdots \circ f}_{n \text{ times}}(x) = \underbrace{f(f(\cdots f}_{n \text{ times}}(x)\cdots)). \tag{2.4}$$

The notation $f^n(x)$ is not to be confused with the n-th power of $f(x)$, which will never be written in this form throughout the book. The convenient notation $f \circ f(x)$ is encountered in many mathematical publications and will be used occasionally in this book.

Fixed points and their stability

It may happen that at some parameter the iteration (2.3) leads to repetition of one and the same number x^*. This is a *fixed point* of the mapping. In general, a fixed point x^* is a solution of the equation

$$x^* = f(\mu, x^*). \tag{2.5}$$

It is a common practice in nonlinear dynamics to check the stability whenever a solution is obtained. To do so we iterate the map in the vicinity of the fixed point by writing

$$x_n = x^* + \epsilon_n,$$

where ϵ_n is a small deviation from the fixed point. Expanding the relation

$$x^* + \epsilon_{n+1} = f(\mu, x^* + \epsilon_n)$$

to first order in ϵ and discarding higher-order terms, we get

$$\left|\frac{\epsilon_{n+1}}{\epsilon_n}\right| = \left|\frac{\partial f(\mu,x)}{\partial x}\right|_{x^*}. \tag{2.6}$$

The iteration converges to the fixed point when the above ratio is smaller than 1:

$$\left|\frac{\partial f(\mu,x)}{\partial x}\right|_{x^*} < 1. \tag{2.7}$$

The most favorable case for stability occurs when the derivative in (2.7) vanishes and the convergence of $\{x_n\}$ happens to be at least quadratic instead of linear. This is the case when the fixed point is located at the critical point $x^* = C$ where the map has a zero slope. It is called a *superstable fixed* point. The condition for a fixed point to be superstable consists in:

$$\left|\frac{\partial f(\mu,x)}{\partial x}\right|_{x^*} = 0. \tag{2.8}$$

Periodic points and their stability

A period n orbit emerges when the iteration loops among n points: $x_{i+n} = x_i$ for all i. Written down explicitly, we have

$$\begin{aligned} x_1 &= f(\mu, x_0), \\ x_2 &= f(\mu, x_1), \\ &\cdots \\ x_{n-1} &= f(\mu, x_{n-2}), \\ x_0 &= f(\mu, x_{n-1}). \end{aligned} \tag{2.9}$$

These n points $x_0, x_1, \cdots, x_{n-1}$ form a period n or nP orbit, or an n-cycle. Using the composite function notation (2.4), we see that each point of the n-cycle is a fixed point of $f^n(\mu, x)$, i.e.,

$$x_i = f^n(\mu, x_i), \quad \text{for} \quad i = 0, 1, \cdots, n-1. \tag{2.10}$$

This allows us to carry over the stability condition (2.7) for a fixed point to a period n orbit. Recalling from calculus the chain rule of differentiation of composite functions, we have the stability condition

$$\left|\frac{\partial f^n(\mu, x_i)}{\partial x}\right| = \prod_{i=0}^{n-1} \left|f'(\mu, x_i)\right| < 1. \tag{2.11}$$

Obviously, this condition remains the same for all points x_i in the n-cycle, i.e., they acquire or lose stability at the same time. A simple fixed point of $f(x)$ corresponds to the case $n = 1$; it is called sometimes a period 1 or $1P$ orbit.

The most favorable case for convergence again takes place when the derivative of $f^n(x)$ in (2.11) vanishes. This happens when the period n orbit contains the critical point C. This is called a *superstable* periodic orbit. The condition for a periodic orbit to be superstable consists in:

$$\left|\frac{\partial f^n(\mu, x_i)}{\partial x}\right| = \prod_{i=0}^{n-1} \left|f'(\mu, x_i)\right| = 0. \tag{2.12}$$

Tangent and period-doubling bifurcations

The marginal case of the stability condition (2.7) or (2.11), when the ratio equals to 1, requires special investigation. This happens at a fixed parameter μ^*. It is proved in the mathematical literature that generally two different kinds of bifurcation, i.e., change of the number and stability of solutions, takes place at the two sides of μ^*. We speak in terms of fixed point only, as a period n orbit is just a fixed point of $f^n(x)$.

1. A tangent bifurcation (called also a saddle node bifurcation) takes place at $\mu = \mu^*$ if

$$\left.\frac{\partial f(\mu, x)}{\partial x}\right|_{\mu^*, x^*} = +1. \tag{2.13}$$

2. A period-doubling bifurcation (called also a pitch fork bifurcation) takes place at $\mu = \mu^*$ if

$$\left.\frac{\partial f(\mu, x)}{\partial x}\right|_{\mu^*, x^*} = -1. \tag{2.14}$$

Without going into details, we formulate two theorems (see, e.g., Guckenheimer [1977]).

Tangent Bifurcation Theorem Suppose the following conditions hold:

1. There is a fixed point at μ^*: $f(\mu^*, x^*) = x^*$.
2. The stability is marginal with $\left.\dfrac{\partial}{\partial x} f(\mu, x)\right|_{\mu^*, x^*} = +1$.
3. $\left.\dfrac{\partial}{\partial \mu} f(\mu, x)\right|_{\mu^*, x^*} \neq 0$.
4. $\left.\dfrac{\partial^2}{\partial x^2} f(\mu, x)\right|_{\mu^*, x^*} \neq 0$.

Then there exists a small region in the μ-x plane, centered at (μ^*, x^*), say, from $\mu^* - \eta$ to $\mu^* + \eta$ along the μ axis and from $x^* - \epsilon$ to $x^* + \epsilon$ in the x direction, such that to one side of μ^*, say, $\mu \in (\mu^*, \mu^* + \eta)$, there are two real solutions to $f(\mu, x) = x$ within $(x^* - \epsilon, x^* + \epsilon)$, one stable and one unstable, whereas to the other side of μ^* there is none.

The situation in the vicinity of a tangent bifurcation point in the μ-x plane is shown in Fig. 2.2 (a). A pair of stable and unstable fixed points emerge "from nowhere", in fact, from a pair of complex-conjugate solutions of the fixed point equation (2.5).

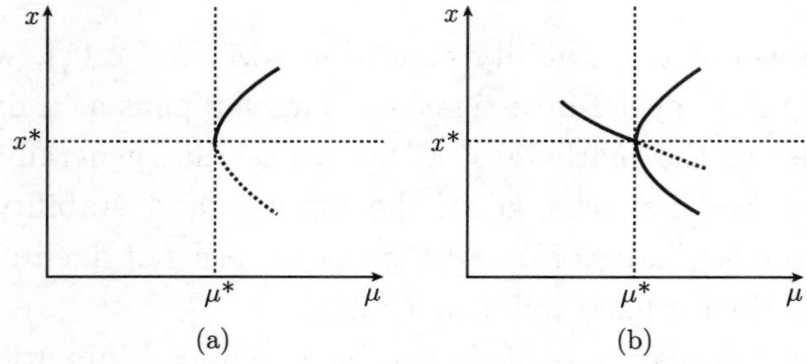

Figure 2.2 Sketch of the vicinity of a bifurcation point in the μ-x plane at a tangent bifurcation (a) and a period-doubling bifurcation (b). Solid line denotes a stable solution, dashed curve—an unstable solution

Period-Doubling Bifurcation Theorem Suppose the following conditions hold:

1. There is a fixed point $f(\mu^*, x^*) = x^*$ in the (μ, x) plane.
2. The stability is marginal with $\left.\dfrac{\partial}{\partial x} f(\mu, x)\right|_{\mu^*, x^*} = -1$.

3. The mixed second derivative $\left.\frac{\partial^2}{\partial x \partial \mu} f(\mu, x)\right|_{\mu^*, x^*} \neq 0$.

4. The Schwarzian derivative $S(f, x)$, is negative at (μ^*, x^*).

Then in the vicinity of the fixed point (μ^*, x^*) there is a small region, say, the box $(\mu^* - \eta < \mu < \mu^* + \eta, x^* - \epsilon < x < x^* + \epsilon)$ in the $\mu - x$ plane, such that to one side of μ^* (depending on the choice of the sign in Condition 3) there is only one stable solution of $f(\mu, x) = x$ in the box, which is certainly a trivial stable solution of $f^2(\mu, x) = x$, but in the other half of the box there are three solutions to $f^2(\mu, x) = x$, two non-trivial stable ones forming a 2-cycle of $f(\mu, x)$ and a trivial unstable one, which is a repetition of the unstable fixed point of $f(\mu, x)$.

Schwarzian derivative

We digress to say a few words on the Schwarzian derivative. A Schwarzian derivative $S(f, x)$ of a function $f(x)$ is the following combination of its derivatives:

$$S(f, x) = \frac{f'''(x)}{f'(x)} - \frac{3}{2}\left(\frac{f''(x)}{f'(x)}\right)^2, \tag{2.15}$$

where f', f'' and f''' are the first, second and third derivative of $f(x)$, respectively. Schwarzian derivative plays an important role in characterizing some metric properties of maps. It appears automatically if one carries out the proof of the period-doubling bifurcation theorem in all its details. Furthermore, maps with a negative Schwarzian derivative on the whole interval may have at most $n + 2$ attracting periodic orbits, where n is the number of extremes in f (Singer [1978]). The number 2 in $n + 2$ takes into accounts the end points of the interval, so at most 2 more orbits may be stable. For the kind of quadratic maps that we are studying it is safe to say that there is at most one stable periodic orbit at each parameter value.

Most maps of physical interest have negative Schwarzian derivatives on the whole interval of definition. For example, the quadratic map and the sine circle map with $b > 1$ (see Section 4.2.2) both have negative Schwarzian derivative on the mapping interval. Therefore, the last condition in the above theorem may often be dropped.

Now back to the theorem. The situation in the vicinity of a period-doubling bifurcation point in the μ-x plane is shown in Fig. 2.2 (b). A stable $1P$ orbit becomes unstable at the bifurcation point and a stable 2-cycle appears. The form of the solution curves explains the alternative name pitchfork bifurcation.

For an elaborated presentation of the proofs we refer to §2.6.2 and §2.6.4 of the book by Hao [B1989].

Graphic representation of the iterations

The iteration process (2.3) is nicely visualized in Fig. 2.3, where a bisector $y = x$ has been drawn in addition to the mapping function $f(x)$. Starting from the initial point x_0, one goes up vertically to locate $x_1 = f(x_0)$. In order to use x_1 as the next input, one goes horizontally to reach the bisector, then turns up or down to locate $x_2 = f(x_1)$. From this construction it is clear that an intersection point of the mapping function with the bisector always represents a fixed point $x = f(x)$.

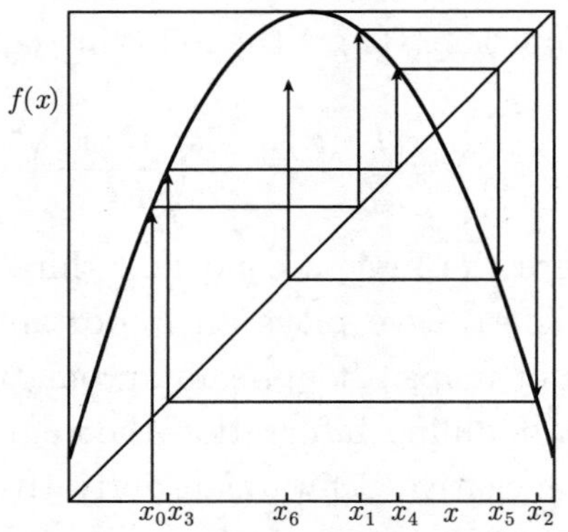

Figure 2.3 Graphic representation of iterations of a map

Partition of the phase space and symbolic sequences

A coarse-grained description means that one is not concerned with the precise value of each x_i. We *partition* the phase space, i.e., the interval I, into a left part L and a right part R. We care only about on which side of C a point x_i falls. If $x_i < C$ we record a letter L, while if $x_i > C$ a letter R. If x_i falls exactly at C, we record a letter C. In this way, from the numerical orbit (2.2) we get a *symbolic sequence* or *symbolic orbit*

$$s_0 s_1 s_2 \cdots s_n \cdots ,$$

where s_i is one of the symbols R, L, or C. For example, the fixed point seen in the right half of Fig. 2.3 has a symbolic sequence R^∞.

It is a good convention to name a symbolic sequence by the initial point which starts the iteration. Thus we may write

$$x_0 = s_0 s_1 s_2 \cdots s_n \cdots . \tag{2.16}$$

In fact, these symbolic sequences involve only two letters R and L, as C may be considered as a limit of either R or L—we will see the precise meaning of this statement in Section 2.8.6. The description using symbolic sequences is, of course, much cruder than a description using all possible numerical orbits (2.2). However, it brings about many simplifications and helps to reveal the essence of the dynamics.

First of all, the description does not need to be one-to-one; many numerical orbits may correspond to the same symbolic sequence. This opens up the possibility of classification of orbits by their equivalent classes: all orbits which correspond to one and the same symbolic sequence are considered equivalent. A simple example elucidates the point better. Take a rather low unimodal map, as shown in Fig. 2.4. Numerically speaking, there are infinitely many different orbits as each initial point x_0 leads to a distinct sequence of numbers. However, it is easy to see that all numerical orbits with $x_0 < C$ lead to the symbolic sequence L^∞, while all that with $x_0 > C$—to the symbolic sequence RL^∞. Despite of an infinity of numerical orbits there are only two or three, if one adds CL^∞, types of symbolic orbits. Thus the dynamics is rather simple. The problem of complexity of dynamics by analyzing symbolic sequences will be elaborated in Chapter 8.

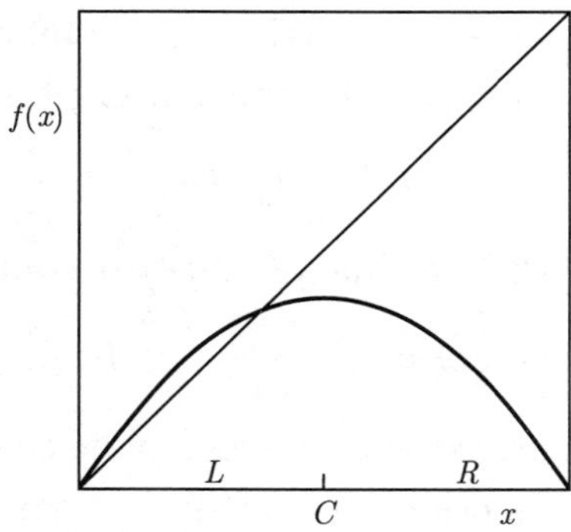

Figure 2.4 A simple map with only two different types of symbolic orbits

Furthermore, symbolic sequences only reflect the stretching and folding action of the mapping function. It does not deal with the concrete functional form of the map. Therefore, conclusions drawn from an analysis of symbolic sequences enjoy great universality. This is why symbolic dynamics is capable to provide a classification scheme and it characterizes the common features of a whole class of systems.

2.1.2 Symbolic Sequence and Functional Composition

Now let us try to reverse the relation $x_n = f^n(x_0)$ in (2.3) to get something like $x_0 = f^{-n}(x_n)$. This cannot be done simply because the inverse of a nonlinear function is multi-valued as a consequence of folding. However, if we do the iteration (2.3) more carefully by adding a subscript s_i to indicate which branch of the mapping function has been used at each step, we will be able to trace the inverse path without ambiguity, or to defold the dynamics. Clearly, the branch $f_{s_i}(x_i)$ is chosen by the argument x_i. Thus we can rewrite (2.3) as

$$\begin{aligned} x_1 &= f_{s_0}(x_0), \\ x_2 &= f_{s_1}(x_1) = f^2(x_0), \\ &\cdots \\ x_n &= f_{s_{n-1}}(x_{n-1}) = f^n(x_0), \\ &\cdots \end{aligned} \tag{2.17}$$

Has this been done, the iterations may be reversed step by step to yield

$$\begin{aligned} x_0 &= f_{s_0}^{-1}(x_1) \\ &= f_{s_0}^{-1}(f_{s_1}^{-1}(x_2)) \\ &\cdots \\ &= f_{s_0}^{-1}(f_{s_1}^{-1}(\cdots f_{s_{n-1}}^{-1}(x_n)\cdots)). \end{aligned} \tag{2.18}$$

To simplify the expressions we will use the same symbol s to denote the branch of the inverse function f_s^{-1} labeled by the subscript s, i.e., let

$$s(y) \equiv f_s^{-1}(y). \tag{2.19}$$

Using this new notation, the last line of (2.18) becomes

$$x_0 = s_0 \circ s_1 \circ \cdots \circ s_{n-1}(x_n). \tag{2.20}$$

This relation is to be compared with the symbolic sequence (2.16). Now each symbol s_i has become a function $s_i(\cdot)$; they appear in the same order as in (2.16). We note that in (2.20) x_0 and x_n are numbers.

Furthermore, we may apply the function $f(x)$ repeatedly to both sides of (2.20) to get

$$\begin{aligned} f(x_0) &= s_1 \circ s_2 \circ \cdots s_{n-1}(x_n), \\ f^2(x_0) &= s_2 \circ s_3 \circ \cdots s_{n-1}(x_n), \\ &\cdots \end{aligned} \tag{2.21}$$

which embody the relation between mapping and shift.

The correspondence of the numerical orbit (2.2) to the symbolic sequence (2.16) and to the functional composition (2.20) is the key message we want to convey to the reader in this Section. We give a simple yet important application of this correspondence.

2.1.3 The Word-Lifting Technique

Anticipating a few later results of applied symbolic dynamics, we indicate that there are three different types of period 5 superstable orbits in an unimodal map, namely, $(RLRRC)^\infty$, $(RLLRC)^\infty$, and $(RLLLC)^\infty$ (see Table 2.2 in Section 2.4). Suppose that we are given a particular unimodal map, say, the map (2.31)

$$x_{n+1} = f(\mu, x_n) = \mu - x_n^2,$$

and are required to determine the location of these periods on the parameter axis. We can proceed as follows.

A periodic orbit $(RLRRC)^\infty$ means, starting from the initial point C, we get the following symbolic sequence

$$C = CRLRRCRLRRC\cdots.$$

We have used the convention of naming a symbolic sequence by its initial point, see (2.16). Replacing the second Cs on the right-hand side with its value, we may stop at the first $RLRRC$ and apply the function $f(x)$ to both sides to get

$$f(C) = R \circ L \circ R \circ R(C), \tag{2.22}$$

see the first line in (2.21). Now the Cs on both sides of (2.22) are numbers, and the symbols R and L are to be understood as inverse mapping functions. For the particular map, $C = 0$, $f(C) = \mu$, and the inverse functions are easily obtained by solving the quadratic equation $y = \mu - x^2$:

$$\begin{aligned} L(y) &= -\sqrt{\mu - y}, \\ R(y) &= \sqrt{\mu - y}. \end{aligned} \tag{2.23}$$

Since everything in (2.22) except for the parameter value is known, it is an equation for the parameter. Writing down explicitly, we have

$$\mu=\sqrt{\mu+\sqrt{\mu-\sqrt{\mu-\sqrt{\mu}}}}. \tag{2.24}$$

In order to solve this equation one may transform it into an iteration (Kaplan [1983]):

$$\mu_{n+1}=\sqrt{\mu_n+\sqrt{\mu_n-\sqrt{\mu_n-\sqrt{\mu_n}}}}. \tag{2.25}$$

The iteration converges quickly for any reasonable initial value, e.g., $\mu_0 = 1.95$, and yields

$$\mu = 1.625\,413\,725\cdots.$$

For the other period 5 orbits $(RL^2RC)^\infty$ and $(RL^3C)^\infty$, one needs merely to change a few signs of the square roots in (2.24).

The form of in Eq. (2.24) may tempt us to get rid of nested square roots by taking squares. In so doing we would get one and the same equation for all three different period 5 orbits, namely,

$$P_5(\mu)=0, \tag{2.26}$$

where we have used the "dark-line" functions $P_n(\mu)$, to be defined by Eq. (2.41) in Section 2.2.3. Its three real roots, located somewhere in between $\mu_\infty = 1.401\cdots$ and $\mu_{\max} = 2$, coincide with what one would get by iterating the (2.25) type equations, but one cannot immediately tell which number corresponds to which period. This does not create a big problem as long as the period under study is short enough.

However, the equation $P_{29}(\mu) = 0$ for period 29 orbits would have 9 256 395 different real roots (see Chapter 7), populated in the same narrow range. At present we do not know any numerical algorithm that is capable to resolve all these roots. The word-lifting technique, to the contrary, can calculate each root with as high precision as one wishes. In other words, symbolic dynamics has overcome the "degeneracy" among various inverse iterations of the same length.

When the inverse functions cannot be written down explicitly, instead of the word-lifting technique one can use a bisection method to locate the orbit.

The bisection method is based on the ordering of symbolic sequences and is not restricted to superstable orbits only. We will discuss it after introducing the ordering rule of symbolic sequences.

2.2 The Quadratic Map

Although the symbolic dynamics that we will develop in this chapter applies to all unimodal maps defined by (2.1), it is instructive to use a particular map to visualize one or another conclusion of the general theory. For this purpose we consider the *quadratic map*. Moreover, here we have a very lucky case in science when a simple one-dimensional model does share some essential properties of higher-dimensional dissipative systems. In fact, the rapid progress of chaotic dynamics since mid 1970s owes a great deal to the quadratic map as a touchstone.

2.2.1 An Over-Simplified Population Model

The quadratic map comes from a simple-minded ecological model. Suppose there is a seasonally breeding insect population in which generations do not overlap. Then the average (or total, or maximal, depending on which quantity is measured) population Y_{n+1} of the next, i.e., $(n+1)$-th, generation will be determined entirely by the population Y_n of the present generation, i.e.,

$$Y_{n+1} = \Phi(Y_n).$$

This is a first order difference equation whose simplest possible form is a linear relation

$$Y_{n+1} = A\,Y_n. \tag{2.27}$$

The linear difference equation (2.27) can be easily solved by letting $Y_n \propto A^n$ to yield

$$Y_{n+1} = Y_0\,A^n,$$

which states that, if, on average, each insect breeds A young ones, then the population will grow exponentially, provided $A > 1$ and will become extinct if $A < 1$.

Simple as it is, this linear equation embodies the "Malthus law of population". In *An Essay on the Principle of Population*[1], after analyzing population

[1] Published by Reeves and Turner, London, 1878, 8th edition.

change in some states in America and Europe, Thomas R. Malthus wrote: "It may safely be pronounced, therefore, that *population when unchecked*, goes on doubling itself every twenty-five years, or *increase in a geometrical ratio* (emphasis original)." In the Malthus law $A = 2$ and the time span from Y_n to Y_{n+1} is 25 years. If we take the initial population Y_0 to be one million, then in less than fourteen 25-year periods, i.e., less than 350 years, Y_{14} would exceed the present population of the Earth. Obviously, the human population cannot be described by such an over-simplified model.

Applied to the insect population with $A > 1$, it would take only a finite number of generations before the globe would be overpopulated solely by this single species. However, new phenomena come into play when Y_n gets large enough: the insects will fight and kill each other for limited food, contagious epidemic may sweep through the population, etc. Either fighting or touching requires the contact of at least two insects, and the number of such events is proportional to Y_n^2 (or, more pedantically, to $Y_n(Y_n - 1)/2$, that makes little difference when Y_n is large). Taking into account this suppressing factor, we can modify Eq. (2.27) into

$$Y_{n+1} = AY_n - B{Y_n}^2. \tag{2.28}$$

Despite its apparently simple form, Equation (2.28) may exhibit quite complicated dynamical behaviour, as Robert May emphasized in his well-read *Nature* paper (May [1976]).

Clearly, one of the two parameters A and B in (2.28) can be scaled out. Usually, one normalizes Y_{n+1} as well and writes (2.28) in one of the following forms,

$$y_{n+1} = \lambda y_n(1 - y_n), \quad y_n \in [0, 1], \quad \lambda \in (2, 4], \tag{2.29}$$

or

$$x_{n+1} = 1 - \mu x_n^2, \quad x_n \in [-1, +1], \quad \mu \in (0, 2]. \tag{2.30}$$

The parameter value $\mu = 0$ is excluded from consideration, as the map then degenerates into a straight line. This minor subtle point has nothing to do with chaotic dynamics, because chaos appears only at $\mu > 1.401\cdots$.

Since both expressions appear frequently in the literature, we write down the corresponding transformation for quick reference. In order to go from (2.29) to (2.30), let

$$y = [(\lambda - 2)x + 2]/4,$$
$$\mu = \lambda(\lambda - 2)/4.$$

The reverse transformation reads

$$\lambda = 1 + \sqrt{1 + 4\mu},$$
$$x = \frac{4y - 2}{\lambda - 2}.$$

Sometimes (2.30) is modified to

$$x_{n+1} = \mu - x_n^2, \quad x_n \in [-\mu, +\mu], \quad \mu \in (0, 2]. \tag{2.31}$$

Throughout this book we shall use either (2.30) or (2.31).

This first order difference equation describes the time evolution of the normalized population x_n. Starting from a number x_n in the interval $I = [-1, 1]$, it generates in a deterministic manner the next number x_{n+1}, located in the same interval I, i.e., the nonlinear transformation $f(\mu, x) = 1 - \mu x^2$ "maps" the interval I into itself. The quadratic map (2.29) is also referred to as the *logistic map*. The functions (2.29) and (2.30) are shown in Fig. 2.1 (a) and (b), respectively.

2.2.2 Bifurcation Diagram of the Quadratic Map

The simplest way to get a feeling on the quadratic map is to visualize it on a computer screen. One covers the parameter range $\mu \in (0, 2]$ by small steps, and calculates, at each parameter value, the iterates of (2.30) starting from $x = 0$ (although other initial values will also do). After throwing away a few hundred points as transients, then the remaining points are displayed on the screen. In this way one gets a *bifurcation diagram* shown in Fig. 2.5.

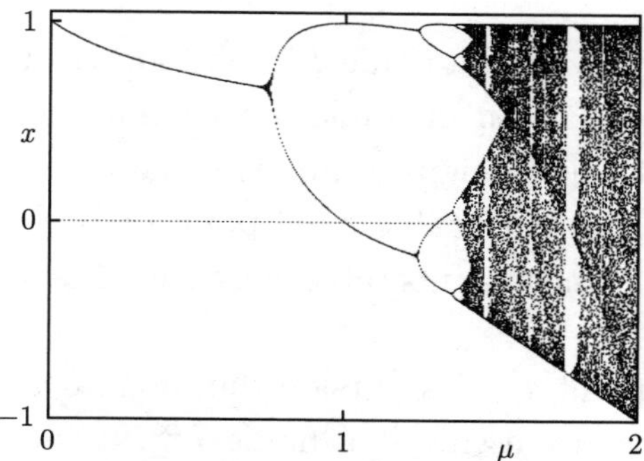

Figure 2.5 A bifurcation diagram of the quadratic map

Fig. 2.5 is a plot of phase space versus parameter space, i.e., μ-x plot. At every fixed parameter value one plots the "limiting set" of the map along the x direction. It is essentially a diagram of attractors, because almost all initial points are attracted to the points shown in the figure, provided a sufficient number of transient points have been discarded. Stable fixed points and periodic points are trivial attractors, while the darkened vertical segments are *chaotic attractors.* Roughly speaking, an attractor is part of the phase space, where the long-time dynamics lives in, no matter how the initial points are specifically chosen from a certain domain.

Now let us have a closer look at the bifurcation diagram Fig. 2.5.

Fixed points

For $\mu \leqslant 0.75$, the limiting set consists of one point. This is the *fixed point* of the mapping. For the quadratic map (2.30), there are, in general, two fixed points. Their parameter dependence is given by

$$x_{\pm}^{*} = \frac{-1 \pm \sqrt{1+4\mu}}{2\mu}. \tag{2.32}$$

The stability condition (2.7), applied to the quadratic map (2.30), reads $|2\mu x^{*}| \leqslant 1$. For the fixed point x_{+}^{*} it yields a stability range

$$-1 \leqslant \sqrt{1+4\mu} - 1 \leqslant 1.$$

Since $\mu \in (0, 2]$, actually we have

$$0 \leqslant \sqrt{1+4\mu} - 1 \leqslant 1. \tag{2.33}$$

The fixed point has a symbolic name R^{∞} and looses stability at $\mu = 3/4$. It further keeps the name as an unstable fixed point. In fact, we will see that a good symbolic dynamics assigns a unique symbolic name to each and every unstable fixed point, while a stable fixed point or period may come into life, sharing a name with another unstable partner, and acquires its own name later.

The other fixed point x_{-}^{*} lies outside the interval $I = [-1, 1]$ until $\mu = 2$, but we may still formally assign it a name L^{∞}. (Note that for the form of (2.29) it is just $y_{-}^{*} = 0$.) It remains unstable on the whole parameter range.

It collides with the chaotic attractor at $\mu = 2$, causing the latter to cease existence beyond $\mu = 2$. This is a *crisis* point, as will be discussed in more detail in Section 2.6.1.

Period-doubling cascade

Right at $\mu = 0.75$ we have marginal stability $f'(\mu, x^*) = -1$ and a period-doubling bifurcation takes place. When μ gets larger than 0.75, the fixed point R^∞ looses stability and a stable 2-cycle comes into play. The orbital points satisfy

$$\begin{aligned} x_2^* &= 1 - \mu(x_1^*)^2, \\ x_1^* &= 1 - \mu(x_2^*)^2, \end{aligned} \tag{2.34}$$

with stability condition

$$-1 \leqslant 4\mu^2 x_1^* x_2^* \leqslant 1. \tag{2.35}$$

Since the fixed points $x_\pm^*$ always contribute to longer periods as trivial solutions, there is no need to deal with the fourth order system (2.34). It is sufficient to solve

$$\frac{f^2(\mu, x) - x}{f(\mu, x) - x} = \mu^2 x^2 - \mu x + 1 - \mu = 0,$$

which yields

$$\begin{aligned} x_1^* &= (1 - \sqrt{4\mu - 3})/(2\mu), \\ x_2^* &= (1 + \sqrt{4\mu - 3})/(2\mu). \end{aligned} \tag{2.36}$$

We see that they become two observable real roots only when $\mu > 0.75$. In the $\mu < 0.75$ region, they make a pair of complex conjugate roots, which cannot be seen in iterations of real numbers.

The stability condition (2.35) leads to

$$-1 \leqslant 4 - 4\mu \leqslant 1,$$

yielding the stability range $(3/4 < \mu < 5/4)$ for the 2-cycle and the superstable parameter $\tilde{\mu}_2 = 1$.

At $\mu = 1.25$ the 2-cycle looses stability and a bifurcation to 4-cycle takes place. This process repeats at closer and closer μ_n, leading consecutively to period 4, 8, $\cdots$, 2^n, $\cdots$ cycles, forming a *period-doubling bifurcation cascade*. The period-doubling cascade reaches an *accumulation point* at $\mu_\infty = 1.40115518909205\cdots$, where the period becomes infinity.

Scaling behavior in the period-doubling cascade

The period-doubling cascade enjoys a number of universal scaling behavior. Although they are associated with metric property of the map, we describe briefly these scalings because symbolic dynamics provides a bridge to the renormalization group equations (see Eq. (2.76) in Section 2.5) which determine the universal scaling constants (exponents).

1. The convergence rate of the successive bifurcation points. We have just seen that in the quadratic map the bifurcation from a fixed point to a 2-cycle occurs at $\mu_1 = 0.75$, the bifurcation from the 2-cycle to a 4-cycle takes place at $\mu_2 = 1.25$, etc. One can proceed further to determine the following bifurcation points μ_n numerically and then explore the way they converge to μ_∞. In fact, the bifurcation parameters μ_n are not the best quantities to deal with, because the iterations get very slow in the vicinity of bifurcation points—the phenomenon of "critical slowing down" (Hao [1981]). However, included in between two successive bifurcation points, there always exists a superstable periodic point $\tilde{\mu}_n$, where the iterations converge much quicker. These superstable $\tilde{\mu}_n$ must converge in the same manner as the μ_n. M. J. Feigenbaum [1978] discovered that they converge geometrically, i.e.,

$$\tilde{\mu}_n = \mu_\infty - \frac{\text{const}}{\delta^n}. \tag{2.37}$$

It is remarkable that

$$\delta = 4.669\ 201\ 609\ 102\ 990\ 67\cdots \tag{2.38}$$

happens to be a "universal" constant for a large class of mappings and even higher-dimensional dissipative systems. The convergence law (2.37) shows that δ may be estimated by looking at the ratio

$$\delta_n = \frac{\tilde{\mu}_{n+1} - \tilde{\mu}_n}{\tilde{\mu}_n - \tilde{\mu}_{n-1}}$$

for several consecutive values of $\tilde{\mu}_n$. Actually, we shall see that δ is only the first, the simplest, and perhaps the most important one among an infinite number of "universal" constants (see the discussion of so-called period-n-doubling sequences in Section 2.5.2).

2. The scaling factor of typical phase space lengths. We have seen the self-similar structure of the bifurcation diagrams Fig. 2.5. Along the parameter axis this structure is characterized by the convergence rate δ. If we take some typical lengths along the x direction, for example, the greatest separation l_n of two nearest periodic points at superstable parameter $\tilde{\mu}_n$ and the smallest separation l'_n at $\tilde{\mu}_n$, then length ratios also approach a definite limit, i.e., both ratios

$$\frac{l_n}{l_{n+1}} \quad \text{and} \quad \frac{l_{n+1}}{l'_{n+1}}$$

approach the constant value (Feigenbaum [1978]):

$$\alpha = 2.502\ 907\ 875\ 095\ 892\ 8 \cdots . \tag{2.39}$$

The above values of α and δ have been taken from Mao and Hu [1988]. We shall see that α also happens to be the most important one of an infinite family of scaling factors (Section 2.5.2).

The most significant contribution of M. J. Feigenbaum [1978, 1979] comprises not only the discovery of these universal constants, but also his device of a universal, renormalization group approach, which reveals the physical meaning of these constants and opens up a way to the calculation of α and δ to high precision, independently of any particular model. Put in a nutshell, appropriate change of length scale and parameter at consecutive period-doubling bifurcations defines a renormalization operation; applied to an arbitrary one-hump function infinitely many times, the renormalization operation produces a universal function $g(x)$—a solution of the following functional equation:

$$\alpha g(\alpha x) = g \circ g(x). \tag{2.40}$$

This is a fixed point in the space of infinitely many one-hump functions. The Eq. (2.40) can be solved only numerically. The constant α obtained together with $g(x)$ is nothing but the scaling factor mentioned above.

One can design a functional iteration process to approach $g(x)$. The eigenvalue of the linearized system determines the convergence rate δ. In fact, there is another universal scaling factor κ which describes the scaling of bifurcation structure with the level of external noise. The dimension D of the attractor at the accumulation point μ_∞ is also an universal constant. The four constants α, δ, κ, and D, play a role, similar to the critical exponents in the theory of continuous phase transitions. Therefore, these constants are also called universal exponents of the period-doubling limit. For details of the renormalization group theory of period-doubling we refer to the book by Hao [B1989]. The most important fact for us is the connection of the functional equation (2.40) with symbolic dynamics. We will return to this point in Section 2.5.2.

If one starts from the maximal parameter $\mu = 2$ and goes downward along the parameter axis, then one encounters first a one-band chaotic region, followed by two, four, $\cdots$ and 2^n-band chaotic regions, n quickly approaching infinity. Therefore, looking upward from smaller to greater parameter values there is a cascade of 2^{n+1} to 2^n band-merging or period-halving cascade. Both period-doubling cascade of periodic orbits and the period-halving cascade of chaotic bands meet at the accumulation point $\mu_\infty = 1.401\,15\cdots$.

We shall see that the band-merging points may be determined precisely by symbolic dynamics.

Periodic windows embedded in chaotic regions

Within a chaotic region, there are many *periodic windows*, i.e., lucid intervals where periodic orbits with their own period-doubling cascade and chaotic bands at smaller scales exist. For example, the most clearly seen window is the period 3, which starts at $\mu = 1.75$. The period 3 orbit then undergoes a period-doubling process, giving birth to periods 3×2, 3×2^2, $\cdots$, 3×2^n, approaching the $3 \times 2^\infty$ limit at a well-definable parameter value. After this period-doubling limit a period-halving cascade of chaotic bands takes place until the last three chaotic bands suddenly merge into the one-band region, leaving a perceptible trace of the three small chaotic bands. The contrast of densities in the one-band region finally disappears at another well-defined parameter. These abrupt changes of the chaotic attractor are examples of *crises* of chaotic attractors.

If one zooms into the smaller chaotic bands, one sees similar embedded periods and their associated chaotic bands at smaller and smaller scales. There are many self-similar structures in the bifurcation diagram.

The rich structure in a bifurcation diagram inspires us to raise many questions. What is the number and systematics of the periodic windows? Can one tell the parameter values of the periodic orbits and the band-merging points? Is it possible to describe the sudden change of the chaotic attractor? It turns out that many, if not all, questions of this kind may be answered by symbolic dynamics.

2.2.3 Dark Lines in the Bifurcation Diagram

In a bifurcation diagram we see many dark lines, either going through the chaotic regions or becoming sharp boundaries of the latter. Can one write down the equations of all the dark lines? It happens to be an easy job (Zeng, Hao, Wang and Chen [1984]).

The map (2.1) is a play of numbers: an initial x_0 produces a long sequence of numbers $\{x_i\}$, $i = 0, 1, 2, \cdots$. Now let us "lift" this iteration of numbers to a recursive definition for a family of composite functions, using the same transformation $f(\mu, x)$. As a function of x, $f(x)$ may have one or more critical points at, say, $x = C$, where its first derivative vanishes. Starting from the critical point C, we define recursively a set of functions $\{P_n(\mu)\}$ of the parameter μ:

$$\begin{aligned} &P_0(\mu) = C, \\ &P_{n+1}(\mu) = f(\mu, P_n(\mu)), \quad n = 0, 1, \cdots . \end{aligned} \tag{2.41}$$

Note that $P_n(\mu)$ is nothing but

$$P_n(\mu) = f^n(\mu, C). \tag{2.42}$$

Our main statement is: the functions $\{P_n(\mu)\}$ give all the dark lines and chaotic zone boundaries seen in the bifurcation diagrams. At the end of this section we will give a simple physical explanation of why the dark-line equations are defined in this way.

In the case of the quadratic map (2.30) we have $C = 0$ and the functions P_n are polynomials of μ:

$$\begin{aligned} P_0(\mu) =& 0, \\ P_1(\mu) =& 1, \\ P_2(\mu) =& 1 - \mu, \\ P_3(\mu) =& 1 - \mu + 2\mu^2 - \mu^3, \\ P_4(\mu) =& 1 - \mu + 2\mu^2 - 5\mu^3 + 6\mu^4 - 6\mu^5 + 4\mu^6 - \mu^7, \\ P_5(\mu) =& 1 - \mu + 2\mu^2 - 5\mu^3 + 14\mu^4 - 26\mu^5 + 44\mu^6 - 69\mu^7 \\ & +94\mu^8 - 114\mu^9 + 116\mu^{10} - 94\mu^{11} + 60\mu^{12} \\ & -28\mu^{13} + 8\mu^{14} - \mu^{15}, \\ \cdots \end{aligned}$$

The curves P_n vs. μ for $n = 0$ to 9 are shown in Fig. 2.6. We see that they do outline the skeleton of the bifurcation diagram Fig. 2.1. We see that the dark lines may either intersect each other or touch each other tangentially. It is easy to realize that once some dark lines intersect or touch at a certain parameter value, there are infinitely many other dark lines going through the intersection or tangency point. The tangent points are associated with superstable periodic orbits, while the intersection points are related to band-merging points, crises of chaotic attractor, or homoclinic or heteroclinic orbits in the map. We will return to these points when we know more about symbolic dynamics (see Section 2.6.2).

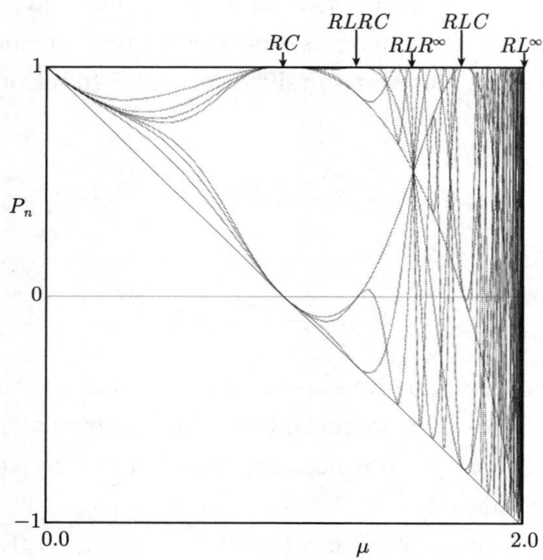

Figure 2.6 Dark lines in the bifurcation diagram of the quadratic map: $P_0(\mu)$ to $P_9(\mu)$

Tangent points between dark lines

Functions from the set $\{P_n(\mu)\}$ touch each other tangentially at well-defined parameter values $\tilde{\mu}$. The tangent points between some dark lines determine the location of certain superstable periodic orbits. In fact, the location of superstable periodic orbits of period n is given by the real roots of the equation

$$P_n(\mu) = C, \tag{2.43}$$

(and hence $P_{kn}(\mu)$ for any positive integer k) within the parameter range $[0, 2]$, or, to be more precise, within the range $[1.401, 2]$, i.e., in the chaotic region after the accumulation point $\mu_\infty = 1.401$ of the Feigenbaum period-doubling cascade.

The Eq. (2.43) may have more than one solutions for $n > 3$. For example, there are 3 different roots for $n = 5$, corresponding to three different types of period 5 orbits in the quadratic map. How many different types of orbits there exist for a given period? This leads to a counting problem. We will give the complete solution of this problem in Chapter 7 not only for the quadratic map, but also for all continuous maps on the interval.

When n is not very large, one can calculate the parameter values for all superstable periods of length n by solving Eq. (2.43). However, as all these roots are populated in the narrow range $[1.401, 2]$, they may hardly be separated numerically. For instance, there are more than one million different roots for $n = 26$ and no known library routines on present-day computers are capable to calculate them with high precision. In this case, a little trick in applied symbolic dynamics—the word-lifting technique introduced in Section 2.1.3 may help us to pick up any superstable period with as high precision as one wishes.

Intersections of dark lines

As Fig. 2.6 shows, the dark lines intersect each other at well-defined parameters. It is remarkable that these intersection points determine the locations of all band-merging points seen in the bifurcation diagram. The most clearly seen band-merging points in Fig. 2.5 are those where a 2^n chaotic band merges into a 2^{n-1} band. In particular, the merging of the two main chaotic bands into a single band occurs at the intersection point $\overline{\mu}$ of all but the first two functions, where

$$P_3(\overline{\mu}) = P_4(\overline{\mu}), \tag{2.44}$$

(and hence $P_3(\overline{\mu}) = P_{3+k}(\overline{\mu}) = C$ for any positive integer k). To the left of $\overline{\mu}$, curves $\{P_n(\mu)\}$ are grouped into two bundles by the parity of their subscripts. For the quadratic map (2.30), from (2.42) one can verify that

$$P_n + P_m = 0 \quad \text{implies} \quad P_{n+1} = P_{m+1}.$$

Specifically for $n = 2$ and $m = 3$, instead of $P_3 = P_4$, one can also solve

$$P_2(\mu) + P_3(\mu) = 2 - 2\mu + 2\mu^2 - \mu^3 = 0$$

to find $\overline{\mu} = 1.543\,689\,013\cdots$.

The merging point of four bands into two is determined by either of the two equalities:

$$P_5 = P_7 \quad \text{or} \quad P_6 = P_8. \tag{2.45}$$

Similarly, to the left of the above band-merging point, the P_n's are grouped according to their subscripts (mod 4), except for the first eight functions which have become the outer boundaries of the chaotic bands. The next merging takes place at the intersection $P_k = P_{k+4}$ for k being either 9, 10, 11 or 12. These relations hold for all unimodal maps. Moreover, the locus of the unstable fixed points (see (2.32)) passes through the first band-merging point (2.44); the loci of the two unstable period 2 orbits (see (2.36)) intersect the next band-merging point (2.45), etc. In other words, the drastic changes of the structure of the chaotic bands may be viewed to be due to the "collision" of unstable orbits with attractors in the phase space. These phenomena have been called the *crisis* of chaotic attractors (Grebogi, Ott and Yorke [1983a]). We shall describe the crises in the general setting of symbolic dynamics in Section 2.6.2.

Other band-merging points can be calculated in the same way. However, the solution of the above equation and other similar equations encounters the same kind of numerical difficulties associated with the calculation of supercritical parameter values from Eq. (2.43). We shall show a much better way out in Section 2.6.2 by extending the word-lifting technique to eventually periodic symbolic sequences.

Dark Lines in maps with multiple critical points

The definition (2.41) applies to any map, whenever an explicit expression of the map is known. If there are more than one critical points C_i, $i = 1, 2, \cdots$, one defines several series of functions $P_n^i(\mu)$, starting from each of the critical point C_i:

$$\begin{aligned} &P_0^i(\mu) = C_i, \\ &P_{n+1}^i(\mu) = f(\mu, P_n^i(\mu)), \quad n = 0, 1, \cdots. \end{aligned} \tag{2.46}$$

The letter μ usually stands for many parameters. A useful example is the sine-square map

$$x_{n+1} = A \sin^2(x_n - B),$$

which has a close relation to optical bistability. For dark lines in the bifurcation diagram of the sine-square map see Zhang *et al.* [1985, 1987]. The symbolic dynamics of the sine-square map will be developed in Section 3.7.

The rainbow: explanation of the dark lines

In order to understand why the recursively defined functions (2.41) describe all the dark lines and band boundaries in the chaotic regions, we digress a little to a high-school physics problem: the rainbow.

If asked to explain the origin of the rainbow, a good student would draw a picture like Fig. 2.7, and then would point out how a droplet works as a prism due to double refraction and single reflection of the light rays at the water-air interfaces. This might be an excellent answer for a middle-school student, but certainly not a complete answer for a graduate student in physics. The point is that the angle of deflection θ of the outgoing ray depends not only on the refraction index n of water, but also on the sight distance (also called the impact parameter) δ of the incoming ray. Since all sight distances from 0 to R (the radius of the droplet) are present in the sunlight, outgoing rays of the same wavelength will spread out, thus different colors will overlap and mix again. Hence it has still to be explained why we can see at all the rainbow.

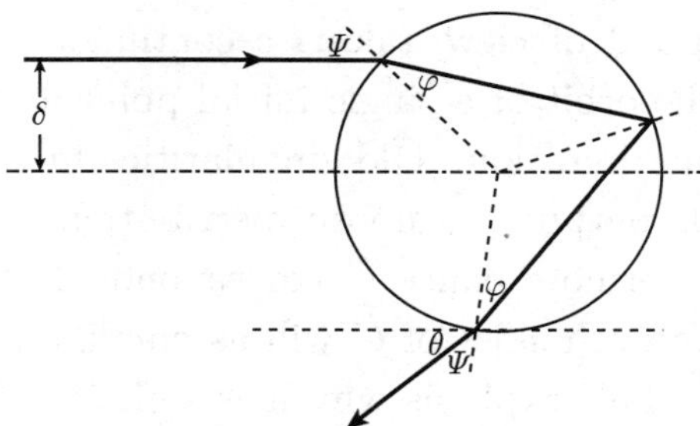

Figure 2.7 Refraction and reflection of a light ray in a water droplet

The correct answer follows from an inspection of the quantitative dependence of θ on n and δ. It is an elementary exercise to derive the formula

$$\theta = 2\arcsin\left\{x\left[\frac{2}{n}\sqrt{(1-x^2)\left(1-\frac{x^2}{n^2}\right)} - 1 + \frac{2x^2}{n^2}\right]\right\}, \tag{2.47}$$

where $x = \delta/R$. The x-θ dependence at fixed n happens to be an unimodal function. Fig. 2.8 is drawn for the refraction index $n = 1.333$, approximately that for water. If we take the incoming rays at equally spaced discrete values of the sight distance δ, then the resulting rays will come out more densely at angles closer to that corresponding to the critical value x_c. If the incoming rays are homogeneously distributed over the interval $x \in (0, 1)$, then the outcoming distribution will display a singularity, i.e., an infinite peak at the maximum of (2.47). In fact, this result gives the rainbow as we see it.[2]

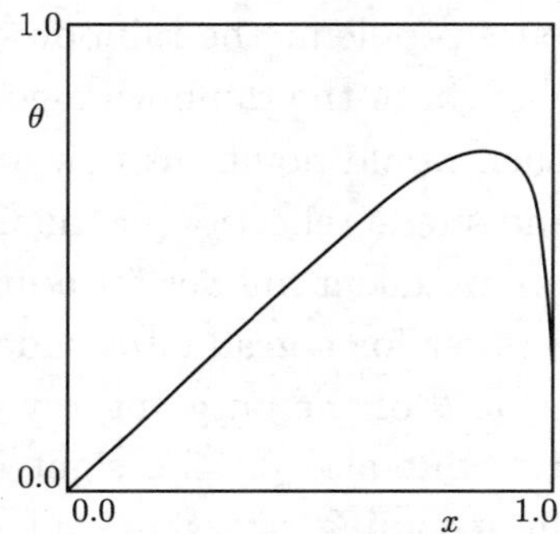

Figure 2.8 The θ-x dependence for $n = 1.333$

In one-dimensional mappings, the amplification near $x = C$ occurs at each iteration, because, being in the chaotic region, we can look at the iterations from an "ensemble" point of view, taking a continuous distribution as input instead of following the orbit of a single initial point and invoking the "rainbow" argument at each iteration. Old singularities in the input distribution will be retained in the output, a smooth distribution near x_c gives rise to a new singularity. Furthermore, since C corresponds to the maximum of the map, the iterates from both sides of C will assemble on the same side of the singularity generated. This explains why it is sufficient to follow the iterates of the critical point C in order to find the location of the dark lines and band boundaries, as well as why there exists one-sided "shadowing" near these lines and boundaries.

We note that the "amplification effect" of a single variable function near its extremes is a common phenomenon. Consider another simple example: a free electron gas. In order to calculate the density of states, one needs merely express the momentum space volume element in terms of the energy

[2] For more on rainbow, see, e.g., H. M. Nussenzveig (1977), *Scientific American*, April, p.116.

$$d^D\mathbf{p} = N(E)dE,$$

where D denotes the dimension of the momentum space. Using the energy-momentum relation $E = p^2/(2m)$, one readily obtains

$$N(E) \propto E^{\frac{D-2}{2}}.$$

For one-dimensional electron gas, a singularity in the density of states occurs at $E = 0$:

$$N(E) \propto \frac{1}{\sqrt{E}}.$$

When applied to one-dimensional crystal, one has in the effective mass approximation

$$N(E) \propto \frac{1}{\sqrt{|E - E_c|}},$$

where E_c is the band edge. This is the same kind of singularity appears in the quadratic map close to the dark lines and zone boundaries (see Section 2.6.2). By the way, an interesting question[3] arises for one-dimensional crystals when the gap between the valence and conduction bands shrinks to zero: what happens to, say, the specific heat or magnetic susceptibility when the singularities at the two energy band edges touch each other? Other examples of the "amplification effect" are the steepest descent method to approximate an integral, the caustics in light or sound propagation, and even the relation between statistical mechanics and thermodynamics.

2.3 Ordering of Symbolic Sequences and the Admissibility Condition

We have seen that many numerical orbits may correspond to one and the same symbolic sequence. On the other hand, different symbolic sequences must correspond to different initial points in the phase space, otherwise it would contradict the fact that the dynamics under study is deterministic. However, an arbitrarily given symbolic sequence, made of Rs and Ls, may not be generated in a unimodal map.

A symbolic sequence is called an *admissible sequence* to a map at a given parameter, if one can choose an initial point to produce a numerical orbit of

[3] The answer was given in B.L. Hao (1961), *Acta Physica Sinica*, **17**, P.505.

the map which leads to the given symbolic sequence, using a given partition of the interval. We need a criterion to check whether a symbolic sequence is admissible or not.

The *admissibility condition* is based on the ordering rule of symbolic sequences and makes use of a few elementary properties of monotone functions. Therefore, we precede the formulation of ordering rule with a discussion of monotone functions.

2.3.1 Property of Monotone Functions

We list a few simple properties of monotone functions, known in college calculus:

1. A monotonically increasing function preserves the order of its arguments, i.e., if $f(x)$ is a monotone increasing function, then

$$x_1 > x_2 \Leftrightarrow f(x_1) > f(x_2).$$

2. A monotonically decreasing function reverses the order of its arguments, i.e., if $f(x)$ is a monotone decreasing function, then

$$x_1 > x_2 \Leftrightarrow f(x_1) < f(x_2).$$

3. The inverse of an increasing (decreasing) function is also increasing (decreasing).

4. The composition $f \circ g(x) = f(g(x))$ or $g \circ f(x) = g(f(x))$ of two increasing or two decreasing functions f and g is increasing.

5. The composition of one increasing and one decreasing function, no matter whichever acts first, is a decreasing function.

The last two statements extend to compositions of a finite number of monotone functions. One counts the number of decreasing functions it contains. If the number is odd, then the composite function is decreasing; if it is even, then the composition is increasing. Therefore, it is convenient to introduce a *parity* for monotone functions: an monotone increasing function has an even parity or parity $+1$, while a decreasing function has an odd parity or parity -1. The parity of a composite function is the product of the parities of its monotone components.

The parity is inherited in the symbols of applied symbolic dynamics, as we always partition the interval in accordance with the monotonicity of the mapping function. Thus for unimodal maps the symbol L and the inverse function $L(y)$ have an even parity, but the symbol R and the inverse function $R(y)$ have an odd parity.

Inasmuch as the parity coincides with the sign of the derivative of the monotone function f, we may as well assign a zero (0) parity to the symbol C in unimodal maps since $df/dx = 0$ at $x = C$. It may only have a formal meaning when $f(x)$ is not continuous at $x = C$.

2.3.2 The Ordering Rule

The fundamental idea underlying the ordering rule of symbolic sequences is the fact that we mentioned in the beginning of this section, namely, two different symbolic sequences must correspond to two different points on the interval. If so, one may assign the natural order of these two points, i.e., their order as real numbers, to the two symbolic sequences they lead to.

In fact, we are going to formulate a general ordering rule for symbolic sequences. The rule works not only for unimodal maps, but also for maps with multiple critical points and/or discontinuities. In order to get a deeper feeling on this important rule of symbolic dynamics, we proceed by looking at a few particular cases in the unimodal map.

First of all, we have a natural order

$$L < C < R, \tag{2.48}$$

which reflects the order of real numbers on the interval I. The simplest case of two different symbolic sequences is $\Sigma_1 = R\cdots$ and $\Sigma_2 = L\cdots$. By definition, the point that corresponds to Σ_1 must locate to the right of C, while the point that gives rise to Σ_2—to the left of C. Therefore, according to the natural order (2.48) we have $\Sigma_2 < \Sigma_1$. If the two symbolic sequences differ only at the second symbol, we can order them by the first two letters according to the following rule

$$LL < LR < RR < RL,$$

which is a simple consequence of the two single monotone functions f_L^{-1} and f_R^{-1}.

In general, the two sequences Σ_1 and Σ_2 may have a common leading string Σ and then differ at the next letter. Suppose we have

$$\begin{aligned}\Sigma_1 &= \Sigma\mu\cdots,\\ \Sigma_2 &= \Sigma\nu\cdots,\end{aligned} \tag{2.49}$$

where $\Sigma = s_0 s_1 \cdots s_n$ and $\mu \neq \nu$. Here s_i, μ, and ν are either R or L. Since $\mu \neq \nu$, they must have been ordered according to the natural order (2.48). Now compare $\Sigma(\mu)$ and $\Sigma(\nu)$, understanding μ and ν as the numbers represented by the subsequences $\mu\cdots$ and $\nu\cdots$, and Σ as the composite function $s_0 \circ s_1 \circ \cdots \circ s_n$. It is clear that the order of μ and ν will be preserved if Σ has an even parity, and the order reversed if Σ has an odd parity. Thus we have

Ordering Rule Given two symbolic sequences $\Sigma_1 = \Sigma\mu\cdots$ and $\Sigma_2 = \Sigma\nu\cdots$ with a common leading string Σ and the next symbol $\mu \neq \nu$, the order of μ and ν in the sense of the natural order (2.48) is the order of Σ_1 and Σ_2 if Σ is even; the order of μ and ν is opposite to that of Σ_1 and Σ_2 if Σ is odd.

This rule works for one-dimensional maps with multiple critical points and discontinuities as well, only the natural order and parity of symbols should be defined in each case. For the unimodal maps we may give a "rule of thumb" for ordering. It comes from the following observation. When comparing two sequences $\Sigma R\cdots$ and $\Sigma L\cdots$, let us call the *principal strings* the strings ΣR and ΣL formed by the common leading string plus the first different letters. We have seen that when Σ is even, the larger sequence Σ_1 has an odd principal string ΣR, whereas when Σ is odd, the larger sequence Σ_2 has an odd principal string ΣL. In both cases the larger sequence has an odd principal string. This observation remains true for the degenerated case of a blank common string, i.e., when the two sequences differ at their first letters. Therefore, we have the following

Ordering Rule (unimodal maps only) When comparing two different symbolic sequences, the larger one always has an odd principal string.

If there is a third symbolic sequence $\Sigma_0 = \Sigma C$ with the same common leading string Σ, then it is easy to see that Σ_0 must be included in between Σ_1 and Σ_2. Therefore, after the order of Σ_1 and Σ_2 has been established by the above rule, the order of Σ_0 is determined naturally.

A BASIC program which compares the order of two given symbolic sequences is given in Program 2.1. In this program the integer variable `P` serves as a parity indicator. Odd parity corresponds to `P = -1`. It changes sign when encountering a letter R (statement `60`).

```
  5  REM PROGRAM 2.1 FOR COMPARING TWO DIFFERENT WORDS
 10  DEFINT I-P
 20  INPUT "THE FIRST WORD IS: ", S1$
 30  INPUT "THE SECOND WORD IS: ", S2$
 40  N=LEN(S1$):  P=1
 50  FOR I=1 TO N: C$=MID$(S1$, I, 1)
 60  IF C$="R" THEN P=-P
 70  IF C$<>MID$(S2$, I, 1) THEN GOTO 90
 80  NEXT I
 90  B$="GREATER":  IF P=1 THEN B$="SMALLER"
100  PRINT S1$;" IS ";B$;" THAN ";S2$;"."
110  STOP: END
```

Program 2.1

The sequence L^∞ is, obviously, the smallest among all possible symbolic sequences, because it always has an even principal string in comparison with any other symbolic sequence. The sequence RL^∞, to the contrary, is the largest among all possible symbolic sequences, since it always has an odd principal string, no matter with which sequence it compares.

We can consider R and L as binary digits 1 and 0. Then L^∞ may corresponds to real number 0, and RL^∞ to 1. There is a one-to-one correspondence between symbolic sequences, made of R and L, and real numbers in the interval $[0, 1]$. However, the ordering differs from that of the usual binary numbers due to the odd parity of R. For more see Section 2.8.6 on metric representations of symbolic sequences.

From the *Ordering Rule* it follows that if two symbolic sequences $\Sigma_1 = \Delta\sigma_1\sigma_2\cdots \equiv \Delta X$ and $\Sigma_2 = \Delta\tau_1\tau_2\cdots \equiv \Delta Y$ both start with an odd string Δ, then $\Sigma_1 > \Sigma_2$ implies $X < Y$. In other words, an odd leading string reverses the order of what follows. In fact, from $\Sigma_1 > \Sigma_2$ it follows that the principal string of Σ_1 must be odd while that of Σ_2 be even. Since Δ is odd, when comparing X with Y, the principal string of X must be even, and that of Y be odd, in consistence with $X < Y$. Note that $\sigma_1\sigma_2\cdots$ and $\tau_1\tau_2\cdots$ may also have a common leading string, i.e., there is no restriction on σ_1 and τ_1, etc., being different. We will often follow this line of reasoning in manipulations with symbolic sequences. To conclude this Section, let us look at a few simple applications of the ordring rule.

Telling the location of orbital points

Suppose a period 5 orbit has the symbolic sequence $(RLRRL)^\infty$, we can determine the order of the five orbital points on the interval. The initial point has a sequence $I_0 = (RLRRL)^\infty$. Its first iterate has the shifted sequence $I_1 = (LRRLR)^\infty$, etc. We get Table 2.1.

Table 2.1 The shift sequences of $(RLRRL)^\infty$ and their order

n	I_n	Order
0	$RLRRL\cdots$	1
1	$LRRLR\cdots$	5
2	$RRLRL\cdots$	3
3	$RLRLR\cdots$	2
4	$LRLRR\cdots$	4

The order has been determined as follows. The sequences I_1 and I_4 start with L, therefore

$$\{I_1, I_4\} < \{I_0, I_2, I_3\}.$$

The principal strings of I_1 and I_4 are LRR and LRL, the latter being odd, thus $I_1 < I_4$. Similarly, we get $I_0 > I_3 > I_2$. Therefore, we have the order $I_0 > I_3 > I_2 > I_4 > I_1$, as indicated in the last column of Table 2.1.

This example addresses ordering of symbolic sequences in the phase space, i.e., on the interval. The next example deals with ordering in the parameter space.

Bisection method to determine parameter of a symbolic sequence

The word-lifting technique, introduced in Section 2.1.3, works well provided the inverse function may be written down explicitly. When this is not the case, it is better to use an even simpler method which is based on the ordering of symbolic sequences. We have in mind the bisection method which works whenever one knows the mapping function; the knowledge of its inverse branches is not required.

We explain this method again on the example of $K = RLR^2C$. Here we regard it as a kneading sequence, which is the sequence of $f(\mu, C)$. More discussion on the kneading sequence will be given in the next subsection. Suppose for certain initially chosen parameters μ_1 and μ_2, we have kneading sequences K_1 and K_2, respectively, and $K_1 < K < K_2$. Take $\mu' = (\mu_1 + \mu_2)/2$ and calculate the corresponding K'. If $K < K'$ we take μ' to be a new μ_2, otherwise take it to be a new μ_1. Proceed further with the bisection until μ is located with the desired precision. A BASIC program implementing this method is listed in Program 2.2.

```
  5  REM PROGRAM 2.2 BISECTION METHOD TO DETERMINE PARAMETERS
 10  CLS: DEFINT I-P
 20  INPUT "THE WORD IS: ", S$:  PRINT
 30  A1=1:  A2=2:  N=LEN(S$):  DA=0.000001:  A=A2
 40  X=1:  S1$=""
 50  FOR I=1 TO N: B$="R":  IF X<0 THEN B$="L"
 60  X=1-A*X*X: S1$=S1+B$
 70  NEXT I
 80  P=1:  FOR I=1 TO N: C$=MID$(S1$, I, 1)
 90  IF C$="R" THEN P=-P
100  IF C$<>MID$(S$, I,1) THEN GOTO 120
110  NEXT I
120  IF P=-1 THEN A2=A ELSE A1=A
130  A=0.5*(A1+A2):  IF A-A1>DA AND A2-A>DA THEN GOTO 40
140  PRINT "THE PARAMETER IS: ";A: STOP: END
```

Program 2.2

Obviously, the bisection method is not limited to the determination of superstable parameters only. Moreover, since ordering rule works for symbolic sequences coming from different points of the phase space, the bisection idea may be adapted to locate the initial point that gives birth to a given symbolic sequence. We will use this approach in the study of two-dimensional maps and ordinary differential equations in later chapters.

2.3.3 Dynamical Invariant Range and Kneading Sequence

Consider a unimodal map, shown in Fig. 2.9. The first and second iterates of C determine an subinterval $[f^2(C), f(C)] \subset I$. If one starts an iteration from a point outside U, then it only takes a finite number of steps to get into U. These finite steps form a trivial transient. Once in U, the orbit can never get out of it, i.e.,

$$f(U) \subset U.$$

The subinterval U is called a *dynamical invariant range.* In chaotic dynamics we are mainly interested in the asymptotic behavior at $t \to \infty$, therefore, only those symbolic sequences which start from points in U will matter.

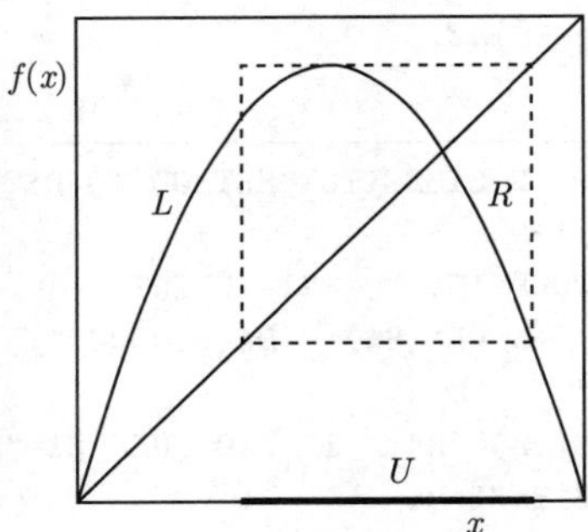

Figure 2.9 The dynamical invariant range $U = [f^2(C), f(C)]$. All points in U have symbolic sequences no larger than the kneading sequence $K = f(C)$

The symbolic sequence that starts from the first iterate of C plays an important role in symbolic dynamics of unimodal maps. It acquires a special name *kneading sequence* (Milnor and Thurston [1977, 1988]). It is often denoted by $K = f(C)$ in accordance with our convention of naming a sequence by its initial point. A unimodal map has only one kneading sequence. It is the largest sequence in U.

Perhaps this is the right place to say a few more about transients. Suppose that we are at a parameter value where there exists, say, a stable periodic orbit. We calculate the orbit, starting from an initial point which is not located exactly on the periodic orbit. If our computer has an infinite precision, we would never reach the periodic regime, always staying in a transient process. Only owing to the finite precision of the computer we will sooner or later get to a final stationary state of periodic motion. However, in a symbolic description we will have a sequence which eventually shows a periodic pattern. Therefore, symbolic dynamics includes the transients naturally. This is yet another advantage of coarse-graining in symbolic dynamics. In the mathematical literature there is a term "eventually periodic sequence" which includes both transients and non-transients. The $\rho\lambda^\infty$ type kneading sequences to be studied later in Section 2.6.2 belong to eventually periodic sequences.

2.3.4 The Admissibility Condition

The admissibility condition is based on the ordering of symbolic sequences and follows from the fact that the kneading sequence K starting from point $f(C)$ is the largest sequence in the dynamical invariant range U.

Given an arbitrary symbolic sequence Σ, we denote by $\mathcal{L}(\Sigma)$ the set of all subsequences that follow a letter L in the given sequence, and by $\mathcal{R}(\Sigma)$ the set of all subsequences that follow a letter R. For example, for the symbolic sequence $\Sigma = (RLRRL)^\infty$ we have

$$\begin{aligned}\mathcal{L}(\Sigma)&=\{(RRLRL)^\infty,(RLRRL)^\infty\},\\ \mathcal{R}(\Sigma)&=\{(LRRLR)^\infty,(RLRLR)^\infty,(LRLRR)^\infty\}.\end{aligned}\tag{2.50}$$

The sets $\mathcal{L}(\Sigma)$ and $\mathcal{R}(\Sigma)$ contain all shifts of Σ, therefore, we will call them the *shifts* of Σ. To be specific, $\mathcal{L}$ and $\mathcal{R}$ are sometimes called L-shifts and R-shifts of Σ, respectively. Using the $\mathcal{L}$ and $\mathcal{R}$ notations, we formulate the admissibility condition as follows.

Admissibility condition A symbolic sequence Σ is admissible to a kneading sequence K if and only if the following conditions are satisfied:

$$\mathcal{L}(\Sigma)\leqslant K \quad \text{and} \quad \mathcal{R}(\Sigma)\leqslant K.\tag{2.51}$$

When the kneading sequence K corresponds to an isolated point on the interval, e.g., K contains C, condition (2.51) will be sharpened to strict inequalities. We shall not repeat this remark when formulating the admissibility conditions for other maps.

Example Taking $K = (RLL)^\infty$, the symbolic sequence $\Sigma = (RLRRL)^\infty$ is admissible, because the sets $\mathcal{L}$ and $\mathcal{R}$, listed above in (2.50), are all smaller than K. However, for the same kneading sequence K, the symbolic sequence $RLRLLLR\cdots$ is forbidden, since we have now

$$K < RLLLR\cdots \in \mathcal{L}(RLRLLLR\cdots),$$

which violates the admissibility condition.

A kneading sequence K itself should satisfy the admissibility condition as well. Since $\mathcal{L}(K)$ and $\mathcal{R}(K)$ contain all shifts $\mathcal{S}^k(K), k = 1,2,\cdots$ of K, the condition (2.51) simply means

$$\mathcal{S}^k(K)\leqslant K, \quad k=0,1,2,\cdots.\tag{2.52}$$

In other words, a kneading sequence must be a *shift-maximal* sequence. When $f(C)$ changes with the parameter, kneading sequence can change. For unimodal maps, a shift-maximal sequence may become a kneading sequence at some parameter value. For more complicated maps there are other conditions to satisfy, see, e.g., Chapter 3 on maps with multiple critical points. Given a kneading sequence, an admissible symbolic sequence always corresponds to an actual orbit of the map, i.e., there exists an initial point leading to that orbit. Correspondingly, non-admissible sequences are sometimes called forbidden words.

An admissible sequence may be larger than the kneading sequence K. For example,

$$RL^3(RLRRL)^\infty > K \equiv (RLL)^\infty.$$

However, at most one shift will make an allowable sequence smaller than the kneading sequence. Physically, it corresponds to a trivial transient from an initial point located outside the dynamical invariant range U, as we have discussed in connection with Fig. 2.9. If we are not interested in trivial transient processes, we can confine ourselves to the dynamical invariant range only. All points in this range have symbolic sequences not larger than the kneading sequence.

Twofold meaning of the admissibility condition

The reader may have noticed the two aspects of the admissibility condition. The kneading sequence K plays the role of parameter for the map. When K is given, the condition (2.51) picks up those symbolic sequences which may be produced by the map by taking suitable initial points. It deals with symbolic sequences in the phase space at fixed parameter value. On the other hand, condition (2.52) works in the parameter space and picks up a symbolic sequence which may become a kneading sequence at some parameter. The parameter value, if necessary, may be calculated by using the word-lifting technique or the bisection method.

Now let us prove two simple assertions related to kneading sequences in unimodal maps.

A. *All kneading sequences, except L^∞ and R^∞, must start with RL.*

Suppose that there are admissible words which do not agree with this assertion, then they must begin with either L^nR $(n \geqslant 1)$ or R^mL $(m \geqslant 2)$.

However, both leading strings violate the shift-maximality condition, since

$$L^nR\cdots < \mathcal{S}^n(L^nR\cdots) = R\cdots\,,$$
$$R^mL\cdots < \mathcal{S}^{m-1}(R^mL\cdots) = RL\cdots\,.$$

Therefore, we have proved the assertion.

B. *A kneading sequence cannot start with a repetition of an odd string except when it is the periodic word made of that string.*

Suppose an admissible word starts with Δ^2, where Δ is an odd string. Denote the length of Δ, i.e., the number of symbols it contains, by $n = |\Delta|$. We can write $K = \Delta\Delta\Gamma\cdots$, where $\Gamma\cdots \neq K$ and $|\Gamma| = n$. Since K is shift-maximal, we must have

$$\Delta\Delta\Gamma\cdots \geqslant \mathcal{S}^n(K) = \Delta\Gamma\cdots\,.$$

However, an odd string reverses the order, so it follows from the above line that $\Delta\Gamma\cdots \leqslant \Gamma\cdots$, i.e., $\Delta \leqslant \Gamma$. Furthermore, we have

$$\Delta\Delta\Gamma\cdots \geqslant \mathcal{S}^{2n}(K) = \Gamma\cdots\,,$$

which yields $\Delta \geqslant \Gamma$. We must admit that $\Gamma = \Delta$. Repeating the arguments, we arrive at

$$K = \Delta^3\cdots = \Delta^4\cdots = \cdots = \Delta^\infty,$$

in contradiction with our assumption that $K \neq \Delta^\infty$. The assertion is proved.

To conclude this Section, a BASIC program which checks the shift-maximality of a given word ending with a letter C is given in Program 2.3.

```
  5  REM PROGRAM 2.3 TO CHECK ADMISSIBILITY
 10  DEFINT I-P: PRINT "PLEASE SET CAPS-LOCK KEY ON."
 20  INPUT "THE WORD TO BE CHECKED IS: ", S$
 30  N=LEN(S$)
 40  FOR I=1 TO N-1:  P=1:  FOR J=1 TO N-1
 50  IF MID$(S$, J, 1)="R" THEN P=-P
 60  IF MID$(S$, J, 1)<>MID$(S$, J+I, 1) THEN GOTO 80
 70  NEXT J
 80  IF P=1 THEN PRINT "THE WORD IS INADMISSIBLE.":  STOP
 90  NEXT I: PRINT "THE WORD IS ADMISSIBLE.":  STOP
100  END
```

Program 2.3

2.4 The Periodic Window Theorem

Analytically, the existence of a periodic window around each parameter value for a superstable orbit follows from continuity considerations in the vicinity of the point where the superstable condition (2.12) holds. Moreover, using the language of symbolic dynamics, we are able to write down the symbolic sequences representing the whole window. The width, i.e., the end points of the window, depend, however, on the mapping explicitly (see Section 2.1.1). Consequently, it cannot be determined from symbolic dynamics alone.

2.4.1 The Periodic Window Theorem

In this Section, we treat merely the symbolic dynamics aspect of the problem, namely, we are going to prove the following

Periodic Window Theorem (Zheng [1989d]) If $\Sigma C \equiv s_0 s_1 \cdots s_n C$ is a kneading sequence, where $s_i = R$ or L, then $(\Sigma t)^\infty$, where $t = R$ or L, are both kneading sequences.

The proof of the Periodic Window Theorem employs a few tricks that are typical in manipulations with symbolic sequences. For this reason, we present it in some detail and advise the reader to go through it carefully.

Proof of the Periodic Window Theorem

In unimodal maps a symbolic sequence being a kneading sequence is equivalent to its shift-maximality. Let us denote $s_0 s_1 \cdots s_n$ by X. We are required to prove that $(Xt)^\infty$ is shift-maximal from the shift-maximality of XC, where t is either R or L. Suppose $Y = \mathcal{S}^k(X)$, where $\mathcal{S}$ is the shift operator (1.8) and $0 < k < |X|$, $|X|$ being the length of X. The shift-maximality of $(Xt)^\infty$ means

$$XtXt \cdots \geqslant YtXt \cdots . \tag{2.53}$$

The shift-maximality of XC implies

$$XC > YC, \tag{2.54}$$

as XC differs from YC. Now, if Y is not a leading string of X, then (2.53) holds as an inequality due to (2.54). If Y is a leading string of X, let us write $X = YsZ$, where s is a single letter and Z is a string which may be blank.

Relation (2.54) now means

$$YsZC > YC. \tag{2.55}$$

There are two ways for this inequality to hold: either Y is even and $s = R$ or Y is odd and $s = L$. In both cases the string Ys is odd and we have to compare $YsZt$ with Yt to check (2.53). If $s \neq t$, (2.53) holds naturally, as an odd principal string Ys is always greater than an even string Yt.

If $s = t$, then both sides of (2.53) starts with one and the same odd leading string Ys. Discarding this common string, we must reverse the inequality to have:

$$ZtXt \cdots \leqslant XtXt \cdots . \tag{2.56}$$

This relation looks the same as (2.53) except that Y has been replaced by Z. From the same kind of reasoning it follows either (2.56), i.e., (2.53), holds as an inequality or Zt is a leading string of Xt. In the latter case, we have

$$Xt = ZtYt = YtZt. \tag{2.57}$$

Inserting this relation into (2.53), we see that (2.53) now holds as an equality. Therefore, we have proved the theorem.

Upper and lower sequences in a periodic window

The Periodic Window Theorem tells us that, whenever there is a superstable orbit described by the symbolic sequence ΣC, one can change the letter C to either R or L to extend ΣC into a window. We denote

$$\begin{aligned}(\Sigma C)_+ &\equiv \max(\Sigma R, \Sigma L),\\ (\Sigma C)_- &\equiv \min(\Sigma R, \Sigma L).\end{aligned} \tag{2.58}$$

The three kneading sequences, describing a window, are ordered as follows:

$$(\Sigma C)_-^\infty < \Sigma C < (\Sigma C)_+^\infty. \tag{2.59}$$

We shall call $(\Sigma C)_-^\infty$ the *lower sequence*, and $(\Sigma C)_+^\infty$ the *upper sequence* of ΣC. It is easy to see that the parity of the basic string, i.e., the unit of repetition, of the lower sequence is always $+1$, whereas that of the upper sequence is -1. When we write a triple $[(\Sigma C)_-, \Sigma C, (\Sigma C)_+]$ to represent a window, the infinite power of the upper and lower sequences are always dropped. The notion of upper and lower sequences will be used in the next section to construct the shortest admissible sequence between two given sequences.

Since the parity of R and L reflects the sign of the derivative of the mapping function on the corresponding branch, we may assign a parity "0" to the letter C since the derivative of $f(x)$ at C is zero. The parities of the triple $(+, 0, -)$, written in the order as they appear in the window, may be called the *signature* of the window.

Kneading sequences for the period-doubling cascade

As the first application of the Periodic Window Theorem we write down the kneading sequences for the main period-doubling cascade. The superstable fixed point C extends to a window (L, C, R), where we have omitted the infinite power in the upper and lower sequences. The leftmost orbit L^∞, as we know, is the smallest among all possible sequences. In the bifurcation diagram we see that a period-doubling bifurcation occurs at the right-hand end of the fixed point window. The bifurcation the parameter is determined by the condition (2.14)

$$\frac{\partial f}{\partial x} = -1,$$

so symbolic dynamics cannot tell us where it exactly is. The knowledge of the mapping function is required, not only the monotonicity of its branches. However, symbolic dynamics does indicate that a bifurcation may take place at the right end of the triple due to its signature $(+, 0, -)$.

We draw a positive conclusion from the negative statement that symbolic dynamics does not know the whereabout of the bifurcation point: the upper kneading sequence R^∞ should remain the same as the parameter passes the bifurcation point. In order to indicate that the period has doubled, we write the sequence as $(RR)^\infty$ after the bifurcation. Further smooth change of the parameter will make an R to become a C at some point, then it goes into L. This means that the next, period 2, window is

$$(RR,\ RC,\ RL),$$

Repeating the above argument, we get the period 4, 8, and so on. They are

$$(RLRL,\ RLRC,\ RLRR),$$
$$(RLRRRLRR,\ RLRRRLRC,\ RLRRRLRL).$$

In general, a periodic window and its period-doubled neighbor are described by

$$
\begin{aligned}
&((\Sigma C)_-, \Sigma C, (\Sigma C)_+) \to \\
&\qquad ((\Sigma C)_+(\Sigma C)_+, (\Sigma C)_+\Sigma C, (\Sigma C)_+(\Sigma C)_-).
\end{aligned}
\tag{2.60}
$$

We see that the rightmost sequence in a window coincides with the leftmost sequence in the next window, if written in their "infinite" form $(RR)^\infty$, $(RLRL)^\infty$, and $(RLRRRLRR)^\infty$, $((\Sigma C)_+)^\infty$, etc., leaving no room for other words to get in between. This remark will become even more clearer later, after we learn how to generate a median word between two given words.

The converse of the periodic window theorem

Suppose $(\Sigma R)^\infty$ or $(\Sigma L)^\infty$ or both are kneading sequences, where Σ is a string made of R and L, can we replace the last R or L by C to get a shift-maximal sequence ΣC? If so, this would provide a converse of the Periodic Window Theorem.

The fact that the converse of the Periodic Window Theorem may not be true can be seen from a counter example. Obviously, the sequence $R^\infty = (RRR)^\infty$ is shift-maximal, so it may become a kneading sequence. However, the sequence $(RRC)^\infty$ is not shift-maximal. Nonetheless, we may formulate the following amended converse of the Periodic Window Theorem. Let s_i, $i = 0, 1, \cdots$ and t be either R or L. If a periodic symbolic sequence $(s_0 s_1 \cdots s_n t)^\infty$ is a kneading sequence and the string $s_0 s_1 \cdots s_n t$ cannot be reduced to the repetition of a shorter string, then the sequence $s_0 s_1 \cdots s_n C$ is shift-maximal, i.e., a kneading sequence candidate.

The proof of this converse theorem goes much like the Periodic Window Theorem. A difference comes only at the last step. If (2.53) holds as an inequality, it implies (2.54). If it holds as an equality, then (2.57) takes place. In the latter case, we might as well let the longer one of Y and Z be Z. Then from (2.57) follows that $Z = YtZ'$. Again from (2.57) it follows that $YtZ't = Z'tYt$. Iterated use of the same argument eventually leads to the conclusion that the strings Zt and Yt must be repetition of a certain basic string. The length of this basic string is the largest common divisor of $|Yt|$ and $|Zt|$. Obviously, when this basic string is of odd parity, one cannot replace its last symbol by C.

2.4.2 Construction of Median Words

An important application of the Periodic Window Theorem is the construction of the shortest kneading sequence between two given kneading sequences.

Suppose that we are given two kneading sequences $\Sigma_1 C < \Sigma_2 C$. It follows from the Periodic Window Theorem that the upper sequence $[(\Sigma_1 C)_+]^\infty$ of the smaller word $\Sigma_1 C$ and the lower sequence $[(\Sigma_2 C)_-]^\infty$ of the larger word $\Sigma_2 C$ are both kneading sequences. They are ordered as

$$\Sigma_1 C < [(\Sigma_1 C)_+]^\infty \leqslant [(\Sigma_2 C)_-]^\infty < \Sigma_2 C. \tag{2.61}$$

If

$$\begin{aligned} [(\Sigma_1 C)_+]^\infty &= \Sigma^* \mu \cdots , \\ [(\Sigma_2 C)_-]^\infty &= \Sigma^* \nu \cdots , \end{aligned} \tag{2.62}$$

where Σ^* is the longest common leading string of the two sequences and $\mu \neq \nu$, then $\Sigma^* C$ is the median word we are looking for.

It is an obvious fact that all symbolic sequences between $\Sigma_1 C$ and $\Sigma_2 C$ must have Σ^* as their common leading string with $\Sigma^* C$ being the shortest one. We have to prove only the shift-maximality of $\Sigma^* C$.

Suppose that $\Sigma^* C$ is not shift-maximal. There must exist a $k \leqslant |\Sigma^*|$ such that $\Sigma^* C < \mathcal{S}^k(\Sigma^* C)$. It is easy to see that one of $\mathcal{S}^k(\Sigma^* \mu)$ and $\mathcal{S}^k(\Sigma^* \nu)$ must be greater than $\mathcal{S}^k(\Sigma^* C)$. Suppose it is $\mathcal{S}^k(\Sigma^* \mu)$ and we are led to the conclusion that

$$\Sigma^* \mu \cdots < \mathcal{S}^k(\Sigma^* C) < \mathcal{S}^k(\Sigma^* \mu \cdots),$$

which contradicts the shift-maximality of $\Sigma^* \mu \cdots$. Therefore, we have proved the shift-maximality of $\Sigma^* C$.

If $[(\Sigma_1 C)_+]^\infty = [(\Sigma_2 C)_-]^\infty$, then no median word exists between $\Sigma_1 C$ and $\Sigma_2 C$. In fact, this is the case when $\Sigma_1 C$ and $\Sigma_2 C$ are adjacent neighbors in a period-doubling cascade.

In the original MSS paper (Metropolis, Stein and Stein [1973]) median words were generated with the help of two auxiliary constructions, called *harmonics* and *anti-harmonics* of symbolic sequences. The anti-harmonics are not shift-maximal themselves. The extension of these notions to maps other than the unimodal map gets very cumbersome, if ever possible (see, e.g., Zeng, Ding and Li [1985, 1988]). Our method, based on the Periodic Window Theorem, deals with shift-maximal sequences all the time. It is easily extensible to maps with multiple critical points in many, if not all, cases. Therefore, we will not discuss the obsolete notions of harmonic and anti-harmonics any more.

A BASIC program to generate a median kneading sequence between two given ones is listed in Program 2.4. The length of the median word is limited by a preset number.

```
  5  REM PROGRAM 2.4 FOR CONSTRUCTION OF MEDIAN WORD
 10  DEFINT I-T: CLS
 20  INPUT "THE SMALLER WORD IS:", S1$
 30  INPUT "THE GREATER WORD IS:", S2$
 40  INPUT "THE MAXIMAL LENGTH OF THE MEDIAN WORD IS:", N
 50  N1=LEN(S1$):  N2=LEN(S2$)
 60  NN=N1-1:  S$=MID$(S1$, 1, NN): P=1 :  GOSUB 200:  S1$=S$
 70  NN=N2-1:  S$=MID$(S2$, 1, NN): P=-1 :  GOSUB 200:  S2$=S$
 80  FOR I=0 TO N-1:  I1=(I MOD N1) + 1:  I2=(I MOD N2) + 1
 90  IF MID$(S1$, I1, 1)<>MID$(S2$, I2, 1) THEN GOTO 120
100  NEXT I
110  PRINT "NO MEDIAN WORD SHORTER THAN "; N; ".":  STOP
120  S$=MID$(S$, 1, I1-1) + "C":  IF I>N1-1 THEN S$=S1$ + S$
130  PRINT "THE MEDIAN WORD IS: "; S$; ".":  STOP
200  FOR I=1 TO NN
210  IF MID$(S$, I, 1) ="R" THEN P=-P;
220  NEXT I
230  C$="R":  IF P=-1 THEN C$="L"
240  S$=S$ + C$:  RETURN
250  END
```

Program 2.4

2.4.3 The MSS Table of Kneading Sequences

We know that among all superstable periodic kneading sequences with period $\leqslant n$ the smallest is C and the largest is $RL^{n-2}C$. These two superstable periodic words generate the first level median word. This median word together with the two previous words generate the second level median words. Carrying out this procedure and keeping only those superstable words whose length does not exceed n, we get a list of superstable periodic kneading sequences up to length n. We call it a *MSS Table* . An infinite collection of all possible superstable periodic kneading sequences will be called a *complete MSS Table*, although one can only generate a finite part of it.

To conclude this Section we list a BASIC program which generates a MSS Table up to period 7 in Program 2.5. Let us make a few remarks on this program. Statements 50 and 60 transform an upper sequence into a lower one. Statements 70 through 100 generate a doubled period. Subroutine, which starts from statement 1000, produces a median word. The string **S$** is dimensioned to 1000 in order to contain any kneading sequence thus produced. This program may easily be modified to generate MSS Table up to a longer period.

```
   2  REM PROGRAM 2.5 TO GENERATE A MSS TABLE
   4  REM S$:  KNEADING SEQUENCE; L: LENGTH OF S$
  10  DIM S$(1000), L(1000)
  30  NMAX=7:  NHALF=3:  IA=1:  IB=1000
  40  S$(1)="L":  L(1)=1:  S$(1000)="RLLLLLR":  L(1000)=7
  50  IF MID$(S$(IA), L(IA), 1)="L" THEN C$="R" ELSE C$="L"
  60  S$(IA)=MID$(S$(IA), 1, L(IA)-1)+ C$
  70  IF L(IA)>NHALF THE GOTO 110
  80  IF C$="L" THEN C$="R" ELSE C$="L"
  90  S$(IA+1) = S$(IA)+MID$(S$(IA), 1, L(IA)-1) + C$
 100  IA=IA+1:  L(IA)=2*L(IA-1):  GOTO 70
 110  S1$=S$(IA): S2$=S$(IB)
 120  FOR M=1 TO 500:  GOSUB 1000
 130  IF ID=2 THEN GOTO 160
 140  IB=IB-1:  L(IB)=K+1:  S$(IB)=SS$:  S2$=SS$
 150  NEXT M
 160  IA=IA+1:  S$(IA)=S$(IB): L(IA)=L(IB): IB=IB+1
 170  IF IB<1001 THEN GOTO 50
 180  FOR I=1 TO IA
 190  PRINT I TAB(8) MID$(S$(I), 1, L(I)-1);"C" TAB(18) L(I)
 200  NEXT I: STOP
1000  L1=LEN(S1$):  L2=LEN(S2$)
1010  FOR K=0 TO NMAX-1
1020  K1=(K MOD L1) + 1:  K2=(K, MOD L2) + 1
1030  IF MID$(S1$, K1, 1)<>MID$(S2$, K2, 1) THEN GOTO 1050
1040  NEXT K: ID=2:  RETURN
1050  SS$=MID$(S1$, 1, K1):  IF K>L1-1 THEN SS$=S1$ + SS$
1060  ID=0:  RETURN: END
```

Program 2.5

Table 2.2 The MSS Table up to period 7. nP stands for period n. nI—for n-piece chaotic band. PDB—peiod-doubling bifurcation cascade

No.	K	Period	Remarks
1	C	1	The superstable fixed point, start of 2^n PDB
2	RC	2	The only $2P$, doubling of C, next in PDB
3	$RLRC$	4	Doubling of R, next in the above PDB
4	RLR^3C	6	$3P$ embedded in $2I$, start of $2 \times 3 \times 2^n$ PDB
5	RLR^4C	7	The 1st $7P$, start of first 7×2^n PDB
6	RLR^2C	5	$5P$ embedded in $1I$, start of 1st 5×2^n PDB
7	RLR^2LRC	7	The 2nd $7P$, start of 2nd 7×2^n PDB

Continued

No.	K	Period	Remarks
8	RLC	3	The only $3P$, start of 3×2^n PDB
9	RL^2RLC	6	Doubling of RL, next in the above PDB
10	RL^2RLRC	7	The 3rd $7P$, start of 3rd 7×2^n PDB
11	RL^2RC	5	The 2nd $5P$ embedded in $1I$, start of 5×2^n
12	RL^2R^3C	7	The 4th $7P$, start of 4th 7×2^n PDB
13	RL^2R^2C	6	The 1st primitive $6P$, start of a 6×2^n PDB
14	RL^2R^2LC	7	The 5th $7P$, start of 5th 7×2^n PDB
15	RL^2C	4	The only primitive $4P$, start of 4×2^n PDB
16	RL^3RLC	7	The 6th $7P$, start of 6th 7×2^n PDB
17	RL^3RC	6	The 2nd primitive $6P$, start of a 6×2^n PDB
18	RL^3R^2C	7	The 7th $7P$, start of 7th 7×2^n PDB
19	RL^3C	5	The last $5P$, start of 3rd 5×2^n PDB
20	RL^4RC	7	The 8th $7P$, start of 8th 7×2^n PDB
21	RL^4C	6	The last $6P$, start of last 6×2^n PDB
22	RL^5C	7	The 9th and last $7P$, start of last 7×2^n PDB

2.4.4 Nomenclature of Unstable Periodic Orbits

Unstable periodic orbits are highly individual objects. Numerically, one can follow an unstable period only within a rather limited time; sooner or later one gets away from it. Symbolically, two different unstable periods should not share the same name. A good symbolic dynamics should be capable to assign a unique name to each and every unstable periodic orbit. This is the first practical criterion to evaluate a symbolic dynamics. Then comes the requirement of ordering, admissibility condition, etc.

A stable periodic orbit in the quadratic map comes into life under the name of someone else. First, at a period-doubling bifurcation the upper sequence $[(\Sigma C)_+]^\infty$ of ΣC continues to be an unstable period, while the new, period-doubled orbit is born with the name $[(\Sigma C)_+(\Sigma C)_+]^\infty$, symbolically identical to the old period. It acquires its own name only after passing the superstable point and keeps this name all the way up to next superstable point on the parameter axis. Second, at a tangent bifurcation a pair of stable and unstable orbits come into existence under the same name. For example, at the beginning of the period 3 window, the unstable and stable orbits are born with the same name $(RLR)^\infty$. The stable period later acquires its own name $(RLL)^\infty$ while the unstable orbit keeps name $(RLR)^\infty$ from the very beginning.

Therefore, the MSS Table also helps us to enumerate all possible unstable periodic orbits up to a certain length. Table 2.3 lists symbolic names of all unstable periodic orbits up to length 7. Only the basic string is written without the ∞ power. A symbol C stands for both R and L, so RLC stands for $(RLR)^\infty$ and $(RLL)^\infty$.

Table 2.3 Nomeclature of unstable periodic orbits in the unimodal map up to period 7. Only the basic strings are written. The symbol C stands for both R and L

Period	Words	Number
1	L, R	2
2	RL	1
3	RLC	2
4	$RLRR, RLLC$	3
5	RLR^2C, RL^2RC, RL^3C	6
6	$RLLRLR, RLR^3C, RL^2R^2C, RL^3RC, RL^4C$	9
7	$RLR^4C, RLR^2LRC, RL^2RLRC, RL^2R^3C$ $RL^2R^2LC, RL^3RLC, RL^3R^2C, RL^4RC, RL^5C$	18

Some remarks are now in order. The fixed point L^∞ is quite specific. In the quadratic map it lies outside the chaotic attractor until the last parameter value $\mu = 2$, where it collides with the attractor, causing the last *crisis* in the system, see Section 2.6.1 for more. This explains why we do not combine it with R^∞ to write a C. In Table 2.3, a word not ending with C comes from period-doubling. Suppose among all period n orbits $P(n)$ come from period-doubling and $M(n)$ come from tangent bifurcation, then the total number of unstable periodic orbits of length n is $P(n) + 2M(n)$. This number is shown in the last column of Table 2.3. The calculation of $P(n)$ and $M(n)$ for any n will be given in Chapter 7.

2.5 Composition Rules

There are many ways to generate new shift-maximal sequences from known ones. Historically, the first such rule is the $*$-composition, introduced by Derrida, Gervois and Pomeau [1978]. The $*$-composition is a powerful tool to describe self-similarity seen in the bifurcation diagram of an unimodal map. It also helps to understand the fine structure of power spectra of some periodic regimes and thus provides a link to experimental observation. The $*$-composition happens to be a particular case of a generalized composition rule to be introduced later in this Section. The generalized composition rule leads naturally to the notion "coarse-grained chaos".

2.5.1 The $*$-Composition

The $*$-composition of a finite sequence P with another sequence $Q = q_1 q_2 \cdots$, which may be infinite, is defined as

$$P * Q = (P * q_1)(P * q_2) \cdots , \tag{2.63}$$

where

$$P * q_i = \begin{cases} (PC)_+, \\ PC, \\ (PC)_-, \end{cases} \quad \text{if} \quad q_i = \begin{cases} R, \\ C, \\ L. \end{cases} \tag{2.64}$$

The definition given here differs slightly from the original one of Derrida, Gervois and Pomeau. It utilizes the notion of upper and lower sequences $(PC)_\pm$ of PC defined in (2.59) and turns out to be more convenient in practice.

Clearly, the $*$-composition preserves parity. That is, $P * R$ is always odd whereas $P * L$ is even. The period-doubling sequence of Feigenbaum, written in terms of the $*$-composition, reads simply

$$C, RC, RLRC, RLR^3LRC, \cdots = \{R^{*n}C\}, \quad n = 0, 1, 2, \cdots , \tag{2.65}$$

where we have used the notation $R^{*n} \equiv (R*)^n$.

We have seen in (2.60) of Section 2.4.1 that a superstable shift-maximal word ΣC has a period-doubled neighbor $(\Sigma C)_+ \Sigma C$. Using the $*$-composition, the latter can be written as

$$(\Sigma C)_+ \Sigma C = \Sigma * (RC). \tag{2.66}$$

In general, there exists a period-doubling sequence $\Sigma * R^{*n}C$ with increasing periods at one side of any superstable kneading sequence ΣC.

In general, the $*$-composition is non-commutative: $P*Q \neq Q*P$. However, it is associative: $(P*Q)*W = P*(Q*W)$, so there is no need to put parentheses in $P * Q * W$. The words P, Q, and W are taken to be finite in the above relations.

The $*$-composition as substitutions

More conveniently, the $*$-composition $P*Q$ may be understood as the following symbol substitutions applied to the word Q:

$$\begin{aligned} R &\to (PC)_+, \\ C &\to PC, \\ L &\to (PC)_-. \end{aligned} \tag{2.67}$$

These substitutions of the $*$-composition preserves order, i.e.,

$$Q_1 < Q_2 \Leftrightarrow P * Q_1 < P * Q_2. \tag{2.68}$$

Consequently, for $Q_1 < Q < Q_2$ and $P * Q_1 < P * Q_2$ there exists $P * Q$ such that $P * Q_1 < P * Q < P * Q_2$; the converse is also true.

The $*$-composition also preserves shift-maximality. Namely, if both Q and PC are kneading sequences, then $P * Q$ is also a kneading sequence. We shall not prove this statement here, as it is a particular case of the generalized composition rule to be discussed later.

Self-similarity in the bifurcation diagram

The self-similarity of the bifurcation diagram that we have seen before reflects the invariance of the complete MSS Table under the $*$-composition. Denote by $[L^\infty, RL^\infty]$ the complete MSS Table from its minimal sequence L^∞ to the maximal sequence RL^∞, and by $R*[L^\infty, RL^\infty]$—the set of symbolic sequences obtained by taking $*$-composition of R with every kneading sequence in the MSS Table, we have

$$R * [L^\infty, RL^\infty] = [(RR)^\infty, RL(RR)^\infty] = [R^\infty, RLR^\infty]. \tag{2.69}$$

It is easy to check that

$$L^\infty < R^\infty < RLR^\infty < RL^\infty.$$

This shows that $[R^\infty, RLR^\infty]$ is a subtable embedded in the complete MSS Table. It starts from the first period-doubling bifurcation point and ends at the point where two chaotic bands merge into one. In the bifurcation diagram it is seen as two copies of the whole diagram, one above the other with a flip.

There is a one-to-one correspondence between this subtable and the complete MSS Table.

We may take one more $*$-composition to get another subtable

$$R * R * [L^\infty, RL^\infty] = [(RL)^\infty, RLRR(RL)^\infty].$$

It is also in one-to-one correspondence with the complete MSS Table. In the bifurcation diagram it is seen as four copies of the whole diagram, stacked one above another in the region between the $2 \to 4$ bifurcation point to the $4 \to 2$ band-merging point. There are infinitely many subtables of this kind, related to period-doublings.

In addition, there exist infinitely many subtables related to tangent bifurcations. For example, we may take the period 3 word RLC to form

$$RL * [L^\infty, RL^\infty] = [(RLR)^\infty, RLL(RLR)^\infty],$$

which also has one-to-one correspondence to the complete MSS Table. It starts from the tangent bifurcation to period 3 and ends at the merging point of the three chaotic bands. In the bifurcation diagram this is seen as three copies of the whole diagram in the period 3 window and its chaotic region.

Within each subtable associated with a tangent bifurcation, there are infinitely many subtables related to both period-doublings and tangent bifurcations. This nested hierarchy repeats *ad infinitum*.

Intra-word self-similarity

What we have just studied addresses the self-similarity of the MSS Table. There exists self-similarity within one and the same word in some limiting regime. For example, at the period-doubling limit $R^{*\infty}$ we have obviously $R * R^{*\infty} = R^{*\infty}$, i.e., $R^{*\infty}$ remains invariant under $*$-composition. Compared with $R * R^{*\infty}$, $R^{*\infty}$ numerically corresponds to keeping one point after every two iterations. These sampled points, after properly rescaled, restore the original orbit. This intra-word self-similarity reflects in the distribution of orbital points at the accumulation point of the period-doubling cascade. As shown in Fig. 2.10, when properly enlarged, a small part of the phase space resembles the whole orbit. To be put in correspondence with the whole, sometimes the enlarged part must be inverted due to the effect of order reversing R. Fig. 2.10 shows such a case.

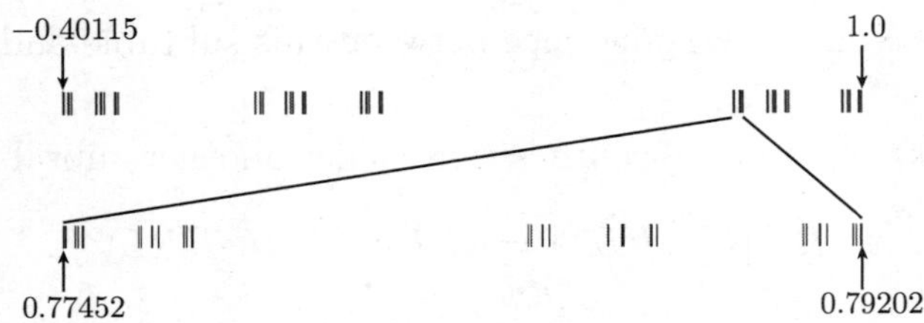

Figure 2.10 Spatial self-similarity in the distribution of orbital points at the accumulation point μ_∞ of the period-doubling cascade

The kneading sequence $R^{*\infty}$ can be compared with a quasi-periodic chain in one-dimensional physics. Suppose R and L are "atoms". Starting from a single "atom" we perform a *doubling transformation* $\mathcal{D}$

$$\mathcal{D}:\begin{cases} R \to RL, \\ L \to RR, \end{cases} \tag{2.70}$$

to get

$$R \to RL \to RLRR \to RLRRRLRL \to \cdots .$$

The infinite limit of these substitutions is a quasi-periodic chain. It has an approximate period 2^n, $n = 1, 2, \cdots$. The kneading sequence at the period-doubling limit is identical to this quasi-periodic chain made of two kinds of "atoms".

One may list infinitely many other quasi-periodic orbits. For instance, from an initial word RLL one gets the period-doubling limit of the period 3 orbit by applying the substitutions (2.70) repeatedly. To the same word RLL one may apply another set of substitutions

$$\begin{aligned} R &\to RLL, \\ L &\to RLR, \end{aligned} \tag{2.71}$$

to get the quasi-periodic limit of the infinite period-tripling sequence. We will return to period-n-tupling sequences later in this Section.

Composite words and fine structure in power spectra

Symbolic sequences that may be obtained by taking $*$-composition of two or more words are called *composite words.* Those that cannot be decomposed by $*$-composition are *primitive words.* A $P * Q$ orbit has both coarse-grained and fine-grained structures, the former being described by P and the latter—by Q. Suppose Q is a periodic orbit of period n_q and the length of PC is n_p. The composite orbit will look like PC globally. However, each point of PC now becomes a small island, whose inner structure resembles the Q orbit. The composite structure of orbits manifests itself in the fine structure of the power spectrum of the data. For example, in Fig. 2.11 we show the power spectra of two different period 6 orbits $RL * RC$ and $R * RLC$.

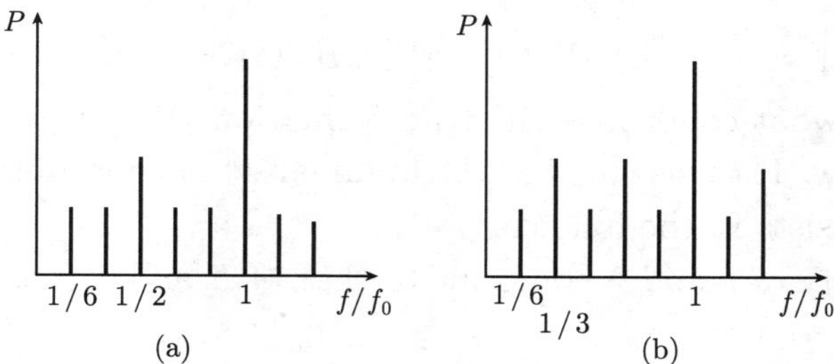

Figure 2.11 Sketch showing fine structure of power spectra for $RL * RC$ (a) and $R * RLC$ (b), where f_0 denotes the fundamental frequency of the system

The fine structure of power spectra is useful in telling the whereabout of periodic orbits in experiments with chaotic systems. For example, in a modern repetition of the Faraday experiment of 1831 it was claimed that deviation from period-doubling sequence was observed (Keolian *et al.* [1981]). Fortunately, the power spectrum of the deviated period was given in the report. It exhibits a clear fine structure of $(RL^6) * RC$ orbit, i.e., an orbit in the two-band region, which should not be confused with the $R^{*4}C$ in the fundamental period-doubling bifurcation cascade.

2.5.2 Generalized Composition Rule

In spite of the usefulness of the $*$-composition in elucidating the self-similarities in the bifurcation diagram and the phase space self-similarity of quasi-periodic orbits, the strict rule of symbol substitutions has limited its application to the analysis of chaotic orbits, intermittency, and other important phenomena. Therefore, one needs to extend the $*$-composition. In what follows we introduce a generalized composition rule, which includes the $*$-composition as a particular case. Postponing the somewhat technical proof of the generalized composition rule to the next Section, which may be skipped at first reading, we first formulate the rule itself and discuss its applications.

The substitutions (2.67), (2.70) and (2.71) mentioned in the last section hint on the consideration of more general symbol substitutions. Let us take a general substitution rule $\mathcal{W}$:

$$\begin{aligned} R \to \mathcal{W}(R) &= \rho \equiv r_1 r_2 \cdots r_u, \\ L \to \mathcal{W}(L) &= \lambda \equiv l_1 l_2 \cdots l_v, \end{aligned} \tag{2.72}$$

where finite strings ρ and λ contain R and L only. A symbolic sequence transforms as

$$\Sigma = s_0 s_1 \cdots s_n \cdots \to \mathcal{W}(\Sigma) = \mathcal{W}(s_0)\mathcal{W}(s_1) \cdots \mathcal{W}(s_n) \cdots . \tag{2.73}$$

We ask under what conditions the transformation $\mathcal{W}$ will preserve order and shift-maximality. It turns out that the minimal set of conditions to be imposed on ρ and λ consists in the following:

1. The parity of ρ and λ conforms to that of R and L, i.e., ρ is odd and λ even.

2. The order of R and L is inherited: $\rho > \lambda$.

3. $\rho|_C$, $\rho\lambda|_C$ and $\rho\lambda^\infty$ should all be shift-maximal sequences. Here $\Sigma|_C$ means replacing the last symbol of Σ by C while keeping others unchanged. We remind that Σ is made of R and L only.

The generalized composition rule states that any kneading sequence except for L^∞ transforms into a kneading sequence by making substitution (2.72) with ρ and λ satisfying the above three conditions. Obviously, the $*$-composition satisfies these three conditions. As compared to the $*$-composition, a distinctive feature of the generalized composition rule consists in that it no longer requires that λ^∞ be shift-maximal.

Let us look at a few examples.

Example 1 Take RL^2RC, one of the period 5 kneading sequences, and replace its last letter C by R and L to get

$$\begin{aligned} \rho &= (RL^2RC)_+ = RL^2RR, \\ \lambda &= (RL^2RC)_- = RL^2RL. \end{aligned} \tag{2.74}$$

These ρ and λ satisfy the conditions of the generalized composition rule and may be applied to RL^2RC itself repeatedly to yield the limit of the period-quintupling sequence. This example may be realized by the usual $*$-composition of RL^2RC with itself; the limiting case is $(RL^2RC)^{*\infty}$. We will discuss period-n-tupling sequences in more detail at the end of this section.

Example 2 The transformation

$$R \to \rho = RLL,$$
$$L \to \lambda = RR,$$

also satisfies the conditions of the generalized composition rule. Applying it to the superstable period 5 kneading sequence RL^2RC and its upper and lower sequences, we get

$$RLL(RR)^2RLLRC, \quad [RLL(RR)^2(RLL)^2]^\infty, \quad [RLL(RR)^2RLLRR]^\infty.$$

They all are shift-maximal, but cannot be obtained by $*$-composition.

Period-n-tupling sequences

Every primitive word WC, describing a periodic orbit born at a tangent bifurcation, gives birth to a period-doubling sequence

$$W * R^{*k}C, \quad k = 0, 1, 2, \cdots .$$

The Feigenbaum period-doubling sequence is the simplest case with W being blank. The next one is the doubling sequence of the most clearly seen period 3 $W = RL$. There are infinitely many such period-doubling cascades. They all share the same scaling property discovered by Feigenbaum, e.g., the same convergence rate $\delta = 4.669\cdots$ and scaling factor $\alpha = 2.5021\cdots$.

Period-doubling sequences, however, are not the only ones that can be selected from the infinitely many periodic windows embedded in the chaotic regime. In fact, there are infinitely many ways to select other sequences having one or another kind of scaling property. In particular, every primitive word WC leads to a sequence

$$W^{*k}, \quad k = 1, 2, \cdots .$$

These are the period-n-tupling sequences, where $n = |WC|$ is the length of the word WC. Unlike the period-doubling sequences, the members of a period-n-tupling sequence are not adjacent in their parameter range. However, they do enjoy some scaling property similar to that of the period-doubling sequences. The scaling property of period-n-tupling sequences was first studied in Zeng, Hao, Wang and Chen [1984] (cf. Chang and McCown [1985]).

A period-n-tupling sequence may be obtained by repeated use of the composition rule

$$\begin{aligned} R &\to (WC)_+, \\ L &\to (WC)_-, \end{aligned} \tag{2.75}$$

of which an example is the substitution (2.74). The infinite limit of this substitution is a symbolic sequence which remains invariant under any further application of the substitution (2.75). Recalling the relation of symbols to the monotone branches of the inverse mapping function, this means that, e.g., in Example 1, with respect to the limiting dynamical regime there exists universal functions g_R and g_L. The result of functional composition $g_R \circ g_L \circ g_L \circ g_R \circ g_R$ is the same as applying a single function g_R, and the result of $g_R \circ g_L \circ g_L \circ g_R \circ g_L$ —the same as g_L. This hints on the existence of a topological conjugacy of the form

$$\begin{aligned} h^{-1} \circ g_R \circ h &= g_R \circ g_L \circ g_L \circ g_R \circ g_R, \\ h^{-1} \circ g_L \circ h &= g_R \circ g_L \circ g_L \circ g_R \circ g_L. \end{aligned} \tag{2.76}$$

Taking the simplest linear choice $h(x) = \alpha x$ and assuming a symmetric form of the universal function, we arrive at a functional equation

$$\alpha^{-1} g(\alpha x) = \underbrace{g \circ g \circ \cdots \circ g}_{n \text{ times}}(x). \tag{2.77}$$

This is the "fixed point" equation in the renormalization group theory of period-n-tupling sequences (Zeng, Hao, Wang, and Chen [1984]). The scaling factor α is obtained as part of the solution together with $g(x)$. The eigenvalue δ of the linearized equation determines the convergence rate. The noise scaling factor κ and dimension D of the limiting set can all be defined. In order to emphasize the fact that these exponents depend on the choice of the primitive word WC, which starts the period-n-tupling sequence, it is better to write α_W, δ_W, κ_W, and D_W. There are infinitely many of them, the Feigenbaum constants are given by the simplest choice of $W = R$. When $n > 3$, the equation has a number of different solutions for one and the same n. The number of solutions is given by the number of different periodic orbits of length n. This discussion will be continued in Section 7.3.1 in Chapter 7 on counting the number of periods.

2.5.3 Proof of the Generalized Composition Rule

This Section sketches the proof of the generalized composition rule. Though instructive for a thorough grasp of manipulating with symbolic sequences, it may be skipped at first reading.

The *Generalized Composition Rule* says that any substitution $\mathcal{W}$ defined in (2.72), which satisfies the three conditions listed in the last Section 2.5.2, transforms a kneading sequence Σ according to (2.73) into another kneading sequence $\mathcal{W}(\Sigma)$. For easier reference we repeat the conditions here:

Condition 1 ρ is odd and λ is even.

Condition 2 $\rho > \lambda$.

Condition 3 $\rho|_C$, $\rho\lambda|_C$ and $\rho\lambda^\infty$ are all shift-maximal.

Before turning to the proof we make a few remarks on the conditions. Conditions 1 and 2 are related to the preservation of order. Since the substitution (2.72) does not change the parity of the leading strings, it follows that

$$\Sigma < \Delta \Longleftrightarrow \mathcal{W}(\Sigma) > \mathcal{W}(\Delta).$$

The $*$-composition clearly satisfies these two conditions.

The Condition 3 guarantees shift-maximality. Obviously, it preserves the shift-maximality of sequences such as R^∞, $(RL)^\infty$, and RL^∞. The $*$-composition naturally satisfies this condition. In contrast with the $*$-composition, now we do not require that λ^∞ is shift-maximal. In other words, the transformation from L^∞ to λ^∞ may violate shift-maximality and we are concerned with the preservation of shift-maximality of all transformed sequences excluding λ^∞. In addition, we have the following observation on Condition 3.

1. The shift-maximality of $\rho|_C$ and $\rho\lambda|_C$ does not necessarily imply the shift-maximality of $\rho\lambda^2|_C$. Example: $\rho = RL$ and $\lambda = LRLRLL$.

2. The shift-maximality of $\rho\lambda|_C$ does not necessarily imply the shift-maximality of $\rho|_C$. Example: $\rho = RLR^3LR$ and $\lambda = LR^2$.

3. The shift-maximality of $\rho|_C$ and $\rho\lambda^\infty$ does not necessarily imply the shift-maximality of $\rho\lambda|_C$. Example: $\rho = RLRR$ and $\lambda = LRLR$.

These observations tell the necessity to include all the three types of sequences in Condition 3.

Since λ^∞ has been excluded and ρ^∞ is shift-maximal from Condition 3, we will consider only the transformation of shift-maximal sequences starting with RL. We first prove that under the above-mentioned conditions symbolic sequences $\rho\lambda^m|_C$ is shift-maximal for any integer $m \geqslant 2$. When $m = 2$ it follows from the shift-maximality of $\rho\lambda|_C$ that if $k \geqslant |\rho\lambda| = u + v$ (u and v are the length of ρ and λ, see (2.72)), then

$$\rho\lambda^2|_C > \mathcal{S}^k(\rho\lambda^2|_C). \tag{2.78}$$

If $k < |\rho\lambda|$, then it follows from the shift-maximality of $\rho\lambda|_C$ that (2.78) still holds as long as

$$(\rho\lambda)_{u+v-k} \neq \mathcal{S}^k(\rho\lambda),$$

where the notation $(P)_k$ means the first k symbols of the string P. If

$$(\rho\lambda)_{u+v-k} = \mathcal{S}^k(\rho\lambda) \equiv \alpha,$$

then it follows from the shift-maximality of $\rho\lambda|_C$ that α must be odd. Now $\rho\lambda^\infty$ and $\mathcal{S}^k(\rho\lambda^\infty)$ must have α as their common leading string. We discard α and compare the remaining sequences, i.e., $\mathcal{S}^{u+v-k}(\rho\lambda^\infty)$ and λ^∞. It follows from the shift-maximality of $\rho\lambda^\infty$ that whenever $\beta \equiv (\mathcal{S}^{u+v-k}(\rho\lambda^\infty))_v \neq \lambda$, one must have $\beta < \lambda$, thus (2.78) holds true. If, however, $\beta = \lambda$, then $\alpha\beta = \alpha\lambda$. Since α is odd and λ is even, $\alpha\beta$ must be odd, so (2.78) holds again. Therefore, we have proved that $\rho\lambda^2|_C$ is shift-maximal.

Now take $\rho\lambda$ as a new ρ. Clearly the three conditions hold true. We can repeat the arguments and prove that $\rho\lambda^3|_C$ is shift-maximal, and so on. Therefore, the shift-maximality of $\rho\lambda^m|_C$ for any $m \geqslant 2$ have been proved. From the above proof it also follows that whenever ρ and λ satisfy the three conditions of the gene-ralized composition rule, ρ and λ^m satisfy these conditions as well. We will make use of this fact in what follows.

Next, we prove that the generalized composition rule preserves shift-maximality for any shift-maximal sequence Σ. Due to Conditions 1 and 2, it follows from $\Sigma \geqslant \mathcal{S}^n(\Sigma)$ that

$$\mathcal{W}(\Sigma) \geqslant \mathcal{S}^{k_0}(\mathcal{W}(\Sigma)),$$

where $k_0 = \sum_{i=0}^{n-1} |\mathcal{W}(s_i)|$. Therefore, it is sufficient to consider the shift-maximality for $k = k_0 + j$, where $0 < j < |\mathcal{W}(s_n)|$. We accomplish the proof by exhausting all possible cases.

Case 1 $\mathcal{S}^n(\Sigma) = s_n s_{n+1} \cdots = RLR\cdots$.

We know from the shift-maximality of $\rho\lambda|_C$ that $(\rho\lambda)^\infty$ is shift-maximal. Introducing the notations $\alpha = \mathcal{S}^j(\rho\lambda)$ and $\beta = (\rho)_j$, if $\rho\lambda \neq \alpha\beta$, one must have

$$\mathcal{W}(\Sigma) > \mathcal{S}^k(\mathcal{W}(\Sigma)). \tag{2.79}$$

Obviously, one cannot have $\rho\lambda = \alpha\beta$, otherwise from the shift-maximality of $\rho\lambda|_C$ it would follow that both α and β should be odd, hence $\alpha\beta$ must be even, in contradiction with the oddness of $\rho\lambda$.

Case 2 $s_n s_{n+1} \cdots = RL^m R \cdots$.

From the shift-maximality of Σ it follows that Σ must start with RL^m, i.e., $\Sigma = RL^m \cdots$. Take λ^m to be a new λ, this case reduces to Case 1.

Case 3 $s_n s_{n+1} \cdots = LR \cdots$.

Denote $\alpha = (\rho\lambda)_{v-j}$ and $\beta = \mathcal{S}^j(\lambda)$. It follows from the shift-maximality of $(\rho\lambda)^\infty$ that if $\alpha \neq \beta$ then (2.79) holds true. If $\alpha = \beta$, then the shift-maximality of $\rho\lambda|_C$ implies that α must be odd. Then from the shift-maximality of $\rho\lambda^2|_C$ it follows

$$\rho\lambda^2 \cdots \geqslant \alpha\lambda \cdots > \alpha\rho \cdots .$$

Here we have used the fact that α is odd and $\rho > \lambda$. Therefore, $\rho\lambda \cdots > \alpha\rho \cdots$, and hence (2.79) holds true.

Case 4 $s_n s_{n+1} \cdots = L^m R \cdots$.

Take λ^m to be a new λ, this case reduces to Case 3.

Case 5 $s_n s_{n+1} \cdots = R^m \cdots, m > 1$.

Denote $\alpha = \mathcal{S}^j(\rho)$ and $\beta = (\rho)_j$. It follows from the shift-maximality of ρ^∞ that if $\rho \neq \alpha\beta$ then (2.79) holds true. However, one cannot have $\rho = \alpha\beta$, otherwise the shift-maximality of $\rho|_C$ would imply that both α and β are odd, in contradiction with the oddness of ρ according to Condition 1. By now all possible cases have been exhausted and we have proved that the generalized composition rule preserves shift-maximality.

Along similar lines one can prove that under the conditions of the generalized composition rule if a sequence Σ is admissible to a kneading sequence K, then the transformed sequence $\mathcal{W}(\Sigma)$ is admissible to the transformed kneading sequence $\mathcal{W}(K)$. This describes the dynamical symmetry of transformation. However, note that admissible sequences need not be always made of ρ and λ. For example, sequence $(RL)^\infty$ admissible to the kneading sequence $\rho\lambda^\infty = RLL(RR)^\infty$ does not consist of RLL and RR.

2.5.4 Applications of the Generalized Composition Rule

We look at a few applications of the generalized composition rule.

Strange repeller behind the stable period 3

The stable period 3 window starts with the kneading sequence $(RLR)^\infty$. Letting $\rho = R$ and $\lambda = LRR$, we may write

$$(RLR)^\infty = R(LRR)^\infty \equiv \rho\lambda^\infty.$$

It may be readily verified that such ρ and λ satisfy the three conditions of the generalized composition rule. Since $\rho\lambda^\infty$ is the greatest sequence written in ρ and λ, any sequence made of ρ and λ will be realizable, i.e., there exists an actual orbit which leads to the sequence. In contradistinction to the coarse-grained chaos to be discussed in the next Section 2.6, where λ^∞ corresponds to an unstable period, here λ^∞ corresponds to a stable period and it attracts almost all initial points on the interval. Although all the symbolic sequences made of ρ and λ still comprises a set of coarse-grained chaotic orbits, they form a strange repeller and may show off only in transient processes.

We can as well write

$$(RLR)^\infty = RL(RRL)^\infty \equiv \rho'\lambda'^\infty,$$

where ρ' and λ' also satisfy the three conditions. One can construct another set of coarse-grained chaotic orbits with them.

"Period 3 implies chaos"

Now it has become a commonplace that whenever there exists in a map a period 3, stable or unstable, there must exist orbits of arbitrary period n. This was what the statement "period three implies chaos" meant partially (Li and Yorke [1975]). With the help of symbolic dynamics we can even name some of these periods. Starting from

$$(RLR)^\infty = R(LRR)^\infty = RL(RRL)^\infty,$$

we can construct admissible sequences consisting of ρ and λ from those of R and L. For example,

$$\begin{array}{ll}
(RL^n)^\infty \to [R(LRR)^n]^\infty & \text{period } 3n+1,\\
(RL^n)^\infty \to [RL(RRL)^n]^\infty & \text{period } 3n+2,\\
(RL^nR)^\infty \to [R(LRR)^nR]^\infty & \text{period } 3n+2,\\
(RL^nRR)^\infty \to [R(LRR)^nRR]^\infty & \text{period } 3(n+1).
\end{array}$$

In this way one can get symbolic sequences of any period; they may be put in correspondence with actual orbits of the same period.

Symbolic dynamics of intermittent chaos

We take the intermittent chaos before the stable period 3 as an example. One can construct kneading sequences $\Delta_m = [R(LRR)^m C]^\infty$, $m = 1, 2, \cdots$, which precede $(RLR)^\infty$ on the parameter axis. It is easy to check that $\Delta_{m+1} > \Delta_m$ and when m goes to infinity, Δ_m approaches $(RLR)^\infty$. In addition, there are kneading sequences of $\rho\lambda^\infty$ type, e.g., $\Sigma_m = R[(LRR)^m RR]^\infty$, $\Sigma_{m+1} > \Sigma_m$, which precede and approach $(RLR)^\infty$ in the limit $m \to \infty$. These sets of coarse-grained chaotic orbits show up as intermittent chaos, the power m describes the length of the "laminar" phase. There is no need to treat intermittent chaos as transient process. Intermittency has been considered with the help of symbolic dynamics also in Refs (Aizawa [1983]; Dias de Deus and Noroula da Costa [1987]).

Non-universal convergence rate of $\rho\lambda^n|_C$ to $\rho\lambda^\infty$

There exist many ways to choose sequences of periodic words which converge to the chaotic limit $\rho\lambda^\infty$. A simplest sequence is $\rho\lambda^n|_C$, $n = 1, 2, \cdots$. We show that the convergence rate is given by the derivative of the unstable periodic orbits λ^∞, taken at the limiting parameter μ_∞ and the unstable periodic point x_∞:

$$\delta = \left.\frac{\partial f^{|\lambda|}}{\partial x}\right|_\infty, \tag{2.80}$$

where $|\lambda|$ denotes the number of letters in the symbolic string λ. In order to derive (2.80) we select a point x in the superstable periodic orbit $\rho\lambda^n|_C$. From the correspondence of symbol to inverse function we have

$$f^{|\rho|}(\mu_n, C) \equiv x_0^{(n)} = \underbrace{\lambda \circ \lambda \cdots \circ \lambda}_{n \text{ times}}(C). \tag{2.81}$$

The first equality

$$x_0^{(n)} = f^{|\rho|}(\mu_n, C) \tag{2.82}$$

leads in the limit $n \to \infty$ to

$$x^\infty = f^{|\rho|}(\mu_\infty, C). \tag{2.83}$$

On the other hand, x^∞ is determined by the fixed point condition

$$x^\infty = \lambda(\mu_\infty, x^\infty),$$

or, written in terms of the map f,

$$x^{\infty} = f^{|\lambda|}(\mu_\infty, x^{\infty}).$$

Subtracting (2.82) from (2.83) and expanding the result near the limit, we get

$$x_0^{(n)} - x^{\infty} = \gamma(\mu_n - \mu_\infty), \tag{2.84}$$

where

$$\gamma = \left.\frac{\partial f^{|\rho|}}{\partial \mu}\right|_{\infty}$$

is a constant independent of n. Applying $f^{|\lambda|}$ to $x_0^{(n)}$, we have

$$x_{|\lambda|}^{(n)} = f^{|\rho\lambda|}(\mu_n, C) = \lambda^{n-1}(\mu_n, C), \tag{2.85}$$

which also goes to the same limit x^{∞} for $n \to \infty$. Similarly, we have

$$x_{|\lambda|}^{(n)} - x^{\infty} = \tilde{\gamma}(\mu_n - \mu_\infty), \quad \tilde{\gamma} = \left.\frac{\partial f^{|\rho\lambda|}}{\partial \mu}\right|_{\infty}. \tag{2.86}$$

We can repeat the discussion with μ_{n+1} as well. Now instead of (2.81) we have

$$f^{|\rho|}(\mu_{n+1}, C) = x_0^{(n+1)} = \underbrace{\lambda \circ \lambda \cdots \circ \lambda}_{n+1 \text{ times}}(\mu_{n+1}, C), \tag{2.87}$$

and

$$x_{|\lambda|}^{(n+1)} = f^{|\rho\lambda|}(\mu_{n+1}, C) = \lambda^{n}(\mu_{n+1}, C). \tag{2.88}$$

Using (2.84) and the above relations, we form two ratios

$$\frac{x^{(n+1)} - x^{\infty}}{x^{(n)} - x^{\infty}} = \frac{x_{|\lambda|}^{(n+1)} - x^{\infty}}{x_{|\lambda|}^{(n)} - x^{\infty}} = \frac{\mu_{n+1} - \mu_\infty}{\mu_n - \mu_\infty}.$$

Now, define the convergence rate of μ_n as

$$\frac{1}{\delta} = \lim_{n \to \infty} \frac{\mu_{n+1} - \mu_n}{\mu_n - \mu_{n-1}} = \frac{\mu_{n+1} - \mu_\infty}{\mu_n - \mu_\infty},$$

we get

$$\begin{aligned}\frac{1}{\delta}=&\lim_{n\to\infty}\frac{x^{(n+1)}-x^{(n+1)}_{|\lambda|}}{x^{(n)}-x^{(n)}_{|\lambda|}}\\ =&\lim_{n\to\infty}\frac{\lambda(\lambda^n(\mu_{n+1},C))-\lambda(\lambda^{n-1}(\mu_{n+1},C))}{\lambda^n(\mu_n,C)-\lambda^{n-1}(\mu_n,C)}\\ =&\lim_{n\to\infty}\lambda'(\lambda^{n-1}(\mu_{n+1},C))\frac{\lambda^n(\mu_{n+1},C)-\lambda^{n-1}(\mu_{n+1},C)}{\lambda^n(\mu_n,C)-\lambda^{n-1}(\mu_n,C)}\\ =&\lambda'(\mu_\infty,x^\infty),\end{aligned}\tag{2.89}$$

which proves (2.80). This δ cannot be universal, as it depends on f. Take, for example, the RL^n sequence. For the quadratic map we have $\mu_\infty=2$, $x^\infty=-1$ and $\delta=4$, whereas for the map $f(\mu,x)=\mu\sin(\pi x)$ we have $\mu_\infty=1$, $x^\infty=0$ and $\delta=\pi$.

This is a right place to compare the above non-universal convergence rate with the universal convergence rate of Feigenbaum. From (2.87) we have, for large n,

$$\Delta_n\equiv\mu_n-\mu_\infty\propto\delta^{-n}\sim\exp(-n\log\delta),$$

Note that here n represents the number of letters or the period. Using n in the same sense, one should write

$$\Delta_n\propto\delta^{-\frac{\log n}{\log 2}}\sim\exp^{-\log n\frac{\log\delta}{\log 2}}$$

for the Feigenbaum sequence. We see that, in general, the non-universal sequence converges much quicker.

2.5.5 Further Remarks on Composition Rules

Although the generalized composition rule includes the $*$-composition as a particular case, it does not exhaust all possible substitution rules $R\to\rho$ and $L\to\lambda$. A simple example is the construction of a Fibonacci sequence of periodic orbits in the unimodal map. Let us look at an over-simplified version of the rabbit population problem. We denote by R an adult rabbit, and by L a little rabbit. Over a certain time period an adult rabbit gives birth to a little rabbit and puts L to the right of R. During the same period a little rabbit grows up to become an adult rabbit. Thus we have the following substitution rule:

$$\begin{aligned}R&\to RL,\\ L&\to R.\end{aligned}\tag{2.90}$$

The substitutions do not satisfy the 3 conditions of the generalized composition rule. Apply (2.90) to an initial letter R, we get

$$\begin{aligned}&R\\&RL\\&RLR\\&RLRRL\\&RLRRLRLR\\&RLRRLRLRRLRRL\\&\cdots\end{aligned} \tag{2.91}$$

As a consequence of violating the 3 conditions of the generalized composition rule, sequences obtained in this way need not be shift-maximal. In fact, the 5th and 6th words in the above list are not shift-maximal. However, if we are interested in a periodic orbit we can always shift its primitive string cyclically to get a shift-maximal one. The situation here is a little simple. It can be proved that the following procedure can easily generate shift-maximal sequences of the above series of sequences. Set $O_0^- = R$ and $O_0^+ = RL$, and denote by $\{O_{n-1}^-, O_{n-1}^+, E_n\}_1^\infty$ the series of shift-maximal sequences. The recursion relations are then

$$E_n = O_n^+ O_n^-, \quad O_{n+1}^- = E_n O_n^+, \quad O_{n+1}^+ = E_n O_{n+1}^-.$$

Written in superstable periods by means of the converse of the periodic window theorem, the series becomes

$$\begin{array}{ll}C & \text{period } 1\\ RC & \text{period } 2\\ RLC & \text{period } 3\\ RLRRC & \text{period } 5\\ RLRRLRRC & \text{period } 8\\ RLRRLRRLRLRRC & \text{period } 13\\ \cdots & \end{array} \tag{2.92}$$

The length of these periods grows as Fibonacci numbers F_n. Fibonacci numbers are defined by the following recursion relation

$$F_{n+1} = F_n + F_{n-1}, \quad n = 1, 2, \cdots, \tag{2.93}$$

with initial numbers $F_0 = 0$ and $F_1 = 1$. We will encounter Fibonacci numbers many times in this book, e.g., in Chapter 4 on circle maps and in Chapter 8 on grammatical complexity of symbolic sequences.

The recursion relation (2.93) is a linear difference equation. We solve it by letting $F_n \propto \omega^n$, just as we did with the simplest difference equation (2.27). This yields a quadratic equation:

$$\omega^2 - \omega - 1 = 0.$$

The two solutions $\omega = (1 \pm \sqrt{5})/2$ lead to the following general expression for the Fibonacci numbers:

$$F_n = \frac{1}{\sqrt{5}}\left[\left(\frac{1+\sqrt{5}}{2}\right)^n + \left(\frac{1-\sqrt{5}}{2}\right)^n\right]. \tag{2.94}$$

We list the first few Fibonacci numbers for future reference:

$$1, 1, 2, 3, 5, 8, 13, 21, 34, 55, 89, 144, 233, 377, 610, \cdots .$$

We note that the substitution (2.90) is not the only way to generate Fibonacci sequence in the unimodal map. A few other methods will be discussed in Chapter 8.

As our last remark we point out that the generalized composition rule themselves are particular case of more general substitutions. The symbols R and L on the left-hand sides of the rule (2.72) may be replaced by two other strings α and β. This leads to so-called homomorphism on free monoid, to be mentioned briefly in Chapter 8, see Section 8.3.2.

All shift-maximal sequences are subdivided into primitive and composite words with respect to the $*$-composition, just as integers are divided into primes and composite numbers. Is it possible to further classify primitive words with respect to generalized compositions? Composite words have a simple manifestation in the fine structure of their power spectra. Does a generalized composition rule which cannot be reduced to an $*$-composition, has any observable manifestation? The generalized composition rules also raise some open problems.

2.6 Coarse-Grained Chaos

The generalized composition rule provides us with a practical definition of chaotic orbit. The sequence $\rho\lambda^\infty$ there has a correspondence with the surjective unimodal map, e.g., the quadratic map (2.30) at $\mu = 2$:

$$x_{n+1} = 1 - 2x_n^2, \tag{2.95}$$

This map is drawn in Fig. 2.12 (a). Fig. 2.12 (b) shows a piecewise linear surjective tent map, to be used later in this section. We start with a summary of what is known for the surjective quadratic map, then proceed to maps with eventually periodic kneading sequences $\rho\lambda^\infty$.

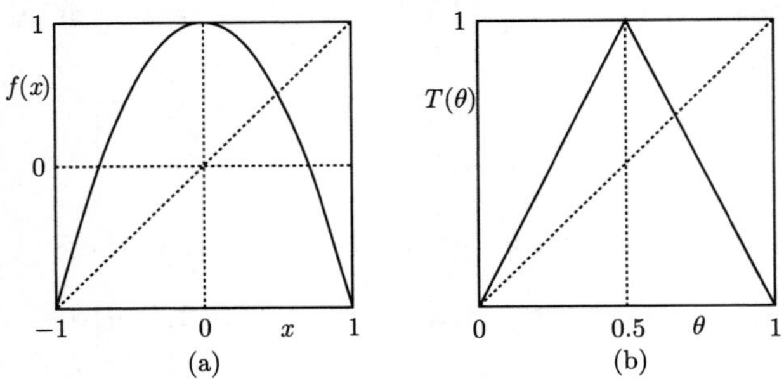

Figure 2.12 The surjective quadratic map (a) and surjective tent map (b)

2.6.1 Chaos in the Surjective Unimodal Map

The maximal sequence in the complete MSS Table is RL^∞. When it becomes the kneading sequence, the mapping is equivalent to the shifts of two-letter symbolic sequences embodied in the surjective tent map, as will be seen in (2.114), Section 2.8.1, with the parameter $\lambda = 2$. In fact, these two maps are *topologically conjugate.* We will explain the topological conjugacy of maps later in this Section. All symbolic sequences, which are not finite sequences appended by RL^∞, are admissible and each sequence has a one-to-one correspondence to a point in the interval. Chaotic nature of this map has been well understood. We summarize here the most important properties, adapted, where necessary, to the quadratic map (2.30) or (2.31).

1. It is a surjective map, i.e., the whole interval I maps *onto* itself. This is clear from the graph of the surjective map, e.g., Fig. 2.14 and does not require any explanation.

2. All symbolic sequences consisting of R and L from the smallest L^∞ to the largest RL^∞ are present except those ending with RL^∞. In other words, for any symbolic sequence, made of L and R and not ending with RL^∞ there is an initial point to produce it. They are as many as real numbers in the interval $[0, 1]$. In fact, this is the symbolic formulation of sensitive dependence on initial conditions.

3. The orbit RL^{∞} itself is a *homoclinic orbit.* The points 1, C and their pre-images comprise the stable set of the unstable fixed point L^{∞}.

4. It is a *crisis point* in the sense of Grebogi, Ott and Yorke [1983a], because the chaotic attractor ceases to exist beyond $\mu = 2$ after colliding with unstable fixed point L^{∞}.

5. It is the band-ending point, or, if one likes, the last, i.e., the $1 \to 0$ band-merging point.

6. There is a continuous distribution $\rho(x)$ for the iterates $\{x_i\}_{i=0}^{\infty}$. For the quadratic map this distribution has been known since 1947 (Ulam and von Neumann [1947]). It is

$$\rho(x) = \frac{1}{\pi\sqrt{1-x^2}}. \tag{2.96}$$

7. It is an intersection point of all $P_n(\mu), n \geqslant 2$. See Section 2.2.3 for the definition of the dark-line functions $P_n(\mu)$.

We explain some of these points in more detail.

Homoclinic orbits in one-dimensional maps

Homoclinic and heteroclinic orbits are key notions in nonlinear dynamics. Especially, the homoclinic and heteroclinic intersections, closely related to these orbits, are the organizing centers of chaotic motion. The French mathematician Henry Poincaré introduced these notions in the 1890s in his study of celestial mechanics (Poincaré [1899]). We explain the essence on the example of a planar system of ordinary differential equations. More will be said in Chapter 5 on symbolic dynamics of two-dimensional maps, see Section 5.2.1.

Generally speaking, the solutions of a planar system of ordinary differential equations

$$\begin{aligned}\frac{dx}{dt} &= g(x,y),\\ \frac{dy}{dt} &= f(x,y),\end{aligned}$$

are given by the family of integral curves passing through various points of the phase plane. However, there may be special solutions at isolated points determined by the equalities

$$\begin{aligned}g(x,y) &= 0,\\ f(x,y) &= 0.\end{aligned}$$

These are fixed points of the differential equations. A stability analysis may be carried out, in principle, for every fixed point. In fact, certain unstable fixed points may be limits of some specific orbits. If from any point of an orbit one reaches the same fixed point in both the $t \to +\infty$ and $t \to -\infty$ limits, the orbit is an *homoclinic orbit*, see Fig. 2.13 for a sketch of the situation.

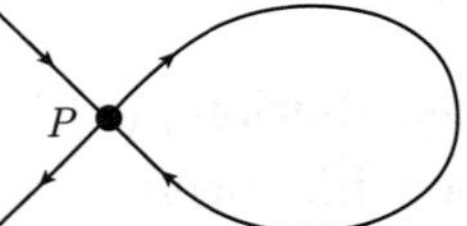

Figure 2.13 A homoclinic orbit in a phase plane (schematic)

In one-dimensional maps the phase space is compressed into a line. The above picture is dimmed by this compression. This explains why the understanding of homoclinic orbits in one-dimensional maps came only in the late 1970s (Block [1978], Misiurewicz [1979]). The reader may find a detailed discussion in the book by Devaney [1986, 1989]. [4] Once understood, it turns out quite simple.

Let us inspect the graph of a surjective quadratic map, Fig. 2.14. Starting from the critical point C as an initial point, it takes two steps to get into the leftmost unstable fixed point L^∞, i.e., the forward orbit is

$$0, 1, -1, -1, -1, -1, \cdots .$$

If one goes backward from the critical point C, then the pre-image $f^{-1}(C)$ consists of two points, the pre-image $f^{-2}(C)$—four points, etc. One can always pick up an infinitely long backward orbit which eventually approaches the fixed point L^∞. In Fig. 2.14, the forward orbit from C is shown by thick lines, and two backward orbits are shown by dashed lines. Therefore, starting from point C or any of its finite pre-images, one reaches the unstable fixed point L^∞ in finite steps forward or in infinite steps backward. These points are then located on homoclinic orbits. In fact, L^∞ in the kneading sequence RL^∞ represents the unstable fixed point while R describes the finite process to reach it. This will become clearer in our discussion on the general kneading sequences $\rho\lambda^\infty$.

[4] The homoclinic orbit discussed here was called "critical homoclinic" by Devaney. He devoted more discussion to homoclinic orbits in expanding maps with $\mu > 2$.

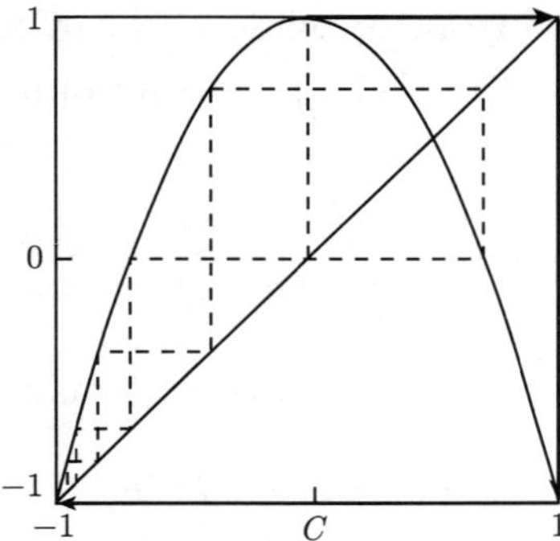

Figure 2.14 Homoclinic orbits in a surjective quadratic map

The surjective tent map

We will use the topological conjugacy of the surjective quadratic map (2.95) and the surjective tent map to derive the density distribution (2.96) for the quadratic map. For convenience sake, we use the following form of the surjective tent map, which is slightly different from the one to be used in Section 2.8:

$$T(\theta) = \begin{cases} 2\theta, & \text{if} \quad 0 \leqslant \theta < 1/2, \\ 2 - 2\theta, & \text{if} \quad 1/2 \leqslant \theta \leqslant 1, \end{cases} \tag{2.97}$$

confining the interval to $[0, 1]$. As shown in Fig. 2.12 (b), the function $T(\theta)$ maps θ from the interval $[0, 1]$ onto itself. It consists of two straight lines and looks like a tent, hence the name tent map.

Before proceeding further, we recollect some relations from elementary trigonometry. First, define a simple function

$$x = h(\theta) \equiv -\cos(\pi\theta), \tag{2.98}$$

where the minus sign is inserted for convenience. The function h maps $\theta \in (0, 1)$ to $x \in (-1, +1)$ monotonically. Therefore, the inverse function is simply

$$\theta = h^{-1}(x) = \frac{1}{\pi} \arccos(-x) = 1 - \frac{1}{\pi} \arccos(x). \tag{2.99}$$

However, for the same range of θ, the function

$$h(2\theta) = -\cos(2\pi\theta) \tag{2.100}$$

has a hump and maps $\theta \in (0, 1)$ twice onto $x \in (-1, +1)$. Consequently, in taking its inverse, one must be careful with the definition of the principal branch of $\arccos(x)$. For instance, while one can annihilate h^{-1} with $h(\theta)$, one cannot do so with $h(2\theta)$. Taking into account the principal branch of $\arccos(x)$, we have

$$h^{-1} \circ h(2\theta) = \begin{cases} 2\theta, & \text{if } \ 0 \leqslant \theta < 1/2, \\ 2 - 2\theta, & \text{if } \ 1/2 \leqslant \theta \leqslant 1. \end{cases} \tag{2.101}$$

Therefore, it turns out that our surjective tent map $T(\theta)$ may be written as

$$T(\theta) = h^{-1} \circ h(2\theta). \tag{2.102}$$

Now return to the surjective quadratic map (2.95). Let us replace x by $h(\theta)$ and look at the result

$$f(x) = f(h(\theta)) = 1 - 2\,\cos^2(\pi\theta) = -\cos(2\pi\theta) = h(2\theta).$$

Applying h^{-1} to both sides of the above relation and taking into account (2.101) and (2.102), we get

$$h^{-1} \circ f \circ h(\theta) = T(\theta), \tag{2.103}$$

or, equivalently, after inserting $\theta = h^{-1}(x)$,

$$f(x) = h \circ T \circ h^{-1}(x). \tag{2.104}$$

If two maps are related by this type of transformation involving a good enough, i.e., continuous and invertible, function h, they are said to be *topologically conjugate*. So the surjective quadratic map is topologically conjugate to the surjective tent map.

Topological conjugacy of maps

The concept of topological conjugacy is very important for the study of maps, because it divides various maps into equivalent classes. We present some more properties of the topological conjugacy relation.

Suppose we have two maps: $f(x)$ maps the interval I into itself, and $g(\theta)$ maps J into itself, where I and J may be different intervals. If there exists a continuous invertible function h such that h transforms J into I and h^{-1} transforms I back into J, and, furthermore

$$f(x) = h \circ g \circ h^{-1}(x) \quad \text{or} \quad g(\theta) = h^{-1} \circ f \circ h(\theta), \tag{2.105}$$

then f and g are topologically conjugate.

It is easy to verify that the conjugacy relation extends also to the iterates $f^{(n)}$ and $g^{(n)}$:

$$\begin{aligned} f^{(n)}(x) &= h \circ g^{(n)} \circ h^{-1}(x), \\ g^{(n)}(\theta) &= h^{-1} \circ f^{(n)} \circ h(\theta). \end{aligned} \tag{2.106}$$

Therefore, an n-cycle of g corresponds to an n-cycle of f and vice versa. Moreover, these cycles enjoy the same stability property. In order to show this, we write

$$f^{(n)} \circ h(\theta) = h \circ g^{(n)}(\theta)$$

and differentiate both sides with respect to θ:

$$\left.\frac{df^{(n)}}{dx}\right|_{x=h} \frac{dh}{d\theta} = \left.\frac{dh}{d\theta'}\right|_{\theta'=g^{(n)}(\theta)} \frac{dg^{(n)}(\theta)}{d\theta}.$$

Now we see that only when θ belongs to an n-cycle of g, i.e., $g^{(n)}(\theta) = \theta$, we have

$$\frac{df^{(n)}(x)}{dx} = \frac{dg^{(n)}(\theta)}{d\theta}.$$

Therefore, from the fact that the tent map does not have any stable periods follows the same conclusion for the logistic map at $\mu = 2$.

Topological conjugacy provides not only a global relation between two maps, but the function $x = h(\theta)$ also establishes a local correspondence between the points in the two intervals. The conservation of probability (or the conservation of number of points) yields a local relation for the invariant distributions

$$\rho_f^*(x)dx = \rho_g^*(\theta)d\theta. \tag{2.107}$$

Here a subscript has been attached to the distribution in order to indicate to which map it belongs. The relation (2.107) permit us to calculate the invariant distribution of one map from that of the other map:

$$\rho_f^*(x) = \rho_g^*(h^{-1}(x)) \left| \frac{dh^{-1}(x)}{dx} \right|. \tag{2.108}$$

Next, we apply (2.108) to the surjective quadratic map (2.95).

The invariant distribution for the surjective quadratic map

The derivative of the surjective tent map

$$\left|\frac{dT(\theta)}{d\theta}\right| = 2$$

is bigger than 1 everywhere, so it does not have any stable periods. Most initial values θ_0 (except for such rational values as 0, 2/3, $\cdots$, which, taken together, form a set of measure zero) will lead to chaotic sequences $\{\theta_n, n = 0, 1, \cdots\}$. How are these points distributed on the interval $(0, 1)$? The answer defines the density of points

$$\rho(\theta) = \lim_{\Delta\theta\to 0}\lim_{n\to\infty} \frac{\text{number of } \theta_n \in (\theta, \theta+\Delta\theta)}{n\Delta\theta}. \tag{2.109}$$

For piecewise linear maps, $\rho(\theta)$ can be calculated exactly (see, e.g., Grossmann and Thomae [1977]). In the case of the tent map (2.101), the result can be anticipated by simple counting arguments to be

$$\rho(\theta) = 1. \tag{2.110}$$

Clearly, this distribution is invariant under the map $T(\theta)$, i.e., it remains the same when θ is replaced by another $\theta' = T(\theta)$. We add an asterisk to denote the invariant distribution: $\rho^*(\theta)$.

Now let g in (2.108) be the tent map $T(\theta)$. From (2.110) we know $\rho_T^* = 1$. Using (2.99) for $h^{-1}(x)$ and differentiating the $\arccos(x)$, we arrive at (2.96)

$$\rho_f^*(x) = \frac{1}{\pi\sqrt{1-x^2}}. \tag{2.111}$$

It is sometimes called the Chebyshev distribution, since it is the weight function in the definition of the Chebyshev polynomials. In fact, the Chebyshev polynomials provide topological conjugacy for a class of piecewise linear maps (Grossmann and Thomae [1977]; Katsura and Fukuda [1985], Fukuda and Katsura [1986]). The $\mu = 2$ logistic map corresponds to the simplest case, namely, the Chebyshev polynomial of order 2.

We take this opportunity to say a few more words on the deterministic nature of chaotic iterations. In principle, the high order iterates of the tent map or the surjective logistic map can all be written down analytically as functions of the initial point. For the tent map, they will be given in Section 2.8.1. The iterates of the $\mu = 2$ logistic map may be expressed via the Chebyshev polynomials (Erber, Johnson and Everett [1981]). If one could fix the initial value and retain the intermediate points with infinitely many digits, an orbit would be determined entirely by the initial value and might be reproduced whenever one wishes. However, due to the instability of all cycles, an orbit will be extremely sensitive to tiny changes in the initial value. On the other hand, it is just this extreme sensitivity to the initial values that guarantees the existence of the invariant distribution. Since one is unable to retain, in practice, the initial point with infinite precision and it is unnecessary, in theory, to bother about single orbits, the invariant distribution (plus some other statistical characteristics) provides us with a satisfactory description of the chaotic motion. In this sense, the output of the $\mu = 2$ logistic map is just as random as any stochastic process with a continuous distribution.

2.6.2 Chaos in $\rho\lambda^\infty$ Maps

For a given unimodal map, if there exists a point on the interval which leads to a symbolic sequence $\rho\lambda^\infty$, where ρ and λ satisfy the three conditions of the generalized composition rule, then a correspondence between ρ, λ and R, L may be established. Any symbolic sequence Σ, included in between L^∞ and RL^∞, will have an image $\mathcal{W}(\Sigma)$ obtained by applying the generalized composition rule $\mathcal{W}$ defined by using these ρ and λ. Using the RL^∞ map as a prototype for chaotic map, we define in this way *coarse-grained chaos*. One may collect a list of properties for the map of the kneading sequence $\rho\lambda^\infty$ point by point in accordance with the surjective map of the kneading sequence RL^∞.

1. All symbolic sequences between λ^∞ and $\rho\lambda^\infty$ are present. They are as many as real numbers in the interval $[0, 1]$, hence sensitive dependence on initial values in this map.

2. The orbit $\rho\lambda^\infty$ itself is a homoclinic orbit. The orbital points corresponding to ρ and their pre-images comprise the stable set of the unstable periodic orbit λ^∞.

3. It is a crisis point where the chaotic attractor collides with the unstable periodic orbit λ^∞.

4. It is a band-merging point.

5. There is a continuous distribution $\rho(x)$, confined to the "sea" of the chaotic attractor, for a proof see Collet and Eckmann [B1980].

6. It corresponds to intersection points of a certain subset of the dark-line functions $P_n(\mu)$.

7. It provides another case when the precise parameter value of the kneading sequence may be calculated by using the word-lifting technique.

8. It is one of the only two classes of regular language associated with the unimodal maps, if symbolic sequences are viewed as words of a formal language, the other case being periodic sequences. This will be explained in Chapter 8 on grammatical complexity of symbolic sequences.

9. When $\rho\lambda^\infty$ is a composite kneading sequence of the type $P * RL^\infty$ there is a locally surjective $|\rho|$-th return map, confined to the subinterval from $|\lambda|^\infty$ to $\rho\lambda^\infty$.

We explain some of these points in more detail.

Homoclinic orbits

An eventually periodic kneading sequence $\rho\lambda^\infty$ implies the existence of a homoclinic orbit. In Fig. 2.15 (a) we draw a quadratic map at $\mu = 1.6437$ when its kneading sequence is RLR^∞. The orbit starting from C gets into the unstable fixed point R^∞ after two iterations represented by RL. This is sometimes called a *Misiurewicz point.* It is clear that the point C and its pre-images are part of the homoclinic orbits.

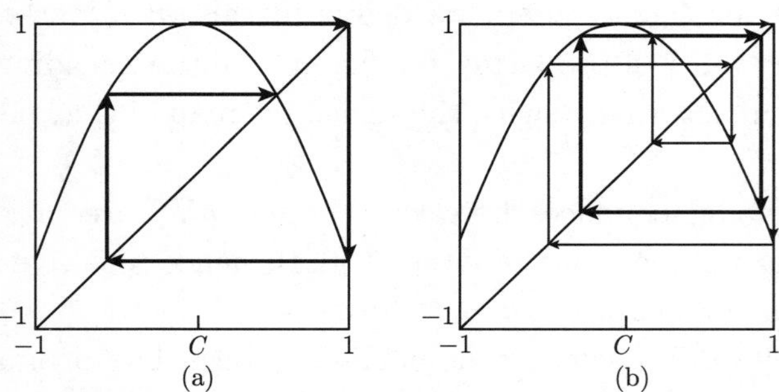

Figure 2.15 Examples of homoclinic orbits. (a) A homoclinic orbit to the unstable fixed point R^∞ at $2 \to 1$ band-merging point RLR^∞. (b) A homoclinic orbit to the unstable 2-cycle $(RL)^\infty$ at $4 \to 2$ band-merging point $RLRR(RL)^\infty$

In general, the λ^∞ represents an unstable periodic orbit of period $|\lambda|$, and ρ describes the finite process to get into this unstable periodic orbit. The point

C and its pre-images are on the homoclinic orbit associated with the unstable periodic orbit λ^∞. Fig. 2.15 (b) shows the forward part of a homoclinic orbit, corresponding to the $4 \to 2$ band-merging point when the kneading sequence is $RLRR(RL)^\infty$. The points C and $f(C) = RLRR(RL)^\infty$ as well as their pre-images lead to the unstable period 2 orbit $(RL)^\infty$ in finite steps.

Sudden change of chaotic attractor—crisis

Applied symbolic dynamics deals with continuity and piecewise monotonicity of the mapping function. It has nothing to do with the stability of the orbit corresponding to a certain symbolic sequence. In order to check the stability of a periodic orbit one has to know the mapping function $f(x)$, since the stability criterion consists in

$$\prod_{i=0}^{n-1} |f'(x_i)| \leqslant 1.$$

In addition, the Schwarzian derivative $Sf(x)$ of the function $f(x)$, see (2.15), depends on the first, second, and third derivatives of f. An unimodal map with negative Schwarzian derivative can have at most one stable periodic orbit at a fixed parameter. It may have none.

If there exists a stable period then it necessarily attracts the critical point C. In other words, the critical point C belongs to the basin of attraction of the stable period. Therefore, if the kneading sequence of the map happens to be $\rho\lambda^\infty$ type and λ^∞ corresponds to a unstable periodic orbit, then the map cannot have any stable period, as the critical point C is attracted to the unstable period. We have just seen that the critical point C is homoclinic to the unstable period. Any infinitesimal deviation will be amplified by the unstable period. Orbital points in numerical simulations will no longer be confined to isolated points; they will fill up intervals. This is how a chaotic attractor appears. When the parameter changes into the value determined by the kneading sequence $K = \rho\lambda^\infty$, the attractor "collides" with unstable periodic orbit, and a sudden change in the structure of attractor occurs, which is called *crisis* by Grebogi, Yorke and Ott [1983a].

For a unimodal map, when the parameter increases from a small value, kneading sequence of the type $\rho\lambda^\infty$ first appears at the accumulation point μ_∞ of the main period-doubling bifurcation cascade. Using the doubling transformation $\mathcal{D}$, introduced in (2.70) in Section 2.5.1, it can be described symbolically as the common limit from both the period 2 and the surjective maps:

$$\lim_{n\to\infty} \mathcal{D}^n(R) = \lim_{n\to\infty} \mathcal{D}^n(RL^\infty) = R^{*\infty}.$$

The two limits in the above equations reflects the twofold nature of the limiting orbit $R^{*\infty}$ of the period-doubling cascade. As we have pointed out in Section 2.5.1, this limit is a quasi-periodic orbit. At the same time, it is the limit of an infinite sequence of $\rho\lambda^\infty$ type kneading sequences. It is just in this sense we say that it signals the beginning of coarse-grained chaos.

Band-merging points

Band-merging points seen in the bifurcation diagram are also crisis points. The main period-doubling cascade starts with a period 1 window (L, C, R). The one-band chaotic region ends at RL^∞, where the chaotic attractor collides with the unstable fixed point L^∞. The next two-band region ends at RLR^∞, where the two pieces of the chaotic attractor meet with the unstable fixed point R^∞. Similarly, the four-band region, corresponding to the period 4 window $(RLRL, RLRC, RLRR)$, ends at $RLRR(RL)^\infty$, where the four pieces of the chaotic attractor meet with the unstable period 2 $(RL)^\infty$ and merge into two bands. In general, a periodic window $((PC)_-, PC, (PC)_+) = (\lambda, \lambda|_C, \rho)$ corresponds to a band-merging point at $\rho\lambda^\infty$, where a $|\rho|$-piece attractor meets with the unstable period λ^∞.

Continuous density distribution of orbital points

When λ is the primitive string of an unstable period, the kneading sequence $K = \rho\lambda^\infty$ is eventually periodic. There is a continuous distribution of orbital points, which resembles the Chebyshev distribution (2.111) in the surjective quadratic map, i.e., with singularity at the edges but smooth in between. As a rule, one cannot write down a closed formula for the distribution.

Locally surjective map

A unimodal map, except for the case when its kneading sequence is RL^∞, cannot be a surjective map. However, for a map whose kneading sequence is of the type $\rho\lambda^\infty = P * RL^\infty$, if one determines its parameter by word-lifting technique and draw the graph of $f^{|\rho|}(x)$, then there is a locally surjective map on the subinterval $[f^{|\rho|+1}(C), f(C)]$ for the quadratic map (2.31). In fact, one can map the two end points of the subinterval to get $|\rho|-1$ more subintervals on each of which there is a locally surjective map. Some of these are formed as a valley instead of a hump. Let us look at two examples.

Example 1 Take the $2 \to 1$ band-merging point. Its kneading sequence is RLR^∞. Word-lifting leads to a par of equations

$$f(C) = R \circ L(\nu),$$
$$\nu = R(\nu).$$

Using the inverse functions of the map (2.31), we have to solve the equations by iteration:

$$\mu_{n+1} = \sqrt{\mu_n + \sqrt{\mu_n - \nu_n}},$$

$$\nu_{n+1} = \sqrt{\mu_n - \nu_n}.$$

Any reasonable initial values satisfying $\nu_0 < \mu_0$, say, $\mu_0 = 2$ and $\nu_0 = 1.95$, lead to the result

$$\mu = 1.543\ 689\ 01 \cdots, \quad \nu = 0.839\ 286\ 75 \cdots.$$

We draw the map $f^2(x)$ at the above parameter μ in Fig. 2.16 (a). In the upper right corner above $x = y = 0.5437$ there is a locally surjective map. We have enlarged it in Fig. 2.16 (b). It looks much like the surjective map in Fig. 2.14 with the difference that it is no longer symmetric with respect to the central vertical. Please note that this is a locally surjective map in $f^2(x)$, not in $f(x)$.

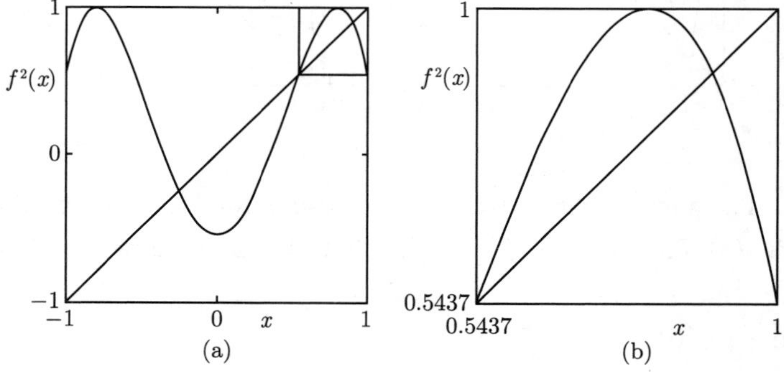

Figure 2.16 The locally surjective map corresponding to the kneading sequence $RL(RR)^\infty$. (a) The function $f^2(x)$. (b) Blow-up of the upper right corner

There is a numerical detail worth mentioning. If we draw the map $f^2(x)$ using the equation (2.31), the μ and ν values given above would determine the interval $[\nu, \mu]$ for the locally surjective map. However, Fig. 2.16 is drawn using the mapping function (2.30). Since x has become x/μ, the interval $[\nu, \mu]$ becomes $[\nu/\mu, 1]$.

Example 2 Our second example deals with the band-merging point of the three chaotic bands in the last part of the period 3 window. It is described by the kneading sequence $RLL(RLR)^\infty$. Word-lifting equations

$$\mu = R \circ L \circ L(\nu),$$
$$\nu = R \circ L \circ R(\nu),$$

yield

$$\mu = 1.790\ 327\ 49\cdots, \quad \nu = 1.745\ 492\ 83\cdots.$$

The function $f^3(x)$ at $\mu = 1.7903$ is drawn in Fig. 2.17 (a). The locally surjective map in its upper right corner can hardly be seen by naked eyes. We blow it up in Fig. 2.17 (b). The starting point of the square $x = y = 0.9749$ is given by the ratio ν/μ. There are two more locally surjective map functions which point downward and are indicated with boxes in Fig. 2.17 (a).

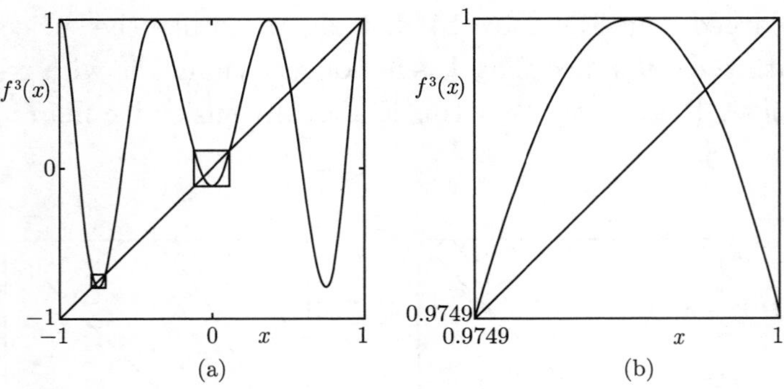

Figure 2.17 The locally surjective map corresponding to the kneading sequence $RLL(RLR)^\infty$. (a) The function $f^3(x)$. (b) Blow-up of the upper right corner

Intersection points of dark lines

We have mentioned in Section 2.2.3 that there are infinitely many intersection points among the dark-line functions $\{P_n(\mu)\}$. These intersections correspond to maps with eventually periodic kneading sequences $K = \rho\lambda^\infty$. To get a more concrete feeling, we draw the dark lines from $P_1(\mu) = 1$ to $P_{12}(\mu)$ in the parameter interval $\mu \in [1.76, 1.86]$ for the quadratic map (2.30) in Fig. 2.18. A bifurcation diagram for the same parameter range is given in Fig. 2.19. A comparison of these two figures shows how the dark lines make deterministic backbones of the chaotic bands.

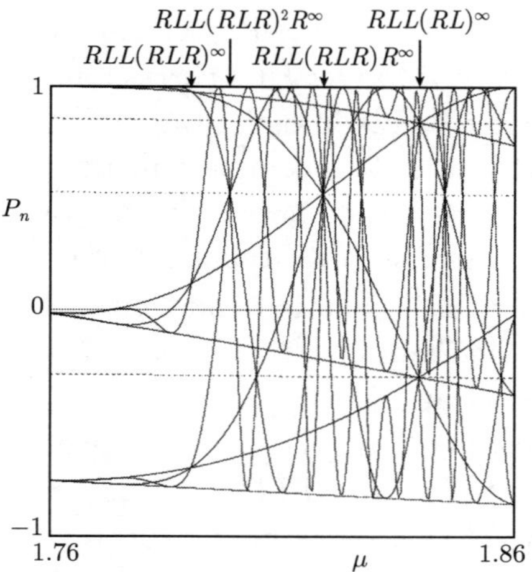

Figure 2.18 The dark lines $P_1(\mu)$ to $P_{12}(\mu)$ in the parameter range, covering approximately the range from RLC to RL^2RC. Dash-dot line—unstable fixed point R^∞. Dash lines—the unstable $(RL)^\infty$ and $(LR)^\infty$ of period 2

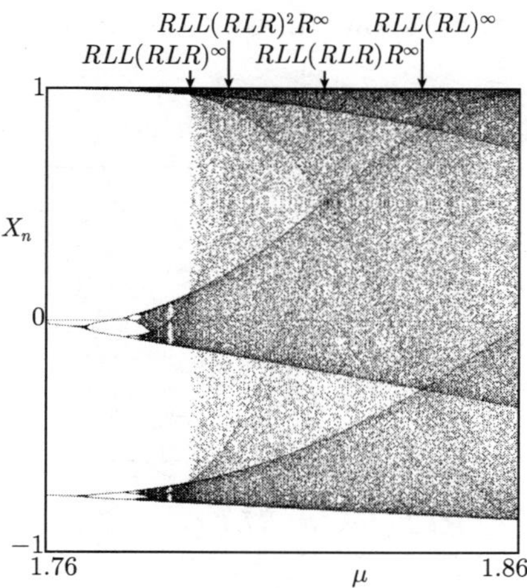

Figure 2.19 A bifurcation diagram in the same parameter range as Fig. 2.18

The parameter range corresponds approximately to that from $K = RLC$ to $K = RLLRC$. One may identify some periodic windows by tangent points between the dark lines, e.g., the RL^2RLC and RL^2RLRC in the MSS Table 2.2.

In Fig. 2.18 also drawn are the locations of the unstable fixed point R^∞ (dash-dot line) and the two unstable period 2 points $(RL)^\infty$ and $(LR)^\infty$ (dash lines), using Eqs. (2.32) and (2.36) in Section 2.2.1. They help us to identify whether the unstable period λ^∞ in $\rho\lambda^\infty$ at an intersection of $P_n(\mu)$ belongs to R^∞ or $(RL)^\infty$. We indicate a few such sequences seen in the figure.

The kneading sequence $RLL(RLR)^\infty$ corresponds to the sudden widening of the three small chaotic bands into the big one-band chaotic region in the bifurcation diagram. It is a fairly incomplete merging, as one can still identify the three chaotic bands from the contrast of the density distribution. This was the first crisis point studied by Grebogi, Yorke and Ott [1983a]. A more complete merging of the three chaotic bands into one-band region takes place at $RLL(RL)^\infty$, where the chaotic attractor collides with the unstable period 2 orbit. This can be seen clearly in Fig. 2.18.

Many series of kneading sequences which converges to $RLL(RLR)^\infty$ can be constructed. For example, the $RLL(RLR)^mR^\infty$ series at the $m \to \infty$ limit. The $m = 1, 2$ members of this series may be seen in Fig. 2.18. However, the $m = 0$ member $RLLR^\infty$ lies beyond the right border of the figure. Another example is the $RLL(RLR)^m(RL)^\infty$ series. The parameter values of all these sequences may be calculated by word-lifting. We show some in the following Table 2.4. The convergence rate of these series is not universal and can be calculated using the method shown in Section 2.5.4.

Table 2.4 Two series of kneading sequences which converge to $RLL(RLR)^\infty$

m	$\rho\lambda^\infty$	μ
0	$RLLR^\infty$	1.892 910 988
1	$RLL(RLR)R^\infty$	1.818 620 134
2	$RLL(RLR)^2R^\infty$	1.798 553 464
3	$RLL(RLR)^3R^\infty$	1.792 694 369
4	$RLL(RLR)^4R^\infty$	1.790 987 279
∞	$RLL(RLR)^\infty$	1.790 327 492
0	$RLL(RL)^\infty$	1.839 286 755
1	$RLL(RLR)(RL)^\infty$	1.804 369 887
2	$RLL(RLR)^2(RL)^\infty$	1.794 415 337
3	$RLL(RLR)^3(RL)^\infty$	1.791 482 606
∞	$RLL(RLR)^\infty$	1.790 327 492

2.7 Topological Entropy

Now we take up briefly the problem of characterizing the chaoticity of the dynamics. There are many characteristics of chaos. The calculation of such strong characteristics as the measure-theoretical (Kolmogoroff-Sinai) entropy or Lyapunov exponents requires detailed knowledge of the mapping function $f(x)$ and thus goes beyond the framework of symbolic dynamics. In spite of the reward that these characteristics are associated with observable chaotic motion in the system, we will not touch on them in this book.

Confining ourselves within reach of symbolic dynamics, we consider one of the weakest characteristics, namely, the topological entropy. Essentially it is associated with counting the number of monotone branches in $f^n(x)$. One starts from an initial point x, the n-th iterate $f^n(x)$ is, of course, unique, due to the fact that the dynamics is deterministic. However, the number of monotone pieces of $f^n(x)$, or the number of minima and maxima of $f^n(x)$, or the number of orbital points of periodic orbits with period not exceeding n, may be quite different for different maps. On the other hand, for one and the same map, these three numbers do not differ much from each other and will be denoted by the same quantity $N(n)$. Intuitively speaking, the faster $N(n)$ grows with n the more "complex" is the dynamics. If the growth of $N(n)$ is slower than an exponential function of n, the dynamics is simple. The topological entropy h measures the exponential growth rate of $N(n)$:

$$h = \lim_{n\to\infty} \frac{\ln N(n)}{n}. \tag{2.112}$$

A rigorous definition of topological entropy makes use of infinite refinement of the partition of phase space by the dynamics, see Section 2.8.5 below. A related problem consists in the following. Given a point such as C, how many ways there exist in order to get to C in n steps or, equivalently, how many pre-images $f^{-n}(C)$ there exist. However, for practical purpose the above formula (2.112) provides an effective way of computing the topological entropy of maps (Collet, Crutchfield and Eckmann [1983]).

It should be made clear that the existence of a positive topological entropy does not guarantee observable chaotic motion. For example, for the super-stable kneading sequence $(RLC)^\infty$ there is an attracting period 3 orbit in the quadratic map; almost all initial values are attracted to this periodic orbit and no chaos shows off. However, the topological entropy acquires a positive value in this periodic window. This follows from the following arguments.

It is easy to see that any symbolic sequence made by concatenations of any number of the two strings R and RL in any order is admissible. We estimate the number $N(n)$ of orbital points with length not exceeding n. In fact, this is a problem of how many ways a boy can jump up a staircase if he jumps either one or two steps at one time. The total number $N(n)$ of combinations that he can make must satisfy the recursion relation

$$N(n+1) = N(n) + N(n-1).$$

The first term on the right-hand side counts for the case when the last jump makes one step and the second term—the case of making two steps. The recursion relation coincides with that of the Fibonacci numbers, see (2.93) in Section 2.5.5. Therefore, we can use the expression (2.94) of the Fibonacci numbers to get

$$N(n) \propto \omega^n, \quad \omega = (1+\sqrt{5})/2, \quad n \gg 1.$$

Finally we get the topological entropy of the period 3 window:

$$h = \ln \omega = 0.481\ 211 \cdots . \tag{2.113}$$

It is definitely a positive number, although the motion is manifestly periodic. The actual chaotic motion occurs as transient chaos when the orbit leaves a strange "repeller". Such chaotic orbits may be observed for finite time span by careful choice of the initial points. They will eventually be attracted to the period 3 orbit.

When the kneading sequence of a unimodal map is either periodic or eventually periodic, the topological entropy can be calculated from the symbolic sequence alone. To be more precise, the topological entropy calculated by using a *generating partition* (see Section 2.8.5 below) coincides with the topological entropy of a piecewise linear tent map. We will devote the next Section 2.8 to the study of piecewise linear maps. The calculation of topological entropy will be continued in Section 2.8.4 when the notions of transfer matrices and characteristic functions are ready.

2.8 Piecewise Linear Maps and Metric Representation of Symbolic Sequences

Since symbolic dynamics reflects general properties of the map such as continuity or discontinuity, increasing (order-preserving) or decreasing (order-reversing) behavior of a certain branch of the map, details of the mapping function, e.g., change of slope within a particular monotone branch, do not matter. Therefore, piecewise linear maps may capture the essence of symbolic dynamics. Furthermore, piecewise linear maps have the merit that much of the study may be carried out analytically. For example, the word-lifting technique works for piecewise linear maps in a very simple manner, as the inverse functions are also linear.

Throughout this book we will often consider piecewise linear counterparts of 1D maps with multiple critical points, of circle maps and of two-dimensional maps. In what follows we start with the simplest case of the tent and the shift maps. This will lead us to a metric representation of symbolic sequences. The admissibility condition of symbolic sequences, reformulated in terms of metric representation, will be of some use in symbolic dynamics of two-dimensional maps.

The shift map leads naturally to so-called λ-expansion of real numbers. This kind of expansions provides a direct method to calculate topological entropy, especially for periodic and eventually periodic kneading sequences.

2.8.1 The Tent Map and Shift Map

In the foregoing sections we have seen that mapping on an interval corresponds to shift of symbolic sequence. For unimodal maps this shift is directly related to a specific piecewise linear map, namely, the tent map, which, in turn, has a close relation with another piecewise linear map—the shift map.

The tent map is a piecewise linear counterpart of the unimodal map. The mapping function

$$x_{n+1} = f(x_n) \equiv 1 - \lambda|x_n| = \begin{cases} 1 + \lambda x_n, & \text{if} \quad x_n \leqslant 0, \\ 1 - \lambda x_n, & \text{if} \quad x_n > 0 \end{cases} \tag{2.114}$$

is shown in Fig. 2.20. The name "tent" comes from its shape.

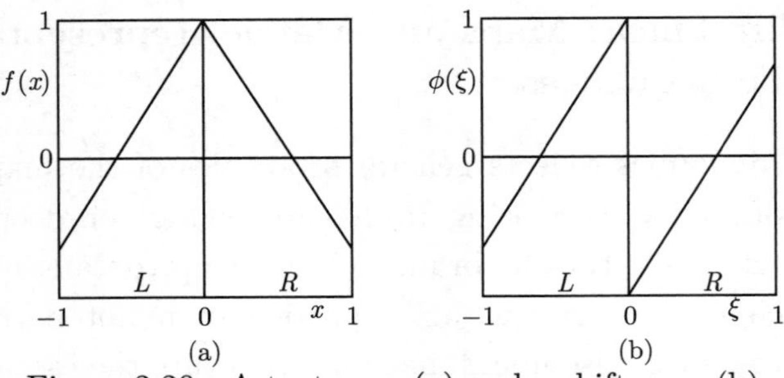

Figure 2.20 A tent map (a) and a shift map (b)

By inverting the decreasing right branch of the tent map with respect to the horizontal axis into an increasing function, we get the shift map, shown in the right part of Fig. 2.20. We write the shift map as:

$$\xi_{n+1} = \phi(\xi_n) \equiv \begin{cases} \lambda\xi_n + 1, & \text{if} \quad \xi_n \leqslant 0, \\ \lambda\xi_n - 1, & \text{if} \quad \xi_n > 0. \end{cases} \tag{2.115}$$

Take an equal value for the initial point $x_0 = \xi_0$, then there is a simple relation between x_n and ξ_n. First, $|x_n| = |\xi_n|$. Second, if in the iterations from x_0 to x_n the right branch appears an even number of times, in other words, among $x_0, x_1, \cdots, x_{n-1}$ there is an even number of points that take a positive value, then the signs of x_n and ξ_n coincide; otherwise, the signs are opposite. Consequently, the symbolic sequence $\sigma_0\sigma_1\cdots$ of ξ_0 may be generated from the symbolic sequence $s_0s_1\cdots$ of x_0 according to the following rule: if the string $s_0s_1\cdots s_{n-1}$ contains an even number of R, then $\sigma_n = s_n$; otherwise $\sigma_n \neq s_n$, i.e., $\sigma_n = L$ when $s_n = R$ and *vice versa*. For example,

$$\begin{aligned} x_0 &: RLRRLRLLR\cdots, \\ \xi_0 &: RRLRRLLLR\cdots. \end{aligned}$$

2.8.2 The λ-Expansion of Real Numbers

We introduce a signature ϵ_i for the the symbol σ_i as follows:

$$\epsilon_i = \begin{cases} +1, & \text{if} \quad \sigma_i = R, \\ 0, & \text{if} \quad \sigma_i = C, \\ -1, & \text{if} \quad \sigma_i = L. \end{cases} \tag{2.116}$$

The mapping function (2.115) may be written as

$$\xi_{n+1} = \lambda\xi_n - \epsilon_n. \tag{2.117}$$

Taking the inverse of the above equation, we get

$$\begin{aligned}\xi_n &= \frac{\epsilon_n}{\lambda} + \frac{\xi_{n+1}}{\lambda} \\ &= \frac{\epsilon_n}{\lambda} + \frac{\epsilon_{n+1}}{\lambda^2} + \frac{\xi_{n+2}}{\lambda^2} \\ &= \cdots \\ &= \sum_{i=1}^{\infty} \frac{\epsilon_{n+i-1}}{\lambda^i}.\end{aligned} \tag{2.118}$$

Therefore, if the symbolic sequence $\sigma_1\sigma_2\cdots$ of ξ_1 does not contain C, then (2.118) gives a λ-expansion for the real number ξ_1 (Derrida, Gervois, and Pomeau [1978]):

$$\xi_1 = \sum_{i=1}^{\infty} \frac{\epsilon_i}{\lambda^i}. \tag{2.119}$$

If the symbolic sequence contains C, then the summation in (2.119) terminates. For example, when $\sigma_n = C$, $\xi_n = \epsilon_n = 0$ and (2.119) becomes

$$\xi_1 = \sum_{i=1}^{n-1} \frac{\epsilon_i}{\lambda^i}. \tag{2.120}$$

Substituting the expansion (2.118) into the iteration (2.117),

$$\xi_{n+1} = \lambda\xi_n - \epsilon_n = \sum_{i=2}^{\infty} \frac{\epsilon_{n+i-1}}{\lambda^i}, \tag{2.121}$$

we see that ϵ_n cancels out. One iteration of the map (2.117) drops the first term in the λ-expansion, i.e., the symbolic sequence shifts one letter leftwards. This is why the map is called a shift map.

2.8.3 Characteristic Function of the Kneading Sequence

Owing to the equivalence of the tent and the shift maps, we will only consider the λ-expansion of ξ from now on. Suppose that a kneading sequence $\sigma_1\sigma_2\cdots$ starts from the critical point $C = 0^-$, where the map has a discontinuity. Since $\xi_1 = \phi(0) = 1$, Eq. (2.119) reads

$$1 = \sum_{i=1}^{\infty} \frac{\epsilon_i}{\lambda^i}. \tag{2.122}$$

In other words, it yields a λ-expansion for the real number 1. We agree that when $\sigma_n = C$ the above expansion is terminated at $i = n$.

Usually the expansion (2.122) is rearranged to define an equation

$$P(\lambda^{-1}) = 1 - \sum_{i=1}^{\infty} \epsilon_i \lambda^{-i} = 0. \tag{2.123}$$

The polynomial $P(t)$ is called the *characteristic function* or *kneading determinant* (Milnor and Thurston [1977, 1988]). This definition applies to all kneading sequences. However, when these sequences are periodic or eventually periodic, there exists a method to write down the final result of summation by simply looking at the symbolic sequence. We will see soon that the smallest real root of (2.123) determines the topological entropy of the map.

2.8.4 Mapping of Subintervals and the Stefan Matrix

So far we have discussed iteration of points. Now let us consider how some subintervals would change under the mapping. We restrict the discussion to mappings which have either a periodic (Σ^∞) or an eventually periodic ($\rho\lambda^\infty$) kneading sequence. For these two types, the orbit of the kneading sequence is made of a finite number of points. These points divide the dynamical invariant range into a finite number of subintervals. One may label these subintervals, say, by $I_j, j = 1, 2, \cdots, j_{\max}$ and write down a transfer matrix, which is often called a Stefan matrix in this special context (Derrida, Gervois, and Pomeau [1978]). We discuss these two types of kneading sequences separately.

The discussion in this section will be continued in Chapter 8, where we will show how the Stefan matrices are related to transfer functions in automaton theory, providing a way to comprehend some infinite limits in the effort to look for more complex languages in the symbolic dynamics of unimodal maps.

Periodic kneading sequences

For the first example, take the superstable kneading sequence $K = RLC$ of period 3. The three orbital points $x_0 = C$, $x_1 = RLC$, and $x_2 = LC$ are visited in the order shown by arrows in Fig. 2.21.

What happens if we choose an initial point other than these periodic orbital points? In general, one would get an infinite aperiodic orbit. It would be "attracted" to the superstable period 3, if one keeps iterating infinitely many times. For any finite number of iterations the sequence is not periodic. Nonetheless, we can tell much about the symbolic structure of these orbits. This can be seen as follows.

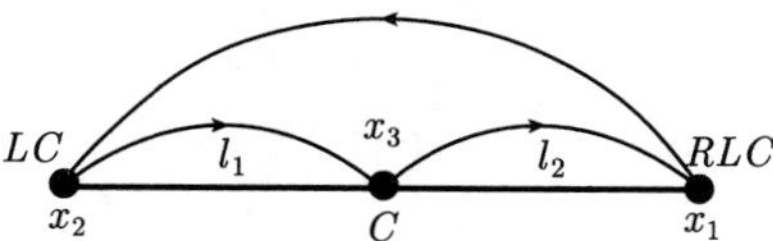

Figure 2.21 The orbital points of period 3 kneading sequence RLC divide the dynamical invariant range into two subintervals I_1 and I_2

The above three orbital points x_0, x_1, and x_2 divide the dynamical invariant range $U = [x_2, x_1]$ into two subintervals $I_1 = (x_2, x_0)$ and $I_2 = (x_0, x_1)$, as shown in Fig. 2.21. It follows from monotonicity and continuity consideration that these subintervals map into each other:

$$f(I_1) = I_2, \quad f(I_2) = (I_1 \cup I_2). \tag{2.124}$$

This result may be represented by a *transfer matrix*:

$$\mathbf{S} = \begin{bmatrix} 0 & 1 \\ 1 & 1 \end{bmatrix}, \tag{2.125}$$

where the rows and columns are labeled by the subscript of subintervals and the matrix elements are defined as

$$\mathbf{S}_{ij} = \begin{cases} 1, & \text{if} \quad f(I_i) \cap I_j \neq \varnothing, \\ 0, & \text{if} \quad f(I_i) \cap I_j = \varnothing. \end{cases} \tag{2.126}$$

The transfer matrix and the kneading sequence are closely related and they characterize the map from different points of view. The leading letter in any symbolic sequence starting from subinterval I_1 or I_2 is L or R, respectively. The transfer matrix shows that a symbol R in any sequence may be followed by either R or L, but a symbol L cannot be followed by another L. This is a strong "grammatical rule": the string LL is forbidden in the map described by the kneading sequence $K = RLC$. We will say more on grammatical aspect of symbolic dynamics in Chapter 8.

Each subinterval I_j may be labeled by a definite string of symbols. Take, for example, the superstable period 5 kneading sequence $K = RLRRC$. Its successive orbital points are obtained by shifting the periodic sequence by one letter at a time, i.e.,

$$\begin{aligned}
x_0 &= CRLRRCRLRR\cdots,\\
x_1 &= RLRRCRLRRC\cdots,\\
x_2 &= LRRCRLRRCR\cdots,\\
x_3 &= RRCRLRRCRL\cdots,\\
x_4 &= RCRLRRCRLR\cdots,\\
x_5 &= x_0.
\end{aligned}$$

These points are ordered in the following way:

$$x_2 < x_0 < x_3 < x_4 < x_1.$$

The dynamical invariant range are now divided into four subintervals, as shown in Fig. 2.22. The label of every subinterval is written under the horizontal line against its name I_j.

We explain how these labels are obtained on the example of $I_2 = (x_0, x_3)$. The two end points of I_2 are C and RRC. Since C is followed by the kneading sequence, these strings may be continued to $CRLRRC\cdots$ and $RRCRLRRC\cdots$. Change the first C in the two sequences into either R or L and order the sequences thus obtained from the smallest to the greatest:

$$\begin{aligned}
&LRLRRC\cdots,\\
x_0 = &CRLRRC\cdots,\\
&RRLRRC\cdots,\\
\cdots&\\
&RRLRLRRC\cdots,\\
x_3 = &RRCRLRRC\cdots,\\
&RRRRLRRC\cdots.
\end{aligned}$$

The longest common string between the two end points is $RRLR$. Any symbolic sequence, which starts from the subinterval must have $RRLR$ as the common leading string. This is why we label I_2 by $RRLR$. Labels for other subintervals are determined likewise. In so doing we have used only the ordering rule and continuity.

By inspecting Fig. 2.22 one easily writes down the transfer matrix

$$\mathbf{S} = \begin{bmatrix} 0 & 0 & 1 & 1 \\ 0 & 0 & 0 & 1 \\ 0 & 1 & 1 & 0 \\ 1 & 0 & 0 & 0 \end{bmatrix}. \qquad (2.127)$$

The non-zero elements of the transfer matrix and the labels of the subintervals tell us that strings $LRRR$, $LRLR$, $RRLR$, $RRRLR$, $RRRR$ and RLR correspond to the 6 non-zero elements $\mathbf{S}_{13}$, $\mathbf{S}_{14}$, $\mathbf{S}_{24}$, $\mathbf{S}_{32}$, $\mathbf{S}_{33}$, and $\mathbf{S}_{41}$, respectively. In this map any symbolic sequence is made by overlapping together strings of the four types LR, $RRLR$, RRR and RLR. After any shifts one of the above 6 strings of non-zero elements must always be the leading string of a subsequence.

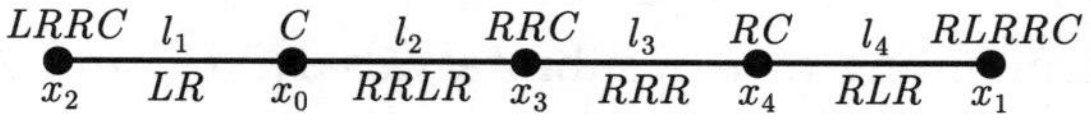

$LRRC$ l_1 C l_2 RRC l_3 RC l_4 $RLRRC$

x_2 LR x_0 $RRLR$ x_3 RRR x_4 RLR x_1

Figure 2.22 The labeling of subintervals for the period 5 superstable kneading sequence $RLRRC$

The above construction may be carried out for periodic kneading sequences which are not superstable. Take, for instance, $K = (RLL)^\infty$, the upper sequence of RLC. The orbit $C(RLL)^\infty$ is now not periodic; it approaches the periodic orbit (x_0, x_1, x_2) in infinitely many iterations. The three periodic points have sequences $L(RLL)^\infty = (LRL)^\infty$, $(RLL)^\infty$, and $(LLR)^\infty$, respectively. There is another point x_0' which is also a preimage of x_1 and has the sequence $R(RLL)^\infty$. By re-defining $I_1 = [x_2, x_0]$ and $I_2 = [x_0', x_1]$, the foregoing discussion for $K = RLC$ carries over easily and the transfer matrix turns out to be the same as (2.125).

Eventually periodic kneading sequences

Eventually periodic kneading sequences also lead to a finite transfer matrix. Two simple examples will elucidate the procedure.

Example 1 The surjective unimodal map with $K = RL^\infty$. The dynamical invariant range is divided into two subintervals, as shown in Fig. 2.23. It looks much like Fig. 2.21 with the difference that there is no closed loop ending at C, but x_2 maps into itself. The transfer matrix is

$$\mathbf{S} = \begin{bmatrix} 1 & 1 \\ 1 & 1 \end{bmatrix}. \tag{2.128}$$

The all-one structure of this matrix tells us that there are no forbidden strings at all; all symbolic sequences are allowed. This, of course, agrees with the admissibility condition with $K = RL^\infty$.

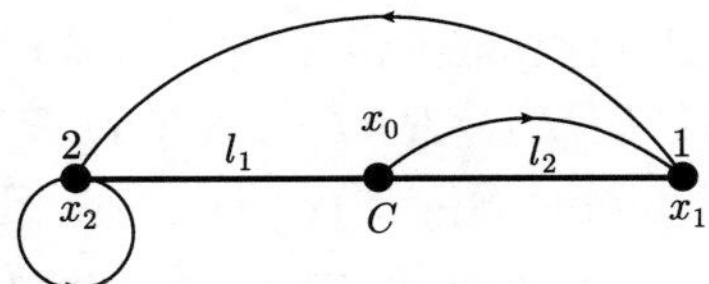

Figure 2.23 The orbital points of the kneading sequence RL^∞ divide the dynamical invariant range into two subintervals I_1 and I_2

Example 2 The 2-band to one-band merging point is given by the kneading sequence $K = RLR^\infty$. The orbital points $x_0 = C$, x_1, x_2, and x_3 are drawn in Fig. 2.24. The mapping of the subintervals I_1, I_2, and I_3 is given by the following transfer matrix:

$$\mathbf{S} = \begin{bmatrix} 0 & 0 & 1 \\ 0 & 0 & 1 \\ 1 & 1 & 0 \end{bmatrix}. \tag{2.129}$$

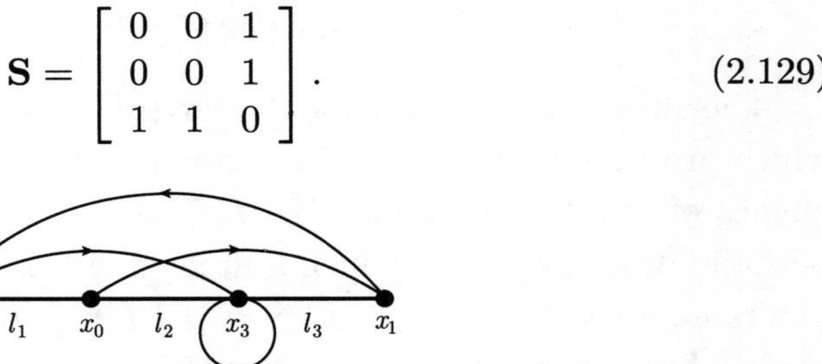

Figure 2.24 Partition of the dynamical invariant range for the kneading sequence $K = RLR^\infty$

Transfer matrix and the characteristic function

The characteristic function $P(t)$ defined in (2.123) is closely related to the transfer matrix. From the definition (2.115) of the shift map it follows that

$$\phi^{(j+1)} = \lambda \phi^{(j)} - \epsilon_j \phi^{(1)}, \tag{2.130}$$

where we have used $\phi^{(1)}(0) \equiv \phi(0) = 1$ and written $\phi^{(j)} \equiv \phi^{(j)}(0)$. If the kneading sequence is a periodic word of length k, one gets a system of homogeneous linear equations from (2.130) by setting $j = 1, 2, \cdots, k-1$ and noting $\phi^{(j+k)}(0) = \phi^{(j)}(0)$. In order to have a non-trivial solution, the determinant of the linear system must vanish:

$$\det \mathbf{A} = \begin{vmatrix} \epsilon_1 - \lambda & 1 & 0 & \cdots & 0 \\ \epsilon_2 & -\lambda & 1 & \cdots & 0 \\ \epsilon_3 & 0 & -\lambda & \cdots & 0 \\ \vdots & \vdots & \vdots & \ddots & \vdots \\ \epsilon_{k-1} & 0 & 0 & \cdots & -\lambda \end{vmatrix} = 0. \tag{2.131}$$

Expanding the determinant by the first column, we get

$$(\epsilon_1-\lambda)\begin{vmatrix} -\lambda & 1 & 0 & \cdots & 0 \\ 0 & -\lambda & 1 & \cdots & 0 \\ 0 & 0 & -\lambda & \cdots & 0 \\ \vdots & \vdots & \vdots & \ddots & \vdots \\ 0 & 0 & 0 & \cdots & -\lambda \end{vmatrix} - \epsilon_2 \begin{vmatrix} 1 & 0 & 0 & \cdots & 0 \\ 0 & -\lambda & 1 & \cdots & 0 \\ 0 & 0 & -\lambda & \cdots & 0 \\ \vdots & \vdots & \vdots & \ddots & \vdots \\ 0 & 0 & 0 & \cdots & -\lambda \end{vmatrix}$$
$$+\epsilon_3 \begin{vmatrix} 1 & 0 & 0 & \cdots & 0 \\ -\lambda & 1 & 0 & \cdots & 0 \\ 0 & 0 & -\lambda & \cdots & 0 \\ \vdots & \vdots & \vdots & \ddots & \vdots \\ 0 & 0 & 0 & \cdots & -\lambda \end{vmatrix} - \cdots = 0.$$

All these $(k-2)\times(k-2)$ determinants may be easily calculated to be powers of $-\lambda$. Finally, we obtain

$$\det \mathbf{A} = (-\lambda)^{k-1}P(\lambda^{-1}), \tag{2.132}$$

where $P(t)$ is the characteristic determinant (2.123).

By definition the Stefan matrix $\mathbf{S}$ transforms the subintervals $\{I_j\}$ into themselves:

$$f(I_i) \subset \bigcup_{j=1}^{k-1} \mathbf{S}_{ij} I_j.$$

Suppose the length of subinterval I_j is l_j. Since the map (2.115) changes l_j to λl_j in one iteration, these l_j must satisfy

$$\sum_j (\mathbf{S}_{ij} - \lambda\delta_{ij}) l_j = 0. \tag{2.133}$$

On the other hand, the quantities $\{l_j\}$ and $\{\phi^{(j)}(0)\}$ differ from each other only by an invertible linear transformation. Therefore, the matrix $\mathbf{S}_{ij} - \lambda\delta_{ij}$ differs from the coefficient matrix of the equations of $\phi^{(i)}(0)$ only by a similarity transformation. Consequently, the characteristic function of the transfer matrix coincides with the coefficient determinant $\det \mathbf{A}$, and is related to the characteristic function of the map in a similar way as (2.132).

Now consider kneading sequences of eventually periodic type $\rho\lambda^\infty$, where λ^∞ represents an unstable period. Let

$$\rho = \sigma_1\sigma_2\cdots\sigma_{m-1}, \quad \lambda = \sigma_m\sigma_{m+1}\cdots\sigma_k.$$

A derivation similar to that leading to (2.131) yields

$$\det \mathbf{A} = \begin{vmatrix} \epsilon_1 - \lambda & 1 & 0 & \cdots & 0 & 0 & \cdots & 0 \\ \epsilon_2 & -\lambda & 1 & \cdots & 0 & 0 & \cdots & 0 \\ \epsilon_3 & 0 & -\lambda & \cdots & 0 & 0 & \cdots & 0 \\ \vdots & \vdots & \vdots & \ddots & \vdots & \vdots & \vdots & \vdots \\ \epsilon_m & 0 & 0 & \cdots & -\lambda & 1 & \cdots & 0 \\ \epsilon_{m+1} & 0 & 0 & \cdots & 0 & -\lambda & \cdots & 0 \\ \vdots & \vdots & \vdots & \vdots & \vdots & \vdots & \ddots & \vdots \\ \epsilon_k & 0 & 0 & \cdots & 1 & 0 & \cdots & -\lambda \end{vmatrix} = 0. \tag{2.134}$$

We have seen that there is a finite number of orbital points for a kneading sequence $\rho\lambda^\infty$, leading to a finite transfer matrix $\mathbf{S}$, with (2.128) and (2.129) being examples. The characteristic function of $\mathbf{S}$ and the determinant of $\mathbf{A}$ contain a common factor, which determines the largest positive eigenvalue λ. It may be checked that the characteristic function of the map satisfies a relation similar to (2.132).

Transfer matrix and topological entropy

We know from Section 2.7 that the total number of orbital points of periods not exceeding n is related to the topological entropy of the map. This number may be calculated from the transfer matrix $\mathbf{S}$ or its equivalents. In fact, from the physical meaning of the transfer matrix $\mathbf{S}$ it follows that the diagonal elements of $\mathbf{S}^n$ correspond to mappings of the subinterval I_j back to itself. Therefore, the trace of $\mathbf{S}^n$ gives the total number $N(n)$ or orbital points of length n. For n big enough, the trace $\mathrm{tr}\mathbf{S}^n$ is determined by λ^n where λ is the *largest* positive eigenvalue of $\mathbf{S}$. In the previous sections we have shown that the transfer matrix $\mathbf{S}$ is equivalent to the coefficient matrix $\mathbf{A}$. Moreover, both $\mathbf{S}$ and $\mathbf{A}$ are closely related to the characteristic function $P(t)$. While formally $P(t)$ is always an infinite series, it may be summed up into a finite expression for periodic and eventually periodic kneading sequences. In these two cases both $\mathbf{S}$ and $\mathbf{A}$ are finite matrices. Since $t = \lambda^{-1}$ the topological entropy is determined by the *smallest* root of $P(t) = 0$.

Therefore, topological entropy may be calculated from the eigenvalue of either $\mathbf{S}$, or $\mathbf{A}$, or $P(t)$. We look at a few examples, including the two considered before.

Example 1 The surjective unimodal map with the kneading sequence RL^∞. The kneading sequence for the corresponding shift map is RR^∞, we have $\epsilon = 1$ for all i. The condition for the determinant $\mathbf{A}$ to vanish reads

$$\det \mathbf{A} = \begin{vmatrix} -\lambda+1 & 1 \\ 1 & -\lambda+1 \end{vmatrix} = 0,$$

coinciding with $\mathbf{S} - \lambda\mathbf{I} = 0$, see (2.128). The largest eigenvalue is $\lambda = 2$, yielding a topological entropy $h = \ln 2$. Of course, one can write the kneading sequence as $R(RR)^\infty$ or $R(RRR)^\infty$, leading to greater determinant $\mathbf{A}$. However, all these determinants contain the factor $\lambda - 2$ which gives the topological entropy.

Example 2 The 2-band to 1-band merging point corresponds to the kneading sequence RLR^∞. The transfer matrix (2.129) has $\lambda_{\max} = \sqrt{2}$, yielding a topological entropy $h = 2^{-1}\ln 2$. The kneading sequence of the corresponding shift map is $R(RL)^\infty$, leading to

$$\det \mathbf{A} = \begin{vmatrix} 1-\lambda & 2 & 0 \\ 1 & -\lambda & 0 \\ -1 & 1 & -\lambda \end{vmatrix} = 0.$$

It has the same $\lambda_{\max} = \sqrt{2}$. In general, the 2^n to 2^{n-1} band merging point has a topological entropy $h = 2^{-n}\ln 2$.

Example 3 The transfer matrix of the period 3 kneading sequence $K = RLC$ was given by (2.125). Its largest eigenvalue leads to the same topological entropy as we have calculated in (2.113).

To conclude this section we point out a direct relation between the characteristic function and the topological entropy. From the definition (2.123) of the characteristic function we know that λ is determined by the root of the characteristic function. The orbital points of period n in the shift map are the intersections of the n-th iterate $\phi^n(\xi)$ with the bisector. Obviously, the function $\phi^n(\xi)$ is also piecewise linear with the slope of each monotone branch being λ^n. The width of each monotone segment is $1/\lambda^n$. When n gets large enough, the number of intersections with the bisector is estimated to be λ^n. Therefore, topological entropy is given by $\ln\lambda$.

2.8.5 Markov Partitions and Generating Partitions

The essence of chaos consists in that a deterministic dynamical system may exhibit irregular behavior which looks so random that cannot be distinguished from a probabilistic random process when a certain degree of coarse-graining is introduced. Symbolic dynamics is an appropriate tool to connect the deterministic and probabilistic aspects. To elucidate the connection we need to explain a few mathematical notions from *Abstract* symbolic dynamics.

Topological Markov chain

Consider an *alphabet* of m letters. These letters may be m symbols or numbers, e.g., $\{1, 2, \cdots, m\}$. Form all possible bi-infinite symbolic sequences

$$\cdots \sigma_{n-2}\sigma_{n-1}\sigma_n\sigma_{n+1}\sigma_{n+2}\cdots,$$

where each σ_i takes one of the m values (symbols). The collection of all such sequences makes a space Σ of symbolic sequences. We single out a subspace of Σ in the following manner.

First, define an $m \times m$ transfer matrix $\mathbf{A} = (a_{ij})$, whose elements are either zero or one. If $a_{ij} = 1$, we say that the transition from the state i to the state j is admissible. Any of the Stefan matrices considered in the previous section has the desired form.

Second, pick up only those symbolic sequences from Σ which are *admissible* by the transfer matrix $\mathbf{A}$, i.e., all pairs of adjacent symbols in a sequence correspond to non-zero elements in $\mathbf{A}$. For example, the presence of $\sigma_n\sigma_{n+1}$ must be guaranteed by $a_{\sigma_n\sigma_{n+1}} = 1$, the presence of $\sigma_{n+1}\sigma_{n+2}$ must be guaranteed by $a_{\sigma_{n+1}\sigma_{n+2}} = 1$, etc.

There is the natural *shift* dynamics on the space Σ of all symbolic sequences embodied in the shift operator $\mathcal{S}$ defined in (1.8). The shift "dynamics" restricted to the subspace of symbolic sequences admissible by the given transfer matrix $\mathbf{A}$ is called a *subshift of finite type* or a topological Markov chain. The name topological Markov chain comes from a comparison with a probabilistic Markov chain, which is defined by a transfer matrix $\mathbf{A}$ whose element a_{ij} takes a probability value $0 \leqslant p_{ij} \leqslant 1$. The elements in a row must satisfy the normalization condition $\sum_j a_{ij} = 1$, as the transition process must keep going without confusion. A probabilistic Markov chain is a typical random process. A topological Markov chain has similar random nature without taking into account the probability of transition. It counts only for "yes" or "no", hence the adjective "topological".

We see that when the kneading sequence is periodic or eventually periodic, it gives rise to a topological Markov chain, defined by the corresponding Stefan matrix. However, the alphabet is not made of the letters R and L, but of the labels of the subintervals. For example, the labels LR, $RRLR$, RRR, and RLR shown in Fig. 2.22 for the periodic sequence $(RLRRC)^\infty$, form the alphabet with $m = 4$. This connection will become even clearer in Chapter 8 when we use the Stefan matrix to construct finite automaton for periodic and eventually periodic kneading sequences.

Markov partitions

The partition of phase space is the starting point of any symbolic dyna-mics. For a compact phase space any partition would lead to some kind of symbolic dynamics, not necessarily interesting or useful. In order to ensure that the symbolic dynamics yields a topological Markov chain, a special kind of partition is required. This is the so-called *Markov partition.* The existence and construction of Markov partitions in general lead to problems of significant mathematical difficulty. The reader may consult Alekseev and Yakobson [1981] or Bedford, Keane and Series [B1991] for the mathematical aspects.

For the unimodal map the requirement of the partition being Markov reduces to the following. First, the subintervals I_j do not overlap with each other. At most they may only have common end points. Second, a subinterval I_j goes into an union of some subintervals under the map f, i.e.,

$$f(I_j) \to \bigcup_k I_k.$$

Therefore, we see the subintervals given by iterates of a periodic or eventually periodic kneading sequences furnish a Markov partition.

Markov partition is not a general requirement in applied symbolic dynamics. The characteristic function or kneading determinant of a map does not require this either. When the partition is Markovian, there are many nice consequences. In particular, the infinite series for the characteristic function can be summed up into a finite expression, whose essential part, i.e., the numerator determining the zeros of the characteristic function is in fact given by the determinant of Stefan or coefficient matrix.

Generating partitions

It was mentioned in Section 2.7 that a rigorous definition of topological entropy requires infinite refinement of the partition by the dynamics. We explain the idea using Fig. 2.25. The shaded broken cake in the squares represents our phase space. A partition V is shown in Fig. 2.25 (a). It has two pieces V_1 and V_2. Another partition H is shown in (b). It has two pieces H_1 and H_2. In both partitions the number of pieces that contain our object is 2. We write $N(V) = N(H) = 2$.

Two partition may be "combined" to get a finer partition. The refined partition obtained by combining V and H is denoted by $V \vee H$ and is shown in Fig. 2.25 (c). Now it has 4 pieces and $N(V \vee H) = 3$.

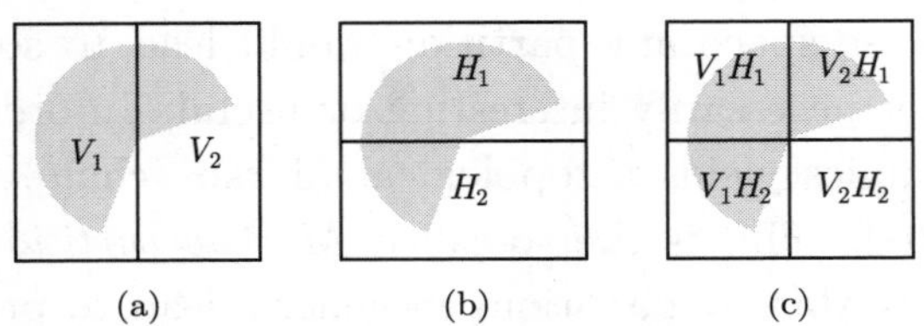

Figure 2.25 Combination of partition V (a) and partition H (b) gives the refined partition $V \vee H$ (c)

For a one-dimensional map f on an interval I we can start with one partition, say, V, and combine it with partition $f^{-1}(V)$, $f^{-2}(V)$, etc., to get finer and finer partitions. In general, $f^{-1}(V)$ contains more pieces than V due to the multi-valuedness of f^{-1}. Take, for example, a surjective unimodal map and the partition V made by its critical point C. It contains two pieces V_1 and V_2, corresponding to L and R. The partition $f^{-1}(V)$ contains 4 pieces, etc.

The topological entropy of f with respect to a "refined-by-dynamics" partition is defined as

$$h(f, V) = \lim_{n \to \infty} \frac{1}{n} \ln \left[N \left(V \vee f^{-1}(V) \vee f^{-2}(V) \vee \cdots \vee f^{-n}(V) \right) \right]. \quad (2.135)$$

The entropy $h(f, V)$ depends on the choice of partition V. The topological entropy of the map f is

$$h(f) = \sup_V h(f, V), \quad (2.136)$$

where the supremum is taken over all possible partitions. If a partition V yields $h(f)$ directly, it is called a *generating partition* or a *generator*. The difficulty in using the definition (2.136) to calculate topological entropy lies in taking the supremum operation or in finding a generating partition. In fact, there is no general method to construct a generating partition. However, maps with a periodic or eventually periodic kneading sequence again distinguish themselves by the readiness of a generating partition (Hsu and Kim [1984, 1985]). In fact, the partition we have been using provides the generating partition.

2.8.6 Metric Representation of Symbolic Sequences

We have mentioned once in Section 1.2.2 that the collection of all possible symbolic sequences forms an abstract space of symbolic sequences. One can introduce additional structure in this space, e.g., distance between any two sequences in order to make it a metric space. In this book we do not need these structures except for a simple metric representation of symbolic sequences. We will use this representation to reformulate the admissibility condition of symbolic sequences. Our ultimate aim is to use the metric representation to formulate admissibility conditions in symbolic dynamics of two-dimensional maps later in Chapter 5. It is with this aim in mind that we put the metric representation in a more general setting.

Consider all possible cases of piecewise linear surjective maps with two monotone branches. There are four cases, shown in Figs. 2.26 (a)–(d). Cases (a) and (c) are the shift and tent maps considered in Section 2.8.1. Case (b) may be called an inverse shift map and Case (d)—an inverse tent map. They exhaust all possible combinations of parities for two symbols.

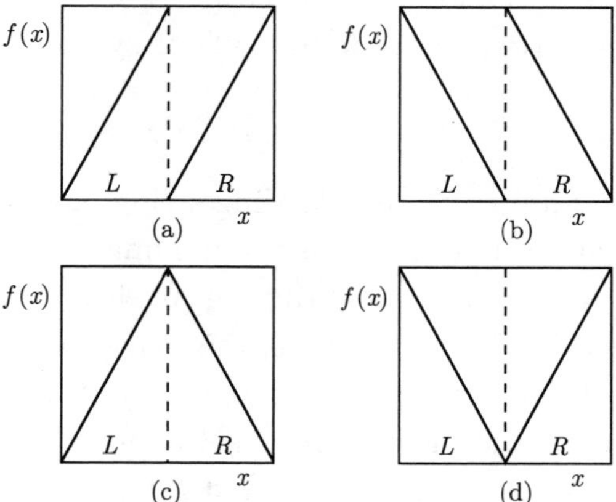

Figure 2.26 Four possible piecewise linear surjective maps with two monotone branches: shift map (a), inverse shift map (b), tent map (c) and inverse tent map (d)

In each case we establish a correspondence between a symbolic sequence

$$\Sigma = \sigma_1 \sigma_2 \cdots$$

and a real number $\alpha(\Sigma) \in [0,1]$:

$$\alpha(\Sigma) = \sum_{i=1}^{\infty} 2^{-i} \mu_i, \tag{2.137}$$

where the coefficient μ_i is determined from the symbol σ_i and, possibly, its preceding symbols in such a way as to keep the ordering of symbolic sequences. In particular, we require that the maximal sequence corresponds to $\alpha = 1$, the minimal sequence—to $\alpha = 0$.

We note that in all cases the natural order

$$L < C_- \leqslant C_+ < R$$

holds. The letter C or $C_\pm$ does not count as an independent symbol. It may always be replaced by an infinite sequence made of R and L.

We consider the four cases separately.

The shift map

The shift map provides the simplest case. By inspecting the graph(see Fig.2.26) (a), we see that the maximal sequence is the upper right fixed point R^∞, and the minimal sequence is the lower left fixed point L^∞. Since both symbols R and L have an even parity, the ordering rule of symbolic sequences may be formulated as

$$\Sigma^* L \cdots < \Sigma^* R \cdots ,$$

where Σ^* denotes any finite common leading string made of R and L.

The sequence that starts from C_+ is the minimal one among all sequences starting with an R. According to the ordering rule it is obtained by appending the minimal sequence L^∞ to R. Likewise, the sequence of C_- is the maximal one among all sequences starting with an L, so it is LR^∞.

The metric representation is nothing but the binary representation of real numbers between 0 and 1, taking R for 1 and L for 0. In other words, for the coefficients μ_i in (2.137) we have

$$\mu_i = \begin{cases} 1, & \text{if} \quad \sigma_i = R, \\ 0, & \text{if} \quad \sigma_i = L. \end{cases} \tag{2.138}$$

It is easy to check that

$$\begin{aligned}&\alpha(R^\infty) = 1,\\&\alpha(C_-) \equiv \alpha(RL^\infty) = \alpha(C_+) \equiv \alpha(LR^\infty) = 1/2,\\&\alpha(L^\infty) = 0.\end{aligned}$$

The inverse shift map

Now both symbols R and L have an odd parity, so the ordering rule is

$$EL\cdots < ER\cdots, \quad OR\cdots < OL\cdots,$$

where E is a common leading string which contains an *even* number of symbols R or L or both; O is a string containing an *odd* number of symbols R or L or both.

The maximal sequence is the sequence $(RL)^\infty$ of period 2; the minimal sequence—the other sequence $(LR)^\infty$ of period 2; C_+ and C_- are $R(RL)^\infty$ and $L(LR)^\infty$, respectively.

To determine the coefficients μ_i in the metric representation (2.137) we put $\epsilon_i = -1$ if $\sigma_i = R$ and $\epsilon_i = 1$ if $\sigma_i = L$. Then we have

$$\mu_i = \begin{cases} 1, & \text{if} \quad (-1)^i\epsilon_i = -1, \\ 0, & \text{if} \quad (-1)^i\epsilon_i = 1, \end{cases} \tag{2.139}$$

and

$$\begin{aligned}&\alpha((RL)^\infty) = 1,\\&\alpha(C_-) \equiv \alpha(R(RL)^\infty) = \alpha(C_+) \equiv \alpha(L(LR)^\infty) = 1/2,\\&\alpha((LR)^\infty) = 0.\end{aligned}$$

The tent map

Only the symbol R has an odd parity, hence the ordering rule:

$$E_RL\cdots < E_RR\cdots, \quad O_RR\cdots < O_RL\cdots,$$

where E_R and O_R are strings containing an *even* and *odd* number of symbol R.

With the same convention $\epsilon_R = -1$ and $\epsilon_L = 1$ we have

$$\mu_i = \begin{cases} 1, & \text{if} \quad \prod_{j=1}^{i} \epsilon_j = -1, \\ 0, & \text{if} \quad \prod_{j=1}^{i} \epsilon_j = 1, \end{cases} \tag{2.140}$$

and

$$\begin{aligned}
&\alpha(RL^\infty) = 1, \\
&\alpha(C) = \alpha(RRL^\infty) = \alpha(LRL^\infty) = 1/2, \\
&\alpha(L^\infty) = 0.
\end{aligned}$$

The inverse tent map

Now only the symbol L has an odd parity, hence the ordering rule

$$E_L L \cdots < E_L R \cdots , \quad O_L R \cdots < O_L L \cdots ,$$

where E_L and O_L are strings containing an *even* and *odd* number of symbol L.

With the same convention $\epsilon_R = -1$ and $\epsilon_L = 1$ we have

$$\mu_i = \begin{cases} 1, & \text{if} \quad (-1)^i \prod_{j=1}^{i} \epsilon_j = 1, \\ 0, & \text{if} \quad (-1)^i \prod_{j=1}^{i} \epsilon_j = -1, \end{cases} \tag{2.141}$$

and

$$\begin{aligned}
&\alpha(R^\infty) = 1, \\
&\alpha(C) = \alpha(RLR^\infty) = \alpha(LLR^\infty) = 1/2, \\
&\alpha(LR^\infty) = 0.
\end{aligned}$$

Admissibility of symbolic sequences

For the surjective maps all symbolic sequences made of two letters R and L are admissible. If a map is not surjective, the transformed interval turns out to be only a part of I; it is an injective, or "into", map. Then not all symbolic sequences made of the two symbols are reproducible by the dynamics. Admissibility condition is needed to tell whether a given sequence is allowed or not.

The metric representation of symbolic sequences allows us to reformulate the admissibility condition. We show this on the example of the tent map $T(\theta)$. Fig. 2.27 (a) shows an injective tent map whose maximal point does not touch the top of the unit square. The kneading sequence K has a metric representation $\alpha(K)$ less than 1, as shown in Fig. 2.27 (b). We ignore the initial points which are greater than K as they only lead to trivial transients.

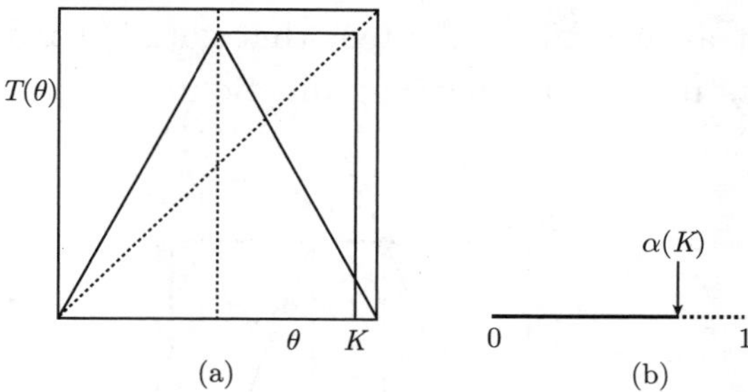

Figure 2.27 The admissibility in terms of metric representation of symbolic sequences. (a) Kneading sequence of the tent map. (b) Forbidden zone in the metric representation

Clearly, if a symbolic sequence Σ is admissible, then any of its "tail" obtained by discarding a number of leading symbols, must be admissible. This means that the metric representation of any shift of Σ must not exceed $\alpha(K)$:

$$\alpha(\mathcal{S}^k\Sigma) \leqslant \alpha(K), \tag{2.142}$$

where $\mathcal{S}$ is the shift operator introduced in Section 1.2.2.

Put in other words, the metric representation of any shift of an admissible symbolic sequence cannot fall in the dashed part of the unit α line shown in Fig. 2.27 (b). In this sense, the dashed part $(\alpha(K), 1]$ of the α line is a *forbidden zone.*

For a general unimodal map one calculates the metric representations of the kneading sequence and the shifts of the sequence under study then compares them as in a tent map. Taking as a simple example, for the very low unimodal map shown in Fig. 2.4, we have $K = L^\infty$ and $\alpha(K) = 0$. Therefore, the forbidden zone extends to the whole unit line except for one point $\alpha = 0$. All other symbolic sequences are inadmissible. As mentioned in Section 2.1.1, the only other kind of symbolic sequence $\Sigma = RL^\infty$ may appear as trivial transient, but we have excluded it from the beginning.

2.8.7 Piecewise Linear Expanding Map

We take the opportunity of studying piecewise linear maps to look at a piecewise linear expanding map. It may be realized by a tent map with slope greater than 2, as shown in Fig. 2.28. The mapping interval is still confined to I. The top central segment of the tent that goes beyond the square $I \times I$ is not considered as part of the mapping function.

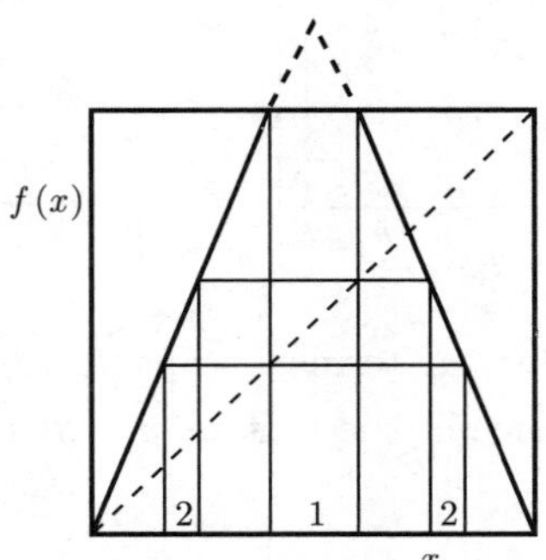

Figure 2.28 An expanding tent map with slope greater than 2

Construct a few auxiliary lines as shown in Fig. 2.28. First draw two verticals from the points where the tent gets out of the top. From the intersection points of these two verticals with the bisector draw horizontal lines, which meet the tent at 4 points. Draw vertical lines from these 4 points and find their intersections with the bisector, and so on and so forth. This construction resembles the construction of inverse paths, shown, e.g., in Fig. 2.14. It may be continued *ad infinitum.*

If an initial point is taken from the subinterval, labeled by the number 1 in the figure, it escapes from I after one iteration. If an initial point is taken from the subintervals, labeled by numbers 2, it escapes from I in two iterations, etc. These "escape regions" are emphasized by thick line segments in the figure. Eventually all points from I will escape except for a collection of infinitely many isolated points which form a Cantor set. When the slope of the tent equals 3, this is the classical one third Cantor set.

Points from the "non-escaping" Cantor set form a *strange repeller.* It is a repeller, because unless a point is located in this set with infinite precision it will eventually be repelled away from I. It is strange, because a Cantor set usually has a fractal dimension.

Points within the Cantor set may be represented by infinite symbolic sequences made of two symbols R and L. This is clear from the above construction. Please note that this is not only a labeling of orbital points, but also an identification of points that do not escape.

3
Maps with Multiple Critical Points

In the last chapter we have studied unimodal maps with one critical point and two monotone branches. The corresponding symbolic dynamics of two letters turns out to be the closest to the simplest dynamics of two symbols described in Section 1.2.2. The two main ingredients are the ordering of symbolic sequences and the admissibility condition based on the ordering. The only topological properties used are the monotonicity and continuity.

There is an infinitely rich variety of nonlinear maps with multiple critical points. By critical points we understand not only maxima and minima of the map, but also discontinuities of the mapping function. Therefore, a critical point is not necessarily connected with vanishing derivative of the map, which matters only in treating metric property of the map. In general, a new critical point brings about a new kneading sequence and an additional monotone branch ushers in a new symbol. Maps with multiple critical points are best parameterized by their kneading sequences.

Most of what we have learned from the symbolic dynamics of unimodal maps may be extended to maps with multiple critical points. Of course, some new features may appear. We precede the case study of maps by a general discussion in Section 3.1.

If one confines oneself to continuous maps, cubic map may be considered next to the simplest. Especially, when the map is anti-symmetric, it reduces to an one-parameter map. The anti-symmetric cubic map is the subject of Section 3.2. Section 3.3 continues the study of the anti-symmetric cubic map by concentrating on the phenomenon of symmetry breaking and symmetry restoration. We will see that symbolic dynamics explains the well-known fact that the symmetry breaking bifurcation always precedes the period-doubling cascade, i.e., period-doublings always take place in asymmetric regime. Symbolic dynamics also provides the selection rule determining which periods are capable to undergo symmetry breaking. Symbolic dynamics explains symmetry restoration as a kind of crisis and helps to locate the parameter precisely. Being performed in symbolic terms, the analysis of symmetry breaking and restoration applies to a much wider class of nonlinear systems with discrete symmetry.

If the map contains one or more discontinuities, the left and right limits of a discontinuity may be treated as critical points. The ordering rule and admissibility conditions may easily be modified to incorporate this case. The presence of discontinuity leads to some new features in the structure of the kneading plane. In particular, there appear so-called *contacts* and *proper contacts*. There are two simple maps with a discontinuity: the gap map and the Lorenz-like map. The gap map is obtained by opening up a gap at the maximal point of the unimodal map. If one further inverses the right branch of the gap map with respect to the horizontal coordinate axis, the Lorenz-like map results. The term Lorenz-like comes from its connection with the geometric Lorenz model which explains the chaotic behavior of the Lorenz model around $r = 28$, see Eqs. (1.3). We will say more in Chapter 6. Both the gap map and the Lorenz-like map have two monotone branches but the critical point splits into two due to the discontinuity. Thus, the symbolic dynamics involves two critical points and two symbols. These cases will be studied in Sections 3.4 and 3.5.

Section 3.6 is devoted to general cubic maps. The generation of kneading sequences and construction of the kneading plane is quite instructive for the understanding of other maps, in particular, for the circle maps to be studied in Chapter 4.

The sine-square map considered in Section 3.7 comes from a model of optical bistable device in the long time-delay limit. Although it is a one-parameter map, it shows clearly why a map with multiple critical points is best parameterized by its kneading sequences. A specific feature of the map consists in there being critical points at the two ends of the interval.

In Section 3.8 we consider a family of maps defined only by their shape but not by any analytical expression. These maps appear in the numerical study of the Lorenz model (1.3) after some manipulation of the first return maps. C. Sparrow anticipated that maps of this type are capable to explain the dynamical behavior of the Lorenz model in a wide parameter range. Hence we suggest the name Lorenz-Sparrow maps. These maps will be studied here *per se* without making connection to the Lorenz equations. We will see that the lacking of analytical expressions does not hinder the construction of symbolic dynamics.

As in Chapter 2, we conclude this chapter by considering piecewise linear maps with multiple critical points, which lead to the kneading determinants and provide an easy way to calculate topological entropy. This is done in Section 3.9.

3.1 General Discussion

We precede the study of various maps with multiple critical points by a general discussion of their common features. A general continuous map with multiple critical points is shown in Fig. 3.1. Starting from the left end of the interval, we label the monotone increasing branches of the map by symbols $L_1, L_2, \cdots$, and that of the decreasing branches by $R_1, R_2, \cdots$. The i-th increasing branch with label L_i has a minimum D_i and a maximum C_i at its left and right ends, respectively. Between the critical points C_i and D_{i+1} there is the i-th decreasing branch label by R_i. The largest and smallest critical points may coincide with end points of the interval.

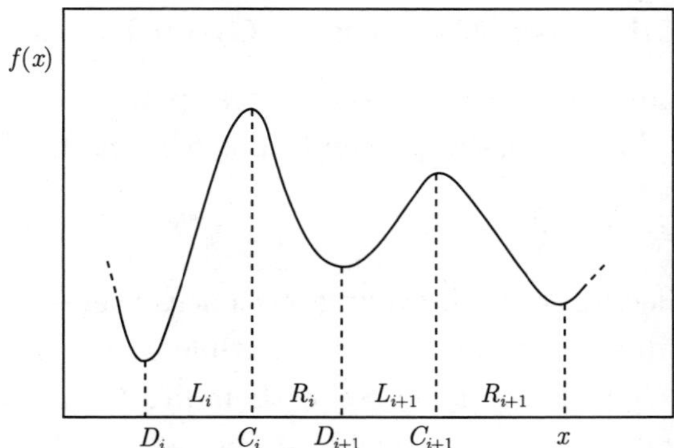

Figure 3.1 A mapping function with multiple critical points

We start with the formulation of the ordering rule of symbolic sequences.

3.1.1 The Ordering Rule

The ordering rule introduced in Section 2.3.2 actually works in all cases when one partition the interval according to the monotone branches of the mapping function. We restate it here for easier reference.

First, we need the notion of parity. For the general map shown in Fig. 3.1, the symbols R_i have an odd parity. The parity of a symbolic string is determined by the total number of R_i it contains.

Second, the ordering of symbolic sequences is based on the natural order of real numbers on the interval. A part of this natural order reads:

$$\cdots < D_i \cdots < L_i \cdots < C_i \cdots < R_i \cdots < D_{i+1} \cdots < \cdots .$$

Third, an even common leading string preserves the order of subsequences following it, while an odd common leading string reverses the order.

Now we are in a position to formulate the following

General Ordering Rule Compare two symbolic sequences Σ_1 and Σ_2:

$$\begin{aligned} \Sigma_1 &= \Sigma^* \sigma \cdots , \\ \Sigma_2 &= \Sigma^* \tau \cdots , \end{aligned} \tag{3.1}$$

where Σ^* is their common leading string and the symbols σ and τ are different. If the parity of Σ^* is even, the order of Σ_1 and Σ_2 is the same as the order of σ and τ as they appear in the natural order. If Σ^* is odd, the order of Σ_1 and Σ_2 is opposite to the order of σ and τ.

3.1.2 Construction of a Map from a Given Kneading Sequence

Denoting the kneading sequences of the critical points C_i and D_i by K_{C_i} and K_{D_i}, respectively. The admissibility conditions for a symbolic sequence Σ read

$$K_{D_i} \leqslant \mathcal{L}_i(\Sigma) \leqslant K_{C_i}, \quad K_{D_{i+1}} \leqslant \mathcal{R}_i(\Sigma) \leqslant K_{C_i}. \tag{3.2}$$

Applied to the collection of all kneading sequences themselves, these relations give the compatibility conditions of a compatible set of kneading sequences.

The kneading sequences of a map with multiple critical points, considered as independent entities, span a *kneading space*. Only those points in the kneading space which meet the compatibility conditions make sense, i.e., correspond to realizable set of parameters for the map. In the case of two parameters these are called *compatible kneading pairs*.

Usually, some generalization of the periodic window theorem holds. Although the phase space is still one-dimensional the dimension of the parameter space is now higher than one. Therefore, multiple-superstable kneading sequences play an especially important role. Compatible kneading pairs determine *joints* in the kneading space. From a joint come out loci of superstable kneading sequences containing a less number of critical point symbols. They are called *bones* of the kneading space. Joints and bones make the *skeleton* of the kneading space. In a high-dimensional parameter space, bones are given by hypersurfaces. However, in problems of physical relevance often a part or a subspace of the kneading space matters; the discussion may usually be simplified.

In this book we will only study maps representable in a kneading plane. For a concrete example of a kneading space, we refer to the symbolic dynamics analysis of a general quartic map (Xie [1994]). In addition, listed in Appendix A there is a C program PERIOD.C to generate all admissible symbolic sequences up to a given length for a given map. It is capable to deal with quite general types of 1D maps with multiple critical points and with discotinuities. However, being primarily designed for counting the number of various periodic orbits in a map, it does not check the compatibility of kneading pairs. Nevertheless we recommend the reader to have a look at this program *after* reading this chapter.

3.2 The Antisymmetric Cubic Map

The anti-symmetric cubic map has only one parameter, as the two critical points are bound to vary together. The symbolic dynamics we are going to develop applies to all continuous maps with two anti-symmetrically located critical points, which divide the interval into three subintervals with one monotonic branch of the mapping function on each subinterval. Nevertheless, it is quite instructive to have a concrete map for comparison. We write this map as (May [1979])

$$x_{n+1} = f(A, x_n) \equiv Ax_n^3 + (1 - A)x_n, \tag{3.3}$$

where $x \in [-1, 1]$ and $A \in (1, 4]$.

The map (3.3) remains unchanged if one reverses the sign of the variable: $x \rightarrow -x$. This is the simplest discrete symmetry a physical system may possess. It is this symmetry that makes the anti-symmetry cubic map an important subject for study, as many higher-dimensional nonlinear models such as the Lorenz equations (Lorenz [1963]) or the double-diffusive system (Knobloch and Weiss [1981]) share the same discrete symmetry. The map (3.3) gives the clue to understand the phenomenon of symmetry breaking and symmetry restoration in dynamical systems with discrete symmetry. In particular, the fact that a symmetry breaking bifurcation must precede the period-doubling bifurcation sequence, the selection rule telling which orbit may undergo symmetry breaking, and the mechanism and location of symmetry restoration point can all be explained concisely in terms of symbolic dynamics.

Before undertaking a detailed study of the symbolic dynamics of the map (3.3), we collect a few facts on the map (3.3) for later reference.

The mapping function (3.3) is shown in Fig. 3.2. It has two critical points

$$C = -\overline{C} = \sqrt{\frac{A-1}{3A}}. \tag{3.4}$$

They divide the mapping interval $[-1, 1]$ into three segments: the *left* segment to the left of $\overline{C}$, the *right* segment to the right of C, and the *middle* segment in between. We label these segments and the monotonic branches of the mapping function on these segments by the letters L, R, and M.

The subinterval U, shown in Fig. 3.2 by a thick line, is the *dynamical invariant range* of the map. Dynamics outside U gives trivial transients, once it enters U it can never get out, as $f(U) \subset U$. Therefore, all the long-time behavior of the dynamics shows itself on U. In what follows we will be interested mainly in the dynamics that takes place on the interval U.

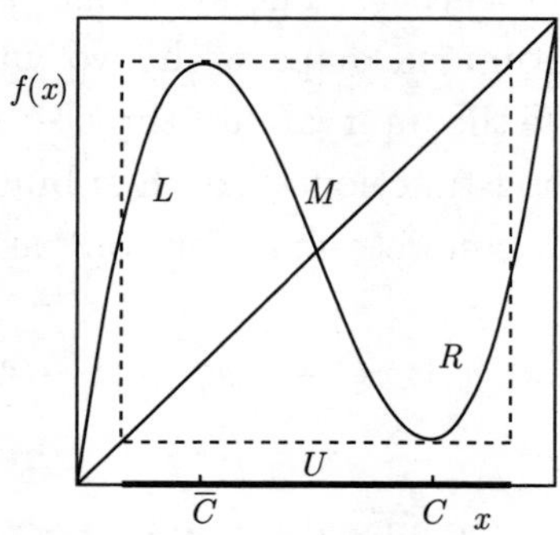

Figure 3.2 The anti-symmetric cubic map. U is the dynamical invariant range

Fixed points and 2-cycle

Three fixed points exist at all parameter values: $x = \pm 1$, always unstable, and $x = 0$, stable for $0 < A < 2$. The M branch comes into being when $A > 1$. At $A = 2$, the $x = 0$ fixed point losses stability and a period-doubling takes place. Owing to the anti-symmetry there are two kinds of period 2 orbits: symmetric orbits, which may be obtained by iterating twice the inversion relation

$$f(A, x^*) = -x^*, \tag{3.5}$$

and asymmetric orbits, which come in pairs

$$x_1^* \to -x_2^* \to x_1^* \quad \text{and} \quad x_2^* \to -x_1^* \to x_2^*. \tag{3.6}$$

We assume that $|x_1^*| \neq |x_2^*|$, otherwise they would degenerate into the symmetric pair.

In order to determine the stability range of the symmetric 2-cycle, we note that each element of the pair $(x^*, -x^*)$ may be viewed as a fixed point of an iteration with inversion, i.e.,

$$x_n = -Ax_n^3 - (1 - A)x_n \, .$$

Its explicit solution yields, besides the trivial fixed point $x^* = 0$,

$$x^* = \sqrt{(A-2)/A}. \tag{3.7}$$

Its stability is determined by

$$s = \left.\frac{\partial f^{(2)}}{\partial x}\right|_{\pm *} = \left.\frac{\partial f}{\partial x}\right|_{\mp *} \left.\frac{\partial f}{\partial x}\right|_{\pm *} = \left(\left.\frac{\partial f}{\partial x}\right|_{*}\right)^2 = (2A - 5)^2 \tag{3.8}$$

(We will write $|_*$ instead of $|_{x^*}$ hereafter). Contrary to an orbit generated by period-doubling, the stability discriminant s can never become negative: $s = 1$ at $A = 2$, as it should be, when the orbit is born due to period-doubling, $s = 0$ when it becomes a symmetric superstable 2-cycle, and again $s = 1$, when it loses stability at $A = 3$. At the last point there occurs a symmetry breaking bifurcation.

In order to calculate the asymmetric 2-cycle, we note that the symmetric 2-cycle still exists when A gets larger than 3, but loses its stability. In order to find the asymmetric orbits, we first exclude the trivial zeros of $f^{(2)} - x$, dividing it by the fixed point relation $f - x$ and the inversion point relation $(f + x)/x$, i.e.,

$$\frac{f^{(2)}(x) - x}{[f(x) - x][f(x) + x]/x} = A\left[A^2x^4 + A(1 - A)x^2 + 1\right]. \tag{3.9}$$

Its two zeros

$$x_\pm^* = \sqrt{\frac{A - 1 \pm \sqrt{(A - 3)(A + 1)}}{2A}} \tag{3.10}$$

determine the two 2-cycles as functions of the parameter A (cf. (3.6)). Their stability range is

$$3 < A < 1 + \sqrt{5},$$

since

$$\left.\frac{\partial f^{(2)}}{\partial x}\right|_{x_1^*} = \left.\frac{\partial f}{\partial x}\right|_{-x_2^*} \left.\frac{\partial f}{\partial x}\right|_{x_1^*} = \left.\frac{\partial f}{\partial x}\right|_{-x_1^*} \left.\frac{\partial f}{\partial x}\right|_{x_2^*}$$
$$= \left.\frac{\partial f^{(2)}}{\partial x}\right|_{x_2^*} = -2A^2 + 4A + 7.$$

The last expression attains -1 at $A = 1 + \sqrt{5}$, where period-doubling into a 4-cycle takes place. Note that both 2-cycles are stable in the above parameter range, each in its own basin. Therefore, only one of them may be observed for a given initial value.

3.2.1 Symbolic Sequences and Their Ordering

We have partitioned the mapping interval according to the monotonic branches of the mapping function, as shown in Fig. 3.2. Any numerical orbit leads to a symbolic sequence after replacing the number by the symbol of the subinterval where the point falls in. Among the three symbols only M now has an odd parity, as the middle branch of the mapping function is monotonically decreasing. We may define the parity of a string made of R, M, and L by the total number of M in it.

The ordering rule of symbolic sequences is formulated as follows. First, there is a natural order $L\cdots < \overline{C}\cdots < M\cdots < C\cdots < R\cdots$, or simply

$$L < \overline{C} < M < C < R. \tag{3.11}$$

Next, consider two symbolic sequences

$$\begin{aligned}\Sigma_1 &= \Sigma^*\mu\cdots,\\ \Sigma_2 &= \Sigma^*\nu\cdots,\end{aligned} \tag{3.12}$$

where Σ^* denotes their common leading string and the symbols $\mu \neq \nu$. Then their ordering is

$$\begin{aligned}\mu > \nu \leftrightarrow \Sigma^*\mu\cdots > \Sigma^*\nu\cdots, &\quad \text{if}\quad \Sigma^* \text{ is even;}\\ \mu > \nu \leftrightarrow \Sigma^*\mu\cdots < \Sigma^*\nu\cdots, &\quad \text{if}\quad \Sigma^* \text{ is odd.}\end{aligned} \tag{3.13}$$

Denote the two kneading sequences by $\overline{K} = f(\overline{C})$ and $K = f(C)$. Due to the anti-symmetry, they are mirror images to each other, i.e., the interchange

$$\mathcal{T}: L \leftrightarrow R, C \leftrightarrow \overline{C}, M \leftrightarrow M, \tag{3.14}$$

leads to $\overline{K} = \mathcal{T}K$. Therefore, there is only one independent kneading sequence. The map may be parameterized by its kneading sequence, i.e., using K or $\overline{K}$ as the parameter.

We note that the interchange (3.14) as well as $K \leftrightarrow \overline{K}$ are parity-preserving transformations. They may be applied to general cubic maps, see, e.g., (3.91) in Section 3.6.3. The anti-symmetry manifests itself in there being only one kneading sequence K and $\overline{K} = \mathcal{T}K$.

Periodic orbits that contain only one of the critical points are asymmetric superstable periods. Their symbolic sequences, containing C or $\overline{C}$ only, are superstable kneading sequences. They exist in pair, being mirror images to each other under the interchange (3.14). Periodic symbolic sequences, which contain both C and $\overline{C}$ are doubly superstable orbits. They must be symmetric orbits of the form $\Sigma C\overline{\Sigma C}$ where Σ and $\overline{\Sigma}$ are mirror images to each other. The simplest doubly superstable orbit is the symmetric period 2 orbit $(C\overline{C})^\infty$.

3.2.2 Admissibility Conditions

Numerical orbits accept certain symbolic sequences. An arbitrarily given symbolic sequence made of R, M and L may not be reproducible by choosing any initial point x_0 at a certain parameter. One needs a criterion to test the admissibility of a symbolic sequence. As we have seen in the quadratic case, the admissibility condition is based on the ordering rule and may be read off from the graph of the mapping function.

For any given symbolic sequence Σ we form three sets of subsequences:

$$\begin{aligned}\mathcal{L}(\Sigma) &= \{\text{subsequences of } \Sigma \text{ that follows any } L \text{ in } \Sigma\};\\ \mathcal{M}(\Sigma) &= \{\text{subsequences of } \Sigma \text{ that follows any } M \text{ in } \Sigma\};\\ \mathcal{R}(\Sigma) &= \{\text{subsequences of } \Sigma \text{ that follows any } R \text{ in } \Sigma\}.\end{aligned}$$

$\mathcal{L}(\Sigma)$, $\mathcal{M}(\Sigma)$ and $\mathcal{R}(\Sigma)$ may be called the *shifts* of L, M and R in Σ, respectively. The general admissibility conditions may be formulated in terms of these shifts by inspecting Fig. 3.2:

$$\begin{aligned}&\mathcal{L}(\Sigma) \leqslant \overline{K},\\ K \leqslant\ &\mathcal{M}(\Sigma) \leqslant \overline{K},\\ K \leqslant\ &\mathcal{R}(\Sigma).\end{aligned} \tag{3.15}$$

The anti-symmetry makes the first and the third inequalities equivalent to each other. If we confine ourselves to the dynamical invariant range $U = [f(C), f(\overline{C})]$, the admissibility conditions may be stated as: for any integer k an admissible sequence Σ must satisfy

$$\overline{K} \geqslant \mathcal{S}^k(\Sigma) \geqslant K, \tag{3.16}$$

where $\mathcal{S}$ is the shift operator, which we first introduced in Section 1.2.2, see (1.8).

Kneading sequences themselves should also satisfy the admissibility conditions. In particular, (3.16) holds for any kneading sequence

$$K \leqslant \mathcal{S}^k(K) \leqslant \overline{K}, \quad K \leqslant \mathcal{S}^k(\overline{K}) \leqslant \overline{K}, \tag{3.17}$$

Again we see the twofold meaning of admissibility conditions. When the kneading sequence K (or $\overline{K}$), or the parameter, is fixed, any admissible symbolic sequence satisfying these relations may be reproduced by choosing suitable initial point x_0. They give the conditions for the orbit to exist in the phase space. When applied to the kneading sequence K itself, they provide the conditions that a given symbolic sequence may serve as a kneading sequence at some parameter value. In the latter sense, conditions (3.17) are often called the consistency conditions for kneading sequences.

In contrast to the unimodal map, where a kneading sequence satisfies the condition of shift-maximality (see Section 2.3.4), the condition (3.17) is stronger, because it requires both shift-maximality for $\overline{K}$ and shift-minimality for its mirror image K.

Given a particular anti-symmetric cubic map, the parameter values for superstable periods and eventually periodic orbits may be determined by using the word-lifting technique or the bisection method. We describe the word-lifting technique on the example of map (3.3).

The word-lifting technique

In order to get the explicit form of the inverse mapping function required by the word-lifting technique, we re-scale the variables in (3.3):

$$\tilde{x} = \frac{x}{2C}. \tag{3.18}$$

Expressed in the new variable, the equivalent map is

$$\tilde{x}_{n+1} = a\, T_3(\tilde{x}_n) = a \cos(3 \arccos \tilde{x}_n), \tag{3.19}$$

where $a \equiv (A-1)/3$ and $T_3(x) = 4x^3 - 3x$ is the Chebyshev polynomial. Taking into account the principal values of the arccos function, we have

$$\begin{aligned} R(y) &= \cos\left(\frac{1}{3}\arccos\frac{y}{a}\right), \\ L(y) &= \cos\left(\frac{1}{3}\arccos\frac{y}{a} + \frac{2\pi}{3}\right), \\ M(y) &= \cos\left(\frac{1}{3}\arccos\frac{y}{a} - \frac{2\pi}{3}\right). \end{aligned} \tag{3.20}$$

The correspondence to the monotonic branches may be checked by looking at the surjective map at $a = 1$. In the new variables the critical points are $\tilde{x} = \pm 1/2$.

Periodic words containing only one of the critical points C or $\overline{C}$ are asymmetric orbits. For instance, the kneading word $K = RMC$ should be lifted to the equation

$$f(C) = R \circ M(C), \tag{3.21}$$

where $R(\cdot)$ and $M(\cdot)$ are defined in (3.20). At the rescaled critical points $\pm 1/2$ the Chebyshev polynomial $T_3(\pm 1/2) = \mp 1$, therefore Eq. (3.21) becomes

$$-a = R \circ M(1/2). \tag{3.22}$$

Similar to what have been done in Section 2.1.3, using the inverse functions (3.20), we transform the above equation into an iteration:

$$a_{n+1} = -\cos\left(\frac{1}{3}\left(\frac{1}{a_n}\cos\left(\frac{1}{3}\arccos\frac{1}{2a_n} - \frac{2\pi}{3}\right)\right)\right), \tag{3.23}$$

we get $A \equiv 3a + 1 = 3.463\ 28$.

Symmetric symbolic sequences must contain both critical points C and $\overline{C}$, hence the name doubly superstable periods. For example, $K = LM\overline{C}RMC$, or equivalently, $\overline{K} = RMCLM\overline{C}$, may be lifted according to one of the half-words:

$$f(\overline{C}) = R \circ M(C) \quad \text{or} \quad f(C) = L \circ M(\overline{C}). \tag{3.24}$$

Owing to the anti-symmetry of the mapping function, these two lifted equations lead to one and the same parameter value. In other words, there is only one independent equation. We have calculated by word-lifting the parameters of all superstable kneading sequences up to period 6 for the map (3.3), and listed them in Table 3.1. We will refer to this table when discussing symmetry

Table 3.1 Superstable kneading sequences for the anti-symmetric cubic map up to period 6. Symmetric sequences are marked with an $*$

No.	Period	$\overline{K}$	A	No.	Period	$\overline{K}$	A
1	2*	$C\overline{C}$	2.5	28	3	$RR\overline{C}$	3.924 990 7
2	2	$R\overline{C}$	3.121 320 3	29	6	$RRMRR\overline{C}$	3.925 457 6
3	4	$RMR\overline{C}$	3.262 878 6	30	6	$RRMRM\overline{C}$	3.935 027 1
4	6	$RMRLR\overline{C}$	3.334 024 1	31	5	$RRMR\overline{C}$	3.940 904 4
5	6*	$RMCLM\overline{C}$	3.345 315 6	32	6	$RRMRL\overline{C}$	3.946 411 0
6	6	$RMMLM\overline{C}$	3.463 283 4	33	6	$RRMML\overline{C}$	3.950 472 1
7	6	$RMMLR\overline{C}$	3.528 227 2	34	5	$RRMM\overline{C}$	3.955 327 4
8	4	$RMM\overline{C}$	3.548 085 8	35	6	$RRMMM\overline{C}$	3.959 701 5
9	6	$RMMMR\overline{C}$	3.565 988 0	36	6	$RRMMR\overline{C}$	3.963 789 8
10	6	$RMMMM\overline{C}$	3.591 181 9	37	4	$RRM\overline{C}$	3.967 540 3
11	5	$RMMM\overline{C}$	3.615 031 9	38	6	$RRMLR\overline{C}$	3.971 091 4
12	5	$RMMR\overline{C}$	3.666 207 0	39	6	$RRMLM\overline{C}$	3.974 519 8
13	3	$RM\overline{C}$	3.700 315 5	40	5	$RRML\overline{C}$	3.977 781 6
14	6	$RMLRM\overline{C}$	3.702 989 4	41	6*	$RRCLL\overline{C}$	3.981 797 4
15	5	$RMLR\overline{C}$	3.733 940 7	42	6	$RRRLL\overline{C}$	3.981 899 0
16	5	$RMLM\overline{C}$	3.775 383 9	43	5	$RRRL\overline{C}$	3.985 488 5
17	6	$RMLMM\overline{C}$	3.790 908 8	44	6	$RRRLM\overline{C}$	3.987 890 5
18	6	$RMLMR\overline{C}$	3.807 368 9	45	6	$RRRLR\overline{C}$	3.990 027 2
19	4*	$RCL\overline{C}$	3.830 811 5	46	4	$RRR\overline{C}$	3.991 930 0
20	4	$RRL\overline{C}$	3.839 894 4	47	6	$RRRMR\overline{C}$	3.993 628 0
21	6	$RRLMR\overline{C}$	3.861 086 0	48	6	$RRRMM\overline{C}$	3.995 129 5
22	6	$RRLMM\overline{C}$	3.873 461 5	49	5	$RRRM\overline{C}$	3.996 426 9
23	5	$RRLM\overline{C}$	3.883 586 0	50	6	$RRRML\overline{C}$	3.997 523 1
24	6	$RRLML\overline{C}$	3.893 355 0	51	6	$RRRRL\overline{C}$	3.998 411 7
25	6	$RRLRL\overline{C}$	3.889 299 2	52	5	$RRRR\overline{C}$	3.999 107 8
26	5	$RRLR\overline{C}$	3.906 906 3	53	6	$RRRRM\overline{C}$	3.999 603 7
27	6	$RRLRM\overline{C}$	3.924 490 1	54	6	$RRRRR\overline{C}$	3.999 900 9

breakings in the map (3.3). Examples of applying word-lifting technique to eventually periodic sequences will be given in Section 3.3 when determining the parameter values for symmetry restoration.

3.2.3 Generation of Superstable Median Words

How to produce *all* admissible superstable periodic symbolic sequences up to a certain length? In particular, how were the superstable kneading sequences listed in Table 3.1 generated? The key point is to generate the shortest superstable periodic sequence between two given ones, i.e., the generation of the median word.

The construction of median words is based on the ordering of symbolic sequences and continuity consideration. We start by looking at an example that makes implicit use of the construction rules which will be formulated and justified in the second part of this Section. Since K is just the mirror image of $\overline{K}$ we will consider $\overline{K}$ only.

An example of constructing median words

Suppose we are given two superstable periods $P_a = R\overline{C} < P_b = RR\overline{C}$ and are required to generate all superstable periods up to a certain length between these two words. We disturb the symbol $\overline{C}$ slightly into its neighboring symbol L or M to extend each of P_a and P_b into an ordered triplet:

$$\begin{aligned} P_a^- &= (RL)^\infty < P_a < P_a^+ = (RM)^\infty, \\ P_b^- &= (RRL)^\infty < P_b < P_b^+ = (RRM)^\infty. \end{aligned}$$

Call P_i^- and P_i^+ the *lower* and *upper* sequence of P_i, $i = a, b$, respectively.

The sequences between P_a and P_b are the lower sequence of the larger P_b and the upper sequence of the smaller P_a:

$$P_a < P_a^+ = (RM)^\infty < P_b^- = (RRL)^\infty < P_b.$$

Inspecting the two sequences symbol by symbol, we see that they differ at the second place. The only critical symbol which may be inserted between the two different symbols is C according to the natural order (3.11). Keeping in mind that $\overline{K}$ corresponds to $\overline{C}$ and taking into account the anti-symmetry, this leads to a doubly superstable word $P_{10} = RCL\overline{C}$. When disturbing C and

$\overline{C}$ in P_{10} to get its lower and upper symmetric sequences, we have to conform with the symmetry, i.e., $(C, \overline{C})$ may be changed either to (R, L) or to (M, M):

$$P_{10}^- = (RMLM)^\infty < P_{10} < P_{10}^+ = (RRLL)^\infty.$$

Comparing P_a^+ with P_{10}^-, we see that the two critical symbols C and $\overline{C}$ may be inserted at the third place to get $P_{21} = RM\overline{C}$ and $P_{20} = RMCLM\overline{C}$, the last C being continued by symmetry to reach $\overline{C}$. Between P_b^- and P_{10}^+ the two critical symbols C and $\overline{C}$ are inserted at the fourth place to get $P_{30} = RRLCLLR\overline{C}$ and $P_{31} = RRL\overline{C}$. We put what has just been described in the following list:

$$\begin{array}{lcl}
P_b & : & RR\overline{C} \\
P_b^- & : & RRLRRLRRL\cdots \\
 & & \cdots \\
P_{30} & : & RRLCLLR\overline{C} \\
 & & \cdots \\
P_{31} & : & RRL\overline{C} \\
 & & \cdots \\
P_{10}^+ & : & RRLLRRLL\cdots \\
P_{10} & : & RCL\overline{C} \\
P_{10}^- & : & RMLMRMLM\cdots \\
 & & \cdots \\
P_{21} & : & RM\overline{C} \\
 & & \cdots \\
P_{20} & : & RMCLM\overline{C} \\
 & & \cdots \\
P_a^+ & : & RMRMRMRM\cdots \\
P_a & : & R\overline{C}
\end{array}$$

Then we extend P_{30}, P_{31}, P_{21}, and P_{20} into lower and upper sequences and compare the neighboring P_i^-, P_j^+ pairs symbol by symbol in order to insert C and/or $\overline{C}$ to form new words. Whenever a ΣC is obtained, it is immediately extended to the symmetric word $\Sigma C \overline{\Sigma C}$, since we are generating kneading sequences ending with $\overline{C}$.

In order to get a deeper feeling, we generate a few more periods in the lower part of the above list. Between P_a^+ and P_{20}^- one may insert only $\overline{C}$ at the fourth place to get $P_{41} = RMR\overline{C}$. There are no median words between P_a and P_{41}, as $P_a^+ = P_{41}^- = (RM)^\infty$. In fact, P_{41} is the doubling of P_a.

Between P_{20}^+ and P_{21}^- one may insert two critical symbols to get $P_{51} = RMM\overline{C}$ and $P_{50} = RMMCLMM\overline{C}$. In between P_{21}^+ and P_{10}^- there exists only one word $P_{60} = RMLCLMR\overline{C}$. Then these words are extended into lower and upper sequences. Further comparison of neighboring sequences will allow us to insert more C and/or $\overline{C}$ to get the next level of median words. A partial list of median words between $P_a = R\overline{C}$ and $P_b = RR\overline{C}$ is given in Table 3.2.

Table 3.2 Some median words between $R\overline{C}$ and $RR\overline{C}$. Symmetric words are marked with an $*$

$RR\overline{C}$			
	$^*RRLCLLR\overline{C}$		
	$RRL\overline{C}$		
$^*RCL\overline{C}$			
		$^*RMLCLMR\overline{C}$	
	$RM\overline{C}$		
		$^*RMMCLMM\overline{C}$	
		$RMM\overline{C}$	
			$^*RMMLCLMMR\overline{C}$
	$^*RMCLM\overline{C}$		
			$^*RMRLMLCLMLRMR\overline{C}$
			$^*RMRLCLMLR\overline{C}$
			$RMRLR\overline{C}$
		$RMR\overline{C}$	
$R\overline{C}$			

Since one keeps inserting the critical symbols $\overline{C}$ and/or C between the already generated sequences and extend them into triplets, the list gets longer and longer. A convenient way of keeping track of the construction process is to use a screen editor on a computer.

Construction rules for median words

Now we are in a position to formulate the rules for generating median words in the anti-symmetric cubic map as follows. Suppose we are given two different kneading sequences $P_b = \Gamma\overline{C} > P_a = \Delta\overline{C}$.

1. First construct the lower sequence $\Sigma_1 = (\Gamma\overline{C})_-^\infty$ of the larger sequence P_b and the upper sequence $\Sigma_2 = (\Delta\overline{C})_+^\infty$ of the smaller sequence P_a. Here the odd string $(\Gamma\overline{C})_+$ and the even string $(\Gamma\overline{C})_-$ denote the larger and smaller strings of ΓL and ΓM.

2. If $\Sigma_1 = \Sigma_2$, then $\Gamma\overline{C}$ and $\Delta\overline{C}$ are neighbors. There is no median word in between. In fact, they are related by period-doubling.

3. If $\Sigma_1 \neq \Sigma_2$, we must have $\Sigma_1 = \Sigma\mu\cdots$, $\Sigma_2 = \Sigma\nu\cdots$, where $\mu \neq \nu$ with $\mu, \nu \in (R, M, L)$, and Σ is a common leading string. Insert whichever of either C or $\overline{C}$ or both intervenes between μ and ν according to the natural order (3.11).

4. Whenever a $\overline{C}$ is inserted, it yields the shortest median word $\Sigma\overline{C}$.

5. Whenever a C is inserted, it is extended to the doubly superstable symmetric word $\Sigma C\overline{\Sigma C}$. In this case there is another median word, namely, the asymmetric superstable word $(\Sigma C)_+\overline{\Sigma C}$.

The median words generated by these rules are of different lengths. One can modify the procedure to generate kneading sequences up to a certain length inclusively. A BASIC program implementing these rules to generate all superstable $\overline{K}$ up to period 6 is listed in Program 3.1, which should be compared with Program 2.5 in Section 2.4.2.

Justification of the construction rules

Readers interested only in using the above rules for practical construction of median words may skip what follows and proceed to the next Section. The justification of these rules goes much like the proof of the Periodic Window Theorem for unimodal maps in Section 2.4.1.

To fix the notation we assume $\Sigma = s_0s_1s_2\cdots s_n$ is a symbolic sequence made of the symbols R, M, and L. In addition, $\Sigma\overline{C}$ satisfies the consistency conditions (3.17). We first formulate a few statements.

A. If the k-th shift of Σ coincides with its leading string, i.e.,

$$\mathcal{S}^k(\Sigma) = s_ks_{k+1}\cdots s_n = s_0s_1\cdots s_{n-k} \equiv \Sigma_{n-k},$$

then there are two possible situations: either Σ_{n-k} is M-even and $s_{n-k+1} \neq L$ or Σ_{n-k} is M-odd and $s_{n-k+1} = L$. Juxtaposing the two sequences

$$\begin{aligned}\Sigma\overline{C} &= \Sigma_{n-k}s_{n-k+1}\cdots s_n\overline{C},\\ \mathcal{S}^k(\Sigma\overline{C}) &= \Sigma_{n-k}\overline{C},\end{aligned}$$

we see that the two possibilities follow from the shift-maximality condition $\mathcal{S}^k(\Sigma\overline{C}) \leqslant \Sigma\overline{C}$ and the natural order (3.11).

```
   2  REM PROG 3.1 SUPERSTABLE K FOR ANTISYMMETRIC CUBIC MAP
   4  REM S$:  KNEADING SEQUENCE; L: LENGTH OF S$
  10  DIM S$(1000), L(1000)
  30  NMAX=6:  IA=1:  IB=1000
  40  S$(1)="RL":  L(1)=2:  S$(1000)="RRRRRL":  L(1000)=6
  50  IF MID$(S$(IA), L(IA), 1)="L" THEN C$="M" ELSE C$="L"
  60  S$(IA)=MID$(S$(IA), 1, L(IA)-1) + C$
  70  S1$=S$(IA): S2$=S$(IB)
  80  FOR M=1 TO 500:  GOSUB 1000
  90  IF ID=2 THEN GOTO 150
 100  IF ID=1 THEN IB=IB-1:  L(IB)=L: S$(IB)=SS$:  S2$=SS$:  GOTO 140
 110  S$(IB-1)=SS$+C2$+MS$+"D":  L(IB-1)=L
 120  S$(IB-2)=SS$+"C"+MS$+"D":  L(IB-2)=L
 130  IB=IB-2:  S2$=SS$(IB)
 140  NEXT M
 150  IA=IA+1:  S$(IA)=S$(IB): L(IA)=L(IB): IB=IB+1
 160  IF IB>1000 THEN GOTO 180
 170  IF MID$(S$(IA), L(IA), 1)="D" THEN GOTO 70 ELSE GOTO 50
 180  K=1:  FOR I=1 TO IA
 190  IF L(I)>6 THEN GOTO 210
 200  PRINT K TAB(8) MID$(S$I), 1, L(I)-1);"D" TAB(18) L(I): K=K+1
 210  NEXT I: STOP
1000  L1=L(IA): L2=L(IB)
1010  SS$=" ":  FOR K=0 TO NMAX-1
1020  K1=(K MOD L1)+1:  K2=(K MOD L2)+1
1030  C1$=MID$(S1$, K1, 1):  C2$=MID$(S2$, K2, 1)
1040  IF(C1$="C" AND C2$="D")OR(C1$="D" AND C2$<>"R")
         THEN C1$="M":  GOTO 1080
1050  IF(C1$="C" AND C2$<>"L") OR (C1$="D" AND C2$<>"R") THEN C1$=C2$
1060  IF(C2$="C" AND C1$<>"L") OR (C2$="D" AND C1$<>"R") THEN C2$=C1$
1070  IF C1$<>C2$ THEN GOTO 1100
1080  SS$=SS$ + C1$
1090  NEXT K: ID=2:  RETURN
1100  L=K+1:  IF C1$="L" THEN SS$=SS$+C1$:  ID=1:  RETURN
1110  IF C2$="L" TEHN SS$=SS$+"M":  ID=1:  RETURN
1120  MS$=" ":  FOR I=1 TO K
1130  C$=MID$(SS$, I, 1)
1140  IF C$="R" THEN C$="L":  GOTO 1160
1150  IF C$="L" THEN C$="R"
1160  MS$=MS$+C$:  NEXT I
1170  L=2*L: ID=0:  RETURN: END
```

Program 3.1

B. The shift-maximality of $\Sigma\overline{C}$ implies shift-maximality of $(\Sigma L)^\infty$ and $(\Sigma M)^\infty$. This statement follows from Statement A and is almost identical to the Periodic Window Theorem of unimodal maps , so we leave it to the reader as an exercise.

C. If $\Sigma\overline{C}$ satisfies the consistency condition (3.17), so does $(\Sigma L)^\infty$ and $(\Sigma M)^\infty$. In other words, the consistency condition

$$(\overline{\Sigma t})^\infty \leqslant \mathcal{S}^k(\Sigma t)^\infty \leqslant (\Sigma t)^\infty, \quad k = 0, 1, \cdots \tag{3.25}$$

holds, where $\mathcal{S}$ is the shift operator and t is either L or M, i.e., one of the neighbors of $\overline{C}$.

We now prove (3.25), starting from (3.17). Since the second inequality in (3.25) follows from **B**, it is enough to prove the first one. Denoting by V the string made of the first k symbols of Σt and writing $\Sigma t = VW \equiv VW_1t$, we are required to prove

$$(\overline{\Sigma t})^\infty \leqslant W(\Sigma t)^\infty. \tag{3.26}$$

In these notations the consistency condition (3.17) means

$$\overline{\Sigma C} \leqslant W_1\overline{C}. \tag{3.27}$$

If W_1 is not a leading string of $\overline{\Sigma}$, the above inequality must hold as $\overline{\Sigma}\cdots <$ $W_1\cdots$, therefore (3.26) holds as an inequality. If W_1 is a leading string of $\overline{\Sigma}$, the consistency condition (3.27) holds as

$$W_1t\cdots \leqslant W_1\overline{C}.$$

This means that either $t = L$ and W_1 is M-even, or $t = M$ and W_1 is M-odd. In both cases $W = W_1t$ is M-even. Denote $\overline{\Sigma t} = WV'$, where V' is of length k, i.e., the same length as V. Since W is even, (3.26) becomes

$$V'(\overline{\Sigma t})^\infty \leqslant (\Sigma t)^\infty. \tag{3.28}$$

Again, if V' is not a leading string of Σ, (3.28) holds as an inequality. If V' is a leading string of Σ, then $V' = V$ as we have denoted $\Sigma t = VW$ in the beginning. This is a special case with $\overline{\Sigma t} = WV$ and (3.28) holds as an equality.

We have proved **C**. It is a common practice in symbolic dynamics that proofs of inequalities require less effort.

D. If both $\Sigma L\cdots$ and $\Sigma M\cdots$ (or $\Sigma R\cdots$) are kneading sequences, so does $\Sigma\overline{C}$. This is clear from continuity consideration.

E. If both $\Sigma M\cdots$ and $\Sigma R\cdots$ are kneading sequences, then the doubly superstable symmetric orbit $(\Sigma C\overline{\Sigma C})^\infty$ and its two "deformed" neighbors $(\Sigma R\overline{\Sigma} L)^\infty$ and $(\Sigma M\overline{\Sigma} M)^\infty$ are kneading sequences.

Denoting the larger of ΣR and ΣM by $(\Sigma C)_+$, and the smaller by $(\Sigma C)_-$, it is easy to see that $(\Sigma C)_+$ is M-even, while $(\Sigma C)_-$ is M-odd. Moreover, $(\Sigma C)_+\overline{\Sigma C}$ is also a kneading sequence. In the next Section we will see that this is nothing but the asymmetric periodic orbit born after symmetry breaking of the symmetric periodic window.

3.3 Symmetry Breaking and Restoration

Symmetry breaking and symmetry restoration are common phenomena in physical systems with a certain kind of symmetry. An equation or a thermodynamic potential may possess a higher symmetry, but a particular solution of the equation or an equilibrium state that minimizes the potential may exhibit only a lower symmetry. However, all asymmetric solutions or states, taken together, restore the original symmetry. In fact, the notion of symmetry breaking has been playing an increasingly important role in the understanding of diverse problems ranging from continuous phase transitions to the origin of the universe.

It is interesting to note that a simple form of symmetry breaking and restoration appears in the bifurcation structure of many dynamical systems. It manifests itself clearly in the bifurcation diagram of the antisymmetric cubic map (3.3), see Fig. 3.3 for the most clearly seen symmetry breaking of the 2-cycle and symmetry restoration in its chaotic band.

Symmetric orbits first undergo symmetry-breaking bifurcation into asymmetric orbits, and then enjoy period-doubling. This has been observed in numerical studies of many systems of ordinary and partial differential equations, and in laboratory experiments as well. Observations have led to the conjecture that symmetry breaking must precede period-doubling, whence follows the expression "precursor" (D'Humieries *et al.* [1982]; Kumar *et al.* [1987]) to period-doubling or "suppression" (Swift and Wiesenfeld [1984]; Wiesenfeld *et al.* [1984]) of period-doubling by symmetry-breaking, etc. The phenomenon was correctly related to the discrete symmetry of the governing equations (D'Humieries *et al.* [1982]), and a heuristic explanation based on bifurcation theory was given by Wiesenfeld *et al.* [1984]. The most complete bifurcation theory analysis applied to the Poincaré mapping of flows appeared in Swift and Wiesenfeld [1984] with the conclusion that a symmetric orbit cannot undergo period-doubling directly except in extraordinary cases.

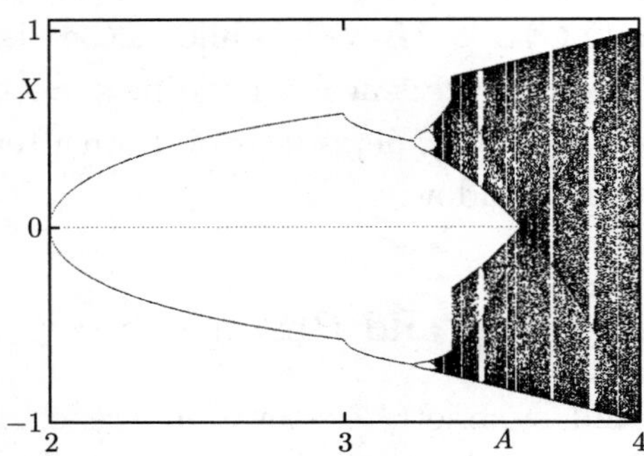

Figure 3.3 A bifurcation diagram of the anti-symmetric cubic map. Note the breaking of the symmetric period 2 and the restoration of symmetry in the chaotic band

In contrast to symmetry breakings, the phenomenon of symmetry restoration has rarely been discussed in the literature. There has been, e.g., cases of mentioning "the symmetry tends to be restored through chaos" (Tomita and Tsuda [1980]) or "a bifurcation back to symmetry" (Knobloch and Weiss [1981]), but nothing was said about the nature of the bifurcation. Chossat and Golubitsky [1988] carried out a numerical study of "symmetry increasing" bifurcations in a family of two-dimensional non-invertible maps without mentioning of symbolic dynamics.

In fact, the bifurcation analysis of symmetry-breakings can be carried out simply and thoroughly for general antisymmetric mappings in much the same way as for tangent and period-doubling bifurcations. Moreover, symbolic dynamics determines the precedence of symmetric orbit over the asymmetric ones, and provides a selection rule to pick up those orbits of even periods that are destined for symmetry breaking. Symbolic dynamics also furnishes a precise description for symmetry restoration. The entire analysis, formulated in the language of symbolic dynamics, applies to differential equations as well, as it will be discussed in connection with symmetry breaking and restoration in the Lorenz equations (Chapter 6).

We perform a symbolic dynamics analysis of symmetry breaking and symmetry restoration on the example of the anti-symmetric cubic map in what follows (Zheng and Hao [1989]). For the reader's reference we formulate a theorem, which we do not need in the sequel, before going to the main theme.

Symmetry-Breaking Bifurcation Theorem Suppose $f(A, x)$ is an anti-symmetric map of the interval I into itself and the following conditions hold:

1. There is an inversion point x^* at the parameter value A^*, i.e., $x^* = -f(x^*)$ at A^*. Then x^* is a fixed point of $f^2(x)$.

2. The stability is marginal with

$$\left.\frac{\partial f}{\partial x}\right|_{x^*} = \left.\frac{\partial f}{\partial x}\right|_{-x^*} = 1.$$

3. The Schwarzian derivative $S(f, x)|_{x^*} < 0$. Most maps of physical interest have negative Schwarzian derivatives on the whole interval I, including the point x^*, so this condition holds automatically.

4. The mixed second derivative of the second iterate does not vanish:

$$\left.\frac{\partial f^2}{\partial x \partial A}\right|_{x^*, A^*} \neq 0.$$

Then on one side of A^* (which side depends on the sign of the above mixed derivative) there exists a stable symmetric 2-cycle which undergoes symmetry-breaking bifurcation at $A = A^*$ into a pair of asymmetric 2-cycles, both stable but observable in their own basins.

The proof is similar to the Tangent Bifurcation and Period-Doubling Bifurcation Theorems formulated in Section 2.1.1, see the Appendix of Zheng and Hao [1989].

3.3.1 Symmetry Breaking of Symmetric Orbits

Symbolic dynamics of the anti-symmetric cubic map has been developed in Section 3.2. Now we apply it to analyze symmetry breakings.

First of all, symmetric orbits of odd period are non-generic, as one has to put one orbital point precisely on the symmetry axis and there are infinitely many perturbations to destroy the situation. Thus only orbits of even period may undergo symmetry breaking, but not all of them are capable to do so.

Next, we look at the simplest superstable $C\overline{C}$ of symmetric period 2. By changing C and $\overline{C}$ into their neighboring symbols in a symmetric way, it extends to the symmetric period 2 window

$$(MM, C\overline{C}, RL). \tag{3.29}$$

This triple has the signature $(+, 0, +)$, so it cannot be followed by a period-doubling bifurcation. In fact, this is the most clearly seen symmetric period 2 window in Fig.3.3. By continuing its upper sequence RL to the lower sequence of the next window, we construct one of the asymmetric period 2 orbits as

$$(RL, R\overline{C}, RM). \tag{3.30}$$

Its mirror image, ordered as minimal sequences,

$$(LR, LC, LM), \tag{3.31}$$

describes the other symmetry-broken orbit. By cyclic shifting, the above window may be transformed to maximal sequences

$$(RL, CL, ML) \tag{3.32}$$

which cannot be brought back to (3.30) by cyclic shifts. Both triples (3.30) and (3.32) have the signature $(+, 0, -)$. Consequently, they are eligible for further period-doubling bifurcation.

Now we consider a general doubly superstable symmetric periodic orbit $\Sigma C\overline{\Sigma C}$. Let us disturb the orbit slightly, giving up the superstability, but retaining the symmetry. The letter C may be changed to either R or M by continuity, and $\overline{C}$ as the mirror image of C to either L or M. The three words, obtained in this way, may be ordered according to the parity of Σ:

$$\begin{aligned}&\Sigma \text{ even} : (\Sigma L\overline{\Sigma}R, \Sigma C\overline{\Sigma C}, \Sigma M\overline{\Sigma}M);\\ &\Sigma \text{ odd} : (\Sigma M\overline{\Sigma}M, \Sigma C\overline{\Sigma C}, \Sigma L\overline{\Sigma}R).\end{aligned}$$

Writing the lower sequences as $[(\Sigma C)_-(\overline{\Sigma C})_+]^\infty$ and the upper sequences as $[(\Sigma C)_+(\overline{\Sigma C})_-]^\infty$, we combine the two cases. Here the basic strings $(\Sigma C)_\pm(\overline{\Sigma C})_\mp$ are necessarily of even parity. This is because the strings $(\Sigma C)_\pm$ and $(\overline{\Sigma C})_\mp$ are mirror images to each other, so their concatenation never contains an odd number of M. At the same time, the string $(\Sigma C)_+$ is even, but $(\Sigma C)_-$ is odd.

In contrast to windows capable of undergoing period-doubling, the window

$$[(\Sigma C)_-(\overline{\Sigma C})_+, \Sigma C\overline{\Sigma C}, (\Sigma C)_+(\overline{\Sigma C})_-] \tag{3.33}$$

has the signature $(+, 0, +)$, therefore it cannot undergo a period-doubling bifurcation requiring a -1 parity. It is easy to check that under the mirror transformation (3.14), i.e., $\mathcal{T} : R \leftrightarrow L,\ M \leftrightarrow M$, the window (3.33) remains unchanged up to a cyclic shift, which is not essential to periodicity. Consequently, the above triple describes the whole window of a symmetric orbit. In (3.33) and hereafter we omit the infinite power, keeping only the basic strings.

This window further undergoes a symmetry-breaking bifurcation, giving birth to an asymmetric window of the same period. However, the bifurcation point cannot be determined from symbolic dynamics as it requires the knowledge of the mapping function. Similarly to what has been done at a period-doubling bifurcation, we use the above negative statement to draw a positive conclusion: since symbolic dynamics does not know where the symmetry breaking bifurcation takes place, the upper sequence of the window (3.33) should go smoothly into the next window, becoming its lower sequence:

$$[(\Sigma C)_+(\overline{\Sigma C})_-, (\Sigma C)_+\overline{\Sigma C}, (\Sigma C)_+(\overline{\Sigma C})_+]. \tag{3.34}$$

The central and upper sequences of the new triple (3.34) may be written down by continuity. Now, taking the mirror image, we will get another triple not equivalent to triple (3.34). It is nothing but the other asymmetric window. If the three words in the original window are ordered as maximal sequences, then the words in the mirror image are ordered as minimal sequences, as the orbit starts from the minimum of the mapping function.

After loosing stability at the symmetry breaking bifurcation point, the symmetric orbit still exist as an unstable period. It will collide with the asymmetric chaotic attractor at the symmetry restoration point. We will discuss this in the next Section.

We have mentioned at the beginning that although symmetry breakings are always associated with orbits of even periods, not all even periods can undergo symmetry breaking. The selection rule follows from the very form $\Sigma C\overline{\Sigma C}$ of the doubly superstable symmetric orbit. There exist as many period $2n$ orbits capable to undergo symmetry breaking as asymmetric orbits ΣC of period n. This leads us to the problem of counting the number of different periodic orbits of a given length. This counting problem has been completely solved not only for the cubic map, but also for arbitrary continuous maps with multiple critical points. We will devote Chapter 7 to this problem. Meanwhile, by making use of the results of Chapter 7, we list the number of superstable orbits that are capable to undergo symmetry breaking in Table 3.3.

Table 3.3 Number of orbits which are capable to undergo symmetry breaking

Period	Number of Orbits	Number of Symmetry Breaking Orbits
1	1	
2	1	1
3	2	
4	5	1
5	12	
6	30	2
8	205	5
10	1 476	12
12	11 070	30

The symbolic names of shorter periods that are capable to undergo symmetry breaking are given in the following examples.

Example 1 There is only one period 4 superstable orbit $RCL\overline{C}$. We write the whole bifurcation structure as

$$(RMLM, RCL\overline{C}, RRLL) \to \begin{cases} (RRLL, RRL\overline{C}, RRLM) \to \cdots, \\ (LLRR, LLRC, LLRM) \to \cdots. \end{cases}$$

Example 2 There are two symmetric orbits among the 32 period 6 windows (Zeng [1985, 1987], Hao and Zeng [1987], and Chapter 7) that allow symmetry-breaking bifurcation: $RMCLM\overline{C}$ and $RRCLL\overline{C}$. Among the 210 period 8 windows, only 5 can undergo symmetry-breaking, namely:

$$RM^2CLM^2\overline{C},\quad RMLCLMR\overline{C},\quad R^2LCL^2R\overline{C},$$
$$R^2MCL^2M\overline{C},\quad R^3CL^3\overline{C}.$$

3.3.2 Analysis of Symmetry Restoration

We have mentioned at the beginning that symmetry restoration has not been analyzed in the literature to date. In fact, symbolic dynamics provides us with a straightforward tool to accomplish this task. We recall that a periodic window with $(+1, 0, -1)$ signature

$$[(\Sigma C)_+(\overline{\Sigma C})_-, (\Sigma C)_+\overline{\Sigma C}, (\Sigma C)_+(\overline{\Sigma C})_+]$$

permits period-doubling. Its corresponding band-merging point is described by the sequence $(\Sigma C)_+(\bar{\Sigma}\bar{C})_+[(\Sigma C)_+(\bar{\Sigma}\bar{C})_-]^\infty$. Applied to the period 2 symmetry-broken window (3.30), we locate the ending of the symmetry-broken regime at

$$RM(RL)^\infty. \tag{3.35}$$

The very form of this sequence tells the nature of the point. It corresponds to the "crisis" created by the collision of the unstable symmetric period 2 orbit represented by $(RL)^\infty$ with the asymmetric chaotic attractor. In other words, the asymmetric chaotic attractor picks up the symmetry from the unstable period 2 orbit and restores the full symmetry.

The parameter value of the symmetry restoration point can be calculated by word-lifting technique from the following pair of "lifted" equations:

$$\begin{aligned} a &= R \circ M(a, z), \\ z &= R \circ L(a, z), \end{aligned} \tag{3.36}$$

where the functions R, M, and L have been defined in (3.20). The iterative solution of (3.36) yields

$$a = 3.360\ 893\ 769\ 096\ 575 \cdots . \tag{3.37}$$

On the other hand, the mirror image of the window (3.30), i.e., the window (3.32), corresponds to an asymmetric chaotic band ending at

$$LM(LR)^\infty. \tag{3.38}$$

Referring to (3.20), it is easy to verify that the parameter value of $LM(LR)^\infty$ is the same as that given by $RM(RL)^\infty$.

The symbolic sequence corresponding to the symmetry restoration point should have, so to speak, a "double personality". Its left half is asymmetric while its right half ought to be symmetric. Indeed, the sequences (3.35) and (3.38) are manifestly mirror images of each other, a property of symmetry-broken pairs. On the other hand, the symmetry has actually been restored, since both sequences appear to be the $m \to \infty$ limit of the following periodic orbits located beyond the symmetry-restoration point (3.37):

$$RM(RL)^m RLM(LR)^m L.$$

These orbits are nothing but the lower sequences of the corresponding window formed by the symmetric periodic orbits

$$RM(RL)^m CLM(LR)^m \overline{C}, \tag{3.39}$$

which are equivalent to the cyclically shifted sequences

$$LM(LR)^m \overline{C} RM(RL)^m C, \tag{3.40}$$

as long as m remains finite. However, the $m \to \infty$ limits of (3.39) and (3.40) approach the asymmetric sequences (3.35) and (3.38), respectively.

Symmetry restoration for higher periods may be analyzed in a similar way. Suppose there is a symmetric orbit $\Sigma C \overline{\Sigma C}$. It extends to the symmetric window:

$$[(\Sigma C)_-(\bar{\Sigma}\bar{C})_+, \Sigma C\bar{\Sigma}\bar{C}, (\Sigma C)_+(\bar{\Sigma}\bar{C})_-], \tag{3.41}$$

where $(\Sigma C)_-$ and $(\bar{\Sigma}\bar{C})_+$ are odd, while $(\Sigma C)_+$ and $(\bar{\Sigma}\bar{C})_-$ even. Let us denote

$$\begin{aligned} \rho &= (\Sigma C)_+, \quad \lambda = \bar{\rho} = (\bar{\Sigma}\bar{C})_-, \\ \mu &= (\Sigma C)_-, \quad \bar{\mu} = (\bar{\Sigma}\bar{C})_+. \end{aligned} \tag{3.42}$$

Then the pair of asymmetric orbits are given by

$$(\rho\lambda,\ \rho\overline{\Sigma C},\ \rho\overline{\mu}) \quad \text{and} \quad (\lambda\rho,\ \lambda\Sigma C,\ \lambda\mu). \tag{3.43}$$

The symmetry restoration point is described by

$$\rho\bar{\mu}(\rho\lambda)^\infty \quad \text{or} \quad \lambda\mu(\lambda\rho)^\infty.$$

They are the $m \to \infty$ limits of the sequence of symmetric orbits

$$\rho\bar{\mu}(\rho\lambda)^m \Sigma C \lambda\mu(\lambda\rho)^m \bar{\Sigma}\bar{C}. \tag{3.44}$$

It is worth mentioning that the above general symmetry breaking and restoration may be obtained from the simplest case associated with the sequence $C\overline{C}$ by making the following substitutions:

$$\begin{aligned} &R \to \rho, \quad L \to \lambda, \\ &C \to \Sigma C, \quad \overline{C} \to \overline{\Sigma C}, \\ &RM \to \rho\overline{\mu}, \quad MM \to \mu\overline{\mu}, \quad LM \to \lambda\mu. \end{aligned} \tag{3.45}$$

All the superstable and symmetry restoration parameters may be calculated by using the word-lifting technique. We have collected a few results in Table 3.4.

Table 3.4 Symmetry breaking and restoration parameters

Period	Σ	Symmetric	Asymmetric	Restoration	δ
2		2.5	3.121 320 343 6	3.360 893 769 1	2.964 55
6	RM	3.345 315 558 4	3.463 283 457 8	3.468 511 941 9	2.839 03
4	R	3.830 811 514 2	3.839 894 487 6	3.844 706 135 3	2.842 05
6	RR	3.981 797 394 8	3.981 899 030 6	3.981 954 188 3	2.820 18

We conclude the analysis of symmetry breakings by making the following remark. In the above discussion attention has been paid to stable periodic orbits. Similar analysis may be carried out for unstable periodic orbits embedded in chaotic attractor to reveal symmetry breakings which may have taken place in the transition to chaos.

3.4 The Gap Map

The gap map is obtained by introducing a discontinuity, or a gap, right at the critical point C in the unimodal map, see Fig. 3.4. The critical point C splits into two: C_+ and C_-, but the monotone branches remain L and R. Therefore, its symbolic sequences consist of two letters R and L.

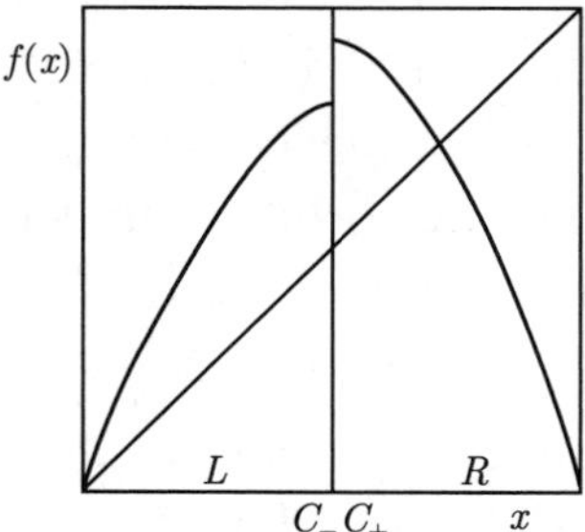

Figure 3.4 The gap map with two monotone branches and split critical points

Numerical study has revealed some "strange" behavior in this map. For example, between neighboring periods of the period-doubling cascade, some new periods may appear (De Sousa Vieira, Lazo and Tsallis [1987]). As a matter of fact, such behavior may be clearly explained by symbolic dynamics (Zheng [1989c]; Zheng and Lu [1991]).

The ordering of symbolic sequences in the gap map remains the same as in the unimodal map. However, due to the splitting of the critical point into two, there appear two independent kneading sequences $K_+ = f(C_+)$ and

$K_- = f(C_-)$. The admissibility condition for a symbolic sequence Σ now reads:

$$\mathcal{L}(\Sigma) \leqslant K_-, \quad \mathcal{R}(\Sigma) \leqslant K_+. \tag{3.46}$$

The meaning of the shift sets $\mathcal{L}(\Sigma)$ and $\mathcal{R}(\Sigma)$ is the same as in the unimodal map, see (2.50) in Section 2.2.3.

There is a simple relation between a periodic kneading sequence of the unimodal map and that of the same length in the gap map, namely, from a kneading sequence $K = \Sigma C$ of the unimodal map one can derive four periodic kneading sequences for the gap map (for period $p \geqslant 3$):

1. $K_- = \max\{\mathcal{L}((\Sigma L)^\infty)\}|_{C_-} = \Sigma C_-$.
2. $K_+ = \max\{\mathcal{R}((\Sigma R)^\infty)\}|_{C_+} = \Sigma C_+$.
3. $K_- = \max\{\mathcal{L}((\Sigma R)^\infty)\}|_{C_-}$.
4. $K_+ = \max\{\mathcal{R}((\Sigma L)^\infty)\}|_{C_+}$.

The notation $\Gamma|_C$ means replacing the last symbol in the string Γ by the symbol C. Postponing the proof to Section 7.5.1, we look at the example of the period 4 kneading sequence $K = RLLC$ of the unimodal map. Its upper and lower sequences $(RLLL)^\infty$ and $(RLLR)^\infty$ lead to two kneading sequences $RLLC_+$ and $RLLC_-$ for the gap map. The maximal member in $\mathcal{L}((RLLR)^\infty)$ is $(RRLL)^\infty$, which gives birth to a kneading sequence $RRLC_-$. The only member in $\mathcal{R}((RLLL)^\infty)$ is $(LLLR)^\infty$, which yields a kneading sequence $LLLC_+$. The last two sequences are not allowed in unimodal maps.

It is instructive to look at these orbits to understand the peculiarity of the gap map. We can find them by using word-lifting technique to a piecewise linear gap map. Let us take the following map:

$$y = \begin{cases} 2Ax, & \text{for} \quad 0 \leqslant x \leqslant 1/2, \\ 2B(1-x), & \text{for} \quad 1/2 \leqslant x \leqslant 1, \end{cases} \tag{3.47}$$

with both parameters $0 < A, B < 1$. The inverse functions are also linear:

$$\begin{aligned} L(y) &= \frac{y}{2A}, \\ R(y) &= 1 - \frac{y}{2B}. \end{aligned} \tag{3.48}$$

Word-lifting of $(RRLC_-)^\infty$ leads to conditions: $A > 0.5$ and

$$B = \frac{1 + \sqrt{2 - 1/A}}{4(1-A)}.$$

The requirement that $B < 1$ further confines the valid values of A to $A < (1+\sqrt{2})/4 = 0.603\ 55$. Such an orbit at $A = 0.6$ is shown in Fig. 3.5 (a).

To see the $(LLLC_+)^\infty$ orbit one must recognize that C_+ picks up the first point on the right branch, then the orbit goes to the left branch. The orbital points are $x_0 = 0.5$, $x_1 = B$, $x_2 = 2AB$, $x_3 = 2Ax_2$. The periodic condition leads to a relation between the two parameters: $16A^3B = 1$. An $(LLLC_+)^\infty$ orbit at $A = 0.6$ is shown in Fig. 3.5 (b).

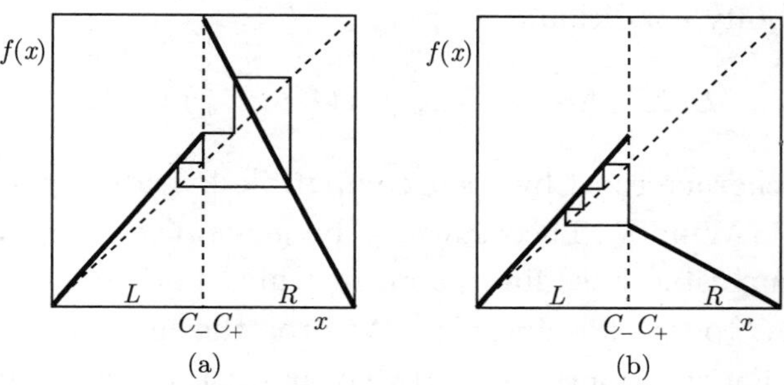

Figure 3.5 Two peculiar period 4 orbits in a piecewise linear gap map. (a) $(RRLC_-)^\infty$; (b) $(LLLC_+)^\infty$

Other peculiar orbits such as those inserted into the period-doubling neighbors (see next Section) may be analyzed in a similar way. For example, the $(RLRLC_+)^\infty$ orbit in between $(RC)^\infty$ and $(RLRC)^\infty$ exists for $0.5 \leqslant B \leqslant 0.709\,822$ and

$$A = \frac{1+\sqrt{2B-1}}{8B(1-B)}.$$

This orbit is shown in Fig. 3.6 at $B = 0.6$.

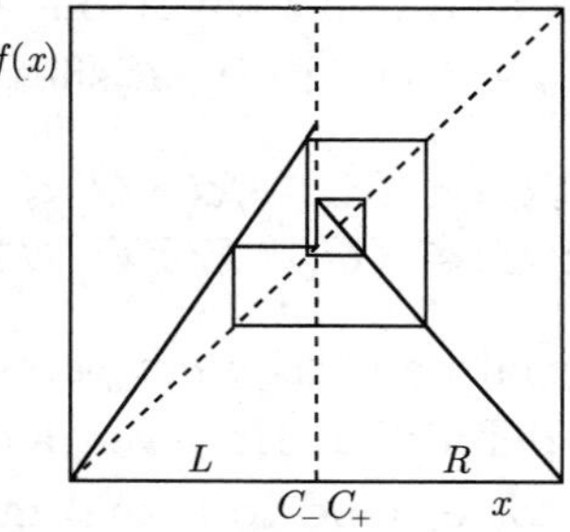

Figure 3.6 Orbit $RLRLC_+$ inserted between the period-doubling neighbors RC and $RLRC$

3.4.1 The Kneading Plane

Clearly, the two kneading sequences K_+ and K_- should satisfy the admissibility conditions at the same time. Only such *compatible kneading pairs* (K_-, K_+) correspond to an actual gap map, i.e., there exist parameters to realize these kneading sequences. The admissibility condition applied to a kneading pair is called compatibility condition:

$$\mathcal{L}(K_-, K_+) \leqslant K_-, \quad \mathcal{R}(K_-, K_+) \leqslant K_+. \tag{3.49}$$

Gap maps are characterized by their compatible kneading pairs.

The Periodic Window Theorem may be generalized to gap maps. Those superstable compatible kneading pairs, in which both sequences contain the critical symbol C, play a special role. (We drop the subscripts $\pm$ when no confusion occurs.) For such pairs, generally speaking, the compatibility conditions (3.49) hold as inequalities.

The two-dimensional parameter plane parameterized by the two kneading sequences is called a *kneading plane*. The kneading plane of the gap map is shown in Fig. 3.7. In this figure K_- increases along the horizontal axis, K_+—along the vertical axis. A vertical upward line corresponds to a particular K_-; a horizontal right-going line—a particular K_+. They intersect only when the kneading pair (K_-, K_+) is compatible. Therefore, at an intersection the kneading pair (K_-, K_+) satisfies

$$\mathcal{L}(K_-, K_+) < K_-, \quad \mathcal{R}(K_-, K_+) < K_+.$$

In contrast to intersections, at a point where one of the kneading lines stops, the kneading pair $(K_-, K_+) = (XC, YC)$ satisfies

$$\begin{aligned} \mathcal{L}(XC) &< XC, \quad \mathcal{R}(XC) \leqslant YC, \\ \mathcal{L}(YC) &\leqslant XC, \quad \mathcal{R}(YC) < YC. \end{aligned} \tag{3.50}$$

We give these points a special name *contact pairs* or *contacts*. Of course, an intersection is a contact by definition. However, a contact is not necessarily an intersection. Intersections correspond to actual maps, but contacts may not. In order for a contact to correspond to an actual map, some more conditions must be satisfied. For example, such conditions could be

$$\mathcal{R}((XL)^\infty) < YC \quad \text{or} \quad \mathcal{L}((YR)^\infty) < XC. \tag{3.51}$$

Here we regard C_- and C_+ as the limit of L and R, respectively. Other conditions could be

$$\mathcal{L}((YLXR)^\infty) < XC \quad \text{or} \quad \mathcal{R}((XRYL)^\infty) < YC. \tag{3.52}$$

Those contacts that correspond to actual maps will be called *proper contacts.*

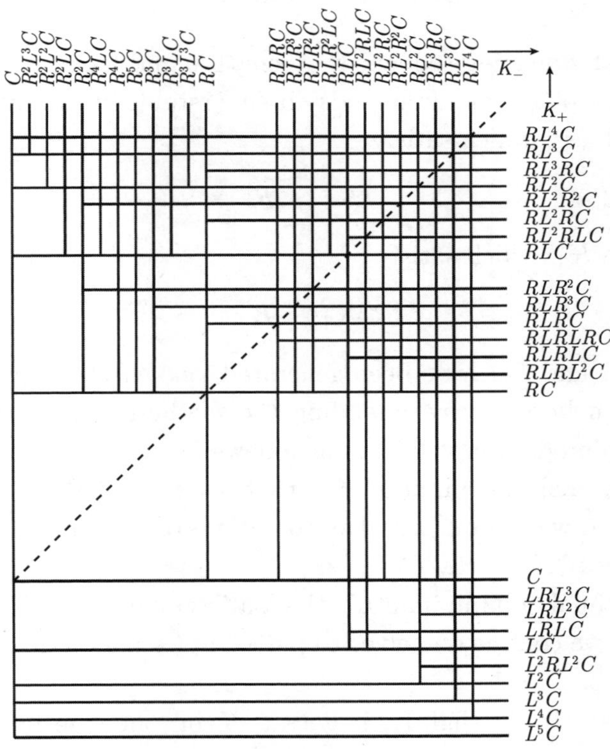

Figure 3.7 The kneading plane for the gap map. Only relative positions make sense

Since the notion of contacts reflects the peculiarity of discontinuity in the mapping function, we compare the definitions once more and give a few examples. Symbolizing the four relations as they stand in (3.50) by a square table, we have

$$\begin{bmatrix} < & < \\ < & < \end{bmatrix} \quad \to \quad \text{intersection,}$$

which corresponds to an actual gap map;

$$\begin{bmatrix} < & \leqslant \\ \leqslant & < \end{bmatrix} \quad \to \quad \text{contact,}$$

where at least one equality should take place, otherwise it is an intersection. We note also that a square table may represent more than 4 relations, as $\mathcal{L}$ and $\mathcal{R}$ are sets of symbolic sequences. A contact may not correspond to an actual gap map; a proper contact does correspond to an actual gap map.

$$\begin{bmatrix} < & \leqslant \\ \leqslant & < \end{bmatrix} + \text{additional conditions} \rightarrow \text{proper contact.}$$

In Fig. 3.7 all kneading sequences of length 6 and less are drawn. In order to construct this plane, one first finds all superstable kneading sequences K_- of length 6 and less, which satisfy

$$\mathcal{L}(K_-) < K_-, \tag{3.53}$$

as well as those K_+ which satisfy

$$\mathcal{R}(K_+) < K_+. \tag{3.54}$$

When these K_- and K_+ have been generated, one checks their compatibility. In fact, this can be done by extending the method of constructing median words in the unimodal map. It goes as follows:

A. The maximal and minimal K_- that satisfy (3.53) are $K_-^{\max} = RL^\infty$ and $K_-^{\min} = C$. If we confine ourselves to periods of a given length, e.g., period 6, we may write $K_-^{\max} = RL^4C$.

B. The maximal and minimal K_+ that satisfy (3.54) are $K_+^{\max} = RL^\infty$ and $K_+^{\min} = L^\infty$. If we confine ourselves to period 6, we may write $K_+^{\max} = RL^4C$ and $K_+^{\min} = L^5C$.

C. If we disturb C_- slightly, it goes to L on one side by continuity; on the other side of the discontinuity it changes to the smallest sequence starting with an R, i.e., $RK_+^{\max}$, the minimal sequence of which is R^2L^4C. Therefore, the upper and lower sequences of $K_- = XC$ may be taken as the greater and smaller of $(XL)^\infty$ and XR^2L^4C.

D. If we disturb C_+ slightly, it goes to R on one side by continuity; on the other side of the discontinuity it changes to the largest sequence starting with an L, i.e., $LK_-^{\max}$. Therefore, the upper and lower sequences of $K_+ = YC$ are defined to be the greater and smaller of $(YR)^\infty$ and $XLRL^4C$.

E. The rest is similar to the unimodal map.

In fact, an effective way of learning the above rules is to look at a program implementing the procedure. A BASIC program which generates all K_- up to length 6 is given in Program 3.2. We leave it to the reader to write a similar program to generate K_+.

```
   2  REM PROG 3.2 SUPERSTABLE K- FOR THE GAP MAP
   4  REM S$:  KNEADING SEQUENCE, L: LENGTH OF S$
  10  DIM S$(1000), L(1000)
  20  NMAX=6:  IA=1:  IB=1000
  30  S$(1)="L":  L(1)=1:  S$(1000)="RLLLLR":  L(1000)=6
  40  S2$=S$(IB): S1$=MID$(S$(IA), 1, L(IA)-1)
  50  IF MID$(S$(IA), L(IA),1)="L"
        THEN S1$=S1$+"RRLLLL" ELSE S1$=S1$+"L"
  60  IF MID$(S2$, L(IB), 1)="R" THEN S2$=S2$ + "RLLLL"
  70  FOR M=1 TO 500:  GOSUB 1000
  80  IF ID=2 THEN GOTO 110
  90  IB=IB-1:  L(IB)=K+1:  S$(IB)=SS$:  S2$=SS$
 100  NEXT M
 110  IA=IA+1:  S$(IA)=S$(IB): L(IA)=L(IB): IB=IB+1
 120  IF IB<1001 THEN GOTO 40
 130  FOR I=1 TO IA
 140  PRINT I TAB(8) MID$(S$(I), 1 L(I)-1):  "C" TAB(18) L(I)
 150  NEXT I: STOP
1000  L1=LEN(S1$):  L2=LEN(S2$)
1010  FOR K=0 TO NMAX -1
1020  K1=(K MOD L1) + 1:  K2=(K MOD L2) + 1
1030  IF MID$(S1$, K1, 1)<>MID$(S2$, K2, 1) THEN GOTO 1050
1040  NEXT K: ID=2:  RETURN
1050  SS$=MID$(S1$, 1, K1):  IF K>L1-1 THEN SS$=S1$+SS$
1060  ID=0:  RETURN: END
```

Program 3.2

3.4.2 Contacts of Even-Odd Type

If in a contact pair $(K_-, K_+) = (EC, OC)$, the strings E and O being *even* and *odd* in the number of Rs, it is called an *even-odd type* contact. If (EC, OC) is an even-odd type contact, then $(EROC, OC)$ and $(EC, OLEC)$ are also even-odd type contacts. We explain here the first one. The proof for the second pair is similar.

From the fact that (EC, OC) is an even-odd type contact it follows:

$$\mathcal{R}(OC) < OC, \quad \mathcal{L}(OC) \leqslant EC < EROC. \tag{3.55}$$

Therefore, we have only to prove that

$$\mathcal{R}(EROC) \leqslant OC, \tag{3.56}$$

and

$$\mathcal{L}(EROC) < EROC. \tag{3.57}$$

For those shifts in $\mathcal{L}(EROC)$ that belong to $\mathcal{L}(OC)$, and for those in $\mathcal{R}(EROC)$ that belong to $\mathcal{R}(OC)$, inequalities (3.56) and (3.57) hold naturally.

Suppose $UOC \in \mathcal{R}(EROC)$, from the contact condition $\mathcal{R}(EC) \leqslant OC$, relation (3.56) holds if U is not a leading string of OC. Otherwise, we write $O = UV$, then from the contact condition $\mathcal{R}(EC) \leqslant OC$, U must be odd. Consequently,

$$VC \in \mathcal{R}(OC) < OC$$

yields (3.56). The only case left is $OC \in \mathcal{R}(EROC)$. Obviously, it goes in accordance with (3.56).

We now prove (3.57). For $U'OC \in \mathcal{L}(EROC)$, if U' is not a leading string of ER, the inequality holds straightforwardly. If $ER = U'V'$, U' must be odd. Relation (3.57) then requires

$$V'OC \in \mathcal{R}(EROC) < OC.$$

Since V' can never be a blank string, the above inequality holds true. We have completed the proof.

It is not difficult to see that the presence of two consecutive contacts $(EROC, OC)$ and $(EC, OLEC)$ implies an intersection $(EROC,\ OLEC)$. Since these two contacts both belong to even-odd type, the process of generating contacts and intersections may be repeated *ad infinitum*. The first few contacts and intersections generated from a given contact (EC, OC) are shown in Fig. 3.8. The shaded region in the figure corresponds to parameter range with zero topological entropy, see Section 3.4.4 below.

The existence of an even-odd type contact (EC, OC) excludes any other contacts in the rectangle demarcated by the four lines $K_- = EC$, $K_- = EROC$, $K_+ = OC$, and $K_+ = OLEC$ in the kneading plane. This can be seen from the following consideration.

Suppose that there is a contact (UC, VC) in the above rectangle, i.e.,

$$EC < UC < EROC, \tag{3.58}$$

$$OC < VC < OLEC. \tag{3.59}$$

Then the following relations should hold:

$$\mathcal{L}(UC, VC) < EROC,$$
$$\mathcal{R}(UC, VC) < OLEC. \tag{3.60}$$

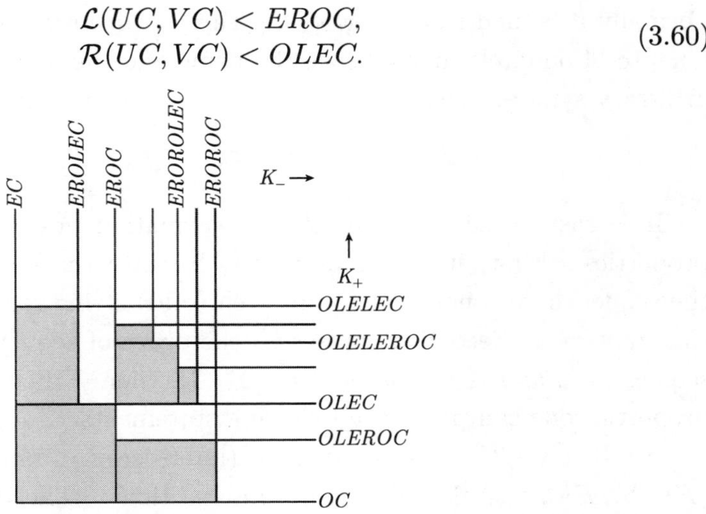

Figure 3.8 An even-odd type contact and its derived contacts and intersections

Relation (3.58) implies that UC must start with an odd string ER, so we can write $UC = ERWC$. From (3.60) we have $\mathcal{R}(UC) \ni WC < OLEC$, hence $EROLEC < UC$, which makes (3.58) stronger to yield $UC = EROLU'C$. It is easy to see that the string U' should satisfy the same condition as UC. This eventually leads to the conclusion that $UC = (EROL)^\infty$, a contradiction. Therefore, such UC does not exist. Along similar lines one may verify the non-existence of VC, hence proves the statement.

3.4.3 Self-Similar Structure in the Kneading Plane

For the unimodal map the composition rules generate kneading sequences from known ones. For the gap map we can find a few rules to transform a compatible kneading pair into a new compatible kneading pair. We discuss two transformations of this kind: the Farey transformation $\mathcal{F}$ which will be studied in more detail in connection with the circle map in the next Chapter, and the doubling transformation $\mathcal{D}$ which we have seen in the study of the unimodal map, see, e.g., (2.70) in Section 2.5.1.

The Farey transformation $\mathcal{F}$

The Farey transformation $\mathcal{F}$ is defined as

$$\mathcal{F}\colon \mathcal{F}R = RL, \mathcal{F}L = L, \mathcal{F}C = C. \tag{3.61}$$

Formally it is identical to one of the Farey transformations to be discussed in Chapter 4 on circle maps, hence the name and the notation $\mathcal{F}$. Applied to an arbitrary symbolic sequence $\Sigma = s_1 s_2 \cdots s_n \cdots$, it results in

$$\mathcal{F}\Sigma = \mathcal{F}s_1\mathcal{F}s_2 \cdots \mathcal{F}s_n \cdots .$$

It is easy to see that the $\mathcal{F}$ transformation possesses the following two properties. First, it preserves parity, because the number of symbol R in the sequence does not change: an even (odd) string remains even (odd) after the transform. Second, it preserves the order of sequences. If two symbolic sequences Σ and Π are ordered as $\Sigma > \Pi$, then $\mathcal{F}\Sigma > \mathcal{F}\Pi$. Using these two properties one can prove the following statements.

A. If (K_-, K_+) is a contact (intersection) and $K_- > K_+$, then $(\mathcal{F}K_-, L\mathcal{F}K_+)$ is also a contact (intersection). The definition (3.61) of $\mathcal{F}$ tells us that the shifts $\mathcal{L}$ and $\mathcal{R}$ of the original sequence will transform as

$$\begin{aligned}\mathcal{R} &\xrightarrow{\mathcal{F}} \tilde{\mathcal{R}} = \{L\mathcal{F}\mathcal{R}\},\\ \mathcal{L} &\xrightarrow{\mathcal{F}} \tilde{\mathcal{L}} = \{\mathcal{F}\mathcal{L}, \mathcal{F}\mathcal{R}\}.\end{aligned}$$

The R and L shifts of the transformed sequence have been denoted by $\tilde{\mathcal{R}}$ and $\tilde{\mathcal{L}}$. Together with the oder-preserving property of $\mathcal{F}$, this proves the statement.

B. If a symbolic sequence Σ is an admissible sequence under a given kneading pair (K_-, K_+), then the transformed sequence $\mathcal{F}\Sigma$ is admissible under the kneading pair $(\mathcal{F}K_-, L\mathcal{F}K_+)$. We omit the proof.

C. The region of $K_+ \leqslant C$ and $C \leqslant K_-$ in the kneading plane can be subdivided into many stripes by various lines $K_+ = L^n C$, as we have seen in Fig. 3.7. Denote by σ_n the stripe outlined by $K_- = C$, $K_+ = L^n C$, and $K_+ = L^{n+1}C$. The stripe σ_n may be generated from σ_0 by applying the $\mathcal{F}$ transformation n times, where σ_0 is the stripe of $K_- > C$ and $LC \leqslant K_+ \leqslant C$, see Fig. 3.9. Therefore, it is sufficient to analyze stripe σ_0 to understand the structure of the whole region $K_+ \leqslant C$.

For an even-odd type contact (EC, OC) in the σ_0 stripe we have

$$\begin{aligned}\mathcal{F}(EROC) &= (\mathcal{F}E)RL(\mathcal{F}O)C = \tilde{E}R\tilde{O}C,\\ L\mathcal{F}(OLEC) &= L(\mathcal{F}O)L(\mathcal{F}E)C = \tilde{O}L\tilde{E}C,\end{aligned} \tag{3.62}$$

where we have introduced $\tilde{E} \equiv \mathcal{F}(E)$ and $\tilde{O} \equiv L\mathcal{F}(O)$. Therefore, the concatenation operations to generate new contacts and the $\mathcal{F}$ transformation are commutable. This commutativity has the consequence that no contact exists in the region demarcated by $K_- = \mathcal{F}(EC)$, $K_- = \mathcal{F}(EROC)$, $K_+ = L\mathcal{F}(OC)$, and $K_+ = L\mathcal{F}(OLEC)$.

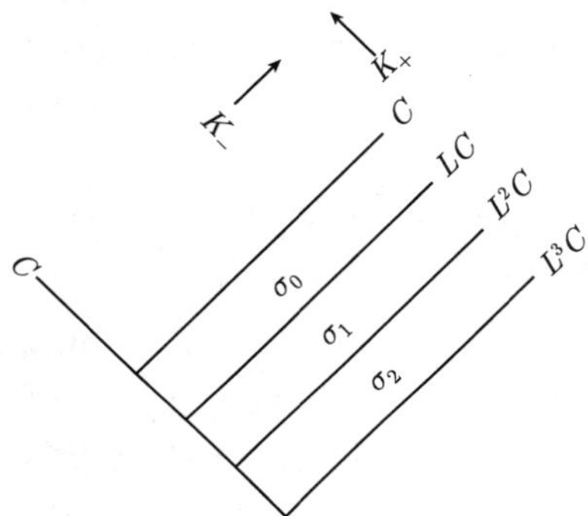

Figure 3.9 The $\mathcal{F}$ transforms σ_0 to σ_1, σ_2, etc

The doubling transformation $\mathcal{D}$

The doubling transformation $\mathcal{D}$ is defined as

$$\mathcal{D}\colon \mathcal{D}R = RL, \mathcal{D}L = RR, \mathcal{D}C = RC. \tag{3.63}$$

In fact, the doubling transformation has played an important role in analyzing the period-doubling bifurcation cascade in the unimodal map, see (2.70) in Section 2.5.1. Similar to $\mathcal{F}$, the transformation $\mathcal{D}$ preserves the parity and order of symbolic sequences. For the shifts $\mathcal{L}$ and $\mathcal{R}$ of any symbolic sequence, we have

$$\begin{aligned} &\mathcal{R} \xrightarrow{\mathcal{D}} \tilde{\mathcal{R}} = \{L\mathcal{D}\mathcal{R}, R\mathcal{D}\mathcal{L}, \mathcal{D}\mathcal{L}\}, \\ &\mathcal{L} \xrightarrow{\mathcal{D}} \tilde{\mathcal{L}} = \{\mathcal{D}\mathcal{R}\}. \end{aligned}$$

Any symbolic sequence X except L^∞ may be denoted as $X = L^k R \cdots$, where $k \geqslant 0$. After the $\mathcal{D}$ transformation it yields

$$R\mathcal{D}X = R^{2k+2}L \cdots < R^{2k+1}L \cdots = \mathcal{D}X.$$

In addition, $\mathcal{D}X$ always starts with R. Therefore,

$$\begin{aligned} &\mathcal{L}(\mathcal{D}K_\pm) = \mathcal{D}\mathcal{R}(K_\pm) < \mathcal{D}K_+, \\ &\mathcal{R}(\mathcal{D}K_\pm) \leqslant \mathcal{D}\mathcal{L}(K_\pm) < \mathcal{D}K_-. \end{aligned}$$

It follows from these relations that if (K_-, K_+) is a contact or an intersection, so does $(\mathcal{D}K_+, \mathcal{D}K_-)$. Note that the two kneading sequences K_- and K_+ have interchanged their position after the transformation.

It is easy to see that under the doubling transformation an even-odd type contact (EC, OC) becomes a contact $(E'C, O'C)$ of the same type with $E' \equiv \mathcal{D}(O)R$ and $O' \equiv \mathcal{D}(E)R$. Moreover, for an even-odd type contact (EC, OC), we have

$$\mathcal{D}(OLEC) = (\mathcal{D}O)RR(\mathcal{D}E)RC = E'RO'C,$$
$$\mathcal{D}(EROC) = (\mathcal{D}E)RL(\mathcal{D}O)RC = O'LE'C.$$

Consequently, the concatenation operations also commute with the $\mathcal{D}$ transformation. Therefore, no contact exists in the region, outlined by $K_- = \mathcal{D}(OC)$, $K_- = \mathcal{D}(OLEC)$, $K_+ = \mathcal{D}(EC)$, and $K_+ = \mathcal{D}(EROC)$.

There exist transformations of stripe-shaped regions under $\mathcal{D}$ in the kneading plane, similar to the $\sigma_0 \to \sigma_i$ transformation discussed above for $\mathcal{F}$. We use Fig. 3.10 to explain it. Denote by Σ_0 the stripe $C \leqslant K_+$ and $C \leqslant K_- \leqslant RC$, and by Δ_0 the stripe $RC \leqslant K_-$ and $C \leqslant K_+ \leqslant RC$. Applying the $\mathcal{D}$ transformation n times to Σ_0 one gets Σ_n, where Σ_n is the stripe outlined by $\mathcal{D}^nC \leqslant K_-$ and $\mathcal{D}^nC \leqslant K_+ \leqslant \mathcal{D}^{n+1}C$ when n is odd, and the stripe $\mathcal{D}^nC \leqslant K_+$ and $\mathcal{D}^nC \leqslant K_- \leqslant \mathcal{D}^{n+1}C$ when n is even. These stripes are shown as shaded rectangles in Fig. 3.10.

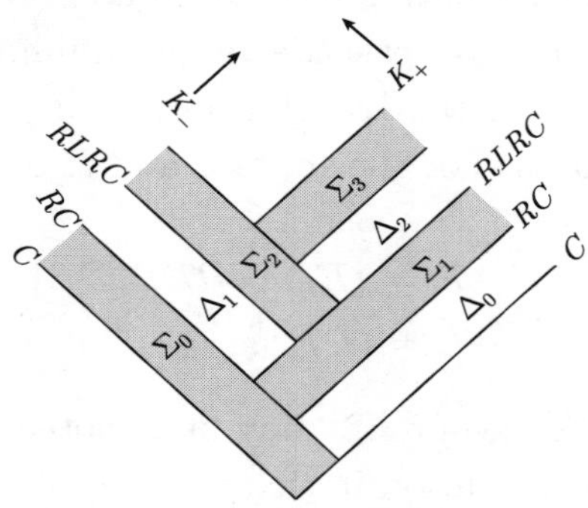

Figure 3.10 The $\mathcal{D}$ transforms Σ_0 to Σ_i and Δ_0 to Δ_i, etc

Shown in the same figure are the transformations of Δ_0 under $\mathcal{D}^n$ into the stripe denoted by Δ_n. When n is odd, these are stripes $\mathcal{D}^{n+1}C \leqslant K_+$ and $\mathcal{D}^nC \leqslant K_- \leqslant \mathcal{D}^{n+1}C$. When n is even, Δ_n are outlined by $\mathcal{D}^{n+1}C \leqslant K_-$ and $\mathcal{D}^nC \leqslant K_+ \leqslant \mathcal{D}^{n+1}C$. The stripes rotate by 90° under each $\mathcal{D}$.

The three sets Δ_n, Σ_n and σ_n of stripes cover the whole kneading plane. The image of σ_0 under $\mathcal{D}$ is the stripe $K_+ \geqslant RC$ and $RC \leqslant K_- \leqslant RLRC$. It contais Δ_1 and a subregion of Σ_1. Therefore, it is enough to understand Σ_0 and Δ_0 in order to comprehend the structure of the whole kneading plane.

3.4.4 Criterion for Topological Chaos

In the parlance of symbolic dynamics topological chaos is the exponential growth of the number of admissible periodic symbolic sequences with their increasing length. For the gap map a simple criterion for the existence of topological chaos may be formulated.

From Fig. 3.8 we see an even-odd type contact (EC, OC) generates a first intersection $(EROC, OLEC)$. Two lines representing the kneading sequences $K_- = EROC$ and $K_+ = OLEC$ pass through this intersection. On each of these two lines there is a second intersection: $(EROC, OLELEC)$ on the K_- line and $(ERO ROC, OLEC)$ on the K_+ line. We will soon prove that at these two secondary intersections there exists topological chaos, i.e., dynamics with a positive topological entropy. Thus this criterion may be briefly stated as "two consecutive intersections imply chaos". A complete statement would be "topological chaos exists at the second intersection on a kneading line passing through a first intersection generated by an even-odd type contact".

We prove the statement for the intersection $(ERO ROC, OLEC)$. The essence of the proof consists in the following.

A. $E_1L \equiv EROL$ and $E_2L \equiv ERO ROL$ are even strings.

B. Any combination of E_1L and E_2L leads to an admissible sequence.

C. Denote the length of the strings E_1L and E_2L by α and β, respectively. Consider periodic sequences made of E_1L and E_2L only. Since one may append E_1L or E_2L to any basic, non-repeating string to get longer admissible periods, the number N_n of periodic sequences of length n and less may be estimated asymptotically for large n as

$$N_n \geqslant N_{n-\alpha} + N_{n-\beta}. \tag{3.64}$$

Since the contribution of admissible periods made of strings other than E_1L and E_2L has not been taken into account, this gives only a lower bound of N_n. From (3.64) one may estimate $N_n \geqslant 2^{n/\alpha}$. Therefore, for the topological entropy we have

$$h(f) = \lim_{n\to\infty} \frac{\ln N_n}{n} \geqslant \frac{1}{\alpha} \ln 2 > 0. \tag{3.65}$$

Now we give the details of the proof. Point **A** is obvious. The key point is **B**. Let s_i stand for E_1L or E_2L. In order to prove that any symbolic sequence $\Sigma = s_1 s_2 \cdots s_n \cdots$ is admissible, we have to check that the shifts $\mathcal{L}(\Sigma)$ and $\mathcal{R}(\Sigma)$ do satisfy the conditions (3.46). For $\mathcal{L}(\Sigma)$ this means

$$\mathcal{L}(\Sigma) \leqslant EROROC = K_-. \tag{3.66}$$

We go through all possible cases. The simplest case is $s_i s_{i+1} \cdots \in \mathcal{L}(\Sigma)$. Since both E_1L and E_2L are less than $EROROC$, the relation

$$s_i s_{i+1} \cdots < EROROC$$

holds naturally.

For other cases of $\mathcal{L}$ we have to prove

$$\mathcal{L}(s_i) s_{i+1} s_{i+2} \cdots \equiv U s_{i+1} s_{i+2} \cdots < EROROC = E_2 C. \tag{3.67}$$

If U is not a leading string of E_2, then it follows from

$$\mathcal{L}(EROROC) < EROROC,$$
$$\mathcal{L}(EROC) < EROROC,$$

that (3.67) is true. If $E_2 = UV$, then the above relations show that U is an odd string and V must be odd. Thus, from $s_{i+1} s_{i+2} \cdots \geqslant (E_1 L)^\infty$, (3.67) holds if

$$\mathcal{L}(EROROC) \ni VC < (E_1 L)^\infty = (EROL)^\infty. \tag{3.68}$$

If $VC \in \mathcal{L}(OC)$, then (3.68) follows from

$$\mathcal{L}(OC) \leqslant EC < ER.$$

If $VC = V'OC$, V' must be even due to V being odd. However, the even string $V' \in \mathcal{L}(EROR)$ can never be a leading string of $EROC$. Therefore, (3.68) holds.

For $\mathcal{R}(\Sigma)$ we have to prove

$$\mathcal{R}(s_i) s_{i+1} s_{i+2} \cdots \equiv U s_{i+1} s_{i+2} \cdots < OLEC. \tag{3.69}$$

Since both $(EROC, OLEC)$ and $(EROROC, OLEC)$ are intersections, we have

$$\mathcal{R}(EROC) < OLEC, \quad \mathcal{R}(EROROC) < OLEC.$$

One has to consider only the case when U is a leading string of $OLEC$. Now U must be an odd string. If we denote $OLE = UV$, then (3.69) is equivalent to

$$s_{i+1}s_{i+2}\cdots > VC \in \mathcal{L}(OLEC),$$

which holds due to $\mathcal{L}(OLEC) < EROC$. So we have proved that any sequence made of E_1L and E_2L is admissible.

The discussion for the other intersection $(EROC, OLELEC)$ goes in a similar way. In fact, by interchanging

$$ER \leftrightarrow OL, \quad EL \leftrightarrow OR, \quad EC \leftrightarrow OC, \quad K_+ \leftrightarrow K_-,$$

one contact transforms into the other.

Now we are in a position to determine the boundaries of topological chaos in the kneading plane. Due to the self-similar structure of the kneading plane, it is enough to analyze the Σ_0 and Δ_0 stripes. When studying the boundaries of topological chaos in Σ_0, it is more convenient to leave the $K_+ < RC$ part of Σ_0 to the analysis of Δ_0. Then in the remainder of Σ_0 the lowest even-odd contact is (C, RC), which is a proper contact, i.e., it corresponds to an actual map.

It is easy to check that the only sequences allowed by (C, RC) are L^∞ and R^∞. One can further generate two contacts (C, RLC) and (R^2C, RC), which allow no sequences other than L^∞ and R^∞. For the intersection (R^2C, RLC) generated from the contact (C, RC), there is only one additional admissible sequence $(RLR)^\infty$ besides L^∞ and R^∞.

If we call (C, RC) a primary contact, then (C, RLC) and (R^2C, RC) are secondary contacts and (R^2C, RLC)—a secondary intersection. In the rectangle formed by these four vertices, the topological entropy of the map is zero. In other words, the rectangle is a zero-entropy region.

One may take these secondary contacts and intersections to be new primary contacts and generate new secondary contacts and intersections from them. They will define new zero-entropy regions. Repeating this process, one may generate a series of contacts and intersections. Among them no two intersections will fall on one and the same K_- or K_+ line. The union of the infinite many zero-entropy regions obtained in this way determines the non-chaotic region in the kneading plane. Its borders give the boundaries of topological chaos.

There is a class of intersections right on the boundaries of topological chaos, which may be generated as follows:

$$\begin{aligned}\cdots \to (EC, OC) &\to (EROC, OLEC)\\ &\to (EROROLEC, OLELEROC) \to \cdots .\end{aligned}$$

It starts with an intermediate intersection (EC, OC). At each step new K_- and K_+ are generated by concatenating old K_- and K_+ with their last symbol properly modified. This process deals with intersections only. A slightly different process may involve contacts as well. For example,

$$\begin{aligned}\cdots \to (EC, OC) &\to (EROC, OC)\\ &\to (EROROC, OLEROC) \to \cdots .\end{aligned}$$

The process of generating boundaries of chaos by mixed use of contacts and intersections may be quite complicated. However, the concatenation at each step is simple.

The above discussion concerns superstable words only. By replacing the last symbol C in a superstable word by R or L one may extend the region of zero-entropy upwards to get the final boundaries of topological chaos.

Now consider the region $C \leqslant K_-$ and $C \leqslant K_+ \leqslant RC$. It includes the stripe Δ_0 and the lower part of Σ_0 left over from the above discussion. In this region one always has $K_+ = R^\infty$, meaning that RR can only be appended by R^∞. When $K_- \leqslant RLC$, the admissible periodic sequences are just R^∞ and L^∞. When $K_- = (RLL)^\infty$, one more admissible sequence $(RLL)^\infty$ is added. When $K_- = RLL(RL)^\infty$, then it is easy to verify that all sequences made of strings RLL and RL are admissible. Moreover, if ignoring a possible leading string RL^k or L^k with $k \geqslant 1$, all admissible sequences, except for R^∞ and L^∞, can only contain these two kinds of strings. Denote

$$\hat{R} = RLL, \quad \hat{L} = RLRL.$$

The limiting word of period-doubling with respect to $\hat{R}$ and $\hat{L}$, namely,

$$\hat{R}\hat{L}\hat{R}^3\hat{L}\hat{R}\hat{L}\cdots = RLLRLRL(RLL)^3RLRLRLLRLRL\cdots,$$

then gives the boundary of topological chaos (Zheng and Lu [1991]).

3.5 The Lorenz-Like Map

Another simple one-dimensional map with a discontinuity originated from the study of chaotic motion in the Lorenz equations, a truncated hydrodynamic model in which Lorenz discovered one of the first strange attractors (Lorenz [1963]). It captures the essence of the so-called geometrical model of Lorenz, hence the name Lorenz-like map (Guckenheimer and Williams [1979]; Williams [1979b]; Arneodo, Collet and Tresser [1981]; Gambaudo, Glendinning and Tresser [1985]). In Chapter 6 we will scrutinize the Lorenz equations by using one- and two-dimensional symbolic dynamics. While the Lorenz-like map, studied in this section, reflects the essence of the dynamics at parameter $r \approx 28$ (see Section 6.3 for the meaning of the parameter), the dynamics in a much wider r-range may be captured by a map with four monotone branches, the Lorenz-Sparrow map to be studied in Section 3.8. For the time being we study the symbolic dynamics of the Lorenz-like map *per se* (Zheng [1990]).

The mapping function $f(x)$ is shown in Fig. 3.11. It has a discontinuity at C and two monotone increasing branches labeled by R and L. If necessary, we shall denote the left and right limits of C by C_- and C_+. The dynamical invariant interval $[f(C_+), f(C_-)]$ is mapped by $f(x)$ to itself.

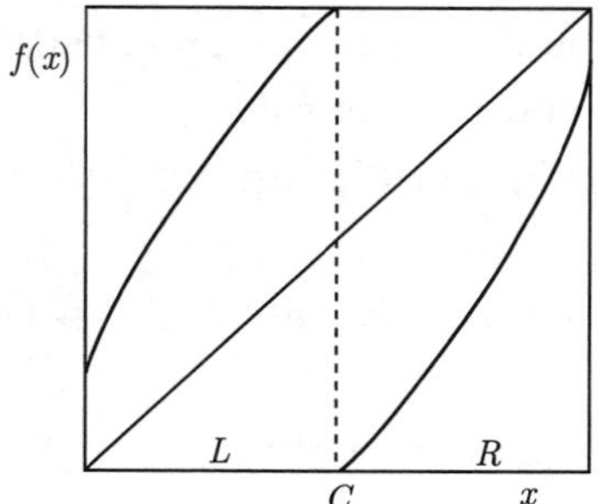

Figure 3.11 The Lorenz-like map

3.5.1 Ordering Rule and Admissibility Conditions

Since both symbols R and L are of even parity, the ordering rule is simply

$$\Sigma L \cdots < \Sigma C \cdots < \Sigma R \cdots , \tag{3.70}$$

where Σ is an arbitrary finite string. The symbolic sequences of $f(C_-)$ and $f(C_+)$ are the kneading sequences K_- and K_+. It is easy to write down the admissibility conditions for a sequence Σ:

$$\mathcal{L}(\Sigma) \leqslant K_-, \quad \mathcal{R}(\Sigma) \geqslant K_+. \tag{3.71}$$

If a pair of kneading sequences (K_-, K_+) corresponds to an actual Lorenz-like map, they themselves should satisfy the admissibility conditions:

$$\mathcal{L}(K_-, K_+) \leqslant K_-, \quad \mathcal{R}(K_-, K_+) \geqslant K_+. \tag{3.72}$$

If so, they are called a *compatible kneading pair*. Continuity consideration still plays an important role at the presence of discontinuity. As usual, "superstable" kneading sequences containing the critical symbol C are crucial in constructing the symbolic dynamics. We have enclosed "superstable" in quotation marks, as the presence of the letter C may have nothing to do with the stability of an orbit. However, we will drop the quotation marks hereafter. For superstable kneading sequences the admissibility condition (3.72) holds as strict inequalities.

3.5.2 Construction of the Kneading Plane

The method of generating superstable kneading sequences K_- and K_+ for the gap map studied in Section 3.4.1 may be extended to the Lorenz-like map as well. We skip the details and list the rules:

A. The maximal and minimal K_- are $K_-^{\max} = R^\infty$ and $K_-^{\min} = C$, respectively.

B. The maximal and minimal K_+ are $K_+^{\max} = C$ and $K_+^{\min} = L^\infty$, respectively.

C. The upper and lower sequences of $K_- = XC$ are $XRK_+^{\min} = XRL^\infty$ and $(XL)^\infty$, respectively.

D. The upper and lower sequences of $K_+ = YC$ are $(YR)^\infty$ and $YLK_-^{\max} = YLR^\infty$, respectively.

E. The rest is similar to the unimodal map.

All possible K_- and K_+ are mirror images to each other, i.e., one set transforms into another by interchanging $R \leftrightarrow L$. By checking the compatibility of the superstable words thus obtained, one constructs the *kneading plane*. The kneading plane showing superstable sequences with length no longer than 6 is given in Fig. 3.12.

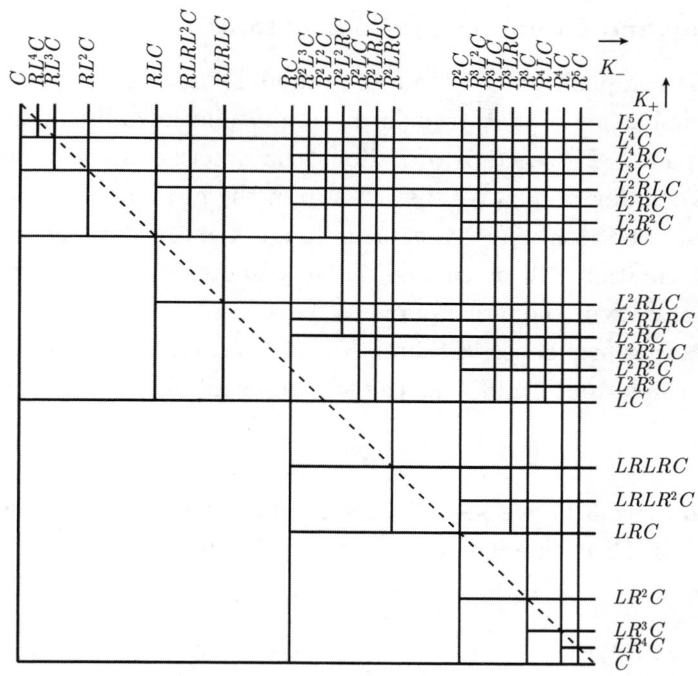

Figure 3.12 Kneading plane of the Lorenz-like map up to period 6

3.5.3 Contacts and Intersections

Similar to the gap map, one can define *contacts* for the Lorenz-like map. A pair of superstable kneading sequences (K_-, K_+) is a contact or contact pair, if they satisfy

$$\begin{aligned} &\mathcal{L}(K_-) < K_-, \quad \mathcal{R}(K_-) \geqslant K_+, \\ &\mathcal{L}(K_+) \leqslant K_-, \quad \mathcal{R}(K_+) > K_+. \end{aligned} \tag{3.73}$$

A contact may not correspond to an actual map. If all relations in (3.73) hold as strict inequalities, the contact is called an *intersection*.

It may be proved that if $(K_-, K_+) = (XC, YC)$ is a contact, then $(XRYC, YC)$ and $(XC, YLXC)$ are also contacts. Furthermore, the pair $(XRYC, YLXC)$ is an intersection. Starting from these contacts and the intersection, one can generate new contacts and intersections. This process may be repeated *ad infinitum*. It also tells about local structure of the kneading plane.

3.5.4 Farey and Doubling Transformations

Similar to the composition rules in the unimodal map, one can find operations in the Lorenz-like map that transform compatible kneading pairs into new compatible pairs. These transformations characterize the self-similar structure of the kneading plane. We discuss two of them, namely, the Farey transformation and the doubling transformation. A more detailed study of the Farey transformation will be carried out in Chapter 4 in connection with the symbolic dynamics of the circle map.

The kneading sequences K_- and K_+ of a general Lorenz-like map have respectively the letter R and L as the leading symbol. Let us write

$$K_- = RJ_-, \quad K_+ = LJ_+. \tag{3.74}$$

By inspecting Fig. 3.11 we see that, if confined to the dynamical invariant range $[f(C_+), f(C_-)]$, one has

$$J_+ \leqslant \mathcal{L} \leqslant K_-, \quad K_+ \leqslant \mathcal{R} \leqslant J_-. \tag{3.75}$$

The Farey Transformations $\mathcal{F}^{\pm}$

The two Farey transformations are defined as

$$\begin{aligned} \mathcal{F}^+ &: R \to R, \quad L \to RL, \quad C \to RC, \\ \mathcal{F}^- &: R \to LR, \quad L \to L, \quad C \to LC. \end{aligned} \tag{3.76}$$

It is not difficult to check that both transformations preserve the order, i.e., for any two symbolic sequences P and Q we have

$$P > Q \Rightarrow \mathcal{F}P > \mathcal{F}Q,$$

where $\mathcal{F}$ may be either $\mathcal{F}^+$ or $\mathcal{F}^-$. For a given kneading pair (K_-, K_+) their shifts $\mathcal{L}$ and $\mathcal{R}$ transform under $\mathcal{F}^+$ as:

$$\begin{aligned} \mathcal{L} &\xrightarrow{\mathcal{F}^+} \tilde{\mathcal{L}} = \{\mathcal{F}^+(\mathcal{L})\} \leqslant \mathcal{F}^+(K_-), \\ \mathcal{R} &\xrightarrow{\mathcal{F}^+} \tilde{\mathcal{R}} = \{\mathcal{F}^+(\mathcal{R}), L\mathcal{F}^+(\mathcal{L})\} \\ &\qquad \geqslant L\mathcal{F}^+(J_+) = \mathcal{S}(\mathcal{F}^+(K_+)), \end{aligned}$$

where $\mathcal{S}$ is the shift operator. Therefore, if (K_-, K_+) is a compatible kneading pair, so does $(\mathcal{F}^+(K_-), \mathcal{S}(\mathcal{F}^+(K_+)))$. Similarly, if (K_-, K_+) is a compatible kneading pair, so does $(\mathcal{S}(\mathcal{F}^-(K_-)), \mathcal{F}^-(K_+))$ under the transformation $\mathcal{F}^-$.

The above proof also shows that if a symbolic sequence Σ is admissible with respect to the kneading pair (K_-, K_+), then $\mathcal{F}(\Sigma)$ is admissible with respect to the transformed kneading pairs.

The Doubling Transformation $\mathcal{D}$

The doubling transformation $\mathcal{D}$ corresponds to the symbol substitutions under a period-doubling bifurcation in a unimodal map. Its definition is

$$\mathcal{D}\colon R \to RL,\ L \to LR,\ C \to C. \tag{3.77}$$

The $\mathcal{D}$ transformation preserves order. The shifts $\mathcal{L}$ and $\mathcal{R}$ of the kneading sequences transform under $\mathcal{D}$ as

$$\begin{aligned} \mathcal{L} \xrightarrow{\mathcal{D}} \tilde{\mathcal{L}} &= \{R\mathcal{D}(\mathcal{L}), \mathcal{D}(\mathcal{R})\} \leqslant R\mathcal{D}(K_-), \\ \mathcal{R} \xrightarrow{\mathcal{D}} \tilde{\mathcal{R}} &= \{L\mathcal{D}(\mathcal{R}), \mathcal{D}(\mathcal{L})\} \geqslant L\mathcal{D}(K_+). \end{aligned}$$

Consequently, the kneading pair transforms into $(R\mathcal{D}(K_-), L\mathcal{D}(K_+))$. A symbolic sequence Σ admissible with respect to the kneading pair (K_-, K_+) transforms into $\mathcal{D}(\Sigma)$, which is an admissible sequence with respect to the transformed kneading pair $(R\mathcal{D}(K_-), L\mathcal{D}(K_+))$.

In addition, denoting by $\overline{\Sigma}$ the mirror image of Σ under the interchange of $R \leftrightarrow L$, we see easily that the mirror image $(\bar{K}_+, \bar{K}_-)$ of a compatible kneading pair (K_-, K_+) is also a compatible pair.

3.6 General Cubic Maps

General cubic maps may serve as a good starting point to learn many new features of maps with more than one critical point. Doubly superstable kneading sequences, which contain the both critical symbols of the map, play a central role in constructing the symbolic dynamics. In the kneading plane they appear as isolated points — "joints". These points may be located precisely by using the word-lifting technique. Four singly superstable kneading sequences grow out from each joint. They are called *bones* . Bones as curves in the kneading plane are also given by word-lifting. Joints and bones make a *skeleton* of the kneading plane. These structures and their relative positions may all be determined by the ordering rule and admissibility conditions.

The symbolic dynamics of cubic maps was first studied by MacKay and Tresser [1987, 1988]. We will put their results in the general setting and notation of this book.

A general cubic map may be represented by the mapping function

$$f(x) = Ax^3 + (1-A)x + B, \tag{3.78}$$

where the constant B, which may take positive as well as negative values, violates the symmetry. When the parameter $A > 1$, the function is shown in Fig. 3.13. The two critical points D and C divide the interval into left, middle, and right segments, which we denote by the symbols L, M, and R, respectively. The mapping function is monotonically increasing on the subintervals L and R, and decreasing on the subinterval M. Therefore, only the symbol M has an odd parity. We say that the cubic map is of $(+,-,+)$ type. The $(-,+,-)$ type map with $f(x) = -Ax^3 + (A-1)x + B$ and $A > 1$ will be discussed in Section 3.6.4 below.

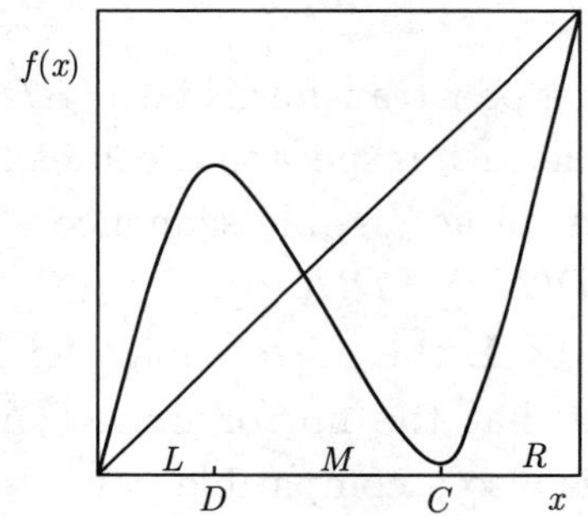

Figure 3.13 A general cubic map of $(+,-,+)$ type

Ordering rule of symbolic sequences

The general ordering rule (3.1) works for any one-dimensional map. In particular, the ordering rule for the general cubic map is identical to that of the anti-symmetric cubic map, see (3.11) to (3.13). In both cases the parity of the common leading string is determined by the number of symbol M representing the only decreasing branch of the map.

Admissibility conditions

In contrast to the anti-symmetric case, now the two kneading sequences $K_C = f(C)$ and $K_D = f(D)$ are no longer dependent. By inspecting the graph of the mapping function, we write down the admissibility conditions for sequence Σ:

$$\begin{aligned}&\mathcal{L}(\Sigma) \leqslant K_D,\\ &K_C \leqslant \mathcal{M}(\Sigma) \leqslant K_D,\\ &K_C \leqslant \mathcal{R}(\Sigma).\end{aligned} \tag{3.79}$$

In general, when K_C or K_D or both are superstable kneading sequences, i.e., are ending with C or D, some of the above relations will hold as inequalities.

Of course, the kneading sequences themselves must satisfy the admissibility conditions (3.79) as well:

$$\begin{aligned}&\mathcal{L}(K_D, K_C) \leqslant K_D,\\ &K_C \leqslant \mathcal{M}(K_D, K_C) \leqslant K_D,\\ &K_C \leqslant \mathcal{R}(K_D, K_C).\end{aligned} \tag{3.80}$$

Then K_C and K_D are compatible to each other and (K_D, K_C) is called a *compatible pair* of kneading sequences or a kneading pair. Obviously, a necessary for compatibility is $K_D \geqslant K_C$.

We again emphasize the twofold meaning of admissibility conditions. Relations (3.79) are conditions for a symbolic sequence Σ to be admissible under a given kneading pair, i.e., they guarantee the reproducibility of Σ in the phase space at a given set of parameters. Relations (3.80) are the conditions for two sequences (K_D, K_C) to become a compatible kneading pair, i.e., they hold at certain combinations of parameters—usually at some points or areas in the kneading plane.

As it is clearly seen in Fig. 3.13, a $(+, -, +)$ type cubic map always has two fixed points L^∞ and R^∞. They provide the minimal and maximal sequences, which may be chosen as K_D or K_C if the compatibility conditions are not taken into account.

3.6.1 Skeleton, Bones and Joints in Kneading Plane

The set of compatible kneading pairs characterizes the essential types of dynamics. When kneading sequences are use as parameters, the parameter plane is called a *kneading plane*. In Fig. 3.14 K_D and K_C are chosen as two coordinates with K_D increasing upward from L^∞ to R^∞ and K_C—downward from L^∞ to R^∞. The qualitative pictures of the map in various parts of the kneading plane are shown schematically.

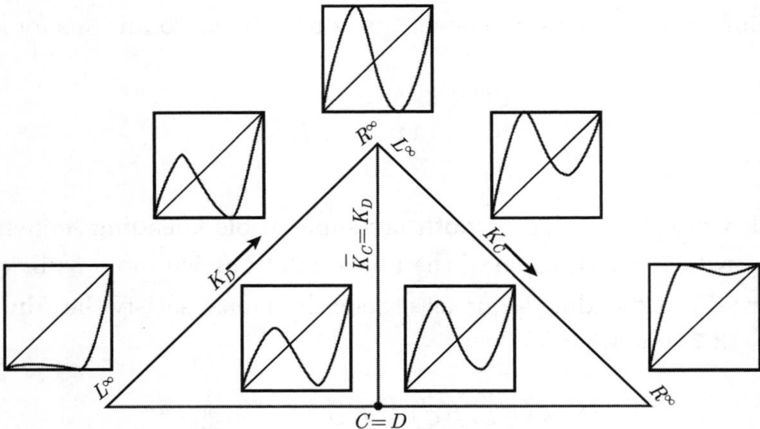

Figure 3.14 The borders of the kneading plane and the mapping function in various parts of the plane (schematic)

Joints in kneading plane

Owing to continuity, superstable kneading sequences provide the clue to understand the structure of the parameter plane. The collection of their loci in the kneading plane is usually called a *skeleton*. Among the superstable kneading sequences the doubly superstable ones, i.e., those that contain both critical symbols D and C, play a key role. These isolated points are called *joints*. A doubly superstable kneading sequence $(XCYD)^\infty$ furnishing a compatible pair by definition means:

$$(K_D, K_C) = (XCYD, YDXC) = [XC, YD], \tag{3.81}$$

where X and Y are strings made of symbols R, M, and L. When dealing with a joint, it is natural to write the two sequences up to the first critical symbol. We use square brackets [,] to distinguish a joint from an ordinary compatible pair.

The location of a given joint for a specific map $f(x)$ may be calculated by using word-lifting of (3.81):

$$f(D) = X(C), \quad f(C) = Y(D).$$

Recollecting that by continuity D goes into R or M, C goes into L or M, and introducing the notations

Next, suppose $U_1 \in \mathcal{L}((YD)_+)$. According to the last inequality of (3.82), $(XC)_+$ cannot be a leading string of U_1. If U_1 is not a leading string of XC, then (3.84) holds. If $(XC)_+ = U_1V_1$, then both U_1 and V_1 are odd, and (3.84) requires

$$V_1(YD)_+XC \in \mathcal{L} < XC,$$

which is alway true since an odd V_1 is never a leading string of XC. No proof is needed for $\mathcal{L}(XC) \in \mathcal{L}$.

By now we have checked all possible cases and finished the proof of statement **A**.

3.6.2 The Construction of the Kneading Plane

The first step to construct the kneading plane consists in generating all possible kneading sequences K_D and K_C. Obviously, they should separately satisfy the necessary conditions

$$\mathcal{L}(K_D) \leqslant K_D, \quad \mathcal{M}(K_D) \leqslant K_D, \tag{3.87}$$

and

$$\mathcal{R}(K_C) \geqslant K_C, \quad \mathcal{M}(K_C) \geqslant K_C. \tag{3.88}$$

(As we have mentioned before, when K_D or K_C corresponds to a *point* on the interval such as a superstable word, these relations may become strict inequalities.)

Generation of kneading sequences

For a general cubic map one may extend the method of generating median words in the unimodal maps to get a K_D satisfying (3.87) or a K_C satisfying (3.88). It goes as follows:

A. The maximal and minimal K_D satisfying (3.87) are $K_D^{\max} = R^\infty$ and $K_D^{\min} = L^\infty$.

B. The maximal and minimal K_C satisfying (3.88) are $K_C^{\max} = R^\infty$ and $K_C^{\min} = L^\infty$.

C. The upper and lower sequences of $K_D = XD$ are $(XD)_+^\infty$ and $(XD)_-^\infty$; the upper and lower sequences of $K_C = YC$ are $(YC)_+^\infty$ and $(YC)_-^\infty$.

D. The upper and lower sequences of $K_D = XC$ are $(XC)_+L^\infty$ and $(XC)_-L^\infty$; the upper and lower sequences of $K_C = YD$ are $(YD)_+R^\infty$ and $(YD)_-R^\infty$.

E. The rest goes just as in the unimodal map.

A kneading plane containing all superstable kneading sequences of length 5 and less is shown in Fig. 3.15.[1] Joints are marked by heavy dots. The dashed diagonal shows the $K_D = \bar{K}_C$ case, i.e., the anti-symmetric cubic map. The two heavy dots on this diagonal represent the doubly superstable period 2 and period 4 sequences.

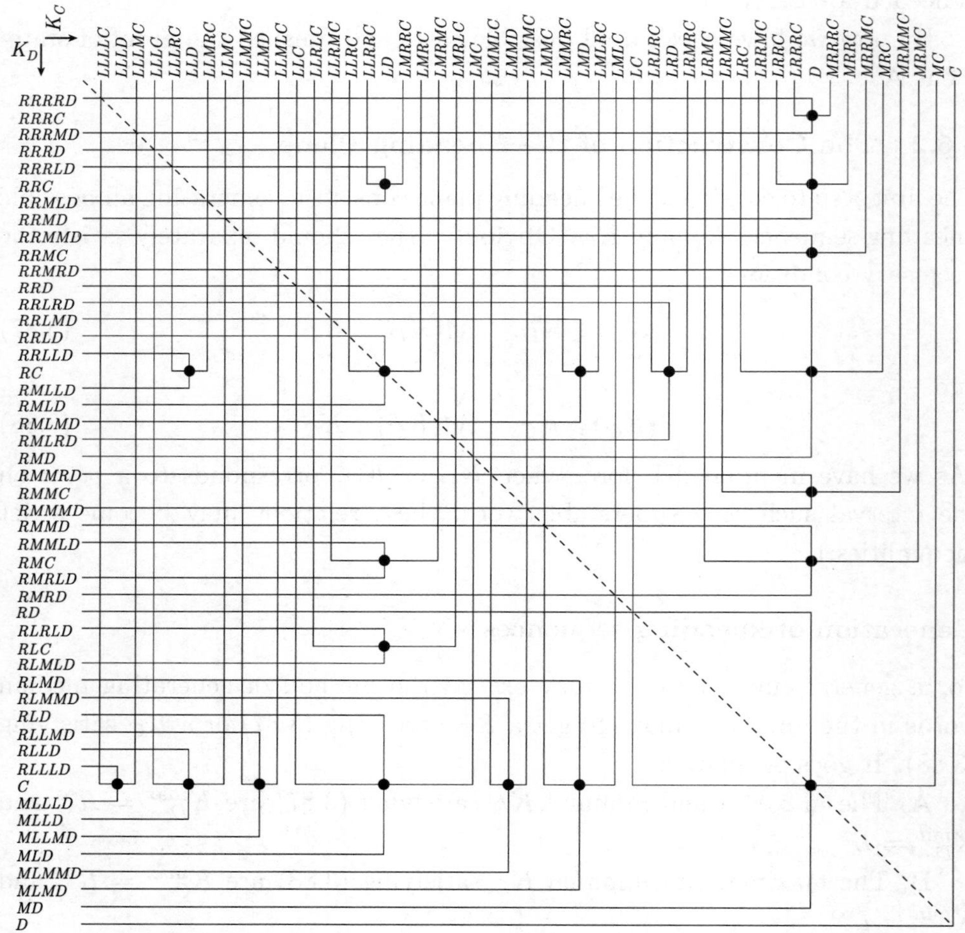

Figure 3.15 The kneading plane of a general $(+,-,+)$ type cubic map containing all superstable kneading sequences of length 5 and less. Joints are marked by heavy dots

[1] This kneading plane was given as Fig. 8 in MacKay and Tresser [1987] without symbolic names and with a period 4 missing.

A BASIC program to generate all superstable kneading sequences K_D up to length 5 is listed in Program 3.3. This program produces all superstable kneading sequences, including doubly superstable ones. Statement 160 turns off the printing of doubly superstable words. One can write a similar program to generate K_C. However, all K_C can be obtained from K_D by taking "mirror images", i.e., interchanging R and L, but keeping M unchanged.

```
   2  REM PROGRAM 3.3 SUPERSTABLE Kd FOR + - + CUBIC MAP
   4  REM S$:  KNEADING SEQUENCE, L: LENGTH OF S$
  10  DIM S$(1000), L(1000)
  20  NMAX=5:  IA=1:  IB=1000
  30  S$(1)="MM":  L(1)=2:  S$(1000)="RRRRL":  L(1000)=5
  40  S2$=S$(IB): S1$=MID$(S$(IA), 1, L(IA)-1)
  50  IF MID$(S$(IA), LA(IA), 1)="L"
         THEN S1$=S1$ + "M" ELSE S1$=S1$ + "L"
  60  IF MID$(S$(IA), LA(IA), 1)="C" THEN S1$=S$(IA) + "LLLL"
  70  IF MID$(S2$, L(IB), 1) = "C" THEN S2$=S2$ + "LLLL"
  80  FOR M=1 TO 500:  GOSUB 1000
  90  IF ID=2 THEN GOTO 130
 100  IB=IB-1:  L(IB)=K+1:  S$(IB)=SS$:  S2$=SS$
 110  IF MID$(S2$, L(IB), 1)="C" THEN S2$=S2$ + "LLLL"
 120  NEXT M
 130  IA=IA+1:  S$(IA)=S$(IB): L(IA)=L(IB): IB=IB+1
 140  IF IB<1001 THEN GOTO 40
 150  K=1:  FOR I=1 TO IA
 160  IF MID$(S$(I), L(I), 1)="C" THEN GOTO 190
 170  S$(I)=MID$(S$(I), 1, L(I)-1) + "D":  K=K+1
 180  PRINT K TAB(8) S$(I) TAB(18) L(I)
 190  NEXT I: STOP
1000  L1=LEN(S1$):  L2=LEN(S2$)
1010  SS$=" ":  FOR K=0 TO NMAX-1
1020  K1=(K MOD L1) + 1:  K2=(K MOD L2) +1
1030  C1$=MID$(S1$, K1, 1):  C2=MID$(S2$, K2, 1)
1040  IF C1$="C" AND C2$<>"L" THEN C1$=C2$
1050  IF C2$="C" AND C1$<>"L" THEN C2$=C1$
1060  SS$=SS$+C1$
1070  IF C1$<>C2$ THEN GOTO 1090
1080  NEXT K: ID=2:  RETURN
1090  IF C1$="L" THEN ID=1:  RETURN
1100  IF C2$="L" THEN SS$=MID$(SS$, 1, K) + "M":  ID=1:  RETURN
1110  SS$=MID$(SS$, 1, K) + "C"
1120  ID=0:  RETURN: END
```

Program 3.3

Compatible kneading pairs

We proceed further to determine compatible (K_D, K_C) pairs. For a given K_D we define

$$J_C = \min\{\mathcal{R}(K_D), \mathcal{M}(K_D), K_D\}, \tag{3.89}$$

then K_D is incompatible with all $K_C > J_C$. In other words, the line representing K_D in the kneading plane stops at the line K_C closest to J_C. Similarly, for a given K_C we define

$$J_D = \max\{\mathcal{L}(K_C), \mathcal{M}(K_C), K_C\}, \tag{3.90}$$

then K_C is incompatible with all $K_D < J_D$. The line representing K_C in the kneading plane stops at the line K_D closest to J_D. Consequently, the intersection points of these two families of lines correspond to compatible kneading pairs.

We have mentioned before that joints are the key points in the kneading plane. For a given joint $(K_D, K_C) = [XC, YD]$, it follows from continuity that there are superstable sequences $(XC)_+YD$ and $(XC)_-YD$ above and below the kneading sequence $K_D = XCYD$. Their representative lines can reach $K_C = YD$ at least. Similarly, above and below the kneading sequence $K_C = YDXC$ there are superstable sequences $(YD)_-XC$ and $(YD)_+XC$; their representative lines can reach $K_D = XC$ at least. Again from continuity we see that these four superstable kneading sequences are followed by non-superstable sequences. For example, above and below $K_D = (XC)_+YD$ there

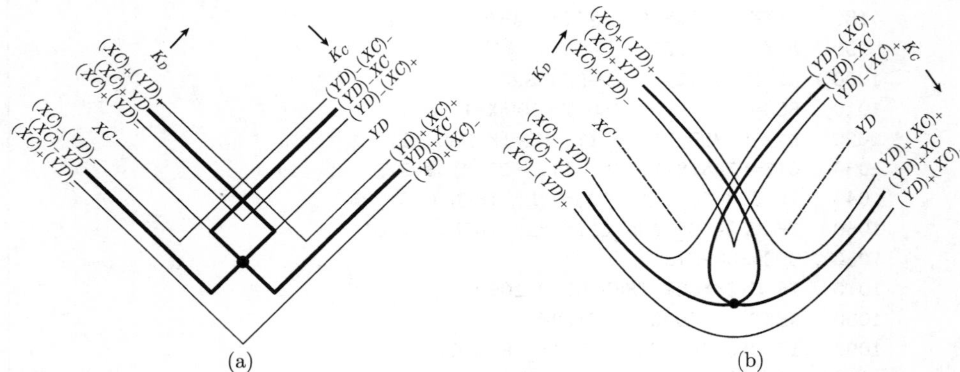

Figure 3.16 Superstable and non-superstable kneading sequences generated from a joint $(K_D, K_C) = [XC, YD]$. (a) Sketch of the kneading plane. (b) Sketch of the parameter plane

are $(XC)_+(YD)_+$ and $(XC)_+(YD)_-$. A series of such superstable and non-superstable kneading sequences born from the joint $(K_D, K_C) = [XC, YD]$ is shown in Fig. 3.16. Fig. 3.16 (a) is a sketch of the kneading plane. Only non-repeating strings of periodic words are indicated. Fig. 3.16 (b) shows how a general parameter plane may look like. Among the non-superstable words those having two + or two − subscripts, e.g., $(XC)_+(YD)_+$ and $(XC)_-(YD)_-$, are strings of odd parity and capable of undergoing period-doubling bifurcations. All other even-parity strings correspond to periods born at tangent bifurcations.

3.6.3 The $*$-Composition Rules

The construction described in the previous two sections meets the practical need of generating admissible symbolic sequences up to a given length. However, theoretically, it is interesting to look for composition rules similar to the $*$-composition of unimodal maps studied in Section 2.5. It turns out that the presence of two critical points leads to two types of $*$-compositions: the "up-star" product $\overline{*}$ and the "down-star" product $\underline{*}$ called by Peng, Zhang and Cao [1998]. In the style and notation of this book it is more convenient to formulate these products as two sets of substitution rules.

Suppose we have a doubly superstable kneading sequence $XCYD$ which leads to a compatible kneading pair $(K_D, K_C) = [XC, YD]$, where [,] is the notation for a joint, and $YD < XC$. We define two sets of substitutions:

$$\mathcal{F}_{\underline{*}}: \begin{cases} L \to (XC)_-(YD)_+, \\ D \to (XC)_-YD, \\ M \to (XC)_-(YD)_-, \\ C \to XC(YD)_-, \\ R \to (XC)_+(YD)_-; \end{cases}$$

$$\mathcal{F}_{\overline{*}}: \begin{cases} L \to (YD)_-(XC)_+, \\ D \to YD(XC)_+, \\ M \to (YD)_+(XC)_+, \\ C \to (YD)_+XC, \\ R \to (YD)_+(XC)_-. \end{cases}$$

As C locates at a minimum, the ordered triple associated with XC is

$$[(XC)_-, XC, (XC)_+] \quad \text{with signature } (-, 0, +).$$

Similarly, near a maximum D the ordered triple is

$$[(YD)_-, YD, (YD)_+] \quad \text{with signature } (+, 0, -).$$

In this way we determine the parity of strings in the above substitutions.

The substitution rules have been written down by taking into account the following requirements:

1. The parity be preserved, e.g., $\mathcal{F}L$ has the same parity as L, etc.
2. The order be preserved, i.e.,

$$\mathcal{F}L < \mathcal{F}D < \mathcal{F}M < \mathcal{F}C < \mathcal{F}R.$$

3. The continuity in the natural order $L < D < M < C < R$ be preserved, e.g., $\mathcal{F}L$ goes continuously to $\mathcal{F}D$ and then to $\mathcal{F}M$ just as L goes to D and then to M.

Similar to the $*$-composition for unimodal maps, the $\overline{*}$ and $\underline{*}$ products are non-commutative, but associative. They preserve order, parity and transform an admissible kneading sequence to another admissible sequence. In addition, they are dual to each other with respect to the parity-preserving interchange $\mathcal{T}$ defined in (3.14) (Peng, Zhang and Cao [1998]):

$$\begin{aligned} \Sigma_1 \overline{*} \Sigma_2 &= \mathcal{T}\{\mathcal{T}\Sigma_1 \underline{*} \mathcal{T}\Sigma_2\}, \\ \Sigma_1 \underline{*} \Sigma_2 &= \mathcal{T}\{\mathcal{T}\Sigma_1 \overline{*} \mathcal{T}\Sigma_2\}. \end{aligned} \tag{3.91}$$

3.6.4 The Down-Up-Down Type Cubic Map

So far we have studied the $(+, -, +)$ type cubic map corresponding to the case $A > 1$ in (3.78). When $f(x) = -Ax^3 + (A-1)x$ with $A > 1$, the graph of the mapping function of the $(-, +, -)$ type may be obtained from Fig. 3.13 by flipping it with respect to the central horizontal line (see Fig. 3.17). The critical points C and D divide the interval into three subintervals R, M, and L as before, but now R and L become monotonically decreasing branches of the map instead of M. When determining the parity of a string, one must count the total number of letters R and L.

The admissibility condition for a sequence Σ now reads:

$$\begin{aligned} &\mathcal{R}(\Sigma) \leqslant K_C, \\ &K_D \leqslant \mathcal{M}(\Sigma) \leqslant K_C, \\ &K_D \leqslant \mathcal{L}(\Sigma). \end{aligned} \tag{3.92}$$

Now the maximal and minimal of both K_C and K_D are $(RL)^\infty$ and $(LR)^\infty$, respectively. They are at the same time the maximal and minimal among all possible sequences.

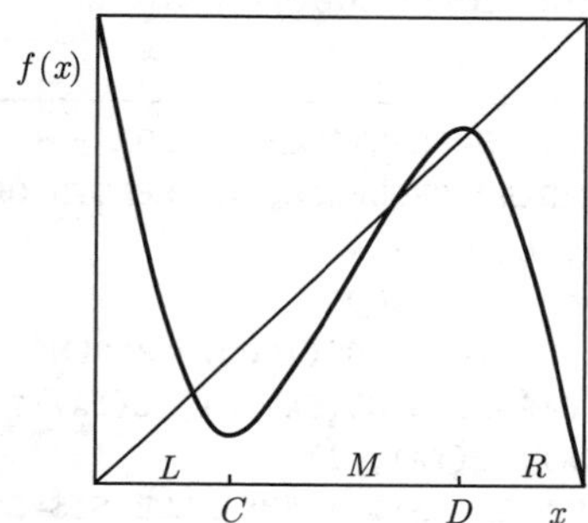

Figure 3.17 A general cubic map of $(-,+,-)$ type

The procedure of generating the kneading plane follows from that for the $(+,-,+)$ type cubic map with minor modification. We list here the BASIC Program 3.4 which generates all K_C up to length 5. Please compare it with Program 3.3 in the last Section. The kneading plane of a $(-,+,-)$ type cubic map up to period 5 is given in Fig. 3.18[2].

We take the opportunity to make two remarks related to one dimensional maps with multiple kneading sequences. First, no matter how many independent kneading sequences there exist, all these sequences may be arranged according to the same ordering rule. In Fig. 3.18 the two sets of kneading sequences, K_C and K_D, are shown ordered along the slanting dashed line in the lower right part of the figure. In principle, kneading planes of other maps with two parameters may all be drawn in this way. Second, we have super-imposed an shaded arrow on the figure in order to say a few words on so-called "anti-monotonicity" in the bifurcation structure of higher-dimensional systems.

Anti-monotonicity in bifurcation diagrams

If an interval map with multiple critical points is parameterized by its kneading sequences and only one parameter is varied, any periodic orbit born at a period-doubling or tangent bifurcation will persist to exist, and only its stabi-

[2] This kneading plane was given as Fig. 12 in MacKay and Tresser [1987] without symbolic names.

lity may change. There is, so to speak, a monotonicity in the overall bifurcation structure: periodic orbits come into life, undergo period-doubling and develop into chaotic bands associated with that period, then these bands merge and, possibly, immerse into higher level chaotic bands.

```
2    REM PROGRAM 3.4 SUPERSTABLE Kc FOR - + - CUBIC MAP
4    REM S$:  KNEADING SEQUENCE, L: LENGTH OF S$
10   DIM S$(1000), L(1000)
20   NMAX=5:  IA=1:  IR=1000
30   S$(1)="LR":  L(1)=2:  S$(1000)="RLRLM":  L(1000)=5
40   S2$=S$(IB): S1$=MID$(S$(IA), 1, L(IA)-1)
45   C$=MID$(S$(IA), L(IA), 1)
50   IF C$="R" THEN S1$=S1$ + "M" ELSE S1$=S1$ + "R"
60   IF C$="D" THEN S1$=S$(IA) + "LRLR"
70   IF ID$(S2$, L(IB), 1) = "D" THEN S2$=S2$ + "LRLR"
80   FOR M=1 TO 500:  GOSUB 1000
90   IF ID=2 THEN GOTO 130
100  IB=IB-1:  L(IB)=K+1:  S$(IB)=SS$:  S2$=SS$
110  IF MID$(S2$, K+1, 1) = "D" THEN S2$=S2$ + "LRLR"
120  NEXT M
130  IA=IA+1:  S$(IA)=S$(IB): L(IA)=L(IB): IB=IB+1
140  IF IB < 1001 THEN GOTO 40
150  K=1:  FOR I=1 TO IA
160  IF MID$(S$(I), L(I), 1) = "D" THEN GOTO 150
170  S$(I)=MID$(S$(I), 1, L(I)-1) + "C":  K=K+1
180  PRINT K TAB(8) S$(I) TAB(18) L(I)
190  NEXT I: STOP
1000 L1=LEN(S1$):  L2=LEN(S2$)
1010 SS$=" ":  FOR K=0 TO NMAX-1
1020 K1=(K MOD L1) + 1:  K2=(K MOD L2) + 1
1030 C1$=MID$(S1$, K1, 1):  C2$=MID$(S2$, K2, 1)
1040 IF C1$="D" AND C2$<>"R" THEN C1$=C2$
1050 IF C2$="D" AND C1$<>"R" THEN C2$=C1$
1060 SS$=SS+C1
1070 IF C1$<>C2$ THEN GOTO 1090
1080 NEXT K: ID=2:  RETURN
1090 IF C1$="R" THEN ID=1:  RETURN
1100 IF C2$="R" THEN SS$=MID$(SS$, 1, K) + "M":  ID=1:  RETURN
1110 SS$= MID$(SS$, 1, K) + "D"
1120 ID=0:  RETURN: END
```

Program 3.4

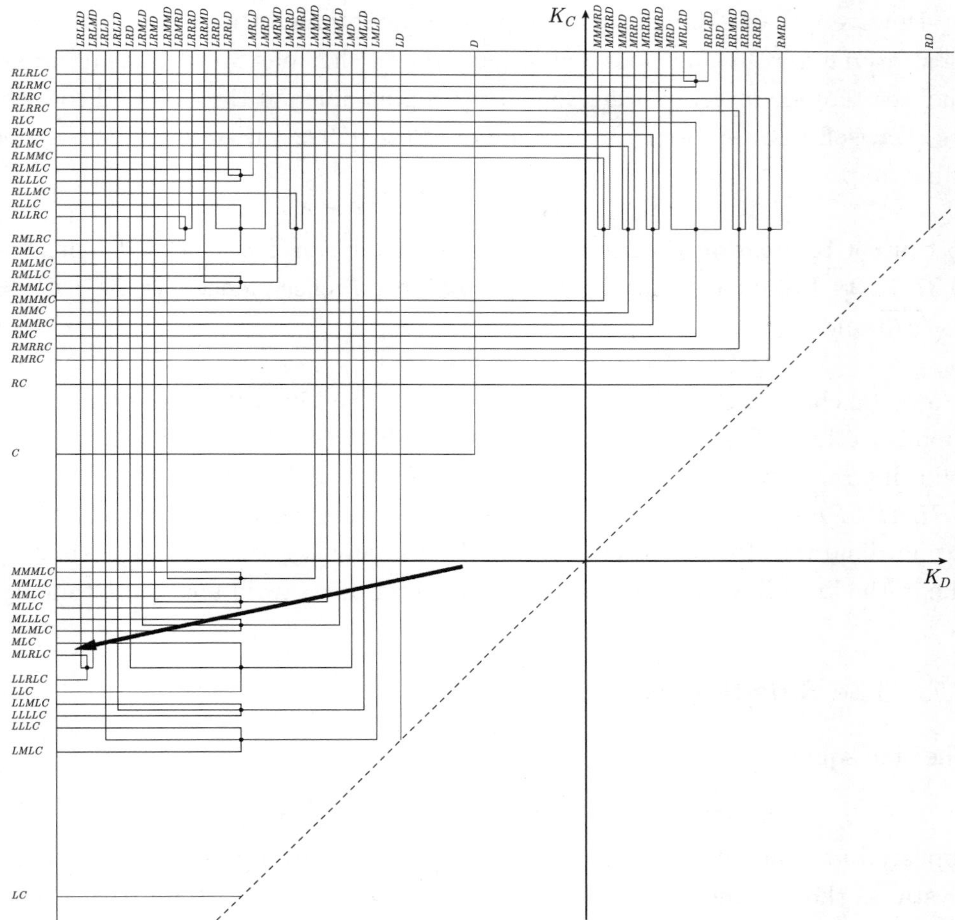

Figure 3.18 The kneading plane of a general $(-, +, -)$ type cubic map containing all superstable kneading sequences of length 5 and less. Joints are marked by heavy dots. For the super-imposed shaded arrow see text

However, if a map is not parameterized by its kneading sequences, deviations from the above-mentioned monotonicity may happen. In particular, periodic orbits and their associated bifurcation structure may be created and then annihilated in a reversed order. In higher-dimensional systems, this may happen at infinitely many parameter values and was called *anti-monotonicity* by Yorke (Kan and Yorke [1990]). An illustration of this phenomenon, among many others, was given in the periodically forced Brusselator in 1982 (Hao and Zhang [1982c]).

However, in one-dimensional maps anti-monotonicity may occur in an almost trivial way if one does not parameterize the map by its kneading sequences. For example, in order to show this phenomenon Dawson and coworkers (Dawson and Grebogi [1991]; Dawson *et al.* [1992]) used a $(-,+,-)$ type cubic map

$$x_{n+1} = f(x) = -x^3 + ax - \lambda$$

to draw a bifurcation diagram in the parameter range $a = 1.2675$ and $\lambda \in [0.37, 1.10]$. Parameterizing this map by its kneading sequences, we have $C_\pm = \pm\sqrt{a/3}$ and

$$K_\pm = f(C_\pm) = \pm 2(a/3)^{3/2} - \lambda.$$

It may be checked that the parameter range of the bifurcation diagram corresponds to the dashed arrow super-imposed on Fig. 3.18, i.e., approximately the following range of kneading pairs: from $(M^3L^3\cdots, L^\infty)$ to $(MLRLLLRLLR\cdots, LRLRLLLR\cdots)$. In this way one can name all the periods seen in the bifurcation diagram. In fact, there are infinitely many ways to choose a slanting line in the kneading plane Fig. 3.18 to encounter or to avoid anti-monotonicity.

3.7 The Sine-Square Map

The sine-square map

$$y_{n+1} = A\sin^2(y_n - B), \quad |y_n - B| \leqslant \pi, \tag{3.93}$$

appeared in a model describing a hybrid optical bistability device using liquid crystal as the nonlinear medium in the cavity (Zhang *et al.* [1983, 1984, 1986, 1988]). There are two parameters in (3.93): A proportional to the intensity of input light, and B depending on the DC bias voltage applied to the cavity. The five critical points, C_i, $i = 1, 2$, and D_i, $i = 1, 2, 3$, divide the interval $(-1, 1)$ into four segments, see Fig. 3.19. On each of these segments the mapping function behaves monotonically and we label the monotone branches by the symbols R, N, M, and L.

The sine-square map was first studied using the above symbolic description without constructing the symbolic dynamics. We distinguish using symbolic description from constructing symbolic dynamics. Labeling the monotone branches and critical points of a map by letters allows one to use the word-lifting technique (see below) to draw the loci of superstable orbits, representing various regimes in the parameter plane. However, without knowing the ordering rule and admissibility conditions, as well as other rules dictated by symbolic dynamics, one can never be confident, for example, that periodic regimes up to some period have been exhaustively listed.

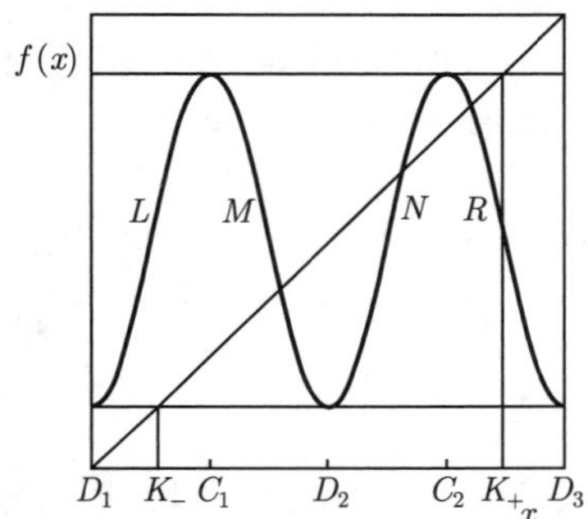

Figure 3.19 The sine-square map

Complete symbolic dynamics of the sine-square map (3.93) was established recently (Xie and Hao [1993]), based on the understanding that the best way to parameterize a map with multiple critical points is to use its kneading sequences as parameters. To this end we introduce a new variable

$$x = \frac{y - B}{\pi} \tag{3.94}$$

to rewrite Eq. (3.93) as

$$x_{n+1} = f(x_n) \equiv (K_+ - K_-)\sin^2(\pi x_n) + K_-, \quad |x_n| \leqslant 1, \tag{3.95}$$

where the two new parameters K_+ and K_- are simply related to the old ones:

$$K_+ = \frac{A - B}{\pi}, \quad K_- = -\frac{B}{\pi}. \tag{3.96}$$

As we will see soon, K_+ and K_- are nothing but the kneading sequences of the map. We will only consider the case $K_+ > K_-$; the opposite case may be treated by a trivial transformation.

3.7.1 Symbolic Sequences and Word-Lifting Technique

As usual, any numerical orbit $x_0\, x_1\, x_2 \cdots x_n\, \cdots$ corresponds to a symbolic orbit, which may be named by the initial point x_0 or by a special notation, say, Σ:

$$x_0 = \Sigma = \sigma_0\sigma_1\sigma_2 \cdots \sigma_n \cdots , \tag{3.97}$$

where σ_i is one of the symbols R, N, M, L, C_1, C_2, D_1, D_2, or D_3.

The symbolic sequence (3.97), truncated at x_n, implies a functional composition

$$x_0 = \sigma_0 \circ \sigma_1 \circ \sigma_2 \circ \cdots \circ \sigma_{n-1}(x_n). \tag{3.98}$$

Now x_0 and x_n are to be understood as numbers, and $\sigma_i(y)$ — the inverse function of the monotone branch labeled by the symbol σ_i. In particular, we have

$$\begin{aligned} R(y) &= 1 - N(y), \\ N(y) &= \frac{1}{2\pi} \arccos \left[1 - \frac{2(y - K_-)}{K_+ - K_-} \right], \\ M(y) &= -N(y), \\ L(y) &= -R(y). \end{aligned} \tag{3.99}$$

To write down these formulae, one must take into account the definition of the proper branches of the inverse trigonometry functions.

Among all the symbolic sequences, superstable periodic sequences and eventually periodic sequences play a special role. A superstable periodic orbit contains at least one of the critical points and represented by, for example, a symbolic sequence $(\Sigma C)^\infty$, where Σ is a string of symbols that do not contain any critical points and C is one of the critical symbols C_i or D_i. Writing this periodic sequence as $C = C\Sigma C \cdots$, applying the map f to the both sides and taking into account (3.98), we get

$$f(C) = \Sigma(C). \tag{3.100}$$

Here Cs on the both sides are a number and Σ must be understood as a composite function. For a given map, everything in Eq. (3.100) is known except the parameter value at which the equality takes place. Therefore, (3.100) is an equation to determine the parameter of the superstable orbit. When there is only one parameter, it yields a number; when there are more parameters, it defines a curve or surface in the parameter space.

Of course, a superstable periodic orbit may contain more than one critical point. For instance, we may encounter $(\Sigma C \Pi D)^\infty$, where C and D are critical points and Σ, Π are strings without critical symbols. In this case, word-lifting leads to a pair of equations

$$\begin{aligned} f(D) &= \Sigma(C), \\ f(C) &= \Pi(D). \end{aligned} \tag{3.101}$$

When there are only two independent parameters, as is the case with our sine-square map, Eq. (3.101) determines an isolated point in the parameter plane.

It corresponds to a doubly superstable periodic orbit and is called a *joint* in the kneading plane. Doubly superstable orbits and joints play an essential role in understanding the parameter space, see Section 3.7.4 below.

Just as in the unimodal map, word-lifting technique also applies to eventually periodic sequences. Suppose that we start from C and get a sequence $\rho\lambda^\infty$, where the strings ρ and λ do not contain critical symbols, then the word $\rho\lambda^\infty$ may be lifted into a pair of equations

$$\begin{aligned} f(C) &= \rho(z), \\ z &= \lambda(z). \end{aligned}$$

We repeat that eventually periodic sequences of type $\rho\lambda^\infty$ also determine the so-called border to chaos. However, we will not elaborate this aspect in connection with the sine-square map.

3.7.2 Ordering Rule and Admissibility Conditions

The general ordering rule (3.1) works for the sine-square map as well. The natural order of symbols is

$$D_1 < L < C_1 < M < D_2 < N < C_2 < R < D_3. \tag{3.102}$$

The parity of a symbolic string is now determined by the total number of symbols M and R.

The sine-square map (3.93) has five critical points, but only two kneading sequences, namely,

$$\begin{aligned} K_+ &= f(C_1) = f(C_2), \\ K_- &= f(D_1) = f(D_2) = f(D_3). \end{aligned} \tag{3.103}$$

The two numbers K_+ and K_- determine a subinterval $[K_-, K_+]$ on the interval of the mapping, see Fig. 3.19. Iterations starting from points within this subinterval will remain in it all the time; those starting from any initial point but outside this subinterval will enter it in finite steps and never get out again. This subinterval is a *dynamically invariant range*. Since we are concerned with long-time behavior of the system, only this invariant range is relevant; dynamics outside it correspond to trivial transients. From now on we will confine ourselves to the dynamically invariant range.

Obviously, an arbitrary string made of the symbols R, N, M, L, C_i, and D_i, may not correspond to a numerical orbit in the sine-square map, no matter how one adjusts the parameters and the initial point. In order to be reproducible by the dynamics, a symbolic sequence Σ must satisfy the admissibility conditions, based on the ordering rule. By inspecting Fig. 3.19, we write down the admissibility conditions for sequence Σ as

$$K_- \leqslant \mathcal{S}^k(\Sigma) \leqslant K_+, \quad k = 0, 1, 2, \cdots, \tag{3.104}$$

where $\mathcal{S}$ is the shift operator introduced in Section 1.2.2. This condition works when the parameters of the map are fixed, i.e., when the kneading sequences are given. The kneading sequences themselves must satisfy the admissibility condition as well. In particular, this means that the sequence K_+ must be shift-maximal:

$$\mathcal{S}^k(K_+) \leqslant K_+, \tag{3.105}$$

while K_- must be shift-minimal:

$$K_- \leqslant \mathcal{S}^k(K_-). \tag{3.106}$$

Any symbolic sequence satisfying (3.105) or (3.106) may be taken as a K_+ or K_-. However, two sequences K_+ and K_- must both satisfy the admissibility condition (3.104) in order to become a *compatible kneading pair* (K_+, K_-) and then correspond to an actual map.

3.7.3 Generation of Kneading Sequences

In order to construct the kneading plane spanned by kneading sequences, the first step consists in generating K_+ and K_- satisfying (3.105) and (3.106) separately. The procedure is similar to what we have done with the general cubic map in Section 3.6.2 when the median words between two given kneading sequences are generated.

We explain the procedure on the example of K_+. From the admissibility condition (3.105) and from Fig. 3.19 it follows that, except $K_+ = D_3$ and $K_- = D_1$, there are no kneading sequences ending with D_3 or D_1. Therefore, we may consider only the critical symbols C_1, C_2, and D_2. Then generation of median words goes as follows (X, Y, and W will denote strings made of non-critical symbols):

A. If a $K_+ = XC_1$ satisfies (3.105), then C_1 may be replaced by either of its neighbors L or M to get the upper and lower sequences $(XC_1)_+^\infty$ and $(XC_1)_-^\infty$, where

$$(XC_1)_+ = \max(XL, XM),$$
$$(XC_1)_- = \min(XL, XM).$$

It is easy to see that $(XC_1)_+$ is odd in the number of R and M.

B. If a $K_+ = YC_2$ satisfies (3.105), then C_2 may be replaced by either of its neighbors R or N to get the upper and lower sequences $(YC_2)_+^\infty$ and $(YC_2)_-^\infty$, where

$$(YC_2)_+ = \max(YR, YN),$$
$$(YC_2)_- = \min(YR, YN).$$

The upper string $(YC_2)_+$ is odd in the number of R and M.

C. If K_+ happens to be WD_2, then from the ordering rule alone, without taking into account the shape of a particular map, it follows that the nearest sequences above and below WD_2 are $(WD_2)_+L^\infty$ and $(WD_2)_-L^\infty$, where

$$(WD_2)_+ = \max(WM, WN),$$
$$(WD_2)_- = \min(WM, WN).$$

Here $(WD_2)_+$ is even in the number of R and M. The sequences $(WD_2)_+L^\infty$ and $(WD_2)_-L^\infty$ are taken to be the upper and lower sequences of WD_2. These infinite sequences are only used to generate short kneading sequences.

Inspection of Fig. 3.19 provides another way of understanding this construction. When the lower critical points D_i touch the bottom, the next step of any iteration that reaches D_2 is falling in the fixed point L^∞.

D. Given two K_+-type kneading sequences, we denote the lower sequence of the greater one by $\Sigma\mu\cdots$ and the upper sequence of the smaller one by $\Sigma\nu\cdots$, where Σ is their common leading string and $\mu \neq \nu$. Suppose $\sigma \in \{C_1, D_2, C_2\}$ is included between μ and ν, then $\Sigma\sigma$ is a median word. Put in other words, a median word starts with the common leading string Σ and ends with any critical symbol that is included between μ and ν. If there are more than one such critical symbols, they are used in the order from μ to ν to get median words ordered in the same way. For example, when $\mu = R$ and $\nu = L$, the ordered median words are ΣC_2, ΣD_2, and ΣC_1. In the opposite case of $\mu = L$ and $\nu = R$, these are ΣC_1, ΣD_2, and ΣC_2.

Now, starting from the maximal $K_+ = D_3$ and minimal $K_+ = L^\infty$ and following the above procedure, we can generate more and more median words.

For K_--type kneading sequences the procedure is similar. Attention must be paid to the upper and lower sequences of XC_1 and YC_2. For instance, the upper and lower sequences of XC_1 are $(XC_1)_+RL^\infty$ and $(XC_1)_-RL^\infty$, respectively.

3.7.4 Joints and Bones in the Kneading Plane

As said before, any symbolic sequence which is shift-maximal or shift-minimal may be taken as a K_+ or K_-. However, two kneading sequences K_+ and K_- should both satisfy the admissibility conditions (3.104) in order to become a compatible kneading pair corresponding to an actual map. A compatible kneading pair is specially denoted as (K_+, K_-). For example, (RLC_1, LLD_2) is a compatible pair, but $K_+ = RLC_1$ and $K_- = MRD_3$ are not compatible.

An important type of compatible pair is made of one and the same doubly superstable kneading sequence. It is enough to write the half-words containing only one critical symbol for each of $K_\pm$:

$$(K_+, K_-) = (YD_iXC_j, XC_jYD_i) = [YD_i, XC_j], \tag{3.107}$$

since by definition D_i is followed by XC_j and C_j — by YD_i. We call it a *joint* and designate it by a pair of square parentheses. Joints correspond to isolated points in the parameter plane, which may be determined from the word-lifted Eq. (3.101).

Using the admissibility conditions of a joint $[YD_i, XC_j]$, one can derive other compatible pairs, e.g., $[(YD)_+XC, (XC)_-YD]$. (The subscripts of C and D have been omitted.) The situation near a joint $[YD, XC]$ on the kneading plane is shown in Fig. 3.20. The joint itself, marked by a heavy dot, is located at the intersection of two thin lines, representing YD and XC. The above-mentioned compatible pair is marked by a circle. There are four

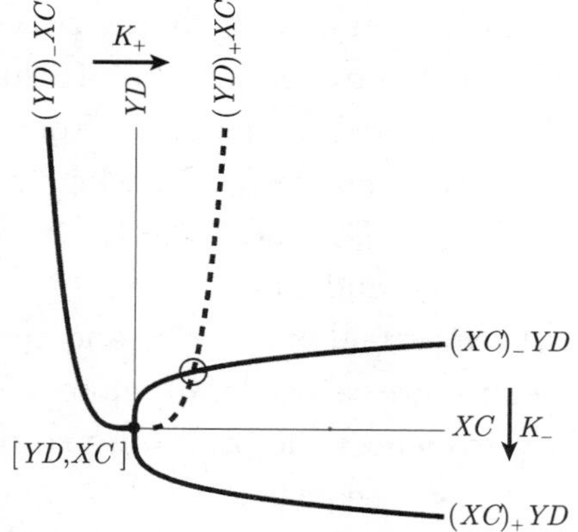

Fig. 3.20 A joint and the superstable kneading sequences ("bones") associated with it

thick lines ("bones") coming out from the joint, they are the loci of the singly superstable kneading sequences. The meaning of the dashed curve will be explained below. Two thick arrows indicate the direction of increase in the ordering of the kneading sequences.

What has just been said applies to joints in other two-parameter maps. For the sine-square map D_3 is the maximal point of the mapping interval. Therefore, there does not exist any $(YD_3)_+$. In this case one may only get three bones from a joint $[D_3, XC]$. The bone shown by a dashed curve in Fig. 3.21 disappears.

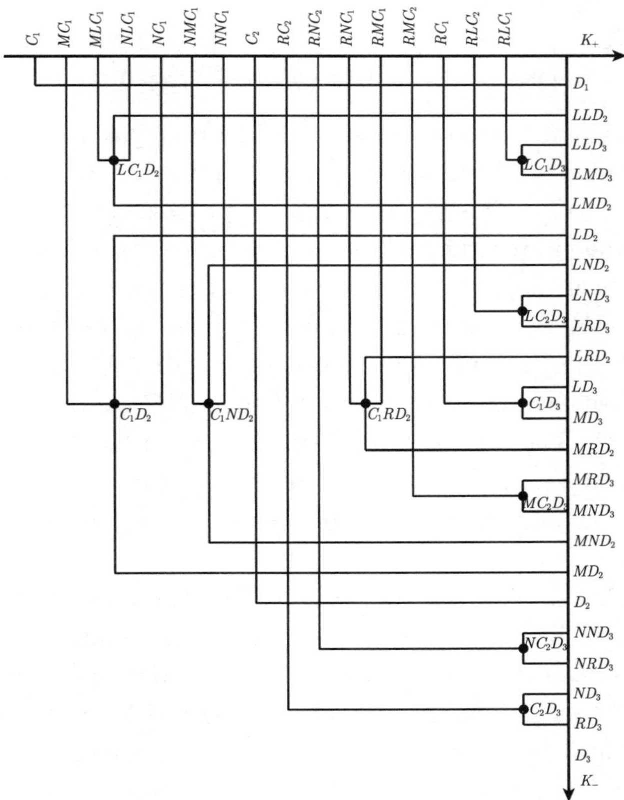

Figure 3.21 Superstable kneading sequences up to period 3. Joints are shown by heavy dots

The collection of joints and bones makes a skeleton in the kneading plane. The skeleton containing all joints and bones up to period 3 is given in Fig. 3.21.

There are ten joints in this figure, shown by solid dots. This figure should be understood schematically: only the relative location of the doubly and singly superstable kneading sequences matter. For any given map of the sine-square type, these curves should be calculated using the word-lifting technique. For example, for the map (3.93) the actual kneading plane is drawn in Fig. 3.24 (a). It is, so to speak, topologically equivalent to Fig. 3.21.

From a given point $[YD, XC]$, one may derive many other joints, for example,

$$[(YD)_+(XC)_+YD, XC], \quad [YD, (XC)_-(YD)_-XC]$$

and

$$[(YD)_+(XC)_+YD, (XC)_-(YD)_-XC].$$

The proof that these are joints goes much like that given in Section 3.6.1, so we skip it. By the way, the first two derived joints describe the period-doubled regimes of $[YD, XC]$. This period-doubling process continues to generate a partially overlapped tree structure. This is similar to other maps with two kneading sequences, e.g., the general cubic map studied in Section 3.6. Only when D_3 is encountered, may some detail disappear, as we have just explained. We call the collection of joints up to a certain length a *joint scheme*. The joint scheme for the sine-square map up to YD and XC containing 4 symbols is given in Fig. 3.22. The ten joints that have appeared in Fig. 3.21 are marked by heavy dots. The joint scheme is a tool to construct joints and to visualize their positions. It embeds in the skeleton graph, but is usually not shown there.

A joint $[YD, XC]$ and a few of its derived joints are shown in Fig. 3.23. In fact, we can tell a few such structures in the joints scheme shown in Fig. 3.22. For example, there is $[D_2, C_1]$ and its derived joints $[D_2, LMC_1]$ and $[NMD_2, C_1]$, as well as $[D_3, C_2]$ and its derived joint $[D_3, NRC_2]$. From every joint in Fig. 3.24 one may derive many longer joints, which will lead to more and more compatible pairs of kneading sequences. They yield a self-similar structure in the kneading plane at finer and finer levels.

Since the complete symbolic dynamics has been constructed, we are confident in having found all the periods. In contrast, in Fig. 3.1 on page 109 of the monograph (Hao [B1989]), seven period 3 superstable kneading sequences were missing.

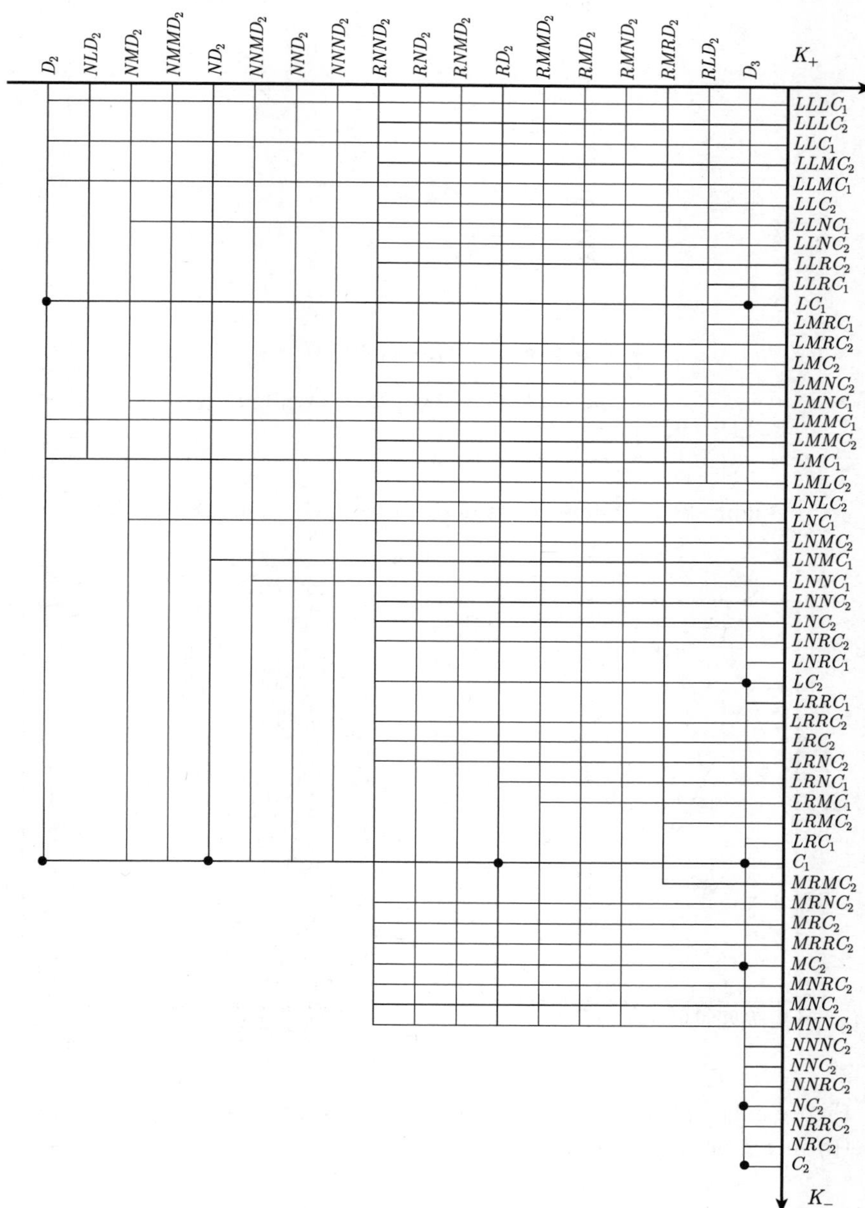

Figure 3.22 Joints scheme of the sine-square map. The 10 joints that have appeared in Fig. 3.21 are marked by solid dots

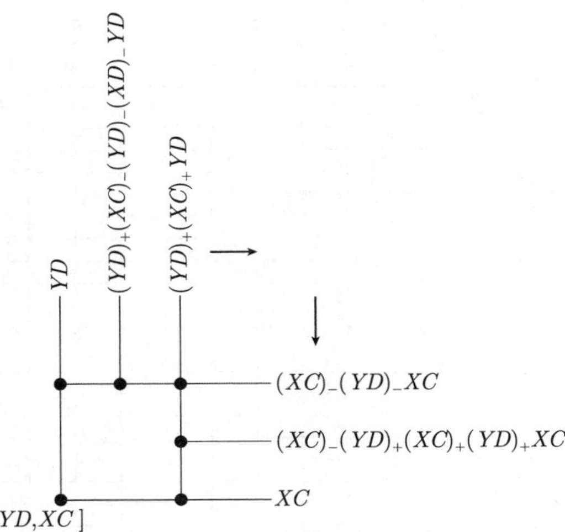

Figure 3.23 Some joints derived from the joint $[YD, XC]$

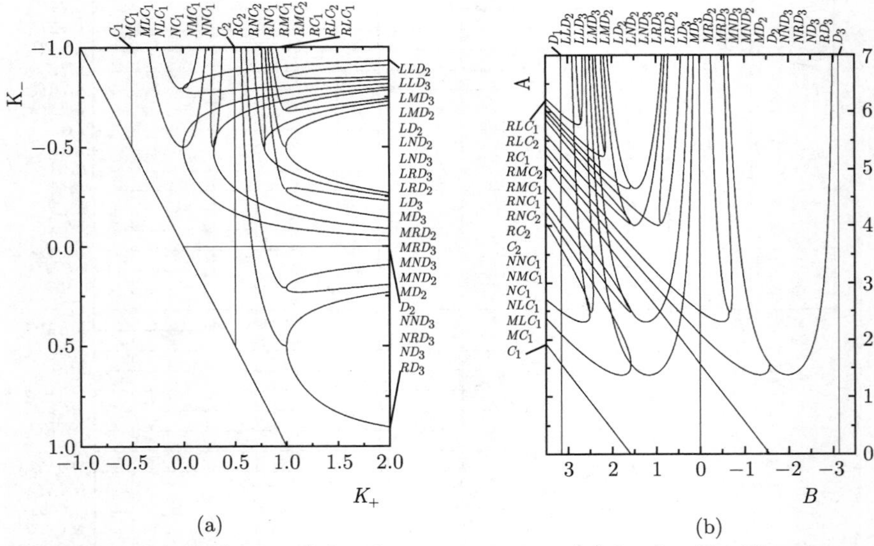

Figure 3.24 Skeleton of the sine-square map. (a) in the K_+-K_- plane. (b) in the original A-B parameter plane

3.7.5 Skeleton of Superstable Orbits and Existence of Topological Chaos

A simple use of Fig. 3.23 or 3.24 is to locate the chaotic region in the parameter plane. To this end one may take the loci of period 3 as indicators since period

3 windows are always embedded in the chaotic regime. To be more precise, one may prove the existence of topological chaos from the joint scheme.

A system exhibits topological chaos if it has positive topological entropy. Perhaps, among various mathematical definitions of chaos, this is the simplest one, although it does not guarantee physically observable chaotic behavior. For our purpose we may define topological entropy as (Crutchfield and Packard [1982])

$$h = \lim_{n\to\infty} \frac{\log N(n)}{n}, \tag{3.108}$$

where $N(n)$ is the total number of periodic points of period n. Now in the spirit of "two crossings imply chaos" (see Section 3.4.4), one can prove that the joints $[YD, XC]$ and

$$[(YD)_+(XC)_+YD, (XC)_-(YD)_-XC]$$

imply topological chaos. The proof goes easily by checking that any symbolic sequence made of only the basic segments $(YD)_+(XC)_-$ and $(YD)_-(XC)_+$ are admissible to the second joint. Assume that the number of periodic points of period n is $N(n)$. Since longer sequences may be obtained from shorter ones by adding either one of the two possible segments, we have

$$N(n) \geqslant 2N(n-\alpha), \tag{3.109}$$

where α is the length of the basic segments. From (3.109) it follows that

$$N(n) \geqslant 2^{n/\alpha}$$

for n big enough. Hence we have an estimate for the topological entropy

$$h \geqslant \frac{1}{\alpha}\log 2 > 0. \tag{3.110}$$

This kind of analysis may be refined to determine the boundary of topological chaos, i.e., the boundaries in the kneading plane where topological entropy begins to acquire a positive value.

To conclude the discussion, we summarize the lessen to be learned from the study of the sine-square map. First, it is insufficient to use symbolic description only. One must construct the entire symbolic dynamics, including the ordering rule, the admissibility condition, the kneading plane, etc. Second, the best way to parameterize a map with multiple critical points is to use its kneading sequences as parameters. An attempt to construct the symbolic dynamics using directly the original map (3.93) would bring about unnecessary complications.

3.8 The Lorenz-Sparrow Maps

In this section we develop symbolic dynamics for a family of maps which are defined only by their shape, but not by any analytical expression. These maps appear in the numerical study of the Lorenz model after some manipulation of the first return maps (Zheng and Liu [1997]). Sparrow anticipated that maps of this type would be "an obvious choice" to explain the dynamical behavior of the Lorenz equations in a wide parameter range (Sparrow [B1982], Appendix J). Hence the suggested name Lorenz-Sparrow maps. Sparrow noted at the same time that a complete kneading sequence theory for these maps "would be quite complicated and would require study of an additional symbolic sequence describing the behavior of successive iterates of one of the turning points of the map $\cdots$". Furnished with our experience on various maps with multiple critical points now we are in a position to accomplish the job and it turns out to be not so formidable.

3.8.1 Ordering and Admissibility of Symbolic Sequences

The shape of a typical Lorenz-Sparrow map is shown in Fig. 3.25. The map has four monotone branches defined on four subintervals labeled by the letters M, L, N, and R, respectively. We will also use these same letters to denote the monotone branches themselves, although we do not have an explicit expression for the mapping function $f(x)$. Among these four branches, R and L are increasing; we say R and L have an even or + *parity.* The decreasing branches M and N have odd or $-$ *parity.* Between the monotone branches there are two "turning points" ("critical points") D and C as well as a "breaking point" B, where discontinuity is present. Any numerical trajectory $x_1x_2\cdots x_i\cdots$ in this map corresponds to a symbolic sequence

$$\Sigma = \sigma_1\sigma_2\cdots\sigma_i\cdots ,$$

where $\sigma_i \in \{M, L, N, R, C, D, B\}$, depending on where the point x_i falls in.

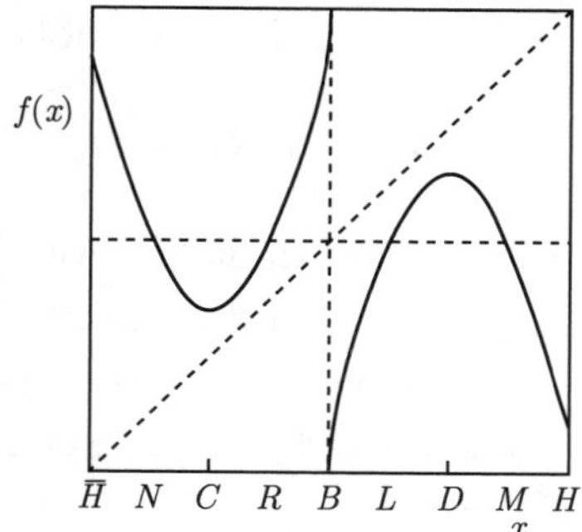

Figure 3.25 A typical Lorenz-Sparrow map

All symbolic sequences made of these letters may be ordered in the following way. First, there is the natural order on the interval:

$$N < C < R < B < L < D < M. \tag{3.111}$$

Next, suppose two symbolic sequences Σ_1 and Σ_2 have a common leading string Σ^*, i.e.,

$$\Sigma_1 = \Sigma^* \sigma \cdots, \quad \Sigma_2 = \Sigma^* \tau \cdots,$$

where $\sigma \neq \tau$. Since σ and τ are different, they must have been ordered according to (3.111). The *ordering rule* is: if Σ^* is even, i.e., it contains an even number of N and M, the order of Σ_1 and Σ_2 is given by that of σ and τ; if Σ^* is odd, the order is the opposite to that of σ and τ. The ordering rule may be put in the following form:

$$\begin{aligned} EN \cdots &< EC \cdots < ER \cdots < EB \cdots \\ &< EL \cdots < ED \cdots < EM \cdots, \\ ON \cdots &> OC \cdots > OR \cdots > OB \cdots \\ &> OL \cdots > OD \cdots > OM \cdots, \end{aligned} \tag{3.112}$$

where E (O) represents a finite string of M, L, N, and R containing an *even* (*odd*) number of letters M and N. We call strings E and O being even and odd, respectively.

In view of its application to the Lorenz equations we are specially interested in a particular class of Lorenz-Sparrow maps that are anti-symmetric with respect to the coordinate inversion $x \to -x$. To incorporate this discrete symmetry, we define a transformation $\mathcal{T}$ of symbols:

$$\mathcal{T}: M \leftrightarrow N,\ L \leftrightarrow R,\ C \leftrightarrow D,\ B \leftrightarrow B. \tag{3.113}$$

Sometimes we distinguish the left and right limit of B, then we add $B_- \leftrightarrow B_+$. We often denote $\mathcal{T}\Sigma$ by $\overline{\Sigma}$ and say Σ and $\overline{\Sigma}$ are mirror images of each other. Symmetry breaking and restoration in the anti-symmetric Lorenz-Sparrow map may be analyzed in a way similar to the antisymmetric cubic map (Section 3.3).

As usual we define kneading sequences as the sequences that start from the next iterate of the turning or breaking points. Insisting on our convention to name a symbolic sequence by the initial point which corresponds to its first symbol, we have two kneading sequences from the turning points:

$$K = f(C), \quad \overline{K} = f(D).$$

With $\overline{K}$ being the mirror image of K, sequence K is taken as the independent one. From the left limit of the point B of discontinuity, we have the second independent kneading sequence $H = f(B_-)$. We may write

$$\begin{aligned} C_- = NK, C+ = RK, B_- = RH, \\ B_+ = L\overline{H}, D_- = L\overline{K}, D_+ = M\overline{K}. \end{aligned} \tag{3.114}$$

As a 1D map with multiple critical points is best parameterized by its kneading sequences, the dynamical behavior of the Lorenz-Sparrow map is entirely determined by a *kneading pair* (K, H). A main task of symbolic dynamics is to determine all symbolic sequences allowed by a given map. In order to formulate the admissibility conditions we need the notion of shift-sets as first introduced in Section 2.3.4. We repeat the definition here. Take a symbolic sequence Σ and inspect its symbols one by one. Whenever a letter M is encountered, we collect the subsequent sequence that follows this M. The set of all such subsequences is denoted by $\mathcal{M}(\Sigma)$ and is called the M-shift set of Σ. Similarly, we define $\mathcal{L}(\Sigma)$, $\mathcal{R}(\Sigma)$ and $\mathcal{N}(\Sigma)$.

Based on the ordering rule (3.112), from (3.114) the *admissibility conditions* follow:

$$\begin{aligned} \overline{H} \leqslant N\mathcal{N}(\Sigma) \leqslant NK, \quad K \leqslant \mathcal{R}(\Sigma) \leqslant H, \\ \overline{H} \leqslant \mathcal{L}(\Sigma) \leqslant \overline{K}, \quad M\overline{K} \leqslant M\mathcal{M}(\Sigma) \leqslant H. \end{aligned} \tag{3.115}$$

Here in the two middle relations we have canceled the leading R or L.

The twofold meaning of the admissibility conditions should be emphasized. On one hand, for a given kneading pair these conditions select those symbolic sequences which may be produced by the dynamics. On the other hand, a kneading pair (K, H), being symbolic sequences themselves, must also satisfy conditions (3.115) with Σ regarded as K and H. Such (K, H) consistent with (3.115) is called a *compatible* kneading pair. The first meaning concerns admissible sequences in the phase space at a fixed parameter set while the second deals with compatible kneading pairs in the parameter space. In accordance with these two aspects there are two pieces of work to be done. First, generate all compatible kneading pairs up to a given length. Second, generate all admissible symbolic sequences up to a certain length for a given kneading pair (K, H).

3.8.2 Generation of Compatible Kneading Pairs

Two kneading sequences K and H must satisfy the admissibility conditions (3.115) in order to become a compatible kneading pair (K, H). This means, in particular, two sequences with $H < K$ cannot make a compatible pair. Moreover, from the admissibility conditions one can deduce that the minimal H compatible with a given K is the following $H_{\min}$:

$$H_{\min} = \max\{\mathcal{R}(K), \mathcal{R}(\overline{K}), \mathcal{N}(K), \mathcal{N}(\overline{K});\ K\},$$

where $\mathcal{R}(K)$ is the R-shift set of K, etc.

According to the ordering rule (3.112) the greatest sequence is $(MN)^\infty$, and the smallest $(NM)^\infty$. Sequence $H = (MN)^\infty$ will be compatible with any K. The admissibility conditions also require that K must be shift-minimal with respect to R and N. Both $(MN)^\infty$ and $(NM)^\infty$ meet this requirement. Taking the extreme sequences $K_1 = (MN)^\infty$, $K_2 = (NM)^\infty$ and $H = (MN)^\infty$, one can generate all compatible kneading pairs up to a certain length by making use of the following propositions. (In what follows $\Sigma = s_0 s_1 \cdots s_n$ denotes a blank or a finite string of M, L, R, and N, and $\mu, \nu \in \{M, L, R, N\}$, $\mu \neq \nu$.)

1. If both $K_a = \Sigma\mu\cdots$ and $K_b = \Sigma\nu\cdots$ are compatible with a given H, then $K = \Sigma\tau$ is also compatible with H, where $\tau \in \{C, B, D\}$ is included between μ and ν, i.e., either $\nu < \tau < \mu$ or $\nu > \tau > \mu$ holds.

2. For $\tau = C$, under the conditions of 1, $K = (\Sigma t)^\infty$ is also compatible with H, where t stands for either R or N.

3. For $\tau = D$, under the conditions of 1, $K = (\Sigma t \overline{\Sigma t})^\infty$ is compatible with H, where t stands for either M or L.

4. For $\tau = B$, under the conditions of 1, $K_- = \Sigma RH$ and $K_+ = \Sigma L\overline{H}$ are both compatible with H.

Without going into the proofs we continue with the construction. Assume that $H = (MN)^\infty$. By means of the above propositions we have the median words $K = D, B, C$ between $K_1 = (MN)^\infty$ and $K_2 = (NM)^\infty$. At this step we have the following sequences listed in ascending order:

$$(NM)^\infty \quad N^\infty, C, R^\infty \quad R(MN)^\infty, B, L(NM)^\infty \\ (LR)^\infty, DC, (MN)^\infty \quad (MN)^\infty.$$

Inside any group centered at C, B, or D there exists no median sequence. Furthermore, no median sequence exists between the group D and $(MN)^\infty$. Taking any two nearby different sequences between the groups, the procedure may be continued. For example, between R^∞ and $R(MN)^\infty$ we have

$$R^\infty \quad RR(MN)^\infty, RB, RL(NM)^\infty \\ (RLLR)^\infty, RDLC, (RMLN)^\infty \quad R(MN)^\infty.$$

This process is repeated to produce all possible K up to a certain length. For each K one determines a $H_{\min}$. In this way we can construct the entire kneading plane for the Lorenz-Sparrow map. In Appendix A, a program PERIOD.C to generate all admissible sequences up to a given length for a given map is explained on the example of the Lorenz-Sparrow map. The reader may consult the program for a deeper understanding. However, the program is written for quite general 1D maps; it does not check the compatibility of kneading pairs.

3.8.3 Generation of Admissible Sequences for a Given Kneading Pair

Given a compatible kneading pair (K, H), one can generate all admissible symbolic sequences up to a given length, e.g., 6. Usually, we are interested in having a list of symbolic names of all short periodic orbits. This can be done by brute force, i.e., first generate all 6^4 possible symbolic sequences then filter them against the admissibility conditions (3.115). In so doing one should avoid repeated counting of words. Therefore, we always write the basic string of a periodic sequence ending with N or R and in the shift-minimal form with respect to N or R. Periodic sequences ending with M or L and written in the shift-maximal form with respect to M or L may be obtained by applying the symmetry transformation $\mathcal{T}$.

However, one can formulate a few rules to generate only the admissible sequences (Zheng [1997c]). These rules are based on continuity in the phase plane and on explicit checking of the admissibility conditions (3.115). To simplify the writing we introduce some notation. Let $\Sigma_n = u_1 u_2 \cdots u_n$ be a finite string of n symbols; let symbols μ, ν, u_i, and s_i, $i = 1, 2, \cdots, n$ be all taken from the set $\{M, L, R, N\}$; and let the symbol τ denote one of $\{C, B, D\}$. Recollect, moreover, that at any step of applying the rules a C at the end of a string is to be continued as CK, a D as $D\overline{K}$, and a B as RH or $L\overline{H}$, see (3.114). We assume that all sequences leading with Σ_n to be considered here are between K and H.

We have the following propositions:

1. If both $\Sigma_n \mu \cdots$ and $\Sigma_n \nu \cdots$ are admissible, then $\Sigma_n \tau \cdots$ is admissible provided τ is include between μ and ν, i.e., either $\nu < \tau < \mu$ or $\nu > \tau > \mu$ takes place.

This proposition is rather obvious. For example, let $u_{i-1} = R$. Both $K \leqslant u_i u_{i+1} \cdots u_n \mu \cdots \leqslant H$ and $K \leqslant u_i u_{i+1} \cdots u_n \nu \cdots \leqslant H$ are satisfied. Thus, $K \leqslant u_i u_{i+1} \cdots u_n \tau \leqslant H$ is also valid since $u_i u_{i+1} \cdots u_n \tau$ is between $u_i u_{i+1} \cdots u_n \mu$ and $u_i u_{i+1} \cdots u_n \nu$.

It is natural that, under the condition of this proposition, if $UC \equiv u_0 u_1 \cdots u_n C$ is admissible, so are URK and UNK. That is, the admissibility of UC always implies that of UtK, where t stands for R and N. Assume $K = s_0 s_1 \cdots s_n \cdots$ and $u_{i-1} = R$. From the admissibility of UC we have $\alpha \equiv s_0 s_1 \cdots s_{n-i+1} \leqslant u_i \cdots u_n t \equiv \alpha'$. If $\alpha \neq \alpha'$, then UtK satisfies the condition due to K. If $\alpha = \alpha'$, then from the admissibility of UC string α must be odd. So, we have to prove $s_{n-i+2} s_{n-i+3} \cdots \geqslant K$, which is, of course, satisfied. Similarly, we can prove that UtK satisfies the condition due to K for $u_{i-1} = N$. The condition due to H can be checked in a similar way to verify that the condition is also satisfied.

2. If $\Sigma_n B$ and $\Sigma_n C$ are admissible then so is $(\Sigma_n R)^\infty$.

As we know, $\Sigma_n B$ and $\Sigma_n C$ imply admissible $\Sigma_n RH$ and $\Sigma_n RK$, respectively. Since for $0 \leqslant k < n+1$ the k-th shift of $(\Sigma_n R)^\infty$ is always between the

k-th shifts of $\Sigma_n RH$ and $\Sigma_n RK$, the admissibility of $(\Sigma_n R)^\infty$ is then guaranteed. Similarly, one can prove that if $\Sigma_n C$ and $\Sigma_n N\overline{H}$ are admissible, so is $(\Sigma_n N)^\infty$.

3. If $\Sigma_n C$ and $\Sigma_n \mu\cdots$ are admissible and, in addition, $\Sigma_n tK < (\Sigma_n t)^\infty < \Sigma_n \mu\cdots$, where $t \in \{R, N\}$, then $(\Sigma_n t)^\infty$ is admissible.

The proof of this proposition is rather trivial, and is then omitted. In contrast with the case considered in the periodic window theorem (for the parameter space), here we regard $\Sigma_n tK$ as the sequence closest to $\Sigma_n C$, and hence closer than $(\Sigma_n t)^\infty$.

Similarly, if $\Sigma_n C$ and $\Sigma_n \nu\cdots$ are admissible, and $\Sigma_n wK > (\Sigma_n w)^\infty > \Sigma_n \nu\cdots$, where $w \in \{R, N\}$, then $(u_0 u_1 \cdots u_n w)^\infty$ is also admissible.

4. If $\Sigma_n \mu\cdots$, $\Sigma_n D$ and $\Sigma_n \nu\cdots$ are admissible and let $\Sigma_n t$ and $\Sigma_n w$ the greater and the smaller of $\Sigma_n L$ and $\Sigma_n M$, respectively, then

$$\Sigma_n t\overline{K} < (\Sigma_n t\overline{\Sigma}_n w)^\infty < \Sigma_n \mu$$

implies the admissibility of $(\Sigma_n t\overline{\Sigma}_n w)^\infty$, and

$$\Sigma_n \nu\cdots < (\Sigma_n w\overline{\Sigma}_n t)^\infty < \Sigma_n w\overline{K}$$

implies the admissibility of $(\Sigma_n w\overline{\Sigma}_n t)^\infty$.

5. If $I_1 = \Sigma_n B$ and $I_2 = \Sigma_n R\cdots$ are admissible and $(\Sigma_n R)^\infty$ is located between $\Sigma_n RH$ and I_2, then $(\Sigma_n R)^\infty$ is admissible. Similarly, If $I_1 = \Sigma_n B$ and $I_2 = \Sigma_n L\cdots$ are admissible and $(\Sigma_n L\overline{\Sigma}_n R)^\infty$ is located between $\Sigma_n L\overline{H}$ and I_2, then $(\Sigma_n L\overline{\Sigma}_n R)^\infty$ is admissible.

Taking K as the lower sequence and H as the upper sequence, we may generate admissible periods by applying the above propositions. We may examine whether a sequence generated in this process is shift-minimal with respect to R or N to avoid redundancy. If it is not shift-minimal, it will make no contribution to the non-redundant list of periods, so can be ignored.

3.8.4 Metric Representation of Symbolic Sequences

To introduce the metric representation we first use $\epsilon = 1$ to mark the even parity of L and R, and $\epsilon = -1$ to mark the odd parity of M and N. Next, a real number $0 \leqslant \alpha(\Sigma) \leqslant 1$ is defined for a symbolic sequence $\Sigma = s_1 s_2 \cdots s_i \cdots$ as

$$\alpha = \sum_{i=1}^{\infty} \mu_i 4^{-i}, \tag{3.116}$$

where

$$\mu_i = \begin{cases} 0, \\ 1, \\ 2, \\ 3 \end{cases} \quad \text{for} \quad s_i = \begin{cases} N, \\ R, \\ L, \\ M \end{cases} \quad \text{if} \quad \epsilon_1\epsilon_2\cdots\epsilon_{i-1} = 1,$$

or

$$\mu_i = \begin{cases} 3, \\ 2, \\ 1, \\ 0 \end{cases} \quad \text{for} \quad s_i = \begin{cases} N \\ R, \\ L, \\ M, \end{cases} \quad \text{if} \quad \epsilon_1\epsilon_2\cdots\epsilon_{i-1} = -1.$$

It is easy to check that

$$\alpha((NM)^\infty) = 0, \quad \alpha((MN)^\infty) = 1.$$

Sometimes for convenience we may set $\alpha(C_\pm) = 1/4$, $\alpha(B_\pm) = 1/2$, and $\alpha(D_\pm) = 3/4$. This will not violate the ordering. In addition, the following relations hold for any symbolic sequence Σ:

$$\alpha(\overline{\Sigma}) = 1 - \alpha(\Sigma). \tag{3.117}$$

One may also formulate the admissibility conditions in terms of the metric representations.

3.8.5 One-Parameter Limits of Lorenz-Sparrow Maps

The family of the Lorenz-Sparrow maps includes some limiting cases.

1. The N branch may disappear, and the minimal point of the R branch moves to the left end of the interval, see Fig. 3.26 (a). This may be described as

$$C = \overline{H} = RK \quad \text{or} \quad H = L\overline{K}. \tag{3.118}$$

It defines the only kneading sequence K from the next iterate of C. In terms of the metric representations the condition (3.118) defines a straight line in the $\alpha(H)$-$\alpha(K)$ plane:

$$\alpha(H) = (3 - \alpha(K))/4. \tag{3.119}$$

We have used (3.117) to derive this relation from (3.118).

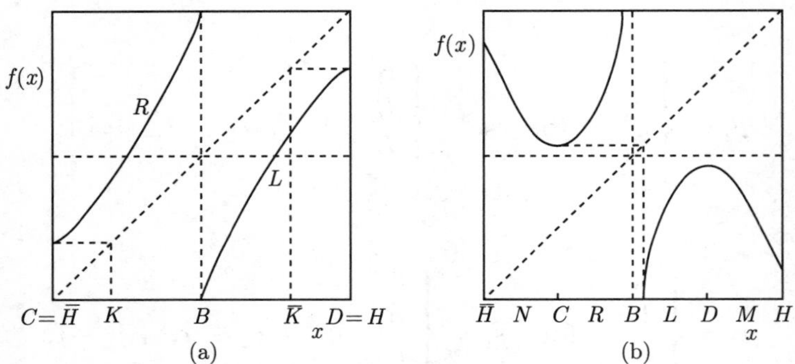

Figure 3.26 One-parameter limits of the Lorenz-Sparrow maps. (a) The $\overline{H}=RK$ limit. (b) The $K=L\overline{H}$ limit. The dynamically invariant set is represented by thick line

2. The minimum at C may rise above the horizontal axis, as shown in Fig. 3.26(b). Now the image of N and R belongs to L and M, and vice versa. The dynamically invariant set consists in two subintervals: $[f^2(C), f(D)]$ and $[f(C), f^2(D)]$, each of which is invariant under the second iterate. Points outside the set on the interval will finally map into the set after a transient. As this invariant set is concerned, the dynamics is described by single kneading sequence $K = f(C)$. If we denote $H = f^2(D)$ instead of $f(B_-)$, we have

$$K = L\overline{H} \quad \text{or} \quad \overline{K} = RH. \tag{3.120}$$

This defines another straight line in the $\alpha(H)$-$\alpha(K)$ plane:

$$\alpha(K) = (3 - \alpha(H))/4. \tag{3.121}$$

These two limiting cases will both be referred to in our study of the Lorenz equations in Section 6.3.

3.9 Piecewise Linear Maps

To conclude this chapter, we examine a piecewise linear model of maps with multiple critical points. The analytical study will lead naturally to the generalization of kneading determinant to maps with multiple critical points. The topological entropy of the map may be calculated from the smallest eigenvalue of the kneading determinant.

3.9.1 Piecewise Linear Maps with Multiple Critical Points

The simplest model of maps with multiple critical points is a piecewise linear map whose monotone pieces have slope $\pm\lambda$. It may be viewed as a physical realization of the symbolic dynamics and may be studied analytically. More important is the fact that it provides a direct mean to calculate the topological entropy of a map.

A piecewise linear continuous mapping function on the interval $[C_0, C_l]$ is given in Fig. 3.27. It has l monotone branches: $I_1, I_2, \cdots, I_l$. The turning points between these branches, i.e., the critical points, are denoted from left to right by $C_1, C_2, \cdots, C_{l-1}$. Write a monotone branch of the mapping function on I_j as

$$x_{n+1} = \lambda\epsilon_i x_n - \delta_i, \tag{3.122}$$

where $\epsilon_i = \pm 1$, the plus and minus signs corresponding to increasing and decreasing branch respectively. Given an initial point x_n, the symbolic sequence is made of the symbols $\{I_j\}$ and $\{C_j\}$. We reverse (3.122) to write

$$x_n = \lambda^{-1} e_n (d_n + x_{n+1}), \tag{3.123}$$

where $e_n = \epsilon_i$ and $d_n = \delta_i$ if $x_n \in I_i$. If x_n happens to coincide with a C_j, one may take either $x_n \in I_j$ or $x_n \in I_{j+1}$. Applying (3.123) repeatedly, we get

$$\begin{aligned} x_n = & \lambda^{-1} e_n d_n + \lambda^{-2} e_n e_{n+1} d_{n+1} \\ & + \cdots + \lambda^{-k-1} e_n e_{n+1} \cdots e_{n+k} d_{n+k} + \cdots . \end{aligned} \tag{3.124}$$

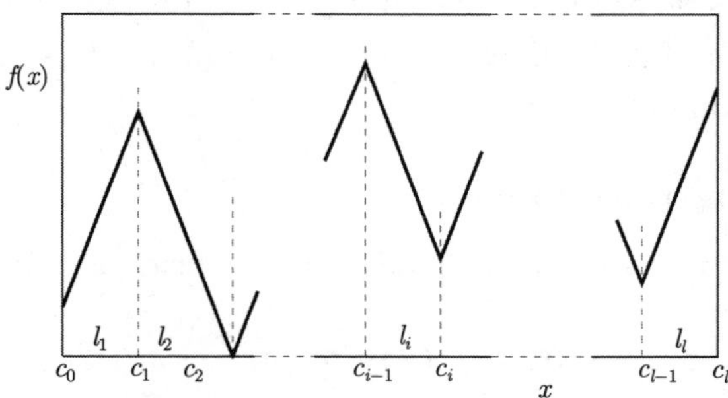

Figure 3.27 A general piecewise linear map with multiple critical points

Without causing confusion, we can denote $I_j \equiv \lambda^{-1}\epsilon_j\delta_j$ by the same symbol I_j. Then (3.124) becomes

$$x_n = I_{j_n} + \lambda^{-1} e_n I_{j_{n+1}} + \cdots + \lambda^{-k} e_n e_{n+1} \cdots I_{j_{n+k}} + \cdots , \tag{3.125}$$

where j_n is one of the integers $1, 2, \cdots, l$, picked up by the actual position of x_n, i.e., by $e_n = \epsilon_{j_n}$ and $d_n = \delta_{j_n}$. As there are only l different I_j, we can rearrange (3.125) into

$$x_n = \Theta_1 I_1 + \Theta_2 I_2 + \cdots + \Theta_l I_l, \tag{3.126}$$

where the Θ_j are power series of λ^{-1} with coefficients being either 0 or ± 1. If we denote $t = \lambda^{-1}$, then (3.126) coincides with the invariant coordinate of x_n, as introduced by Milnor and Thurston [1977, 1988].

We have not fixed the origin of x in the above description. A translation of the origin by d corresponds to the transformation

$$x_n \to x_n + d, \quad d_k \to d_k + (\lambda e_k - 1)d. \tag{3.127}$$

Performing this transformation in both sides of (3.124) and collecting the terms proportional to d, we get

$$1 = \lambda^{-1} e_n(\lambda e_n - 1) + \lambda^{-2} e_n e_{n+1}(\lambda e_{n+1} - 1) + \cdots. \tag{3.128}$$

Re-arranging terms and using the same notations Θ_j as in (3.126), Eq. (3.128) may be written as

$$\sum_{j=1}^{l}(1 - \lambda^{-1}\epsilon_j)\Theta_j = 1. \tag{3.129}$$

We note that the same kind of manipulation may be carried out for the difference of two expansions of points x_n and x_m. Then instead of 1 we would get a zero on one side of Eq. (3.128) and (3.129). Of course, there will be some other polynomials of λ^{-1} instead of Θ_j. We will use this trick to derive the kneading determinant.

3.9.2 Kneading Determinants

The kneading sequence K_i corresponding to the critical point C_i may be written as

$$K_i = f(C_i) = \sigma_{i_1}\sigma_{i_2}\cdots, \tag{3.130}$$

where a symbol σ_{i_k} belongs to the set $\{I_j\}$ (or $\{C_j\}$ if it falls on a turning point). Since $K_i = f(C_i)$, we can apply (3.122) and (3.124) to both left and right limits of C_i, i.e., to $C_i^{\pm}$, to get one and the same expansion on the other side:

$$\begin{aligned} \lambda_{i+1}C_i^{+} - \delta_{i+1} &= \lambda^{-1} e_{i_1} d_{i_1} + \cdots + \lambda^{-k-1} e_{i_1} e_{i_2} \cdots e_{i_k} d_{i_k} + \cdots, \\ \lambda_i C_i^{-} - \delta_i &= \lambda^{-1} e_{i_1} d_{i_1} + \cdots + \lambda^{-k-1} e_{i_1} e_{i_2} \cdots e_{i_k} d_{i_k} + \cdots. \end{aligned} \tag{3.131}$$

Formally we can consider the right-hand side of the above equations as expansion of $K_i^{\pm}$, though K_i^+ and K_i^- are the same.

If the k-th symbol in K_i happens to be C_j, then when $\epsilon_i e_{i_1} \cdots e_{i_{k-1}} = \pm 1$ holds, all terms corresponding to $C_i^{\pm}$ in the above equations are either $\lambda^{-k}\epsilon_i \cdots \epsilon_j \delta_j$ or $\lambda^{-k}\epsilon_i \cdots \epsilon_{j+1}\delta_{j+1}$.

Summing up the two equations in (3.131) and noticing that $\epsilon_{i+1} = -\epsilon_i$, we have

$$\begin{aligned} 0 &= -\lambda\epsilon_i(C_i^+ - C_i^-) \\ &= \delta_i + \delta_{i+1} + 2\lambda^{-1}e_{i_1}d_{i_1} + 2\lambda^{-2}e_{i_1}e_{i_2}d_{i_2} + \cdots \\ &\equiv M_{i_1}\delta_1 + M_{i_2}\delta_2 + \cdots + M_{i_l}\delta_l. \end{aligned} \tag{3.132}$$

In the above expansion the coefficients of the negative powers of λ can only be ± 2.

By shifting the origin one can always put one of the δ_j to zero. Suppose that $\delta_l = 0$. Write down the expansion (3.132) for all $(l-1)$ kneading sequences, we get a system of $(l-1)$ homogeneous equations for the $(l-1)$ quantities δ_j. In order to allow for a non-trivial solution the characteristic determinant must vanish, leading to

$$D_M = |M_{ij}|_{(l-1)\times(l-1)} = 0. \tag{3.133}$$

The smallest positive root λ^{-1} of Eq. (3.133) determines the topological entropy $h = \ln\lambda$ of the map. The determinant D_M may be identified with the *kneading determinant* of the map.

In general, M_{ij} is a $(l-1)\times l$ matrix. We introduce another matrix N_{ij} by defining

$$N_{ij} = -\epsilon_i\epsilon_j M_{ij}. \tag{3.134}$$

It is the Milnor-Thurston matrix (Milnor and Thurston [1977, 1988]), often encountered in the literature on symbolic dynamics of one-dimensional maps.

From (3.132) it is easy to get

$$\nu_i \equiv C_i^+ - C_i^- = \sum_j N_{ij}I_j. \tag{3.135}$$

Recollecting the derivation leading to (3.129), we have

$$\sum_j (1-\lambda^{-1}\epsilon_i)N_{ij} = 0, \tag{3.136}$$

which shows that the l columns of the matrix N_{ij} are linearly dependent. Denote by D_k the determinant of the square matrix obtained by deleting the k-th column from N_{ij}, it follows from matrix theory that the quantity

$$D = (-1)^{k+1} D_k / (1 - \lambda^{-1} \epsilon_k) \tag{3.137}$$

does not depend on the subscript k. It is called the Milnor-Thurston *kneading determinant.* The kneading determinant D is related simply to the determinant D_M introduced in (3.133):

$$D = (-1)^{l-1} D_M / (1 - \lambda^{-1} \epsilon_l). \tag{3.138}$$

The determinants D and D_M are entirely equivalent for the calculation of the topological entropy.

4
Symbolic Dynamics of Circle Maps

Circle maps transform a circle into itself. They provide another important paradigm in nonlinear dynamics. They give clues to the understanding of chaotic behaviour in a much broader class of nonlinear systems, namely, those involving two or more competing frequencies. In fact, circle maps furnish the simplest model of coupled nonlinear oscillators, yet they exhibit a variety of new phenomena, e.g., mode-locking and many new forms of transitions from quasi-periodic regimes to chaos. These phenomena are much more frequently encountered in higher dimensional systems and are either absent or not so transparent to be readily seen in one-dimensional maps of the interval that we have studied in the preceding chapters.

The importance of circle maps consists of the fact that it highlights the physics of chaos as one of the typical regimes of nonlinear oscillations that cannot be reduced to periodic or quasi-periodic motions. The "physics" here is centered around the phenomena of resonance and their nonlinear generalizations—mode-locking or frequency-locking and the destruction of these locked regimes. Although an elaborated description of nonlinear oscillations inevitably involves nonlinear differential equations, the main ingredients are present in circle maps and can be studied in a fairly elementary manner. Moreover, many circle maps are very nice mathematical creatures in their pre-chaotic regimes: they are diffeomorphisms, i.e., continuous and invertible transformations of a circle into itself. Circle maps are in their own rights rich subjects; there exists a wide mathematical literature on them. However, in accordance with the goal of this book, we shall confine ourselves mainly to symbolic dynamics of these maps, which is a natural extension of that for interval maps with multiple critical points and discontinuities.

Symbolic dynamics of circle maps has been studied from various viewpoints by Piña [1986], Veerman [1986], Zeng and Glass [1989], and Alseda and Ma nosas [1990]. Our presentation is mainly based on Zheng [1989b, 1991a, 1994].

This Chapter is organized as follows. Section 4.1 discusses the "physics" associated with circle maps. We will emphasize the difference between linear and nonlinear oscillations and see why nonlinear oscillators, including circle maps as the simplest realization of such systems, are widely used in modeling natural processes. Section 4.2 is a general discussion of the "geometry" of circle maps. The important notion of the *lift* and the *rotation number* of circle maps will be introduced. This Section ends with the simplest case of circle maps, namely, a linear map describing rigid rotations, called also a "bare" circle map in the sense that there are no nonlinear "interaction" or "coupling" terms. Simple as it is, it provides a standard ordering of orbits. This well-ordering of orbits will figure in the discussion of more complicated circle maps as a reference framework. The well-ordering of orbits is closely related to the ordering of rational numbers. Thus the study of circle maps is involved in some specific arithmetic of rationals. This is embodied in various representations of the so-called Farey tree of rational numbers. Section 4.3 introduces these representations for use in the rest of the Chapter. Section 4.4 concentrates on well-ordered orbits. The study of composition rules that form new well-ordered orbits from old ones takes a significant part of this Section. These rules are known under the name Farey transformations. Section 4.5 studies circle maps with non-monotone lift. After formulating the ordering rule and admissibility condition we will show that the existence of any periodic sequence implies the existence of a well-ordered periodic sequence with the same rotation number. This again illustrates the importance of well-ordered orbits in understanding the dynamics of circle maps. Section 4.6 is devoted to the structure of Arnold tongues, the generation of kneading sequences, and the construction of kneading plane for circle maps. Doubly superstable kneading sequences again play an essential role. They form "joints" and from joints come out "bones" of singly superstable kneading sequences. The last Section 4.7 makes use of piecewise linear circle maps to derive the analog of kneading determinants for circle maps. As usual, they are related to topological entropy of the maps.

4.1 The Physics of Linear and Nonlinear Oscillators

We begin the discussion of circle maps with a qualitative comparison of linear and nonlinear oscillators without the use of the language of differential equations. A simple linear oscillation is described by $A\sin(\omega t + \phi)$. Among the three characteristics of this oscillation: the amplitude A, the phase ϕ and the frequency ω, only the last, i.e., the frequency, is an intrinsic property of the oscillator. The amplitude and phase are determined by initial conditions. There are no transients in such simple linear oscillations without friction or external forcing; a pendulum would start swinging from the point where it has been released and return to this point indefinitely many times, provided the initial conditions have put it in the linear, i.e., the small amplitude regime.

If two linear oscillators are coupled, they "interact" only under one condition, namely, that their frequencies must be the same: $\omega_1 = \omega_2$, otherwise they will just go their own way without influencing each other. Recall a middle school physics demonstration: two pendulums hanging on a beam will interchange energy only when they are of the same length. Even in this case, the transfer of energy from one pendulum to another and *vice versa* becomes possible only due to small nonlinearities inevitably present in the system. At resonance, the amplitude of oscillations will grow without limits in the simple linear theory; in practice, it is bounded by dissipations and nonlinearities. Although dissipations can be included in oscillator models in a linear manner, e.g., by a friction term $\gamma\dot{x}$ proportional to the friction coefficient γ and velocity $\dot{x}$, they originate from interactions, i.e., nonlinearities, among a huge number of degrees of freedom. When friction is included, the resonance amplitude will be restricted to a finite peak and acquires a finite width. This width smears the sharp resonance condition and causes interactions under near-resonance conditions. Nevertheless, we repeat that friction is essentially a nonlinear and many-body effect.

There are many types of nonlinear oscillators, a frequently encountered type being a limit cycle oscillator. Contrary to a linear oscillator, the frequency, amplitude and phase of a limit cycle oscillator are closely related intrinsic properties of the system which eventually reach a steady oscillatory state of motion, independent of the initial conditions. If we "couple" two nonlinear oscillators, e.g., by making the state of one oscillator dependent on the amplitude of the other, then in most cases they will try "to sing in unison":

either the frequency of the stronger one takes over, or they compromise by finding a common frequency—usually at a fixed ratio of the two "free" frequencies. This is the phenomenon of *frequency-locking* or *mode-locking* . Linear oscillators simply cannot do this, because they are not capable of changing their own frequency. Even coupled, nonlinear oscillators cannot always reach a compromise at a common periodic regime, whence quasi-periodic and chaotic motions come into play. Chaos is just one frequently encountered regime of nonlinear oscillation, which cannot be reduced to simple periodic or quasi-periodic motion.

The nice properties of limit cycle oscillators of retaining their individuality, on the one hand, and of being entrained by external periods, on the other hand, have made them good candidates for the modeling of biological processes. Indeed, many biological rhythms, in the first place, must be an intrinsic property of a species independent of initial conditions or small changes in the environment, and, at the same time, they should be in harmony with the rhythms of Nature (seasons, day and night, etc.). These facts might explain why researchers in the life sciences have paid so much attention and contributed so widely to the study of circle maps (see, e.g., the nice book by Glass and Mackey [B1988]).

4.2 Circle Maps and Their Lifts

The phase space of a circle map is a circle. Usually, a full rotation along the circle is taken to be 1 instead of 2π or 360°. Therefore, a general circle map is given by

$$x_{n+1} = f_{a,b}(x_n) = kx_n + a + bg(x_n) \pmod 1, \tag{4.1}$$

where k is an integer, $g(x)$ a periodic function of period 1:

$$g(x+1) = g(x),$$

and the modulus operation (mod 1) means discarding the integer part but keeping the fraction part of the numerical result of the preceding expression.

The integer k in (4.1) is called *degree* or *topological degree* of the map. When $a = 0$ the integer k gives the number of pre-images of the point $x = 0$. Degree 0 and 1 maps are the most frequently encountered cases in practice. In this chapter we will study the $k = 1$ case only.

Let us look at two examples of circle maps.

4.2.1 The Rigid Rotation—Bare Circle Map

It is quite instructive to commence our study of circle maps with the simplest case of a rigid rotation, i.e., a linear map without nonlinear terms

$$x_{n+1} = x_n + a \pmod 1. \tag{4.2}$$

In Fig. 4.1 a linear circle map is given by two segments of straight lines above and below the bisector and a periodic orbit starting from $x = x_0$ is shown.

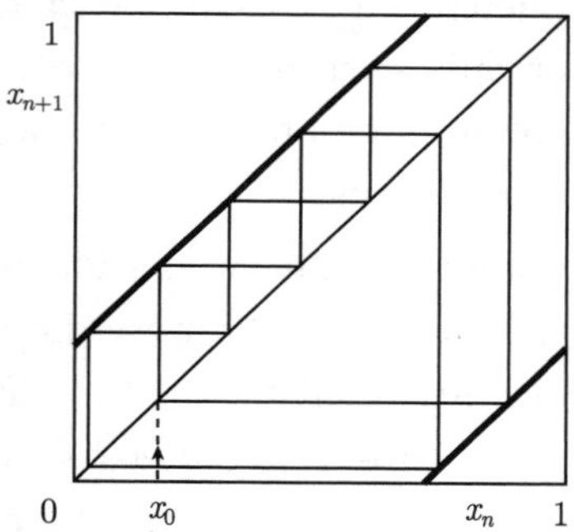

Figure 4.1 The bare circle map at $a = 2/7$ with a periodic orbit shown

The n-th iterate of this simple linear map can be written down easily as

$$x_n = x_0 + na \pmod 1.$$

Therefore, if $a = p/q$ happens to be the ratio of two coprime integers p and q, then qa drops out when taking the modulus and we have a q-cycle with

$$x_q = x_0.$$

Notice that in the figure of the bare circle map q is the total number of points, while p is the number of points located on the lower branch of the map. For instance, in Fig. 4.1, we have a periodic orbit at $a = 2/7$.

If we wish to retain the cumulative effect of the rotations, we can drop the modulus (mod 1) and write

$$x_{n+1} = x_n + a.$$

In general, when a circle map is drawn without taking (mod 1), it extends to the whole $f(x)$-x plane instead of being confined to the unit square. This is called a *lift* of the circle map. We shall define the lift in general terms later. The lift of the bare map is an ever ascending straight line. When confined to the unit square, each of the p points on the lower branch results from a single (mod 1) operation. It will be quite useful to become accustomed to going back and forth between these two representations.

We see that when $a = p/q$, there are q points on the circle, and they are visited in p full turns. Parameter a is called the (bare) *rotation number*. Moreover, for the rigid rotation any point belongs to a cyclic orbit. In the case of a general circle map (see below), the rotation number no longer enters into the original map explicitly and has to be calculated especially, and at a rational rotation number only certain points will fall on cyclic orbits. The dependence of the rotation number on the parameters is one of the central problems in the study of circle maps.

When a is an irrational number, i.e., it cannot be represented as a ratio of two integers, we will never obtain a finite periodic cycle from the bare circle map (4.2); the successive iterates will fill up the circle without ending or repeating themselves. We say that we have a *quasi-periodic motion* characterized by the irrational rotation number a. Quasi-periodic motion is a qualitatively new phenomenon that comes into play whenever two or more incommensurable basic frequencies are present in the system. In the case of (4.2), these frequencies are 1 and an irrational a.

Now we have a complete picture of the bare circle map in its parameter space $a \in [0, 1)$. When a varies from 0 to 1, the map (4.2) shows a periodic cycle at each rational a, and it exhibits quasi-periodic motion at all other values. Since rational numbers form only a countably infinite set of measure zero in the unit interval, and irrationals fill up the line with measure 1, periodic cycles are exceptions, and quasi-periodic motion is the general rule. When one picks up an a at random, there is an overwhelming chance that a quasi-periodic orbit will be obtained.

The bare circle map is rather important for at least two reasons. Firstly, inclusion of a nonlinearity in the map (4.2) tends to enhance the possibility of mode-locking, increasing thereby the chance of a periodic regime and giving it a positive measure in the parameter space. We will see this on the example of the sine circle map (4.3). Secondly, the bare circle map serves as a reference for the ordering of orbits in more general circle maps. The point is that, whenever the map $f(x)$ remains monotonically increasing in x, i.e., the slope df/dx remains positive, the order of the visits along the unit circle in such maps is the same as that in the bare circle with the same rotation number, although the distance between points may vary. If two or more initial points are chosen, then their relative order will be preserved at subsequent iterations. This is, of course, a consequence of the monotonicity of the map. We shall call this kind of order *well-ordering*, and the corresponding map—an order-preserving map. Not-well-ordered orbits are sometimes said to be *ill-ordered.* Since the orientation of the circle may be determined by following a small "vector" from one point to its neighboring point, order-preserving also means orientation-preserving.

4.2.2 The Sine-Circle Map

A common way of adding interaction into the bare circle map is to take the $g(x)$ in (4.1) to be a sine function, which makes it a *sine circle map*

$$x_{n+1} = f(x_n) = x_n + a + \frac{b}{2\pi}\sin(2\pi x_n) \pmod 1, \tag{4.3}$$

where the sine term is clearly a nonlinear periodic function. The new parameter b characterizes the strength of the coupling to the nonlinearity.

The previous periodic regime of the bare map at each rational point $a = p/q$ will be widened into a finite interval of a at nonzero b. This is a new phenomenon called *frequency-locking* or *mode-locking.* The nonlinear oscillator represented by $f(x)$ locks into the p/q regime instead of exhibiting its own a of the bare linear map. Fig. 4.2 gives a schematic presentation of these *mode-locking tongues* or *Arnold tongues* (see Section 4.2.5 below) in the a-b parameter plane. Only the wide ones among the infinite many tongues are shown. The structure of the Arnold tongues may be analyzed by using symbolic dynamics, see Section 4.6.1.

When b is small, mode-locking tongues with different p/g do not overlap. Therefore, at least for sufficient small b, the systematics of periodic regimes in the nonlinear map (4.3) will be given by that of the bare circle map.

We collect a few more facts on the sine circle map. First, the horizontal line $b = 1$ in the a-b parameter plane demarcates regions of different qualitative behavior of the map. When $b < 1$ the mapping function $f(x)$ is monotone increasing with a discontinuity caused by taking (mod 1). When $b = 1$ there appears an inflection point, signaling the end of monotonicity. When $b > 1$ the function $f(x)$ will have a minimum and a maximum. Locally it looks like a cubic map. Second, for $b \geqslant 1$ the function $f(x)$ has a negative Schwarzian derivative (see (2.15)) on the whole mapping interval $[0, 1]$.

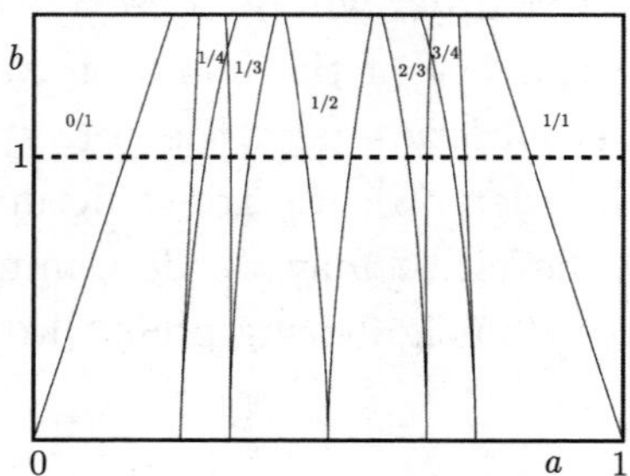

Figure 4.2 Sketch of mode-locking tongues in the parameter plane of a circle map

4.2.3 Lift of Circle Maps

A unit circle may be "unwrapped" into an infinite straight line, repeating itself between consecutive integers. Put in another way, the infinite real line covers a unit circle infinitely many times. A circle map may be "lifted" to a map of the real line. This is obtained from (4.1) by dropping the modulus operation:

$$t_{n+1} = F_{a,b}(t_n) = kt_n + a + bg(t_n), \quad t_n \in (-\infty, \infty). \tag{4.4}$$

When the argument takes its value from the whole real line, we write t instead of x, which takes value from the unit interval with its two end points identified. The map $F(t)$ is called a *lift* of $f(x)$.

The lift $F(t)$ satisfies the relation

$$F(t+1) = F(t) + k,$$

for all t with the integer k being the degree of the map.

To make the relation between a circle map $f(x)$ and its lift $F(t)$ clearer, we introduce a "covering map" $\pi : \mathcal{R} \to \mathcal{S}^1$ which maps the real line $\mathcal{R}$ to the circle $\mathcal{S}^1$:

$$\pi(t) = e^{2\pi i t},$$

then the circle map $f(x)$ and its lift $F(t)$ satisfy

$$\pi \circ F = f \circ \pi. \tag{4.5}$$

Since $\pi(t)$ defines a many-to-one correspondence, the above relation does not imply that f and F are topologically conjugate.

The dynamical behavior of circle maps with monotone and non-monotone lift differs significantly. Circle maps with monotone lift have been fully understood. A circle map with monotone lift is an invertible map. If it does not have periodic points, it is topologically equivalent to a rigid rotation with the same rotation number.

The border separating non-decreasing lift from non-monotone lift in the parameter space is sometimes called a *critical line.* For the sine circle map it is given by the horizontal line $b = 1$. In more general cases the critical line has to be determined numerically. Maps with parameters below the critical line are *subcritical maps*, while those above the critical line are *supercritical maps.* We will rarely use these terms.

Circle maps with non-monotone lift exhibit much more complicated dynamics. Nonetheless, they are "bounded" by maps with non-decreasing lift from above and from below in the sense of the construction shown in Fig. 4.3 (Chenciner, Gambaudo and Tresser [1984]; Boyland [1986]). In this figure a non-monotone part of $F(t)$ is drawn for a unit interval as a solid curve. There are, of course, infinite repetitions of such part along the whole t-line. In Fig. 4.3 two lines are labeled by P and Q. They are tangent to the maximum or minimum of $F(t)$ and intersect the latter at some point. One can define two non-decreasing functions

$$F_+(t) = \sup_{t' \leqslant t} F(t'), \quad F_-(t) = \inf_{t' \geqslant t} F(t'). \tag{4.6}$$

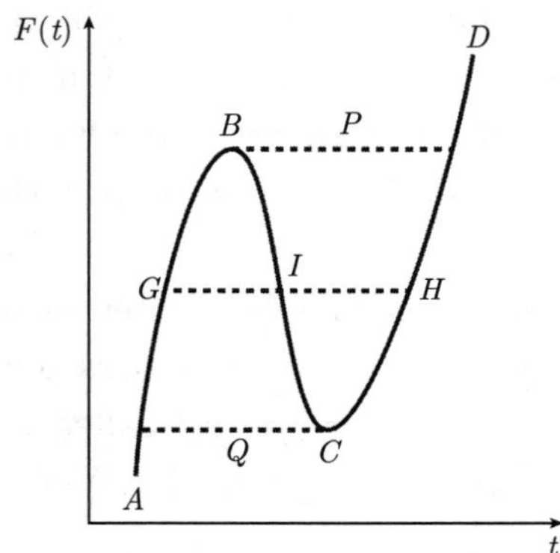

Figure 4.3 Construction of non-decreasing functions from a non-monotone lift

In Fig. 4.3 the function $F_+(t)$ is given by the curve $AGBPD$ and $F_-(t)$ by the $AQCHD$ curve. One always has

$$F_-(t) \leqslant F(t) \leqslant F_+(t), \quad \forall t.$$

One may also construct many non-decreasing functions between F_- and F_+. One example is given by the curve $AGIHD$ in Fig. 4.3. We will make use of this construction in what follows.

4.2.4 Rotation Number and Rotation Interval

The notion of "rotation number" plays a key role in the study of circle maps. It was introduced by Poincaré in 1885. It gives the average rotation rate of the map, and is a dynamical invariant of circle maps with non-decreasing lift.

We use the lift F of a circle map f to calculate the average rotation per iteration. Take a point t and the corresponding point x on the circle. Iterate it n times to get $F^n(t)$; the rotation of consecutive iterations accumulates due to the absence of taking (mod 1). Therefore, one can define

$$\rho(x, f) = \rho(t, F) = \lim_{n\to\infty} \frac{F^n(t) - t}{n}. \tag{4.7}$$

If the limit exists, it defines the *rotation number*, called also a *winding number*.

In general, the rotation number $\rho(x, F)$ depends on the orbit. However, when the lift $F(t)$ is a non-decreasing function, all orbits have one and the same rotation number and it may be called the rotation number of the map — $\rho(f)$. This is the case in the sine circle map (4.3) when the parameter $b < 1$.

The rigid rotation (4.2) is used as a reference to define a *well-ordered orbit* or *well-ordering* for other circle maps. If the orbital points of an orbit with rotation number ρ in a circle map are located in the same order along a circle as that of a rigid rotation with the same rotation number, it is called a well-ordered orbit. All orbits in a circle map with non-decreasing lift are well-ordered.

In a circle map with non-monotone lift, different orbits may have different rotation numbers. All possible rotation numbers $\rho(x, f)$ forms a non-empty closed set, usually an interval $\rho(t) = [\rho_1, \rho_2]$ called a *rotation interval* (Newhouse, Palis and Takens [1983]; Ito [1981][1]). When there exists a rotation

[1] The 1981 Ito's paper followed a preprint of Newhouse, Palis and Takens that finally appeared in 1983.

interval, there must exist orbits without a rotation number, i.e., orbits for which the limit in (4.7) does not exist. The existence of a rotation interval implies the existence of positive topological entropy, but the converse is not necessarily true. Therefore, rotation interval is always associated with topological chaos, but chaos may exist when there is only a unique rotation number. We will illustrate some of these points by way of symbolic dynamics in the subsequent sections.

Intuitively speaking, when an initial point maps successively into its images along the circle, if it always moves forward in one sense without marking time or jumping back, the motion is regular (periodic or quasi-periodic). Whenever it starts to hesitate or to step back, owing to the non-monotonicity in $f(x)$, the well-ordering is no longer preserved and one is approaching a chaotic regime. However, ill-ordering alone does not necessarily mean chaos, since period-doubling already violates the well-ordering. This may be seen from the period 4 orbit $(RLRC)^\infty$ in the main period-doubling cascade of unimodal maps, if one identifies the two end points of the mapping interval I.

Now recall the functions $F_\pm$ constructed in Fig. 4.3. Being non-decreasing functions, they have well-defined rotation numbers, $\rho(F_+)$ and $\rho(F_-)$. It is easy to see that the rotation number of the non-monotone lift bounded by F_+ and F_- is included between the two rotation numbers:

$$\rho(F_-) \leqslant \rho(t, F) \leqslant \rho(F_+).$$

Thus, we may write

$$\rho(t, F) \in [\rho(F_-), \rho(F_+)].$$

This notation includes the case when the interval degenerates into a point.

In fact, one may construct an infinite family of non-decreasing functions, each included between F_- and F_+, as the one represented by the curve $AGIHD$ in Fig. 4.3. For any number ρ within the above rotation interval, there exists a function in the family that has ρ as its rotation number. Suppose that the function $AGIHD$ has a rotation number ρ. If ρ is a rational number, then there exists a well-ordered orbit which has no orbital point within the interval GH. If ρ is irrational, then the orbit of G or H does not have any point that falls in the interval GH. However, the orbits of G and H are also orbits of $F(t)$. Therefore, for any ρ within the rotation interval there exists a well-ordered orbit with that rotation number.

4.2.5 Arnold Tongues in the Parameter Plane

Equipped with such notions as rotation number and rotation interval, we are in a right position to look at the structure of the parameter plane. All the maps which have at least one orbit with a certain rational rotation number r form a tongue-shaped region in the parameter plane. We call this region a r-Arnold tongue and say that any map which has an orbit with the rotation number r belongs to this r-Arnold tongue. Due to the existence of rotation interval, different Arnold tongues may overlap.

If all orbits of a map have the same rational rotation number r, the map is said to be frequency-locked to rotation number r. Frequency-locked region is a subregion of the Arnold tongue. If there exists a stable well-ordered orbit with rotation number r, the map is said to be r-stable.

The notion of frequency-locking and Arnold tongue may be extended to maps with an irrational rotation number ω. Below the critical line, the frequency-locked region reduces to a line. It opens up to an Arnold tongue on crossing the critical line (Boyland [1986]).

What has been said is illustrated in Fig. 4.4. In the figure A labels Arnold tongue, L means frequency-locking, and S stands for stable region. Generally speaking, the rotation interval of a map is a continuous function of the parameters. Whenever the rotation interval does not shrink to a point, there must exist chaotic orbits—rotational chaos (Boyland [1986]). However,

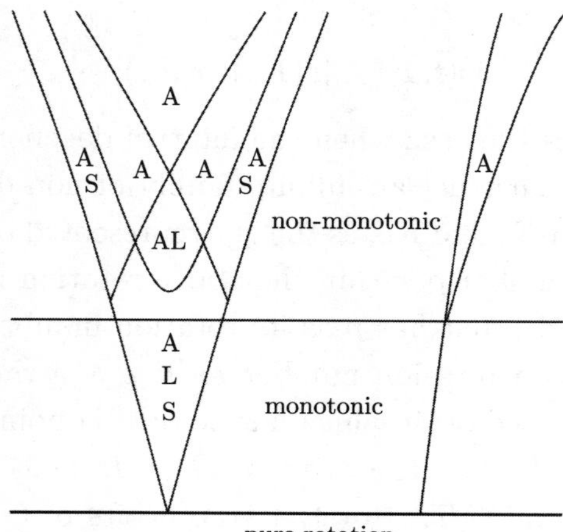

Figure 4.4 Arnold tongues in the parameter plane: A-Arnold tongue; L-frequency-locked region; S-stable region

even when a map has only a unique rotation number in a frequency-locked region, it is still possible to have chaotic orbits (MacKay and Tresser [1986]).

4.3 Continued Fractions and Farey Tree

It is quite curious that the development of nonlinear dynamics has related some well-known constructions in elementary number theory to physical reality. In Chapter 7 we will apply some number theory functions such as the Euler and Möbius functions to the counting of periodic orbits in one-dimensional maps. The study of circle maps makes frequent use of continued fraction representation of real numbers and the arrangement of real numbers into a tree structure, so-called Farey tree. Several notions such as the Farey addresses and Farey matrices are related to the Farey tree representation. In order not to interrupt the narration in the subsequent sections, we will now make a detour to recollect a few useful notions.

4.3.1 Farey Tree: Rational Fraction Representation

For real numbers on the unit interval $[0, 1]$, besides their positions on the interval as their natural order, there are other ways of ordering them. One of these ways, namely, the Farey construction, has attracted much attention due to a recent "experimental" observation that between two frequency-locked regimes described respectively by rational rotation numbers p/q and p'/q', the most easily observable, i.e., the widest in the parameter space, period would often be given by the Farey composition

$$\frac{p}{q} \oplus \frac{p'}{q'} = \frac{p+p'}{q+q'}. \tag{4.8}$$

Applying this composition rule to the two extremes of the unit interval $0/1$ and $1/1$, one gets $1/2$. Re-applying the Farey construction to all adjacent pairs, we arrive at the *Farey tree* shown in Fig. 4.5.

After excluding the two "ancestors" $0/1$ and $1/1$ at the top level of the binary tree, there is $2^0 = 1$ member, namely $1/2$, at the zeroth level. On the k-th level there are 2^k members $\{x_i = p_i/q_i\}$, the sum of which is

$$\sum_{i=0}^{2^k-1} x_i = 2^{k-1}, \quad \forall\, k > 0.$$

All rational numbers at the same level are ordered in the usual sense, increasing monotonically from left to right. Moreover, the values of two Farey members located symmetrically with respect to the central vertical yield the sum 1, being p/q and $1 - p/q = (q - p)/q$. Geometrically, as rotation numbers, they differ only in the sense of rotation.

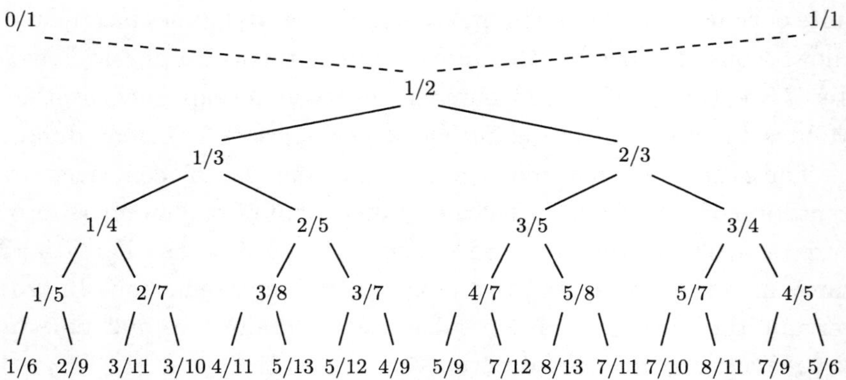

Figure 4.5 The Farey tree: rational fraction representation

Since the Farey tree plays a key role in understanding the ordering of periods in circle maps, we will present a few alternative representations, which may be more useful in one or another context.

4.3.2 Farey Tree: Continued Fraction Representation

Any rational number may be converted into a unique, simple *continued fraction* by applying the division algorithm usually used to find the greatest common divisor of two integers. Take, for example, the fraction 3/11, we have

$$\frac{3}{11} = \frac{1}{\dfrac{11}{3}} = \frac{1}{3 + \dfrac{2}{3}} = \frac{1}{3 + \dfrac{1}{1 + \dfrac{1}{2}}} \equiv [3, 1, 2].$$

A *simple* continued fraction has only 1's in the successive numerators. In general, we write

$$\frac{p}{q} = \frac{1}{a_1 + \dfrac{1}{a_2 + \dfrac{1}{a_3 + \dfrac{1}{a_4 + \cdots}}}} = [a_1, a_2, a_3, a_4, \cdots]. \tag{4.9}$$

The integer a_i is called the i-th partial quotients of the continued fraction.

Given a real number $\rho \in (0,1)$, we can define these partial quotients as a sequence of positive integers $\{a_i\}_1^\infty$ through a sequence of real numbers $\{\rho_i\}_0^\infty$ in the following recursive manner. Let $\rho_0 = \rho$, then

$$a_{i+1} = \left[\frac{1}{\rho_i}\right], \quad \rho_{i+1} = \frac{1}{\rho_i} - a_{i+1}, \tag{4.10}$$

where $[x]$ denotes the largest integer not exceeding x. Therefore, we have

$$\begin{aligned} \rho = \rho_0 &= \frac{1}{a_1 + \rho_1} = \frac{1}{a_1 + \cfrac{1}{a_2 + \rho_2}} \\ &= \frac{1}{a_1 + \cfrac{1}{a_2 + \cfrac{1}{a_3 + \cdots}}} \\ &= [a_1, a_2, a_3, \cdots]. \end{aligned} \tag{4.11}$$

Thus we have obtained a continued fraction expression for the real number ρ. A rational number is expressed by a finite (terminate) continued fraction and an irrational number by a non-terminate continued fraction.

For a finite continued fraction it follows from the definition that

$$[a_1, a_2, \cdots, a_{n-1}, a_n, 1] = [a_1, a_2, \cdots, a_{n-1}, a_n + 1]. \tag{4.12}$$

In order to make our notation unique, we will always use the first expression. Moreover, we will write the first quotient as $a_1 + 1$, singling out a 1 for convenience, i.e., our standard shorthand for a finite continued fraction is

$$[a_1 + 1, a_2, \cdots, a_n, 1]. \tag{4.13}$$

Golden mean and Fibonacci numbers

The truncations of an infinite continued fraction give a systematic approximation to an irrational number by rational fractions. Take, for example,

$$g = [1, 1, 1, 1, \cdots].$$

The continued fraction leads to a closed equation $g = 1/(1+g)$, which has the *golden mean* $g = \dfrac{\sqrt{5}-1}{2}$ as a solution. The finite truncations of g are:

$$\begin{aligned} &[1,1] = 1/2, \\ &[1,1,1] = 2/3, \\ &[1,1,1,1] = 3/5, \\ &[1,1,1,1,1] = 5/8, \\ &\cdots \\ &[\underbrace{1,1,\cdots,1,1}_{n \text{ times}}] = F_n/F_{n+1}, \\ &\cdots \end{aligned} \tag{4.14}$$

where F_n are the Fibonacci numbers defined recursively by

$$F_{n+2} = F_n + F_{n+1}, \tag{4.15}$$

with the initial numbers $F_0 = 0$ and $F_1 = 1$.

The Farey tree shown in Fig. 4.5 can be transformed into a continued fraction representation, as shown in Fig. 4.6. Now, at the k-th level of the tree, the Farey members are different partitions of the number $k+2$, i.e.,

$$\sum_i a_i = k + 2.$$

One can obtain readily from the continued fraction representation of the Farey tree the rule for the generation of the next level members. Any member (mother) in the tree gives birth to two members (daughters) at the next level according to the following two different rules.

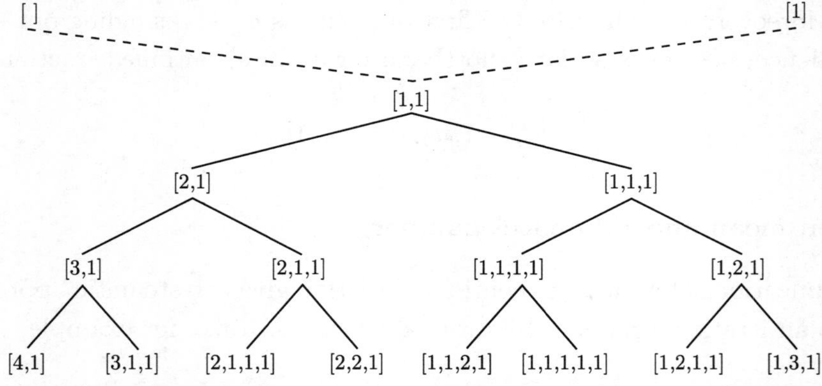

Figure 4.6 The Farey tree: continued fraction representation

Rule 0 $[a_1+1,\cdots,a_n-1,1] \rightarrow [a_1+1,\cdots,a_n,1]$,
where the length of the daughter equals that of the mother, i.e., the length increment is 0.

Rule 1 $[a_1+1,\cdots,a_n,1] \rightarrow [a_1+1,\cdots,a_n,1,1]$,
where the length of the daughter is longer than that of the mother by an increment 1.

These rules are used between any two levels starting from $k=1$ in an alternating way, e.g., from the 2nd to the 3rd level the rules

$$0\ 1\ \ 1\ 0\ \ 0\ 1\ \ 1\ 0$$

are used to the four 2nd level members, with Rule 0 being used at the two flanks of the tree. Written on the branches of the tree and read from the top ($[1,1]$), these 0's and 1's assign a unique binary number to each Farey member. In this way one could have given a binary number representation of the Farey tree. However, we prefer an equivalent, more convenient representation using so-called *Farey addresses* (Ostlund and Kim [1985]).

Before turning to the Farey address, let us point out another regularity in applying Rules 0 and 1 to generate the tree. Count the number of partial quotients in the continued fraction representation of the Farey tree in Fig. 4.6. If it is an even number, we say that the Farey member has an even parity; otherwise it has an odd parity. Then Rule 0 generates a daughter whose parity is always the same as her mother's, while Rule 1 generates a daughter with different parity.

4.3.3 Farey Tree: Farey Addresses and Farey Matrices

Instead of assigning 0 and 1 to the branches of Farey tree in an alternating way, we assign -1 and 1 in a fixed manner as shown in Fig. 4.7: a -1 is always put on the branch connecting a mother to the smaller daughter, and

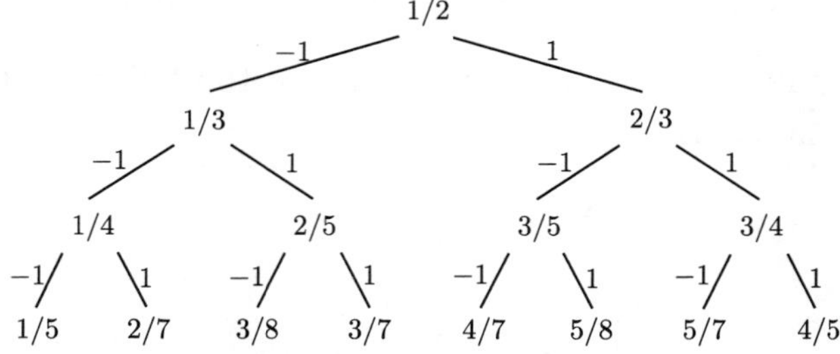

Figure 4.7 The assignment of Farey addresses

$a+1$ on the other branch. Take, for example, 2/7. Starting from 1/2, one goes through three consecutive branches, labeled by -1, -1, and 1 to reach 2/7. These three labels define the Farey address $\langle\bar{1}\bar{1}1\rangle$ of 2/7. We have written -1 as $\bar{1}$ for compactness and introduced the delimiters $\langle\cdots\rangle$ to distinguish a Farey address.

Formally a *Farey address* of a rational number ρ may be defined as

$$\rho = \langle b_{11}b_{12}\cdots b_{1a_1}b_{21}\cdots b_{2a_2}\cdots b_{n1}\cdots b_{na_n}\rangle, \tag{4.16}$$

where a_i is the partial quotients in the continued fraction representation (4.13) of ρ except for a difference 1 in a_1, and $b_{ij} = \pm 1$ is determined by its first index i only:

$$b_{ij} = (-1)^i. \tag{4.17}$$

Take $2/7 = [3,1,1]$ again for example. Recollecting the standard notation (4.13), we have $a_1 = 2$ and $a_2 = 1$. Therefore, $\langle b_{11}b_{12}b_{21}\rangle = \langle\bar{1}\bar{1}1\rangle$. Therefore,

$$2/7 = [3,1,1] = \langle\bar{1}\bar{1}1\rangle.$$

Representing the mth $\bar{1}$ or 1 in a Farey address by a single letter b_m, we write the Farey address of a rational number as

$$p/q = \langle b_1 b_2\cdots b_k\rangle \equiv \langle\beta\rangle, \tag{4.18}$$

where p and q are coprime integers. Now we introduce two *Farey matrices*

$$\mathcal{F}_1 = \begin{bmatrix} 0 & 1 \\ -1 & 2 \end{bmatrix}, \quad \mathcal{F}_{\bar{1}} = \begin{bmatrix} 1 & 0 \\ 1 & 1 \end{bmatrix}. \tag{4.19}$$

The rational number (4.18) may be represented as

$$\begin{bmatrix} p \\ q \end{bmatrix} = \mathcal{F}_{b_1}\mathcal{F}_{b_2}\cdots\mathcal{F}_{b_k} \equiv \langle\beta\rangle \circ \begin{bmatrix} 1 \\ 2 \end{bmatrix}. \tag{4.20}$$

Using the Farey matrices (4.19) one may prove the following Farey transformations of real numbers:

$$\begin{aligned} \langle 1\beta\rangle &= \frac{1}{2-\langle\beta\rangle} > \langle\beta\rangle, \\ \langle\bar{1}\beta\rangle &= \frac{\langle\beta\rangle}{1+\langle\beta\rangle} < \langle\beta\rangle. \end{aligned} \tag{4.21}$$

Moreover, real numbers may be ordered by their Farey addresses as follows:

$$\langle \beta 1 \cdots \rangle > \langle \beta \rangle > \langle \beta \bar{1} \cdots \rangle, \tag{4.22}$$

where β is an arbitrary string of 1 and $\bar{1}$. Relation (4.22) shows that the order of real numbers agrees with the lexicographical order of their Farey addresses.

We leave the proof of the relations (4.20) through (4.22) to the end of this section. Many readers may wish to skip the technical details of the proofs and turn directly to the use of these relations in developing the symbolic dynamics of the circle map. Before doing so, let us consider another way of introducing the Farey addresses.

We notice that among the two daughters one is greater and the other is smaller than the mother (in the sense of the natural order of numbers between 0 and 1). An inspection of the continued fraction representation of the Farey tree (Fig. 4.6) shows that the smaller daughter always has an even number of partial quotients or an even parity, while the greater daughter has an odd parity, independently of the number of partial quotients of the mother.

Now let us formulate the rules for finding the parents of a Farey member from its own continued fraction representation. It is easy to see that a member $[a_1+1, a_2, \cdots, a_n, 1]$ comes from the Farey composition of the following two members:

$$\begin{array}{ll} \text{“mother”} & [a_1+1, a_2, \cdots, a_n - 1, 1], \\ \text{“father”} & [a_1+1, a_2, \cdots, a_{n-1}, 1], \end{array}$$

the latter of which is located at some higher level than the “mother”. In fact, one can forget the “fathers” at all and trace back only along the maternal line. If the continued fraction under consideration has an even parity it must be a smaller daughter, then there is a label -1 on the connection to her mother; otherwise there is a 1. Tracing back from the given member up to the top member $[1,1]$ step by step and recording the label encountered at each step, we have the encoding scheme, e.g., for $3/11 = [3,1,1,1]$:

$$[3,1,1,1] \overset{-1}{\rightleftarrows} [3,1,1] \overset{+1}{\rightleftarrows} [3,1] \overset{-1}{\rightleftarrows} [2,1] \overset{-1}{\rightleftarrows} [1,1]. \tag{4.23}$$

Reading the labels from the top $[1,1]$ downwards, we find the Farey address $\langle \bar{1}\bar{1}1\bar{1} \rangle$ of the Farey member 3/11. In this way we may give the “address representation” for the Farey tree.

4.3.4 More on Continued Fractions and Farey Representations

Owing to the importance of continued fraction and Farey representations of rational numbers in the study of circle maps, we collect here some formulas and proofs for reference. This subsection may be skipped at first reading.

Continued fraction representation of numbers

We need a few more formulas related to the continued fraction representation (4.11). First, we recollect the shorthand for the same continued fraction:

$$\rho = [a_1, a_2, \cdots] = \frac{1}{a_1 +} \frac{1}{a_2 +} \frac{1}{a_3 +} \cdots . \tag{4.24}$$

Let $p_{-1} = q_0 = 1$ and $q_{-1} = p_0 = 0$, we can define a sequence $\{p_n, q_n\}$ recursively as

$$\begin{aligned} p_n &= a_n p_{n-1} + p_{n-2}, \\ q_n &= a_n q_{n-1} + q_{n-2}. \end{aligned} \tag{4.25}$$

We have by induction

$$\rho = \frac{p_n + p_{n-1}\rho_n}{q_n + q_{n-1}\rho_n}, \tag{4.26}$$

where ρ_n is from the sequence of real numbers defined in (4.10), and

$$q_n p_{n-1} - p_n q_{n-1} = (-1)^n. \tag{4.27}$$

The last relation (4.27) shows that p_n and q_n are coprime. By solving (4.26) we get

$$\rho_n = -\frac{p_n - \rho q_n}{p_{n-1} - \rho q_{n-1}}. \tag{4.28}$$

When $\rho = p_n/q_n$ we have $\rho_n = 0$. Therefore,

$$\frac{p_n}{q_n} = [a_1, a_2, \cdots, a_n].$$

It follows from (4.28) that $\rho - p_n/q_n$ changes sign upon increment of n by 1 and $p_0/q_0 = 0 < \rho$. Therefore, $p_n/q_n < \rho$ for even n and $p_n/q_n > \rho$ for odd n. In other words, ρ is always kept between p_n/q_n and p_{n+1}/q_{n+1}. Consequently, from (4.27) it follows

$$\left|\frac{p_n}{q_n} - \rho\right| + \left|\frac{p_{n+1}}{q_{n+1}} - \rho\right| = \left|\frac{p_n}{q_n} - \frac{p_{n+1}}{q_{n+1}}\right| = \frac{1}{q_n q_{n+1}}.$$

One may rewrite (4.25) in matrix form as

$$\begin{bmatrix} p_{n+1} & p_n \\ q_{n+1} & q_n \end{bmatrix} = \begin{bmatrix} p_n & p_{n-1} \\ q_n & q_{n-1} \end{bmatrix} \begin{bmatrix} a_{n+1} & 1 \\ 1 & 0 \end{bmatrix}.$$

We have accordingly

$$\begin{aligned} \begin{bmatrix} p_{n+1} & p_n \\ q_{n+1} & q_n \end{bmatrix} &= \begin{bmatrix} p_{n-1} & p_{n-2} \\ q_{n-1} & q_{n-2} \end{bmatrix} \begin{bmatrix} a_n & 1 \\ 1 & 0 \end{bmatrix} \begin{bmatrix} a_{n+1} & 1 \\ 1 & 0 \end{bmatrix} \\ &= \begin{bmatrix} 0 & 1 \\ 1 & 0 \end{bmatrix} \begin{bmatrix} a_1 & 1 \\ 1 & 0 \end{bmatrix} \begin{bmatrix} a_2 & 1 \\ 1 & 0 \end{bmatrix} \cdots \begin{bmatrix} a_{n+1} & 1 \\ 1 & 0 \end{bmatrix}. \end{aligned}$$

Therefore, finally we have

$$\begin{bmatrix} p_n \\ q_n \end{bmatrix} = \begin{bmatrix} 0 & 1 \\ 1 & 0 \end{bmatrix} \begin{bmatrix} a_1 & 1 \\ 1 & 0 \end{bmatrix} \begin{bmatrix} a_2 & 1 \\ 1 & 0 \end{bmatrix} \cdots \begin{bmatrix} a_n & 1 \\ 1 & 0 \end{bmatrix} \begin{bmatrix} 1 \\ 0 \end{bmatrix}.$$

This matrix form of continued fractions turns out to be quite convenient in the derivation of some relations.

Farey product and Farey sum

First, recollecting the Farey representation (4.18) of a rational number, we have the simplest Farey representation

$$\langle\,\rangle = \frac{1}{2}. \tag{4.29}$$

Next, we get by induction

$$\langle 1^k \rangle = \frac{k+1}{k+2}, \quad \langle \bar{1}^k \rangle = \frac{1}{k+2}. \tag{4.30}$$

Here 1^k and $\bar{1}^k$ denote k consecutive 1 and $\bar{1}$, respectively.

We further define a *Farey product* $\odot$ by the following concatenation

$$\langle \alpha \rangle \odot \langle \beta \rangle = \langle \alpha\beta \rangle. \tag{4.31}$$

For two rational numbers p_1/q_1 and p_2/q_2 we define a *Farey sum* $\oplus$ as

$$\langle \beta_1 \rangle \oplus \langle \beta_2 \rangle = \frac{p_1+p_2}{q_1+q_2}. \tag{4.32}$$

According to the definition (4.20) of product of Farey matrices, the Farey product and sum correspond to product and sum of Farey matrices. Therefore, it is easy to prove the following associative relation:

$$\langle\alpha\rangle \odot (\langle\beta_1\rangle \oplus \langle\beta_2\rangle) = (\langle\alpha\rangle \odot \langle\beta_1\rangle) \oplus (\langle\alpha\rangle \odot \langle\beta_2\rangle). \tag{4.33}$$

Furthermore, it follows from (4.30) and (4.32) that

$$\langle\ \rangle \oplus \langle 1\bar{1}^k\rangle = \langle 1\bar{1}^{k+1}\rangle, \quad \langle\ \rangle \oplus \langle \bar{1}1^k\rangle = \langle \bar{1}1^{k+1}\rangle. \tag{4.34}$$

The above two relations may be combined to

$$\langle\ \rangle \oplus \langle b\bar{b}^k\rangle = \langle b\bar{b}^{k+1}\rangle, \tag{4.35}$$

where b and $\bar{b}$, taken from ± 1, must be different. Take $b = 1$ for example. Since

$$\langle 1\bar{1}^k\rangle = [1, 1, k, 1] = \frac{k+2}{2k+3},$$

we have

$$\langle\ \rangle \oplus \langle 1\bar{1}^k\rangle = \frac{1}{2} \oplus \frac{k+2}{2k+3} = \frac{(k+1)+2}{2(k+1)+3} = \langle 1\bar{1}^{k+1}\rangle.$$

Equivalence of continued fraction and Farey representations

Now we are well prepared to prove the equivalence of the Farey Matrix representation (4.20) to the continued fraction representation (4.11) of a rational number. It goes by induction. First, it follows directly from (4.20) that

$$\langle\ \rangle = [1,1] = \frac{1}{2}, \quad \langle 1\rangle = [1,1,1] = \frac{2}{3}, \quad \langle\bar{1}\rangle = [2,1] = \frac{1}{3}. \tag{4.36}$$

Next, suppose the equivalence has been proved for some Farey representation $\langle\beta b\rangle$ of finite length, we prove that the equivalence holds for Farey representations of length increased by one, i.e., it holds for $\langle\beta bb\rangle$ and $\langle\beta b\bar{b}\rangle$. We consider the two cases separately.

A. Proof for $\langle\beta b\bar{b}\rangle$. Suppose

$$\langle\beta b\rangle = [a_1+1, a_2, \cdots, a_n, 1] \equiv \frac{p_{n+1}}{q_{n+1}},$$

$$\langle\beta\rangle = [a_1+1, a_2, \cdots, a_n] \equiv \frac{p_n}{q_n},$$

then it follows from (4.25) that

$$\frac{p_n}{q_n} \oplus \frac{p_{n+1}}{q_{n+1}} = [a_1+1, a_2, \cdots, a_n, 1, 1]. \tag{4.37}$$

However, according to (4.33) we have

$$\langle\beta\rangle \oplus \langle\beta b\rangle = \langle\beta\rangle \odot (\langle\ \rangle \oplus \langle b\rangle) = \langle\beta\rangle \odot \langle b\bar{b}\rangle = \langle\beta b\bar{b}\rangle. \tag{4.38}$$

The equivalence is proven.

B. Proof for $\langle\beta bb\rangle$. If $\beta b = b^k$, then the equivalence of $\langle b^{k+1}\rangle$ to the corresponding continued fraction may be readily proved.

If $\langle\beta b\rangle = \langle\beta'\bar{b}b^k\rangle$, we suppose

$$\begin{aligned} \langle\beta b\rangle &= \langle\beta'\bar{b}b^k\rangle = [a_1+1, a_2, \cdots, a_n, 1] \\ &= [a_1+1, a_2, \cdots, a_n+1] \equiv \frac{p_n}{q_n}, \\ \langle\beta'\bar{b}\rangle &= [a_1+1, a_2, \cdots, a_{n-1}] \equiv \frac{p_{n-1}}{q_{n-1}}. \end{aligned} \tag{4.39}$$

From (4.25) we have

$$\frac{p_{n-1}}{q_{n-1}} \oplus \frac{p_n}{q_n} = [a_1+1, a_2, \cdots, a_n+1, 1]. \tag{4.40}$$

However, according to (4.33) we have

$$\langle\beta'\rangle \oplus \langle\beta'\bar{b}b^k\rangle = \langle\beta'\rangle \odot (\langle\ \rangle \oplus \langle\bar{b}b^k\rangle) = \langle\beta'\bar{b}b^{k+1}\rangle = \langle\beta bb\rangle, \tag{4.41}$$

which proves the equivalence.

Ordering of real numbers in Farey representation

From the definition of Farey matrices we have

$$\begin{aligned} \langle\bar{1}\beta\rangle &= \langle\bar{1}\rangle \odot \langle\beta\rangle = \frac{\langle\beta\rangle}{1+\langle\beta\rangle} < \frac{1}{2}, \\ \langle 1\beta\rangle &= \langle 1\rangle \odot \langle\beta\rangle = \frac{1}{2-\langle\beta\rangle} > \frac{1}{2}. \end{aligned} \tag{4.42}$$

Therefore, for any $\langle\alpha\rangle$ and $\langle\beta\rangle$ one always has

$$\langle 1\alpha\rangle > \langle\bar{1}\beta\rangle.$$

In addition, it may be checked by using (4.42) that if $\langle\beta_1\rangle > \langle\beta_2\rangle$, then

$$\langle\bar{1}\beta_1\rangle > \langle\bar{1}\beta_2\rangle, \quad \langle 1\beta_1\rangle > \langle 1\beta_2\rangle,$$

i.e., the Farey matrices preserve order. Consequently, considering $\langle\ \rangle = 1/2$, we have the Farey ordering of real numbers:

$$\langle\beta 1\cdots\rangle > \langle\beta\rangle > \langle\beta\bar{1}\cdots\rangle. \tag{4.43}$$

Mirror image of a Farey representation

One may get a "mirror image" $\langle\bar{\beta}\rangle$ from $\langle\beta\rangle$ by interchanging $1 \leftrightarrow \bar{1}$. One can prove by induction that

$$\langle\beta\rangle + \langle\bar{\beta}\rangle = 1. \tag{4.44}$$

This is nothing but the simple fact that symmetrically located members on the same level of a Farey tree sums up to one. The proof starts from the trivial relation $\langle\ \rangle + \langle\ \rangle = 1$. If $\langle\beta\rangle = p/q$ and $\langle\bar{\beta}\rangle = (q-p)/q$, then

$$\langle 1\beta\rangle \begin{bmatrix} 1 \\ 2 \end{bmatrix} = \langle 1\rangle \begin{bmatrix} p \\ q \end{bmatrix} = \begin{bmatrix} q \\ 2q-p \end{bmatrix},$$

$$\langle\bar{1}\bar{\beta}\rangle \begin{bmatrix} 1 \\ 2 \end{bmatrix} = \langle\bar{1}\rangle \begin{bmatrix} q-p \\ q \end{bmatrix} = \begin{bmatrix} q-p \\ 2q-p \end{bmatrix}.$$

Therefore,

$$\langle 1\beta\rangle + \langle\bar{1}\bar{\beta}\rangle = 1.$$

4.3.5 Farey Tree: Symbolic Representation

We shall try to introduce symbolic systematics for all possible orbits in the circle mapping and to establish some links to the symbolic dynamics for the mappings of the interval that we have discussed in Chapter 3. Since the rational rotation numbers represented in the Farey tree will play an essential role even for the systematics of chaotic orbits, we start with replacing each rational fraction in the tree by a symbolic word. In order to obtain some feeling for the connection to mappings, we draw the graph of a general circle map in Fig. 4.8 with a period 2 orbit shown explicitly. This is a periodic orbit in the 1/2 mode-locked tongue. If we label the upper left branch by the letter L and the lower right branch—by R, this orbit may be represented by the word

RL. Therefore, the orbit of rotation number 1/2 (with its continued fraction representation [2]) has a symbolic word RL. In contradistinction to unimodal maps, now both letters represent monotone ascending branches. This remains true as long as we deal with subcritical circle maps (see Section 4.2.3). Above the critical line, there appears one or more descending branches in the map, which may require additional letters, say, M (see Fig. 4.12 in Section 4.5).

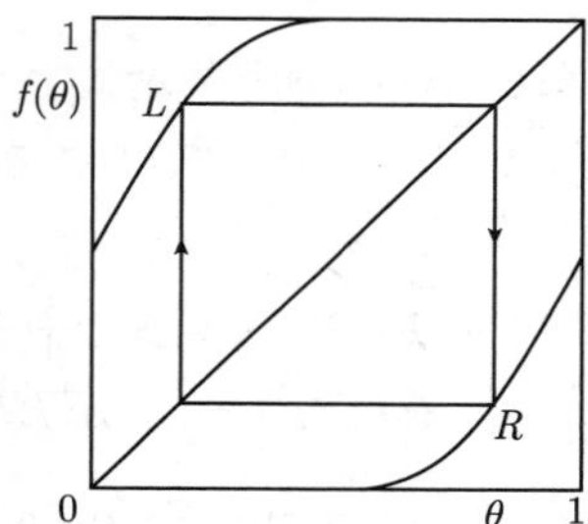

Figure 4.8 A period 2 orbit in circle map

The symbolic representation of any Farey member may be obtained from its Farey address by applying the Farey transformation to the top of the tree, i.e., the word RL (Zheng [1989b]). We will define Farey transformations as substitution rules applied to the letters R and L, analogously to the generalized composition rules discussed in Section 2.5.3.

There are two *Farey transformations*:

$$\begin{aligned}\mathcal{F}_1 &: R \to R, \quad L \to RL,\\ \mathcal{F}_{\bar{1}} &: R \to LR, L \to L.\end{aligned} \tag{4.45}$$

Here, instead of $\mathcal{F}_{-1}$, we have written it as $\mathcal{F}_{\bar{1}}$. In general, we will write $\mathcal{F}_\epsilon$ where ϵ may be ± 1. Suppose we have a word

$$\Sigma = s_0 s_1 \cdots s_n,$$

then its Farey transformation is

$$\mathcal{F}_\epsilon(\Sigma) = \mathcal{F}_\epsilon(s_0)\mathcal{F}_\epsilon(s_1)\cdots\mathcal{F}_\epsilon(s_n).$$

Now the rule to generate the symbolic representation W of a Farey member from its Farey address $\langle b_0 b_1 \cdots b_n \rangle$ can be formulated as follows:

$$W = \mathcal{F}_{b_0}\mathcal{F}_{b_1}\cdots\mathcal{F}_{b_n}(RL). \tag{4.46}$$

The following working procedure may be adopted:

1. Transform the given rational fraction into its continued fraction representation by using the division algorithm.

2. Write the Farey address directly from the continued fraction representation.

3. Apply the Farey transformation (4.46) to the Farey address to get the word W.

A few simple examples are given as follows.

Example 1 The fraction 2/7 has a Farey address $\langle\bar{1}\bar{1}1\rangle$. Its symbolic representation is

$$\begin{aligned} W_{\frac{2}{7}} &= \mathcal{F}_{\bar{1}}\mathcal{F}_{\bar{1}}\mathcal{F}_{1}(RL) = \mathcal{F}_{\bar{1}}\mathcal{F}_{\bar{1}}(RRL) \\ &= \mathcal{F}_{\bar{1}}(LRLRL) = LLRLLRL. \end{aligned}$$

Example 2 We have calculated the Farey address of 3/11 to be $\langle\bar{1}\bar{1}1\bar{1}\rangle$ (see Eq. (4.23)). Consequently,

$$\begin{aligned} W_{\frac{3}{11}} &= \mathcal{F}_{\bar{1}}\mathcal{F}_{\bar{1}}\mathcal{F}_{1}\mathcal{F}_{\bar{1}}(RL) \\ &= LLRLLLRLLRL = L^2RL^3RL^2RL. \end{aligned}$$

As we shall see later on, the Farey transformation is useful for general derivations. Fig. 4.9 is the top part of the symbolic representation of the Farey tree.

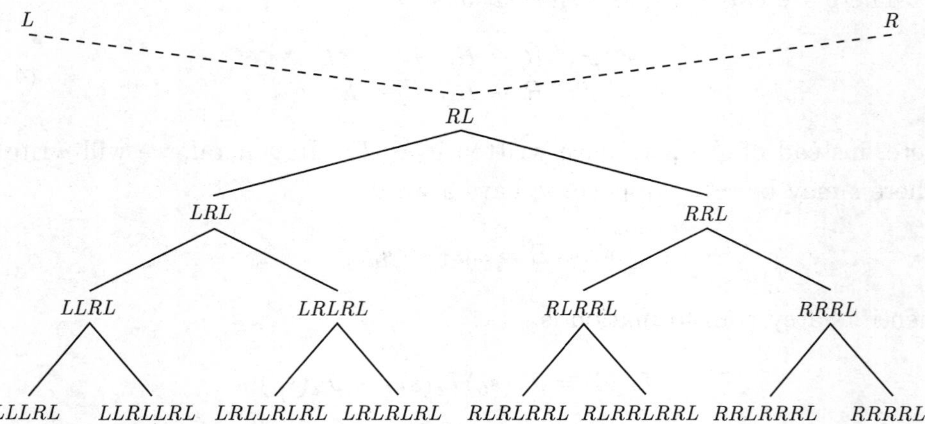

Figure 4.9 The Farey tree: symbolic representation

If only a single rational fraction is under consideration, then there exists a simple graphic method to find the symbolic sequence without knowing its

Farey address. Given a fraction p/q, first draw a circle and distribute q points on the circle more or less evenly, then draw a roughly vertical line dividing the circle into right and left halves, leaving p points in the right half. Each point in the right half carries a letter R, and each point in the left half — a letter L. Starting from the lowest point in the left half, go around the circle counterclockwise, and skip $p-1$ points at each step. After q steps, one has visited all the points on the circle and returned to the starting point. One then gets the desired symbolic word. Fig. 4.10 show two examples of assigning symbolic words to rational fractions. It is readily checked that the symbolic words thus obtained coincide with those built by applying the transformations (4.45) according to the Farey addresses.

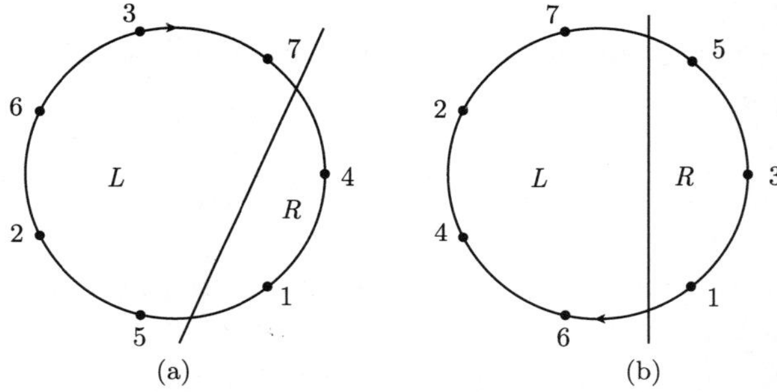

Figure 4.10 Assignment of symbolic sequence to rational rotation numbers. (a) $\rho = 2/7$. (b) $\rho = 3/7$

4.4 Farey Transformations and Well-Ordered Orbits

We have mentioned in Section 4.2.4 that if the lift of a circle map is non-decreasing, then all its orbits are well-ordered and they have one and the same rotation number. Now we study this simple case.

The mapping function of a monotone circle map looks like what is shown in Fig. 4.11. It has two monotonically increasing branches labeled by L and R and a discontinuity at point C. Any orbit corresponds to a symbolic sequence made of R and L. Each appearance of R in the sequence signals that the orbit has made one complete rotation. Therefore, the rotation number may be calculated from the symbolic sequence as

$$\rho = \lim_{n\to\infty} \frac{n_R}{n}, \tag{4.47}$$

where n_R is the number of Rs in the first n symbols of the sequence.

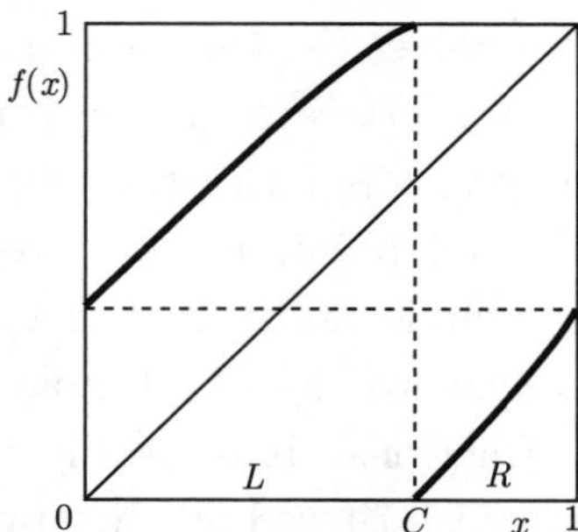

Figure 4.11 Mapping function of a circle map with non-decreasing lift

From Fig. 4.11 and the natural order $L < R$ follows the ordering rule of symbolic sequences

$$s_1 s_2 \cdots s_n L \cdots < s_1 s_2 \cdots s_n R \cdots , \tag{4.48}$$

where s_i stands for either R or L. Moreover, the shift sets of a symbolic sequence Σ satisfy

$$\mathcal{L}(\Sigma) \leqslant \mathcal{R}(\Sigma). \tag{4.49}$$

The shift set $\mathcal{L}(\Sigma)$ contains all subsequences of Σ that follow a symbol L; $\mathcal{R}(\Sigma)$ is analogous. We have first introduced them in Section 2.3.4, see, e.g., (2.50).

4.4.1 Well-Ordered Symbolic Sequences

Symbolic sequences are called *well-ordered sequences* if they are made of R and L only and satisfy condition (4.49). Well-ordered sequences correspond to well-ordered orbits.

Take, for example, sequences with rotation number $\rho = 1/2$. The symbolic sequence $(RL)^n R(RL)^\infty$ is well-ordered, as

$$\mathcal{R}((RL)^n R(RL)^\infty) \leqslant (RL)^\infty \leqslant \mathcal{L}((RL)^n R(RL)^\infty).$$

However, the sequence $(R^2L^2)^\infty$ is not well-ordered, because

$$\mathcal{R}((R^2L^2)^\infty) \ni (RL^2R)^\infty > (LR^2L)^\infty \in \mathcal{L}((R^2L^2)^\infty).$$

The seemingly innocent condition (4.49) for well-ordering has significant consequences. For instance, it follows from (4.49) that

1. A well-ordered symbolic sequence with rotation number ρ between $(n-1)/n$ and $n/(n+1)$ must be a sequence which contains only strings $R^{n-1}L$ and R^nL, or a shift of such a sequence.

2. A well-ordered symbolic sequences with rotation number ρ between $1/(n+1)$ and $1/n$ must be a sequence which contains only strings RL^{n-1} and RL^n, or a shift of such a sequence.

Since the proofs are similar, we treat the first one. Obviously, any symbolic sequence made of strings $R^{n-1}L$ and R^nL only must have a rotation number between $(n-1)/n$ and $n/(n+1)$. Suppose that a sequence contains another string $R^{n+k}L$, $k \geqslant 1$, then

$$\mathcal{R} \ni R^nL\cdots > R^{n-1}L\cdots \in \mathcal{L}.$$

On the other hand, if a symbolic sequence contains a string $R^{n-1-k}L$, then

$$\mathcal{R} \ni R^{n-1}L\cdots > R^{n-1-k}L\cdots \in \mathcal{L}.$$

Both relations contradict the well-ordering condition (4.49).

4.4.2 Farey Transformations as Composition Rules

For well-ordered symbolic sequences there exists a simple composition rule — the *Farey transformation* $\mathcal{F}_1$ and $\mathcal{F}_{\bar{1}}$. They are defined by (4.45) and has been introduced in Section 4.3.5. These transformations preserve the well-ordering of symbolic sequences. The proof goes as follows.

First, we have

$$\begin{aligned}\mathcal{F}_1(R\cdots) &= RR\cdots, \quad \mathcal{F}_1(L\cdots) = RL\cdots, \\ \mathcal{F}_{\bar{1}}(R\cdots) &= LR\cdots, \quad \mathcal{F}_{\bar{1}}(L\cdots) = LL\cdots.\end{aligned}$$

Furthermore, for any symbolic sequences P and Q, either well-ordered or ill-ordered, we can check that

$$P > Q \Rightarrow \mathcal{F}(P) > \mathcal{F}(Q), \tag{4.50}$$

where $\mathcal{F}$ denotes either $\mathcal{F}_1$ or $\mathcal{F}_{\bar{1}}$. Relation (4.50) shows that the Farey transformations preserve order.

Now we prove that the Farey transformations preserve well-ordering as well. The arguments will be given for $\mathcal{F}_1$ only; for $\mathcal{F}_{\bar{1}}$ the rephrasing is easy. Suppose that the shift sets $\mathcal{L}$ and $\mathcal{R}$ of a well-ordered symbolic sequence transform into $\tilde{\mathcal{L}}$ and $\tilde{\mathcal{R}}$, then we should have

$$\begin{aligned} \mathcal{L} \xrightarrow{\mathcal{F}_1} \tilde{\mathcal{L}} &= \{\mathcal{F}_1(\mathcal{L})\}, \\ \mathcal{R} \xrightarrow{\mathcal{F}_1} \tilde{\mathcal{R}} &= \{\mathcal{F}_1(\mathcal{R}), L\mathcal{F}_1(\mathcal{L})\}. \end{aligned} \tag{4.51}$$

It follows from the definition of $\mathcal{F}_1$ that the first symbol of $\mathcal{F}_1(\mathcal{L})$ must be R; thus $\mathcal{F}_1(\mathcal{L}) > L\mathcal{F}_1(\mathcal{L})$. From the well-ordering of the original sequence it follows that $\mathcal{L} \geqslant \mathcal{R}$. Since $\mathcal{F}_1$ preserves the order, we have $\mathcal{F}_1(\mathcal{L}) \geqslant \mathcal{F}_1(\mathcal{R})$. Therefore, $\tilde{\mathcal{L}} \geqslant \tilde{\mathcal{R}}$.

From the definition of $\mathcal{F}_{\bar{1}}$ follows that

$$\begin{aligned} \mathcal{L} \xrightarrow{\mathcal{F}_{\bar{1}}} \tilde{\mathcal{L}} &= \{\mathcal{F}_{\bar{1}}(\mathcal{L}), R\mathcal{F}_{\bar{1}}(\mathcal{R})\}, \\ \mathcal{R} \xrightarrow{\mathcal{F}_{\bar{1}}} \tilde{\mathcal{R}} &= \{\mathcal{F}_{\bar{1}}(\mathcal{R})\}. \end{aligned} \tag{4.52}$$

Let $\mathcal{F}$ stand for either $\mathcal{F}_1$ or $\mathcal{F}_{\bar{1}}$. Relations (4.51) and (4.52) tell us that the following equalities hold:

$$\begin{aligned} \max\{\tilde{\mathcal{R}}\} &= \mathcal{F}(\max\{\mathcal{R}\}), \\ \min\{\tilde{\mathcal{L}}\} &= \mathcal{F}(\min\{\mathcal{L}\}). \end{aligned} \tag{4.53}$$

If one requires only the order-preserving and well-ordering, one of the possible choices for Farey transformations would be:

$$\begin{aligned} &\mathcal{F}_1' : R \to R, \quad L \to LR, \\ &\mathcal{F}_{\bar{1}}' : R \to RL, L \to L. \end{aligned} \tag{4.54}$$

However, such transformations other than (4.45) do not satisfy (4.53), which makes them less convenient in practice. Therefore, we will never use the transformations (4.54).

4.4.3 Extreme Property of Well-Ordered Periodic Sequences

Among all periodic symbolic sequences made of letters R and L and having one and the same rotation number, the well-ordered sequences distinguish others by being the extreme in the following sense. Collect the $\max\{\mathcal{R}\} \equiv \mathcal{R}_{\max}$ of all such sequences, and denote $\mathcal{R}_{\max}$ of a well-ordered sequence by $\mathbf{R}^{\circ}_{\max}$, then the $\mathbf{R}^{\circ}_{\max}$ is the minimal. Likewise, $\mathbf{L}^{\circ}_{\min}$ from the well-ordered sequence is the maximal among all $\min\{\mathcal{L}\} \equiv \mathcal{L}_{\min}$. Since $R\mathcal{R}_{\max}$ and $L\mathcal{L}_{\min}$ are respectively the shift-maximal and shift-minimal sequences of a periodic sequence, the aforementioned property may be concisely stated as "the well-ordered periodic symbolic sequence possesses the minimal right-extreme and the maximal left-extreme".

Now we prove this statement by reduction to absurdity. Those readers who are not interested in technicalities may skip the proof and proceed to the next subsection. Suppose that there exists an ill-ordered periodic symbolic sequence I, whose $\max\{\mathcal{R}\} = \mathcal{R}_{\max}$ is the minimal among all periodic symbolic sequences with the same rotation number. Since I is not well-ordered, for it we must have $\min\{\mathcal{L}\} = \mathcal{L}_{\min} < \mathcal{R}_{\max}$. Therefore, there exist strings P and Q satisfying

$$\mathcal{R}_{\max} = (PLQR)^{\infty}, \quad \mathcal{L}_{\min} = (QRPL)^{\infty},$$

with

$$PLQR > QRPL. \tag{4.55}$$

Here $PLQR$ is the basic string of the periodic sequence I. Denote the lengths of various strings by $p = |P|$ and $q = |Q|$. String P must end with R. Otherwise $\mathcal{L}_{\min}$ would not be $(QRPL)^{\infty}$. Similarly, the string Q must end with L. In addition, we recollect the obvious fact that $R\mathcal{R}_{\max} = (RPLQ)^{\infty}$ is a shift-maximal sequence.

With all these ingredients ready, considering all possibilities case by case, we proceed further with the proof.

Case 1 Suppose that PL satisfies

$$\mathcal{S}^k(PL) < RPL, \quad \text{for} \quad 0 \leqslant k \leqslant p, \tag{4.56}$$

under the action of the shift operator $\mathcal{S}$, then one can construct a new periodic symbolic sequence $I' = (PLRQ)^{\infty}$, whose shifts will be all less than $R\mathcal{R}_{\mathrm{Max}}$. This is done as follows. From (4.56) it follows

$$\mathcal{S}^k(PLRQ) < (RPLQ)^{\infty}, \quad \text{for} \quad 0 \leqslant k \leqslant p. \tag{4.57}$$

From (4.55) we have

$$RQPL < RQRPL < RPLQR. \tag{4.58}$$

This means that (4.57) holds for $k = p + 1$ as well. Furthermore, from the shift-maximality of $R\mathcal{R}_{\max}$ we have

$$(RPLQ)^{\infty} > \mathcal{S}^{j}((QRPL)^{\infty}) > \mathcal{S}^{j}((QPLR)^{\infty}), \quad 0 \leqslant j \leqslant q. \tag{4.59}$$

So far we have considered all possible shifts of sequence I'. Therefore, this new sequence has a smaller $\mathcal{R}_{\max}$, in contradiction with our original assumption that the sequence I has the minimal $\mathcal{R}_{\max}$ among all periodic sequences with the same rotation number.

Case 2 String PL does not satisfy (4.56). This means that after some shifts the tail string of PL coincides with a leading string of RPL, i.e., the leading string of RPL forms its own tail. Suppose the shortest leading string of RPL coinciding with its tail is RX, then there are again two possibilities:

A. $RPL = RXYRX$.

B. $RPL = RURVU'L$, where $X = URV = VU'L$.

In addition, from the fact that $(RPLQ)^{\infty}$ is a shift-maximal sequence and RX is the shortest string causing the coincidence we have

$$\mathcal{S}^{k}(X) < RX, \quad 0 \leqslant k < |X|, \tag{4.60}$$

where, as usual, $|X|$ denotes the length of the string X. For occurrence A the relation (4.55) leads directly to

$$RXYRXQR > RQRXYRX. \tag{4.61}$$

We can construct a new symbolic sequence $J = (RQRXYX)^{\infty}$, and prove that its shifts are all less than $R\mathcal{R}_{\max}$. In fact, from the shift-maximality of $R\mathcal{R}_{\max}$ and (4.61) we have

$$R\mathcal{R}_{\max} > \mathcal{S}^{k}(RQRXYRX) > \mathcal{S}^{k}(RQRXYXR)$$

when $0 \leqslant k \leqslant |RQRXY|$. On the other hand, when $0 \leqslant k < |X|$, it follows from (4.60) that

$$R\mathcal{R}_{\max} > \mathcal{S}^{k}(XRQRXY).$$

Therefore, we have checked all the shifts of J and have shown that this new sequence would have a smaller $\mathcal{R}_{\max}$, in contradiction with our original assumption that I has the smallest $\mathcal{R}_{\max}$ among all periodic symbolic sequences with the same rotation number.

For case B we may construct a new sequence $J' = (RQRUVU'L)^\infty$. It is easy to show that for $0 \leqslant k \leqslant |RQRU|$ and $0 \leqslant h \leqslant |VU'L|$, we have respectively

$$R\mathcal{R}_{\max} > \mathcal{S}^k(RQRURVU'L) > \mathcal{S}^k(RQRUVU'L)$$

and

$$R\mathcal{R}_{\max} > \mathcal{S}^h(XRQRU) = \mathcal{S}^h(VU'LRQRU).$$

These inequalities mean that the sequence J' would have a smaller $\mathcal{R}_{\max}$, in contradiction with our original assumption. Therefore, we have excluded all possibilities that a sequence other than the well-ordered sequence would have the minimal $\mathcal{R}_{\max}$. In other words, we have proved that the well-ordered sequence has the smallest $\mathbf{R}^\circ_{\max}$ among all periodic symbolic sequences with the same rotation number.

Of course, by using similar arguments one can also prove that the well-ordered sequence has a greatest $\mathbf{L}^\circ_{\min}$ among all periodic sequences with the same rotation number. However, this conclusion may be reached by symmetry consideration as follows. A "mirror" sequence $\overline{I}$ is obtained from the original sequence I by interchanging $R \leftrightarrow L$. The order of two mirror sequences is opposite to the original order, i.e.,

$$I > J \Rightarrow \overline{I} < \overline{J}.$$

Therefore, the mirror sequence of a well-ordered sequence remains well-ordered. However, the mirror of $\mathcal{L}_{\min}$ becomes $\mathcal{R}_{\max}$ of the mirror sequence with a changed rotation number. This finishes the proof.

The $\mathcal{R}_{\max}$ of a non-periodic sequence cannot be the minimal. Suppose the opposite is true, i.e., $\mathcal{R}_{\max} = PRQ$ is minimal, where PRQ is non-periodic. By definition of $\mathcal{R}_{\max}$ we have $\mathcal{R}_{\max} > \mathcal{R}(Q)$, which means that $\mathcal{R}(Q)$ would provide an even smaller $\mathcal{R}_{\max}$, a contradiction.

We repeat the main result: $\mathbf{R}^\circ_{\max}$ ($\mathbf{L}^\circ_{\min}$) is the minimal (maximal) among $\max\{\mathcal{R}\}$ ($\min\{\mathcal{L}\}$) of all periodic and non-periodic symbolic sequences with the same rotation number.

4.4.4 Generation of $\mathbf{R}^\circ_{\max}$ and $\mathbf{L}^\circ_{\min}$

Different well-ordered symbolic sequences may have different rotation numbers. Denote the extreme sequences of the well-ordered periodic sequence with rotation number $r = p/q$ by

$$\mathbf{R}^\circ_{\max,r} = \min\{\max\{\mathcal{R}\}\}, \quad \mathbf{L}^\circ_{\min,r} = \max\{\min\{\mathcal{L}\}\},$$

we show how to generate these extreme sequences for different rotation number r. We will use the Farey address notation and their manipulation introduced in Section 4.3.3.

First of all, it is easy to check that for rotation number $1/2 = \langle\,\rangle$ we have

$$\mathbf{R}^\circ_{\max,1/2} = (LR)^\infty, \quad \mathbf{L}^\circ_{\min,1/2} = (RL)^\infty. \tag{4.62}$$

For other rotation numbers the corresponding extreme sequences may be generated according to the Farey address of the rotation number as follows:

$$\begin{aligned} \mathbf{R}^\circ_{\max,r} &= \mathcal{F}_{b_1}\mathcal{F}_{b_2}\cdots\mathcal{F}_{b_k}((LR)^\infty) \equiv \langle\beta\rangle \circ (LR)^\infty, \\ \mathbf{L}^\circ_{\min,r} &= \mathcal{F}_{b_1}\mathcal{F}_{b_2}\cdots\mathcal{F}_{b_k}((RL)^\infty) \equiv \langle\beta\rangle \circ (RL)^\infty. \end{aligned} \tag{4.63}$$

From (4.53) it is clear that they do correspond to the extreme sequences of the new well-ordered sequences. We need only to prove that the rotation number is just r. To this end we consider how the rotation number is changed under the Farey transformations. Suppose that before the transformation the the basic string of a periodic sequence is of length q; it contains p letters R and the rotation number is r. After the transformation these numbers become $\tilde{q}$, $\tilde{p}$, and $\tilde{r}$. From the definition (4.45) of the Farey transformation we have

$$\begin{aligned} \mathcal{F}_1 &: \tilde{q} = 2q - p, \tilde{p} = q, \tilde{r} = \frac{1}{2-r}, \\ \mathcal{F}_{\bar{1}} &: \tilde{q} = q + p, \ \tilde{p} = p, \tilde{r} = \frac{r}{1+r}. \end{aligned} \tag{4.64}$$

Compare the expression of $\tilde{r}$ with the transformation (4.21) of a real number $\langle\beta\rangle$ under the Farey transformations, we see that they change in the same way. This proves that the rotation number resulted from the transformation (4.63) is $r = \langle\beta\rangle$.

Take the rotation number $2/7 = \langle\bar{1}\bar{1}1\rangle$ as an example. From (4.63) we have

$$\begin{aligned} \mathbf{R}^\circ_{\max,2/7} &= \langle\bar{1}\bar{1}1\rangle \circ (LR)^\infty = \langle\bar{1}\bar{1}\rangle \circ (RLR)^\infty \\ &= \langle\bar{1}\rangle \circ (LRLLR)^\infty = (L^2RL^3R)^\infty. \end{aligned}$$

Similarly, one gets $\mathbf{L}^\circ_{\min,2/7} = (L^2RL^2RL)^\infty$.

Denote by 1^k and $\bar{1}^k$ the segments of k consecutive 1 and $\bar{1}$ in a Farey address, respectively. It is easy to show that

$$\begin{aligned}\langle 1^k\rangle \circ (RL) = R^{k+1}L, \quad \langle 1^k\rangle \circ (LR) = R^k LR,\\ \langle \bar{1}^k\rangle \circ (RL) = L^k RL, \quad \langle \bar{1}^k\rangle \circ (LR) = L^{k+1}R.\end{aligned} \tag{4.65}$$

Obviously, for an arbitrary Farey address $\langle\alpha\rangle$ one can always write

$$\langle\alpha\rangle \circ (RL) = L^{i_1}R^{i_2}L\cdots, \quad \langle\alpha\rangle \circ (LR) = R^{j_1}L^{j_2}R\cdots, \tag{4.66}$$

where i_1 and j_1 are non-negative integers, and i_2 and j_2 positive non-zero integers. Therefore, we have

$$\begin{aligned}\langle 1\alpha\rangle \circ (LR) = (RL)^{j_1}RR\cdots > (RL)^\infty,\\ \langle \bar{1}\alpha\rangle \circ (RL) = (LR)^{i_1}LL\cdots < (LR)^\infty.\end{aligned}$$

This leads to the following general ordering relations:

$$\begin{aligned}\langle\beta 1\alpha\rangle \circ (RL)^\infty &> \langle\beta 1\alpha\rangle \circ (LR)^\infty\\ &> \langle\beta\rangle \circ (RL)^\infty > \langle\beta\rangle \circ (LR)^\infty\\ &> \langle\beta\bar{1}\alpha'\rangle \circ (RL)^\infty > \langle\beta\bar{1}\alpha'\rangle \circ (LR)^\infty,\end{aligned} \tag{4.67}$$

where $\langle\alpha'\rangle$ is just another arbitrary Farey address. Relations (4.67) show the consistency of the following ordering:

The ordering of well-ordered symbolic sequences as represented by the order of their corresponding $\mathbf{R}^\circ_{\max}$.

The ordering of well-ordered symbolic sequences as represented by the order of their corresponding $\mathbf{L}^\circ_{\min}$.

The ordering of the Farey addresses of rotation numbers.

The ordering of rotation numbers themselves.

In addition, one can prove by induction

$$\begin{aligned}(\langle\beta\rangle \circ RL)(\langle\beta 1\bar{1}^k\rangle \circ RL) = \langle\beta 1\bar{1}^{k+1}\rangle \circ RL,\\ (\langle\beta\rangle \circ RL)(\langle\beta\bar{1}1^k\rangle \circ RL) = \langle\beta\bar{1}1^{k+1}\rangle \circ RL.\end{aligned} \tag{4.68}$$

In fact, the string RL in the above lines may be replaced by LR or any other string to make the relations still hold. This means two Farey transformations may be concatenated to yield another Farey transformation.

Finally, we point out that since a well-ordered periodic sequence can only be composed either from the strings R^nL and $R^{n+1}L$ or from the strings RL^m

and RL^{m+1}, one can always apply to it the inverse transformation of a Farey transformation. The inverse transformation necessarily produces again a well-ordered sequence. One can further apply inverse transformations until $(RL)^\infty$ or $(LR)^\infty$ is reached at a certain step. Therefore, a well-ordered periodic sequence is uniquely specified by its rotation number.

4.5 Circle Map with Non-Monotone Lift

We already know that the lift of the sine circle map

$$\theta' = \theta + a + \frac{b}{2\pi}\sin(2\pi\theta) \pmod 1 \tag{4.69}$$

becomes non-monotonic when the parameter $b > 1$. We will tackle this problem with a minor restriction. Namely, we confine ourselves to the case when there is only one pre-image to $x = 0$. However, the method employed in this section may be applied to analyze more general cases.

4.5.1 Symbolic Sequences and Their Continuous Transformations

The shape of a mapping function is shown in Fig. 4.12 (a). The pre-image of $x = 0$ is the point d where the function experiences a apparent discontinuity. The function has a minimum at s and a maximum at g. (d, s, and g are the mnemonic for "discontinuity", "small", and "great", respectively.) The origin $x = 0$ is put neither at the minimum nor at the maximum. The three points s, d, and g divide the interval $[0, 1)$ into four segments, labeled by the symbols M, L, R, and N. Symbolic sequences of orbits are made of these four letters, including s, d, and g as limits of the four letters.

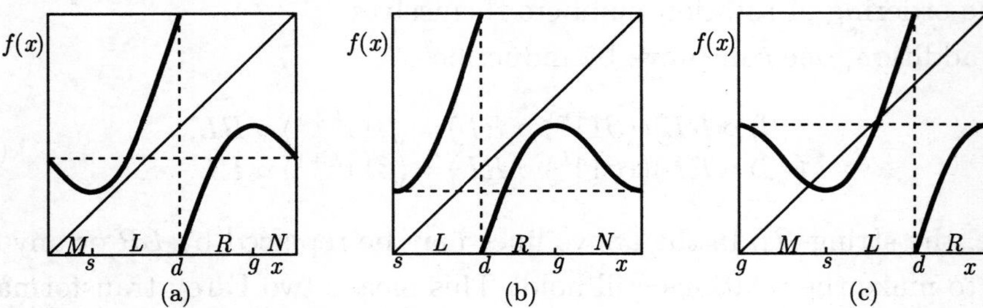

Figure 4.12 Mapping function of a circle map with non-monotone lift. (a) General case. (b) Case with the origin put at the minimum. (c) Case with the origin put at the maximum

Usually, one can translate the x-axis to place the origin at the minimum s, as shown in Fig. 4.12 (b). Now symbolic sequences may only contain three letters L, R, and N. There is another possibility of using three letters, namely, the case shown in Fig. 4.12 (c) where the origin is put at the maximum g. The letters are M, L, and R. In the last case, in order to ensure the order $s < d < g$, one can take $(0, 1]$ to be the mapping interval.

The dynamics is independent of the point where we cut the circle to flatten it. As long as these three equivalent ways of putting the origin all exist and the origin has only a unique pre-image, then symbolic sequences of one and the same orbit, obtained in these cases, are related by continuity. We explain this as follows.

Take a symbolic sequence I made of four letters as given in Fig. 4.12 (a). If a symbol R in I is followed by the minimal member $\min\{\mathcal{R}\}$ of the shift set $\mathcal{R}$, we call it the minimal R and denote it by $R^<$. Similarly, denote the minimal M by $M^<$, the maximal L by $L^>$, and the maximal N by $N^>$. Their following subsequences are $\min\{\mathcal{M}\}$, $\max\{\mathcal{L}\}$, and $\max\{\mathcal{N}\}$, respectively. When we change the origin continuously, the following transformation of symbol pairs may take place in I:

$$R^<M^< \to LN, \quad L^>N^> \to RM. \tag{4.70}$$

These transformations connect symbolic sequences of the same orbit when changing different sets of symbols from those of Fig. 4.12 (b) to those of (c).

When parameters of a map vary continuously, a specific orbit may also change continuously. If its symbolic sequence changes in accordance with the transformation (4.70), then the topology of the orbit may be just the same. In addition, continuous move of initial point of an orbit in a fixed map may cause the following symbolic change:

$$XLN\cdots \to Xd\cdots \to XRM\cdots$$

or

$$XLN\cdots \leftarrow Xd\cdots \leftarrow XRM\cdots ,$$

where X is some finite string.

For a general four-letter symbolic sequence one can extend the two-letter rotation number formula (4.47) to

$$\rho = \lim_{n\to\infty} \frac{n_R + n_N}{n}, \tag{4.71}$$

where n in the denominator is the length of a finite leading string, n_R and n_N are the number of letters R and N in this leading string. Obviously, the continuous transformation (4.70) does not change rotation number.

4.5.2 Ordering Rule and Admissibility Condition

For the four-letter map shown in Fig. 4.12 (a) one can extend the ordering rule of interval maps with multiple critical points as follows. The symbols N and M are odd, R and L are even. The natural order is

$$M < L < R < N. \tag{4.72}$$

If the common leading string of two symbolic sequences is even, then the order of the two sequences is the natural order of their first different symbols; otherwise, it is the opposite of the latter.

The kneading sequences of the map are given by the first iterate of the minimum point s and the maximum point g: $K_s = f(s)$ and $K_g = f(g)$. By inspecting Fig. 4.12 (a) the admissibility condition for a symbolic sequence Σ reads:

$$\mathcal{R}(\Sigma) \leqslant K_g, \quad K_s \leqslant \mathcal{M}(\Sigma) \leqslant \mathcal{N}(\Sigma) \leqslant K_g, \quad K_s \leqslant \mathcal{L}(\Sigma). \tag{4.73}$$

Of course, the kneading sequences K_s and K_g themselves should satisfy the admissibility condition (4.73) as well. We shall call two kneading sequences K_s and K_g satisfying (4.73) a *compatible kneading pair* or simply a *kneading pair* and denote it by (K_s, K_g).

4.5.3 Existence of Well-Ordered Symbolic Sequences

An important statement consists in the following: the existence of any periodic symbolic sequence implies the existence of a well-ordered periodic symbolic sequence with the same rotation number (Bernhardt [1982]). This statement can be proved as follows.

First, if the given symbolic sequence consists of letters R and L only, then the result of the last section implies

$$\mathbf{R}^{\circ}_{\max} < \mathcal{R}_{\max} \leqslant K_g, \quad \mathbf{L}^{\circ}_{\min} > \mathcal{L}_{\min} \geqslant K_s, \tag{4.74}$$

where $\mathcal{R}_{\max}$ and $\mathbf{R}^{\circ}_{\max}$ denote the $\max\{\mathcal{R}\}$ of the given sequence and of the well-ordered sequence, respectively. The meaning of $\mathcal{L}_{\min}$ and $\mathbf{L}^{\circ}_{\min}$ is similar.

Therefore, the well-ordered sequence satisfies the admissibility condition for sure. Put in other words, the well-ordered sequence is allowed by the same map.

When the given sequence contains letters N and M, we write

$$\max\{\mathcal{R},\mathcal{N}\} = X_1t_1X_2t_2\cdots X_nt_n\cdots \equiv I \leqslant K_g, \tag{4.75}$$

where string X_i contains only R and L or simply a blank string, and t_i stands for N or M. From the admissibility condition (4.73), we have

$$X_it_iX_{i+1}t_{i+1}\cdots \leqslant I \leqslant K_g, \quad \text{for} \quad i>1. \tag{4.76}$$

In the case of $X_1t_1 = L^kM$, all the X_it_i must be L^hM with $h \leqslant k$. The well-ordered periodic sequence L^∞ with the same rotation number is then allowed since L^∞ is greater than any $\mathcal{L}$ and $\mathcal{M}$. We can then assume that X_1 contains R. We generate $(X_it_i)'$ from X_it_i as follows. If $t_i = N$, then $(X_it_i)' = X_iR$. In the case of $t_i = M$, if X_i contains R, we replace its last R by L, and t_i with R; otherwise, just replace t_i by L. From (4.76) we have $X_1t_1 \geqslant X_it_i$ for $i>1$. As for the relation between X_1t_1 and X_it_i there are four possibilities:

A. $X_1t_1 = X_it_i$.

B. $X_1 > X_i$.

C. X_1 is a leading string of X_i and $t_1 = N$.

D. X_i is a leading string of X_1 and $t_i = M$.

It follows that

$$(X_it_i)' < X_1t_1$$

is always true except for the case when $X_jt_j = L^kM$ for some j and at the same time $X_1t_1 = L^{k+1}Xt_1$ with X being a string of R and L only. In this case we must have

$$Xt_1 > X_1t_1 \geqslant X_{j+1}t_{j+1},$$

which guareantees

$$I = X_1t_1\cdots > X_jt_j\cdots .$$

Let us denote by I' the sequence $(X_1t_1)'(X_2t_2)'\cdots$. Thus, we have for $i \geqslant 1$

$$\mathcal{R} \ni (X_it_i)'(X_{i+1}t_{i+1})'\cdots < I. \tag{4.77}$$

For other $\mathcal{R}(I')$ we consider $X_i t_i$ of the form $X_i = X_{i1} R X_{i2}$. The case when X_{i2} contains no R and $t_i = M$ has been considered in the above. In the other cases we have

$$(X_i t_i)' = X_{i1} R (X_{i2} t_i)'.$$

Since I is $\max\{\mathcal{R},\mathcal{N}\}$, we have

$$X_{i2} t_i X_{i+1} t_{i+1} \cdots < I.$$

The argument for deriving (4.77 now leads to

$$\mathcal{R} \ni (X_{i2} t_i)'(X_{i+1} t_{i+1})' \cdots < I.$$

We have considered all $\mathcal{R}(I')$, so

$$\mathbf{R}^{\circ}_{\max} < \mathcal{R}(I') < I \leqslant K_g.$$

The mirror image T of I may be generated from I by interchanging R with L and M with N. By taking the mirror image, it can be proved that

$$\mathbf{L}^{\circ}_{\min} \geqslant \min\{\mathcal{L}(I')\} > I \geqslant K_s.$$

Finally, the admissibility of the well-ordered periodic sequence with the same rotation number is verified.

4.5.4 The Farey Transformations

The Farey transformations (4.45) involving two letters R and L may be readily extended to circle maps with non-monotone lift that involve letters M and N as well:

$$\begin{aligned} \mathcal{F}_1 &: R \to R, L \to RL, N \to N, M \to RM, \\ \mathcal{F}_{\bar{1}} &: R \to LR, L \to L, N \to LN, M \to M. \end{aligned} \tag{4.78}$$

Clearly, both transformations $\mathcal{F}_1$ and $\mathcal{F}_{\bar{1}}$ do not change parity. Moreover, they both preserve order, i.e., for any two ordered symbolic sequences $P > Q$,

$$P > Q \Rightarrow \mathcal{F}(P) > \mathcal{F}(Q), \tag{4.79}$$

where $\mathcal{F}$ stands for either $\mathcal{F}_1$ or $\mathcal{F}_{\bar{1}}$.

Furthermore, a compatible kneading pair (K_s, K_g) transforms into a compatible pair $(\mathcal{F}K_s, \mathcal{F}K_g)$. We prove this for $\mathcal{F}_1$ only, as the proof for $\mathcal{F}_{\bar{1}}$ goes in a similar manner. Under the transformation $\mathcal{F}_1$ we have

$$\begin{aligned} &\mathcal{L} \xrightarrow{\mathcal{F}_1} \tilde{\mathcal{L}} = \{\mathcal{F}_1(\mathcal{L})\}, \\ &\mathcal{M} \xrightarrow{\mathcal{F}_1} \tilde{\mathcal{M}} = \{\mathcal{F}_1(\mathcal{M})\}, \\ &\mathcal{N} \xrightarrow{\mathcal{F}_1} \tilde{\mathcal{N}} = \{\mathcal{F}_1(\mathcal{N})\}, \\ &\mathcal{R} \xrightarrow{\mathcal{F}_1} \tilde{\mathcal{R}} = \{\mathcal{F}_1(\mathcal{R}), L\mathcal{F}_1(\mathcal{L}), M\mathcal{F}_1(\mathcal{M})\}. \end{aligned} \tag{4.80}$$

We know from the definition (4.78) that the transformed $\mathcal{F}_1(I)$ of any sequence I must start with either R or N, therefore

$$\begin{aligned} \min\{\tilde{\mathcal{L}}, \tilde{\mathcal{M}}\} &= \mathcal{F}_1(\min\{\mathcal{L}, \mathcal{M}\}), \\ \max\{\tilde{\mathcal{N}}, \tilde{\mathcal{R}}\} &= \mathcal{F}_1(\max\{\mathcal{N}, \mathcal{R}\}). \end{aligned} \tag{4.81}$$

Accordingly, we have

$$\begin{aligned} \min\{\mathcal{L}, \mathcal{M}\} \geqslant K_s \Rightarrow \min\{\tilde{\mathcal{L}}, \tilde{\mathcal{M}}\} \geqslant \mathcal{F}_1(K_s) = \tilde{K}_s, \\ \max\{\mathcal{N}, \mathcal{R}\} \leqslant K_g \Rightarrow \max\{\tilde{\mathcal{N}}, \tilde{\mathcal{R}}\} \leqslant \mathcal{F}_1(K_g) = \tilde{K}_g, \end{aligned} \tag{4.82}$$

which proves the statement. Furthermore, it is clear from the proof that if a symbolic sequence is allowed by a compatible kneading pair, then the transformed sequence is also allowed by the transformed kneading pair.

4.5.5 Existence of Symbolic Sequence without Rotation Number

We show the existence of symbolic sequences without a definite rotation number by giving an explicit example. Let the compatible kneading pair be $K_g = (RL)^\infty$ and $K_s = (LRL)^\infty$. Their compatibility may be checked easily. It can be verified that any symbolic sequence made of strings RL and RLL is allowed by this compatible pair.

We concatenate strings $(RL)^{n_{2k}}$ and $(RLL)^{n_{2k+1}}$ from n_1 to n_m to form a finite string Σ_m. There may be two kinds of string Σ_m depending on the integer m being even or odd:

$$\Sigma_m = \begin{cases} (RLL)^{n_1}(RL)^{n_2}(RLL)^{n_3}\cdots(RL)^{n_{m-1}}(RLL)^{n_m}, & m \text{ odd}, \\ (RLL)^{n_1}(RL)^{n_2}(RLL)^{n_3}\cdots(RLL)^{n_{m-1}}(RL)^{n_m}, & m \text{ even}. \end{cases}$$

There are many ways to choose these n_k in order to prevent the rotation number ρ from approaching a definite limit as $m \to \infty$. For example, we choose them by the following recursion relation:

$$n_1 = 1, \quad n_{k+1} = \left[\frac{k+3}{4}\right] \sum_{i=1}^{k} n_i, \tag{4.83}$$

where $[x]$ means taking integer part of x. Depending on whether we look at $\lim_{m\to\infty} \rho$ with an odd or even m, the ratio for ρ will approach either 1/2 or 1/3. Therefore, the infinite sequence Σ_∞ does not possess a definite limit. The detailed proof of this fact makes use of the following easily verifiable inequalities:

$$\begin{aligned} &\frac{1}{2} > \frac{i+j}{3i+2j} > \frac{1}{2} - \epsilon, \quad \text{if} \quad j \geqslant \frac{i}{4\epsilon}, \\ &\frac{1}{3} < \frac{i+j}{2i+3j} < \frac{1}{3} + \epsilon, \quad \text{if} \quad j \geqslant \frac{i}{9\epsilon}, \end{aligned} \tag{4.84}$$

and then take $j = n_{2k+1}$, $i = n_{2k}$ and $\epsilon = 1/(2k)$. The idea may be applied to other maps with a rotation interval.

4.6 Kneading Plane of Circle Maps

We have seen that the main objects in the parameter plane are the Arnold tongues of various rotation numbers. Since the kneading plane is just another way of parameterizing a map, we first study the structure of Arnold tongues in the kneading plane. All Arnold tongues with rational rotation numbers have a similar structure, so we will take the $\rho = 1/2$ tongue for a detailed study.

4.6.1 Arnold Tongue with Rotation Number 1/2

We have just proved in the previous section that if there exists a symbolic sequence with rotation number 1/2, there must be a well-ordered periodic sequence of $\rho = 1/2$. Therefore, the necessary and sufficient condition for the existence of symbolic sequences of rotation number 1/2 consists in

$$K_s \leqslant \mathbf{L}^{\circ}_{\min,1/2} = (RL)^\infty, \quad K_g \geqslant \mathbf{R}^{\circ}_{\max,1/2} = (LR)^\infty. \tag{4.85}$$

Consequently, the kneading sequences $K_s = (RL)^\infty$ and $K_g = (LR)^\infty$ make the outer boundaries of the $\rho = 1/2$ Arnold tongue in the kneading plane.

When the lift of the circle map is monotone, K_s and K_g may be understood as $\mathcal{L}_{\min}$ and $\mathcal{R}_{\max}$, respectively. Since $\langle\bar{1}1^\infty\rangle\circ(RL)^\infty = (LR)^\infty$, if $K_s < (LR)^\infty$

there must exist a large enough m such that

$$\langle \bar{1}1^m \rangle \circ (RL)^\infty \geqslant K_s,$$

which means that there would exist a rotation number less than 1/2. Similarly, if $K_g > (RL)^\infty$, then there must exist a large enough n such that

$$\langle 1\bar{1}^n \rangle \circ (LR)^\infty \leqslant K_g,$$

which implies the existence of rotation number greater than 1/2. Therefore, the inner borders of the frequency-locked region within the $\rho = 1/2$ Arnold tongue must be $K_s = (LR)^\infty$ and $K_g = (LR)^\infty$.

The structure of the Arnold tongue in the kneading plane is shown schematically in Fig. 4.13. The rectangle area labeled by L designates the frequency-locked regime. A dot "•" in the figure denotes the location of rotation number 1/2 and the horizontal error bar shows the possible rotational interval with respect to "•" in different areas of the kneading plane.

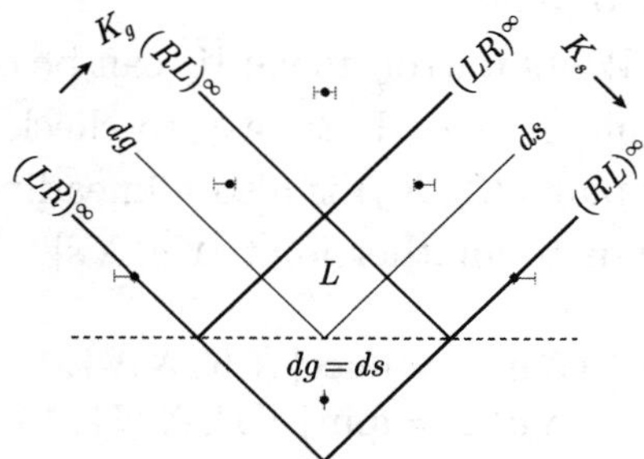

Figure 4.13 The $\rho = 1/2$ Arnold tongue in the kneading plane

The outer boundaries of the Arnold tongue correspond to tangent bifurcations, where the symbolic sequences of the critical points are $g(LR)^\infty$ and $s(RL)^\infty$; their degenerated forms are $(RL)^\infty$ and $(LR)^\infty$, respectively. On the other hand, at the boundaries of the frequency-locked area within the tongue the symbolic sequences of the critical points are $g(RL)^\infty$ and $s(LR)^\infty$, which have the degenerated forms $R(RL)^\infty$ and $L(LR)^\infty$, respectively. This shows that they correspond to homoclinic orbits. On the critical border line which divides the region with monotone and non-monotone lift, there is a point where $dg = ds$, or to be more precise, $d_- g = d_+ s$. The kneading sequences $K_g = dg$ and $K_s = ds$ change into each other continuously at this point.

The above discussion applies to rotation number $r = \langle b_1 b_2 \cdots b_k \rangle \equiv \langle \beta \rangle$ as well. In order to see the structure of an r-Arnold tongue, it is enough to understand R and L in Fig. 4.13 as $\langle \beta \rangle \circ R$ and $\langle \beta \rangle \circ L$, respectively.

4.6.2 Doubly Superstable Kneading Sequences: Joints and Bones

Just as interval maps with multiple critical points, the doubly superstable kneading sequences play a crucial role in elucidating the structure of the kneading plane of circle maps.

A pair of doubly superstable kneading sequences

$$(K_s, K_g) = (XgYs, YsXg) = [Xg, Ys]$$

is called a *joint.* We use square brackets $[\,,]$ to distinguish a joint from an ordinary compatible kneading pair, as we did in Section 3.6.1.

It can be proved by using the definition (4.78) of the Farey transformations that if $[Xg, Ys]$ is a joint, so does $[\mathcal{F}(Xg), \mathcal{F}(Ys)]$. Moreover, $[\mathcal{F}_{\bar{1}}(Xg), Ys]$ and $[Xg, \mathcal{F}_1(Ys)]$ are also joints.

From a given string W its mirror image $\overline{W}$ can be obtained by interchanging $R \leftrightarrow L$, $N \leftrightarrow M$, and $s \leftrightarrow g$. It is easy to check that the mirror image $(\overline{K}_g, \overline{K}_s)$ of a kneading pair (K_s, K_g) is also a kneading pair. Furthermore, a joint $[Xg, Ys]$ corresponds to another joint $[\overline{Y}g, \overline{X}s]$.

Let us denote

$$\begin{aligned}(Xg)_+ &= \max\{XR, XN\},\\ (Xg)_- &= \min\{XR, XN\},\\ (Ys)_+ &= \max\{YL, YM\},\\ (Ys)_- &= \min\{YL, YM\}.\end{aligned}$$

From the existence of a joint $[Xg, Ys]$ one can further prove the following:

A. There are compatible kneading pairs $(XgYs, YsXg)$, $((Xg)_-Ys, (Ys)_+Xg)$, and $((Xg)_+(Ys)_-, (Ys)_-(Xg)_+)$.

B. There are joints $[(Xg)_-(Ys)_-Xg, Ys]$, $[Xg, (Ys)_+(Xg)_+Ys]$, and $[(Xg)_-(Ys)_-Xg, (Ys)_+(Xg)_+Ys]$.

The proof is much similar to that for the derived kneading pairs in a general cubic map in Section 3.6.1. We leave it to the reader as an exercise (see Zheng [1991a]).

The above results are shown schematically in Fig. 4.14 (a). The loci of singly superstable kneading sequences $K_s = Us$ and $K_g = Vg$ in the parameter plane are called a *skeleton*. A simple skeleton corresponding to the the joints shown in Fig. 4.14 (a) is given in Fig. 4.14 (b).

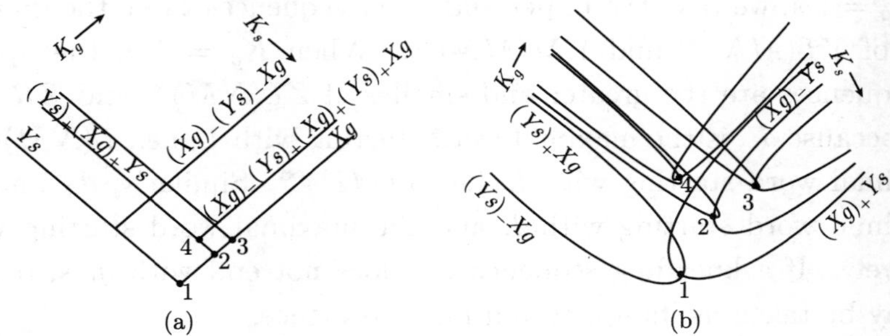

Figure 4.14 Joints and kneading pairs generated from a joint. (a) The joint $[Xg, Ys]$ and some joints derived from it. (b) Some simple skeleton corresponding to the joints shown in (a). Four numbered joints in (a) correspond to those with the same numbering in (b)

4.6.3 Generation of Kneading Sequences K_g and K_s

So far we have described the structure of a kneading plane in the vicinity of a given joint. In order to construct the whole kneading plane we have to generate all possible kneading sequences up to a certain length and then to form their compatible pairs. For the circle map one may extend the method of generating median words discussed in connection with maps of the interval to get all possible superstable K_s and K_g. We take the kneading sequences K_g as an example.

It follows from the admissibility condition (4.73) that any candidate for K_g must be shift-maximal with respect to the symbols R and N. The maximal K_g written with the four letters L, M, N, and R is $(NM)^\infty$. It is also the maximal sequence among all symbolic sequences. The sequence $(MN)^\infty$ may be taken as the minimal K_g if the additional condition $\mathcal{M} < \mathcal{N}$ is ignored. Therefore, all other K_g must be located between these maximal and minimal words.

In the process of generating median words, symbol d plays a role similar to s and g. Since the kneading sequence K_g comes from the iterate of g, the upper and lower sequences of $K_g = Xg$ are $(Xg)_\pm^\infty$, where

$$\begin{aligned}(Xg)_+ &= \max\{XR, XN\},\\ (Xg)_- &= \min\{XR, XN\}.\end{aligned}$$

However, for sequences ending with s or d we must consider sequences which are closest to s or d, i.e., those corresponding to $s_\pm$ or $d_\pm$. To be more specific, when $K_g = Ys$, we take the upper and lower sequences to be the greater and smaller of $YL(MN)^\infty$ and $YM(MN)^\infty$. When $K_g = Zd$, the upper and lower sequences are the greater and smaller of $ZL(NM)^\infty$ and $ZR(MN)^\infty$. This is because d_- is the maximal word starting with L, i.e., $L(NM)^\infty$; s_+ is the minimal word starting with L, i.e., $L(MN)^\infty$. Similarly, d_+ and s_- are the minimal word starting with R and the maximal word starting with M, respectively. If a kneading sequence K_g does not end with g, s, or d, then itself may be taken as its upper and lower sequence.

Once we have defined all upper and lower sequences, the method of generating all median words goes just like what we have done for interval maps, see, e.g., Sections 3.6.1 or 3.7.3. For two given kneading sequences K_g, compare the lower sequence of the greater one with the upper sequence of the smaller. If they coincide, then the two given sequences are neighbors, i.e., there is no median word in between. If not, let their common leading string be J. Append to J whichever of the letters s, d and g is between the next different symbols according to the natural order

$$M < s < L < d < R < g < N.$$

Denote any one of these appended letters by $\sigma \in \{s, d, g\}$, then $J\sigma$ is a median word.

Example 1 Take $K_{g_1} = (NM)^\infty$ and $K_{g_2} = (MN)^\infty$. Both words do not end with s, d, or g. The common leading string is empty. The letters s, d, and g included between the first different letters N and M, each provides a median word.

Example 2 Next, take $K_{g_1} = d$ and $K_{g_2} = s$. Comparing their upper and lower sequences $d_- = L(NM)^\infty$ and $s_+ = L(MN)^\infty$, we see that the common leading string is L. Inserting the "critical" letters in the order g, d, and s, we get three median words Lg, Ld, and Ls.

The median words ending with d are intermediate products. We need them to generate all superstable kneading sequences, but they themselves are not counted in the final list. Repeat the above procedure, we may generate all kneading sequences K_g up to a given length. All K_g of length 3 and less generated in this way are listed in Table 4.1.

Table 4.1 Superstable kneading sequences K_g of length 3 and less

No.	Sequence	No.	Sequence	No.	Sequence
1	NMg	11	RLg	21	Ls
2	NMs	12	RLs	22	LMs
3	Ns	13	Rs	23	LMg
4	NLs	14	RMs	24	s
5	NLg	15	RMg	25	MMg
6	NRs	16	LNs	26	MMs
7	NRg	17	Lg	27	Ms
8	Ng	18	LRs	28	MLs
9	g	19	LLg	29	MLg
10	RRs	20	LLs	30	Mg

Once all superstable K_g have been obtained, the simplest way to get the superstable K_s consists in taking the mirror images of K_g, because these K_s must be shift-minimal words with respect to L and M.

4.6.4 Construction of the Kneading Plane

Now we are in a position to construct the kneading plane. We arrange the K_g along straight lines pointing to the "northwest", and the K_s along straight lines pointing to the "northeast"; the K_g increases as one goes from "southwest" to "northeast", while K_s increases as one goes from "northwest" to "southeast", as shown in Fig. 4.15. Consequently, a pair of (K_s, K_g) corresponds to a point in the kneading plane. One may say that this point has K_s and K_g as its coordinates.

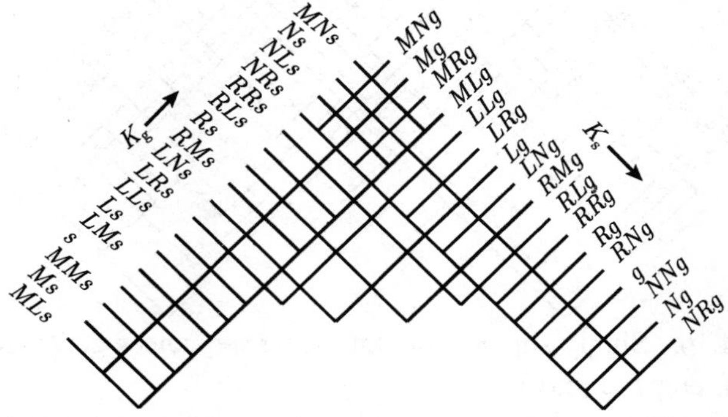

Figure 4.15 Joints scheme formed by doubly superstable kneading sequences $XgYs$ with length of Xg and Ys not exceeding 3

The K_s coordinate of the end point of a K_g line is determined by the $\min\{\mathcal{L}, \mathcal{M}, K_g\}$ of that K_g. The K_g coordinate of the end point of a K_s line is determined by the $\max\{\mathcal{R}, \mathcal{N}, K_s\}$ of that K_s. Therefore, whenever two straight lines have an intersection, the following relations hold at the intersection:

$$\begin{aligned}
&K_g \geqslant K_s,\\
&\min\{\mathcal{L}(K_g, K_s), \mathcal{M}(K_g, K_s)\} \geqslant K_s,\\
&\max\{\mathcal{R}(K_g, K_s), \mathcal{N}(K_g, K_s)\} \leqslant K_g,\\
&\mathcal{M}(K_g, K_s) \leqslant \mathcal{N}(K_g, K_s).
\end{aligned} \tag{4.86}$$

These relations show that an intersection corresponds to a compatible kneading pair. Conversely, a compatible kneading pair corresponds to an intersection in the kneading plane.

In Fig. 4.15 we have drawn all joints formed by superstable kneading words not exceeding length 3. This is a *joints scheme* as we have called it in Fig. 3.22. All intersections in this figure are joints. One can generate all singly superstable kneading sequences from these joints to get the kneading plane. Fig. 4.16 shows all singly superstable kneading sequences generated from joints with Xg and Ys of length 2 and less, i.e., from part of the joints shown in Fig. 4.15. The $\rho = 1/2$ Arnold tongue has been indicated by four unlabeled thin lines in this figure. They are to be compared with what has been shown in Fig. 4.16.

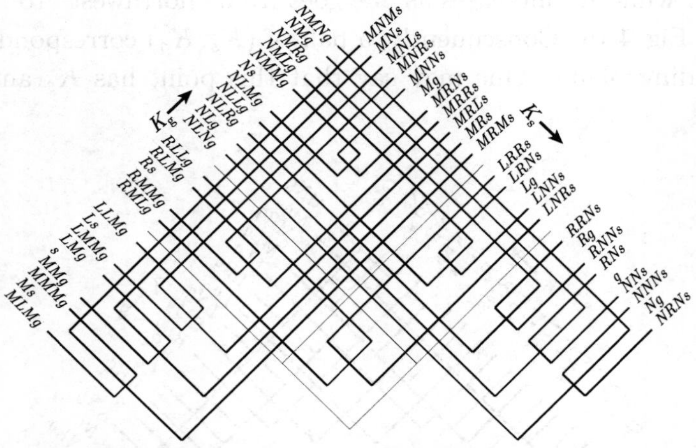

Figure 4.16 Singly superstable kneading sequences generated from joints of length 2 and less

4.7 Piecewise Linear Circle Maps and Topological Entropy

As we have shown many times in this book, piecewise linear maps have the merit that many instructive results may be obtained analytically. It is especially easy to construct a circle map with a given kneading sequence using a piecewise linear mapping function. We precede this construction by looking at the piecewise linear analog of the sine circle map (4.87).

4.7.1 The Sawtooth Circle Map

Replacing the $\sin(2\pi\theta)$ in (4.3) or (4.69) by a properly normalized sawtooth function, we get a piecewise linear circle map

$$\theta_{n+1} = \theta_n + A - Bg(\theta_n) \pmod{1}, \tag{4.87}$$

where

$$g(\theta) = (-1)^k(4\theta - 2k) \quad \text{for} \quad \theta \in [(2k-1)/4, (2k+1)/4].$$

The function $g(\theta)$ is shown in Fig. 4.17. It has been shown by Yang and Hao [1987] that the boundaries of the frequency-locked tongues in the A-B parameter plane are all given by rational fractions of the form

$$A = \frac{P(B)}{Q(B)},$$

where $P(B)$ and $Q(B)$ are polynomials of B. A striking feature of the structure in the parameter plane is that all Arnold tongues of rotation numbers p/q with $q > 3$ become "sausages"; their widths shrink to zero at well-defined values of B. We refer the interested reader to the original paper (Yang and Hao [1987]) and turn to the symbolic dynamics of piecewise linear circle maps.

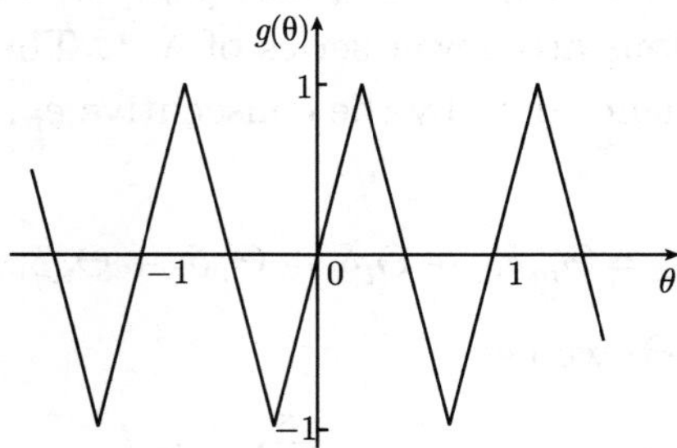

Figure 4.17 The sawtooth function

4.7.2 Circle Map with Given Kneading Sequences

The derivation below goes in parallel to what has been described in Section 3.8. However, we present it from scratch in order to make this chapter self-content.

In order to construct a circle map with given kneading sequences we take a mapping function as shown in Fig. 4.18 with linear segments having the same absolute value of slope λ. (Now the positive and negative slopes of $g(\theta)$ in (4.87) should have different absolute values.) Equality of the function at the two end points of the interval ensures the continuity of the lift. The minimum s, maximum g, discontinuity d, and labels of the monotone branches M, L, R, and N remain the same as in previous sections. Within a monotone branch the mapping function can be written as

$$x' = \lambda\epsilon_i x - \delta_i, \tag{4.88}$$

where the subscript i is one of m, l, r, and n, corresponding to M, L, R, and N in the symbolic sequence; $\epsilon_i = \pm 1$ reflects the parity of the symbol, i.e., increasing or decreasing behavior of the monotone branch. Here R and L have parity $+1$, while M and N have -1. Due to the continuity of the lift from (4.88) we derive the following relations between various δ_i:

$$\begin{aligned}\delta_m &= \lambda + \delta_n,\\ \delta_r &= 1 + \delta_l.\end{aligned} \tag{4.89}$$

Suppose that there is a point x_n, which has a symbolic sequence $s_n s_{n+1}\cdots$. It follows from (4.88) that

$$\begin{aligned}x_n &= \lambda^{-1}e_n d_n + \lambda^{-1}e_n x_{n+1}\\ &= \lambda^{-1}e_n d_n + \lambda^{-2}e_n e_{n+1}d_{n+1} + \cdots\\ &\quad + \lambda^{-k-1}e_n\cdots e_{n+k}d_{n+k} + \cdots,\end{aligned} \tag{4.90}$$

where $e_k \in \{\epsilon_i\}$, $d_k \in \{\delta_i\}$. Since the d_n enter every term linearly, we can collect the coefficients of δ_m, δ_l, δ_r, and δ_n into Θ_m, Θ_l, Θ_r and Θ_n, respectively. These coefficient functions are power series of λ^{-1}. They are entirely fixed by the given symbolic sequence, i.e., by the consecutive e_n, e_{n+1}, etc. Using these notations, we have

$$x_n = \Theta_m\delta_m + \Theta_l\delta_l + \Theta_r\delta_r + \Theta_n\delta_n. \tag{4.91}$$

From the definition of Θ_i we get

$$\Theta_m + \Theta_l + \Theta_n + \Theta_r = \sum_{k=0}^{\infty}\lambda^{-k-1}e_n\cdots e_{n+k}. \tag{4.92}$$

Since in the expansion (4.90) a d_j is always associated with the last e_j in the product of e_i, we may multiply a factor λe_j to each term to cancel the factor e_j, using $e_j^2 = 1$. Collecting again the coefficients of δ_i, we arrive at

$$\begin{aligned}\lambda\Theta_m\epsilon_m\delta_m + \lambda\Theta_l\epsilon_l\delta_l + \lambda\Theta_r\epsilon_r\delta_r + \lambda\Theta_n\epsilon_n\delta_n \\ = d_n + \sum_{k=1}^{\infty}\lambda^{-k}e_ne_{n+1}\cdots e_{n+k-1}d_{n+k},\end{aligned}$$

hence

$$-\lambda\Theta_m + \lambda\Theta_l + \lambda\Theta_r - \lambda\Theta_n = 1 + \sum_{k=0}^{\infty}\lambda^{-k-1}e_n\cdots e_{n+k}. \tag{4.93}$$

Subtraction of (4.92) from (4.93) yields

$$(\lambda-1)(\Theta_l+\Theta_r) - (\lambda+1)(\Theta_m+\Theta_n) = 1. \tag{4.94}$$

This identity may also be obtained from the freedom of shifting the origin of x, i.e., by applying the following translation to both sides of (4.91):

$$x_n \to x_n + d, \quad \delta_i \to \delta_i + (\lambda\epsilon_i - 1)d. \tag{4.95}$$

When the symbolic sequence of a point x_n contains only the letters R and L, the expansion (4.91) acquires an extremely simple form. In fact, by using (4.89) and (4.94) we have

$$x_n = \Theta_l\delta_l + \Theta_r\delta_r = \Theta_r + \frac{\delta_l}{\lambda-1}. \tag{4.96}$$

We note that the identity (4.94) is derived from the expansion of one point x_n. If we take the difference of the expansions for two points, in the right-hand side of (4.94) we would get a zero instead of 1. We will make use of this remark in what follows.

4.7.3 Kneading Determinant and Topological Entropy

Denote the kneading sequence of the minimum s as $K_s = \sigma_1\sigma_2\cdots$. Since $K_s = f(s_\pm)$, the point x_1 corresponding to σ_1 is reached from both s_+ and s_-. We write down the expansion (4.90) for both $f(s_+)$ and $f(s_-)$:

$$\begin{aligned}\lambda s_+ - \delta_l &= \lambda^{-1}e_1d_1 + \lambda^{-2}e_1e_2d_2 + \cdots, \\ -\lambda s_- - \delta_m &= \lambda^{-1}e_1d_1 + \lambda^{-2}e_1e_2d_2 + \cdots.\end{aligned} \tag{4.97}$$

Taking the sum of the two expansions, we get

$$\begin{aligned}\lambda(s_+ - s_-) &= \delta_l + \delta_m + 2\lambda^{-1}e_1 d_1 + 2\lambda^{-2}e_1 e_2 d_2 + \cdots \\ &\equiv A^s_m\delta_m + A^s_l\delta_l + A^s_r\delta_r + A^s_n\delta_n.\end{aligned} \tag{4.98}$$

The coefficient functions A^s_i are polynomials of λ^{-1}. An argument similar to that in derivation of (4.94) leads to

$$(1+\lambda)(A^s_m + A^s_n) + (1-\lambda)(A^s_l + A^s_r) = 0, \tag{4.99}$$

telling the linear dependence of A^s_m, A^s_n, A^s_l and A^s_r. This linear dependence is equivalent to saying that the physics does not change under the translation (4.95) of the origin.

Similarly, from $f(g_\pm) = K_g$ we get

$$\begin{aligned}\lambda(g_+ - g_-) &= \delta_r + \delta_n + 2\lambda^{-1}e'_1 d'_1 + 2\lambda^{-2}e'_1 e'_2 d'_2 + \cdots \\ &\equiv A^g_m\delta_m + A^g_l\delta_l + A^g_r\delta_r + A^g_n\delta_n,\end{aligned} \tag{4.100}$$

where e'_i and d'_i correspond to δ_i of the i-th symbol in the sequence K_g. Moreover, these A^g_m, A^g_l, A^g_r and A^g_n are linearly dependent, satisfying a relation similar to (4.99).

In what follows we choose the origin to make $\delta_n = 0$ for convenience. From (4.89), (4.98) and (4.100), finally it follows that

$$\begin{aligned}(A^s_l + A^s_r)\delta_l + \lambda A^s_m + A^s_r &= 0, \\ (A^g_l + A^g_r)\delta_l + \lambda A^g_m + A^g_r &= 0.\end{aligned} \tag{4.101}$$

Viewed as two linear equations for one and the same variable δ_l, this system may have a solution only at a vanishing determinant

$$A \equiv \begin{vmatrix} A^s_l + A^s_r & \lambda A^s_m + A^s_r \\ A^g_l + A^g_r & \lambda A^g_m + A^g_r \end{vmatrix} = 0. \tag{4.102}$$

The determinant A is nothing but the *kneading determinant* of the circle map. The smallest positive root λ^{-1} of Eq. (4.102) determines the topological entropy of the piecewise linear circle map: $h = \ln\lambda$.

When both kneading sequences K_s and K_g do not contain the symbol M, the kneading determinant acquires a more symmetric form

$$\begin{vmatrix} A^s_l & A^s_r \\ A^g_l & A^g_r \end{vmatrix} = 0. \tag{4.103}$$

We remind that the coefficient functions $A^{s,g}_{l,r}$ are polynomials of λ^{-1}, therefore (4.103) is an equation for λ^{-1}.

4.7.4 Construction of a Map from a Given Kneading Sequence

Now we have been prepared to construct a piecewise linear map (4.88) which produces the given kneading sequences.

Let us first fix the unit square in Fig. 4.18 with respect to the infinite horizontal and vertical axes. Suppose that it occupies $[-c, 1-c]$ in both directions, with the end points c and $1-c$ identified. The M and L branches meet at $x = s$, yielding $\delta_l - \delta_m = 2\lambda s$. The R and N branches meet at $x = g$, yielding $\delta_r - \delta_n = 2\lambda g$. The L branch reaches $1-c$ and the R branch reaches $-c$ at $x = d$, yielding $1 - 2c = 2\lambda d - \delta_l - \delta_r$. Putting the origin in such a way that $\delta_n = 0$ and making use of (4.89), we get

$$\begin{aligned} 2\lambda s &= \delta_l - \lambda, \\ 2\lambda g &= \delta_l + 1, \\ \lambda d &= \delta_l + 1 - c. \end{aligned} \tag{4.104}$$

When the kneading sequences are given, the slope λ is calculated from the kneading determinant equation $A = 0$. Then δ_l may be found from (4.101). The locations of s and g are completely determined from the first two equations of (4.104), but the third equation in (4.104) only gives a relation between c and d.

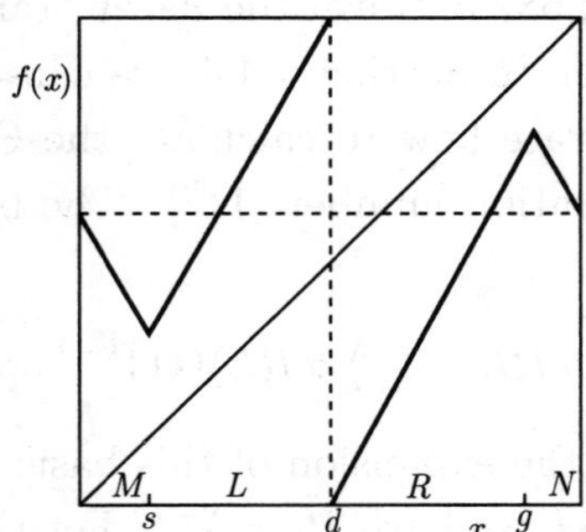

Figure 4.18 The mapping function of a piecewise linear circle map

In order to fix the map, we proceed as follows. The extreme position of g would put the maximum at the top of the unit square, corresponding to a discontinuity at d_1 on the L branch, which yields

$$\lambda d_1 - \delta_l = \lambda g - \delta_r.$$

Similarly, the extreme position of s would put the minimum at the bottom of the unit square, corresponding to a discontinuity at d_2 on the R branch, which yields

$$\lambda d_2 - \delta_r = \lambda s - \delta_l.$$

The actual discontinuity d may be put anywhere between d_1 and d_2. Suppose d is taken to be the middle point $(d_1 + d_2)/2$, we get

$$d = \frac{1}{2}(s+g), \quad c = \frac{1}{2}(1+\lambda). \tag{4.105}$$

4.7.5 Rotation Interval and Well-Ordered Periodic Sequences

A well-ordered periodic symbolic sequence can only contain the letters R and L. Therefore, it follows from the simplified expansion (4.96) that we have to deal with Θ_r only.

For a well-ordered periodic symbolic sequence the extreme members $\mathbf{L}^{\circ}_{\min}$ and $\mathbf{R}^{\circ}_{\max}$ of the corresponding shift sets must satisfy (4.74). We rewrite it here for easier reference:

$$\mathbf{R}^{\circ}_{\max} \leqslant K_g, \quad \mathbf{L}^{\circ}_{\min} \geqslant K_s. \tag{4.106}$$

From these relations one can find the upper and lower bounds of the rotation interval from K_s and K_g.

Using the relations (4.68) between the Farey transformation and the concatenation operation given in Section 4.4.4, we can calculate $\mathbf{L}^{\circ}_{\min}$ and $\mathbf{R}^{\circ}_{\max}$ recursively. We demonstrate how to calculate the Θ_r corresponding to $\mathbf{L}^{\circ}_{\min}$ on the example of the rotation number $\langle 1\bar{1}^k\rangle$. The basic string of $\mathbf{L}^{\circ}_{\min}$ under this rotation number is

$$\langle 1\bar{1}^k\rangle \circ RL = (\langle\ \rangle \circ RL)(\langle 1\bar{1}^{k-1}\rangle \circ RL). \tag{4.107}$$

Denote by P_{1k} the Θ_r in the expansion of this basic string. For the Θ_r in the expansion of $\langle\ \rangle \circ RL = RL$ we have $P_0 = \lambda^{-1}$. For the Θ_r in the expansion of $\langle 1\rangle \circ RL = RRL$ we have $P_{10} = \lambda^{-1} + \lambda^{-2}$. From (4.107) we have the recursion relations

$$P_{1k} = P_0 + \lambda^{-2} P_{1,k-1},$$

with $P_0 = \lambda^{-1}$ and $P_{10} = \lambda^{-1} + \lambda^{-2}$. Once the P_{1k} has been calculated, we have finally

$$\mathbf{L}^{\circ}_{\min,\langle 1\bar{1}^k\rangle} = \frac{\delta_l}{\lambda - 1} + \frac{P_{1k}}{1 - \lambda^{-2k-3}}. \tag{4.108}$$

Here we have used the fact that the period of a well-ordered symbolic sequence with rotation number $\langle 1\bar{1}^k \rangle$ is $2k+3$.

In principle, from the above P_0 and P_{10} as well as $P_{\bar{1}0} = \lambda^{-2}$ which corresponds to $\langle \bar{1} \rangle \circ RL = LRL$ we can calculate $\mathbf{L}^{\circ}_{\min}$ for any rotation number. Likewise, from $Q_0 = \lambda^{-2}$ corresponding to $\langle\ \rangle \circ LR$, $Q_{10} = \lambda^{-1} + \lambda^{-3}$ corresponding to $\langle 1 \rangle \circ LR = RLR$, and $Q_{\bar{1}0} = \lambda^{-3}$ corresponding to $\langle \bar{1} \rangle \circ LR = LLR$ we can calculate $\mathbf{R}^{\circ}_{\max}$ for any rotation number. However, in practice it is usually more convenient to generate first the $\mathbf{L}^{\circ}_{\min}$ and $\mathbf{R}^{\circ}_{\max}$ from the Farey address of the rotation number, then check them against (4.106) to determine the rotation interval.

5
Symbolic Dynamics of Two-Dimensional Maps

From now on we will study the symbolic dynamics of two-dimensional maps and apply what we learn to ordinary differential equations (ODEs) in Chapter 6. We emphasize that a thorough understanding of symbolic dynamics of 2D maps is crucial for the study of ODEs, as the essential dynamical behavior of ODEs may be captured as Poincaré maps in various sections. On the other hand, two-dimensional maps arise as models in many physical problems. Higher dimensional flows may be visualized as two dimensional mappings by taking Poincaré sections. Therefore, 2D maps deserve detailed study on their own.

The success of symbolic dynamics of one-dimensional maps has been largely based on the nice ordering property of real numbers, which is lacking in higher dimensions. This has hindered the development of symbolic dynamics of two-dimensional maps. Although symbolic dynamics of one-dimensional maps is topological in nature, it has a flavor of an algebraic approach, since we have been mostly dealing with symbols and combinatorics. The partition of the phase interval and assignment of symbols comes out so naturally that the extension to two-dimensional dynamics does not seem straightforward.

In order to solve the central problem of symbolic dynamics—the partition of phase plane, assignment of symbolic sequences, their ordering and admissibility condition, etc., a better understanding of the geometric aspects of the dynamics is essential. In addition, the geometric picture of chaotic motion, suppressed to a large extent in one dimension, is better exposed in a two-dimensional setting. To this end an essential step forward was the suggestion of Grassberger and Kantz [1985] to determine the partition line of the phase plane by connecting "primary" homoclinic tangencies between the invariant, stable and unstable, manifolds of the fixed points in the attractor. This construction has been further extended by invoking tangencies between the forward contracting foliations (FCFs) and backward contracting foliations (BCFs), i.e., the dynamical foliations of the phase plane. The FCFs and BCFs need not be invariant manifolds themselves, but they contain the invariant manifolds as subsets. Conceptually, the tangencies between FCFs and BCFs are much more general than the homoclinic tangencies. Operationally, this extension also facilitates the determination of partition lines, as the need of keeping on the invariant manifolds drops out.

In fact, the topological and geometrical approaches to chaos are closely related and their combination furnishes the most rigorous, but often not practical, definition for chaos. Nowadays, researchers in the physical sciences feel the necessity to become acquainted with a number of notions that several decades ago belonged to the realm of mathematics. The whole geometric approach is now backed by a well-shaped mathematical theory of dynamical systems. However, a thorough presentation of the geometrical approach is not the task of this book. We shall only touch a few geometric notions inasmuch as it is needed in developing symbolic dynamics. At the same time, we shall not pretend to mathematical rigor and shall continue to insist on an intuitive and practical way of presentation. For those readers who wish to learn more we recommend the excellent book by Thompson and Stewart [B1986].

In the construction of symbolic dynamics for two-dimensional maps there is a number of technical details. Without proper understanding of these details it is impossible to make symbolic dynamics a practical tool. On the other hand, we do not want the essence of 2D symbolic dynamics be dimmed by these details. Therefore, we organize this Chapter as follows.

Section 5.1 is a general discussion of an ideal situation that we would wish to have in order to construct symbolic dynamics in two dimensions. Section 5.2 shows that our ideal can indeed be realized by the dynamics itself, i.e., by the dynamical foliation of the phase plane. We precede the study of dynamical foliations by a discussion of invariant manifolds, which are subsets of the dynamical foliations. Intricate foldings and intersections of the stable and unstable manifolds are essential for understanding the origin of chaotic motion in dynamical systems and in terms of symbolic dynamics they acquire a concise and workable description.

All what is discussed in general terms in Sections 5.1 and 5.2 will be worked out explicitly on the example of two piecewise linear maps, the Tél map (Section 5.3) and the Lozi map (Section 5.4). The partition of the phase space for the both maps is given by the definition of the maps. Furthermore, the construction of symbolic dynamics in these two cases, carried out to a large extent analytically due to the piecewise linear nature of the maps, is quite instructive for understanding more complicated maps where numerical work is inevitable. In particular, the Lozi map provides the clues and hints for the Hénon map, which is the subject of Section 5.5.

The above three maps all lead to symbolic dynamics of two letters. The dissipative standard map, incorporating features of the circle map, requires more than two symbols to encode the trajectories. Nevertheless, the symbolic dynamics may be constructed. This is studied in Section 5.6. We will refer to this map when studying the symbolic dynamics of the periodically forced Brusselator.

There are a few cases of conservative systems where the idea of symbolic dynamics still works. For example, in the stadium billiard problem, one can encode the orbits by symbolic names, and work out the ordering rule and admissibility conditions for symbolic sequences as well. This makes the content of Section 5.7.

All sections on concrete examples are written independently from each other so the reader may consult any section to get a feeling of how 2D symbolic dynamics works, ignoring safely occasional references to other maps. However, in order to master a really working knowledge of 2D symbolic dynamics it is advised to go through more than one example.

5.1 General Discussion

What do we mean by constructing a good symbolic dynamics for a dynamical system from a practical point of view? First, one must partition the phase space properly in order to assign a unique symbolic name to each periodic orbit. As this can be checked for periodic orbits up to a certain length, there is a good hope that the partition may be used to name other orbits and put them into a certain classification scheme. Second, symbolic sequences thus obtained should be ordered properly. Based on the ordering rule, one would be able to formulate conditions to check whether a symbolic sequence may be produced by the dynamics or not, i.e., to formulate admissibility conditions for symbolic sequences. Third, equipped with ordering rule and admissibility conditions of symbolic sequences we will be able to predict and locate periodic orbits up to a given length, to indicate the structure of some chaotic orbits, and to study bifurcations in the parameter space as well. Once the above requirements are met, we will have a fairly good understanding of the topological aspects of the dynamics.

The very idea of symbolic dynamics as a way of coarse-grained description applies to dynamics in any dimension. However, as we have seen in the previous chapters, the construction of symbolic dynamics for one-dimensional maps essentially depends on the nice ordering property of real numbers. In particular, the admissibility conditions are based on ordering of symbolic sequences. The ordering, in turn, is made possible by the fact that all points on an one-dimensional line are ordered as naturally as real numbers.

In two and higher dimensions this nice ordering no longer exists. Nonetheless, a two-dimensional phase plane may be decomposed by two families of one-dimensional curves and one may order curves from one family along a curve of the other family, if these two families of curves intersect each other transversally. The transversal intersections may become tangencies in some parts of the phase plane. We will see that these tangencies determine the location of partition lines in the phase plane.

In order to see how this program is carried out, let us first imagine an ideal situation we would wish to have. In the next Section we will show that such a situation may indeed be realized by the dynamics itself.

5.1.1 Bi-Infinite Symbolic Sequences

The necessity of dealing with bi-infinite symbolic sequences is a characteristic feature of two- and higher-dimensional symbolic dynamics. To be more specific, let us consider a two-dimensional map of the following "triangle" form:

$$\begin{aligned} x_{n+1} &= f(a, x_n) + by_n, \\ y_{n+1} &= x_n, \end{aligned} \tag{5.1}$$

where $f(a, x)$ is a nonlinear function with parameter a and a second parameter b couples the variables x and y.

The Jacobian matrix of the map (5.1)

$$\mathbf{J} = \begin{bmatrix} \dfrac{\partial f}{\partial x} & b \\ 1 & 0 \end{bmatrix}, \tag{5.2}$$

in general, has a non-zero determinant $\det \mathbf{J} = -b$. The map may then be iterated backward as well as forward. Consequently, for a given point in the phase plane there is a bi-infinite numerical trajectory, which, once a certain partition of the phase space is given, would correspond to a bi-infinite symbolic sequence

$$\cdots s_{n-2}s_{n-1} \bullet s_n s_{n+1} s_{n+2} \cdots , \tag{5.3}$$

where we have put a dot "•" to denote the present time instant, and s_i represents a symbol describing where the i-th orbital point falls with respect to the partition.

Historically, the first much-studied two-dimensional map, proposed by M. Hénon (Hénon [1976], Hénon and Pomeau [1977]), corresponds to the choice of $f(a, x)$ to be a quadratic map:

$$f(a, x) = 1 - ax^2. \tag{5.4}$$

M. Lozi [1978] introduced a piecewise linear counterpart of the Hénon map, replacing the parabola (5.4) by a tent :

$$f(a, x) = 1 - a|x|. \tag{5.5}$$

An even simpler map is obtained by substituting the tent map (5.5) with a shift map (Tél [1983]):

$$f(a, x) = ax - \operatorname{sgn}(x), \tag{5.6}$$

where $\operatorname{sgn}(x)$ denotes the sign function of x, i.e.,

$$\operatorname{sgn}(x) = \begin{cases} 1, & \text{if } \ x > 0, \\ -1, & \text{if } \ x < 0. \end{cases} \tag{5.7}$$

For the last two piecewise linear maps the x interval may be partitioned using two letters R and L, according to whether x is greater or less than 0. Owing to the second equation in (5.1) this partition carries over to y according to whether y is greater or less than 0.

Take (x_n, y_n) to be the present point, by iterating (5.1) we get a numerical trajectory

$$\cdots,(x_{n-2},y_{n-2}),(x_{n-1},y_{n-1}),(x_n,y_n),(x_{n+1},y_{n+1}),(x_{n+2},y_{n+2}),\cdots. \quad (5.8)$$

Using $y_{n+1} = x_n$, one can keep only one variable for each orbital point to write

$$\cdots x_{n-2}x_{n-1} \bullet x_n x_{n+1} x_{n+2} \cdots \quad (5.9)$$

for the same orbit, where a dot "•" is put between $x_{n-1} = y_n$ and x_n. This means that iterating the map backward or forward once respectively corresponds to a leftward or rightward shift of the present dot "•". Assigning R to $x > 0$ and L to $x < 0$, we get a symbolic orbit like (5.3).

The semi-infinite symbolic sequence

$$\bullet\, s_n s_{n+1} s_{n+2} \cdots$$

is called a *forward symbolic sequence* (FSS), while the semi-infinite sequence

$$\cdots s_{n-2}s_{n-1} \bullet$$

is called a *backward symbolic sequence* (BSS). Shorthand notations $\bullet P$, $Q\bullet$, etc., will be often used for these symbolic sequences. Contrary to the one-dimensional (1D) cases where critical points divide the interval into subintervals, the partition of the two-dimensional phase space is provided by lines of partition. Generally, symbolic sequence $\bullet P$ or $Q\bullet$ corresponds to a line.

5.1.2 Decomposition of the Phase Plane

By taking point (x, y) as the initial point, its forward orbit corresponds to a forward sequence $\bullet P$. Generally, the points which share the same forward sequence $\bullet P$ form a line segment. Similarly, the points which share the same backward sequence $Q\bullet$ also form a line segment. Point (x, y) is the intersection of the two lines corresponding to its forward and backward sequences.

There are two families of one-dimensional curves: one corresponds to forward sequences and the other to backward sequences. Let us call them *family F* and *family B*, respectively. Suppose that we can decompose the phase plane with either the curves of family F or those of B. Curves from one family may intersect some curves from the other in a transversal manner. Then curves from one family, say family F, may be ordered along a curve of family B by using the ordering property of points on a one-dimensional line, and, similarly, curves of B be ordered on a curve of F. Although all points on a segment of a curve in the family B have one and the same backward symbolic sequence, the forward symbolic sequences of these points are, in general, different. (It may happen that an area of the phase plane, instead of a line, has one and the same forward symbolic sequence. A simple example is a basin of attraction in a one-dimensional map. This will not bring about any difficulty as it will become clear in the subsequent sections.) Then these different points on the segment from B may be labeled by their corresponding forward symbolic sequences. In this way an order can be assigned to these forward sequences and hence to their corresponding curve segments, e.g., $\bullet P_1 > \bullet P_2 > \bullet P_3$, etc. This situation is shown in Fig. 5.1.

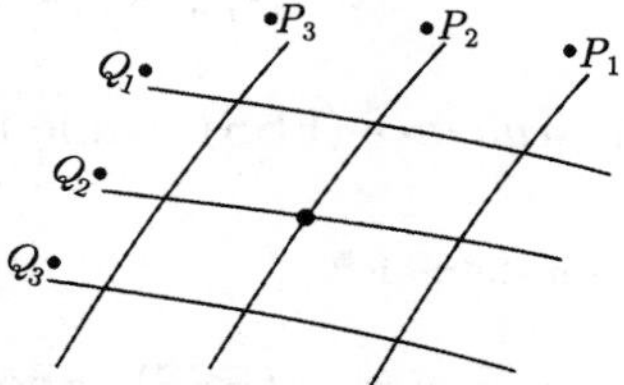

Figure 5.1 Decomposition of the phase plane using two families of curves

Similarly, points on a segment of F have one and the same forward symbolic sequence. They may be labeled by their corresponding backward symbolic sequences and the latter gain an order from the order of those points on the segment, e.g., $Q_3\bullet < Q_2\bullet < Q_1\bullet$, etc., as shown in Fig. 5.1.

Once this decomposition has been carried out, we can assign a two-sided symbolic sequence to each point in the phase plane. For example, for the central point in Fig. 5.1 we have $Q_2 \bullet P_2$. Sequences $Q_2\bullet$ and $\bullet P_2$ provide the "symbolic coordinates" of the point.

5.1.3 Tangencies and Admissibility Conditions

Generally, it is not always possible to have transversal intersections between the two families of curves. Situation shown in Fig. 5.2 may happen, when a curve, say, from family B touches a curve from family F tangentially and bends or folds back. Whether a tangency may happen or not depends, of course, on the dynamics itself. Whenever this happens, there are two immediate consequences.

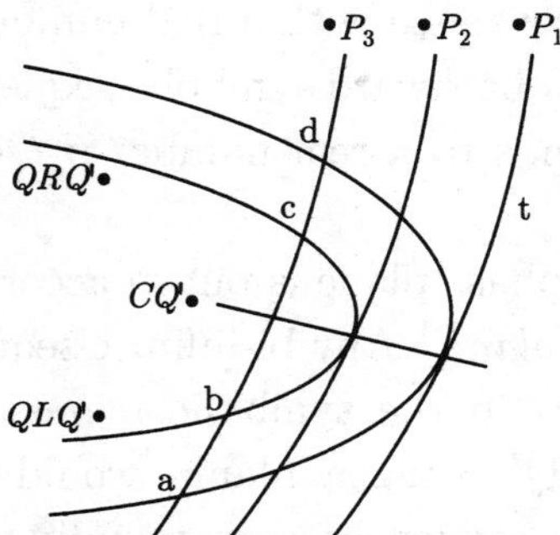

Figure 5.2 A tangency between the two families of curves and a pre-image of the partition line passing through the tangency

1. In Fig. 5.2 two continuous pieces from the family B are shown, which intersect with the line $\bullet P_3$ of the family F forming four crossings labeled with a, b, c and d. Points b and c on the same piece must have different backward sequences, so must points a and d. Otherwise we would not have a consistent ordering of the four crossings on $\bullet P_3$. Thus, the piece btc has to be cut in order to assign different backward sequences to b and c. The tangency point t is where the curves passing through b and c meet, hence provides a cutting point. A general form of backward sequenes for the four crossings a, b, c and d could be, say, $Q_1RQ'\bullet$, $QRQ'\bullet$, $QLQ'\bullet$, and $Q_1LQ'\bullet$, respectively (The only alternative is obtained by interchanging R with L in the four sequences). The *partition line* passing through the tangency shown in the figure may denote by $CQ'\bullet$.

2. Furthermore, the existence of a tangency implies that both the $QR\bullet$ and $QL\bullet$ curves, and all curves from family B which are located to the one side of the tangency, cannot intersect with segments in family F located on the other side of the tangency. In particular, in the situation shown in Fig. 5.2, both symbolic sequences $QRQ' \bullet P_1$ and $QLQ' \bullet P_1$ do not exist in this dynamics. Therefore, a tangency leads to an *admissibility condition* for a two-dimensional map.

5.1.4 Admissibility Conditions in Symbolic Plane

The admissibility condition, imposed by a tangency, may be better visualized by using the metric representations of the symbolic sequences introduced in Section 2.8. Suppose each forward symbolic sequence $\bullet P$, from the minimal to the maximal, is put in accordance with a real number $\alpha(\bullet P)$ with α changing from 0 to 1. Similarly, each backward symbolic sequence $Q\bullet$, from the minimal to the maximal, corresponds to a real number $\beta(Q\bullet)$, which changes from 0 to 1.

In this way the whole phase plane is put in accordance with a unit square, which is called a *symbolic plane.*[1] Any bi-infinite sequence $Q \bullet P$ is represented by a point $(\alpha(\bullet P), \beta(Q\bullet))$ in the symbolic plane. The tangency discussed above, in the case with Q' being a blank, would look like what is shown in Fig. 5.3. All symbolic sequences corresponding to points in the shaded rectangle demarcated by the four straight lines $\alpha = \alpha(\bullet P_2)$, $\alpha = 1$, $\beta = \beta(QR\bullet)$, and $\beta = \beta(QL\bullet)$, cannot occur in the given dynamics. Therefore, it determines a *forbidden zone* (FZ), which belongs to point $QC \bullet P$ on partition line $C\bullet$. Each point on $C\bullet$ with different forward sequence gives a FZ. Of course, all images and preimages of any FZ are forbidden by the dynamics.

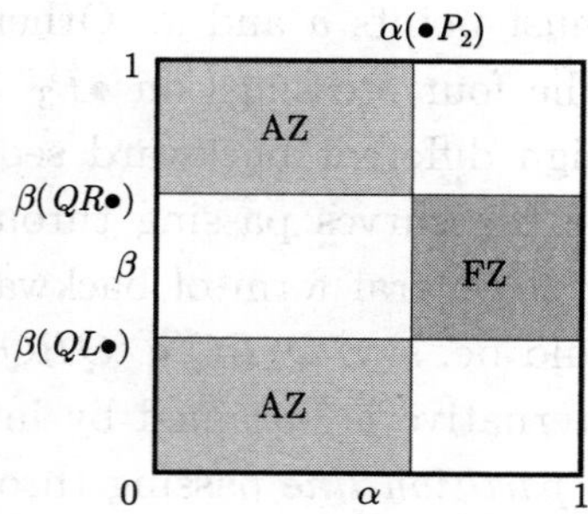

Figure 5.3 Symbolic plane of a two-dimensional map: the situation of one tangency $QC \bullet P_2$. FZ—forbidden zone, AZ—allowed zone, see the text

In fact, as shown in Fig. 5.2, when a tangency takes place, other tangencies may occur between nearby curves from the two families. A *partition line* connects together these tangent points. Each tangency gives rise to a forbidden zone like the one shown in Fig. 5.3. The union of all forbidden zones determines the admissibility of any symbolic sequence. Taken together, these FZs determine a *fundamental forbidden zone* (FFZ).

[1] The notion of a symbolic plane was first introduced in Cvitanović, Gunaratne and Procaccia [1988].

Given an admissible bi-infinite symbolic sequence, naturally, all its shifts must not be forbidden. Calculate the metric representations of all these shifts, *none* of them may fall in any forbidden rectangle determined by tangencies. (When all shifts of a given sequence are examined for its admissibility, we need consider only FZs given by $C\bullet$, safely ignoring images and preimages of FZs.) In Fig. 5.3 there are two rectangles labeled by allowed zones (AZ). For certain two-dimensional maps, AZs and FZs are exclusive in the sense that AZs contain no points of FZs. (Generally, AZs contain points of images or preimages of FZs while FZs contain no real orbit points.) If the representative points for *all shifts* of a symbolic sequence fall in a finite number of AZs, then it is admissible.

We call the reader's attention on the *all or none* alternatives in formulating the admissibility condition. In order to be an admissible sequence, *none* of its shifts should fall in the FFZ or *all* its shifts do fall in AZs. Therefore, in practice, it is much easier to exclude a symbolic sequence on the basis of a single tangency than accept one. Obviously, the admissibility of many sequences may not be decided by the single tangency under consideration. If some shifts fall in the rectangles other than those labeled by AZ or FZ, one has to invoke more tangencies to make the decision.

In the context of admissibility conditions tangencies play a role similar to kneading sequences in one-dimensional maps while a partition line plays the role of a critical point. However, there is a significant difference one should bear in mind. While for a 1D map with multiple critical points there is only a finite number of kneading sequences, for a 2D map there are infinitely many tangencies, hence infinite many "kneading sequences" at a fixed set of parameters. In practice, however, one can only be interested in symbolic sequences of finite length, then usually only a finite number of tangencies will matter.

In the next Section we will show that all what has been said is not a wishful thinking, but is provided by a realistic construction dictated by the dynamics. The two families of curves, family F and family B, correspond to the two kinds of dynamical foliations of the phase plane: the forward contracting foliations (FCFs) and the BCFs.

5.2 Invariant Manifolds and Dynamical Foliations of Phase Plane

How can one decompose the phase plane with two families of foliations, using the dynamics under study? We need to fix our notations first.

Consider a 2D map T:

$$T\colon \begin{cases} x_{n+1} = f(x_n, y_n), \\ y_{n+1} = g(x_n, y_n). \end{cases} \tag{5.10}$$

We suppose that the determinant of the Jacobian matrix of T

$$\mathbf{J} = \begin{bmatrix} \dfrac{\partial f}{\partial x} & \dfrac{\partial g}{\partial x} \\ \dfrac{\partial f}{\partial y} & \dfrac{\partial g}{\partial y} \end{bmatrix}$$

does not vanish in the phase plane, so the map (5.10) may be iterated both forward and backward. We note that the linearized dynamics near any point of a trajectory is given by $\mathbf{J}$ taken at that point. In particular, the dynamical behavior in a small neighborhood of that point is given by the eigenvalues of $\mathbf{J}$. We will be interested only in maps with real eigenvalues.

Moreover, we will assume that the absolute value of one eigenvalue, λ_u, is greater than 1 and that of the other eigenvalue, λ_s, is less than 1. Locally speaking, at each point of the phase plane there is a stretching (unstable) direction along the eigenvector of λ_u and a contracting (stable) direction along the eigenvector of λ_s.

5.2.1 Stable and Unstable Invariant Manifolds

A two-dimensional map realizes a transformation of the plane into itself. Objects in the plane that remain unchanged under the transformation are called invariant. Obviously, fixed points are invariant under the mapping. Sinks and sources are isolated invariant points, as any neighboring point will get closer to or farther away from the fixed point when the transformation is applied repeatedly. Saddle points distinguish themselves significantly in this respect: they sit at the intersection of two invariant curves.

Indeed, a small deviation from the saddle along the eigenvector $\vec{\xi}$ corresponding to the eigenvalue whose absolute is larger than 1 (denoted by λ_u

hereafter) only stretches by a factor λ_u. It remains in the same or opposite direction, depending on whether the sign of λ_u is positive or negative. Similarly, a deviation along the other eigenvector $\vec{\eta}$ corresponding to $|\lambda_s| < 1$ remains in the contracting direction, possibly with a sign change. The stretching direction is called the *unstable direction* (or the *outset*) of the saddle, while the contracting direction—the *stable direction* (the *inset*) of the saddle. What has been said holds in the linearized regime, as long as the map may be replaced by the tangent map acting on small deviations from the saddle.

What happens with these small invariant straight lines when they get further away from the saddle? In principle, they can develop into invariant curves called in general *invariant manifolds;* these curves may either terminate at other fixed points or continue to wander in the plane.

There are at least two methods for the visualization of these invariant manifolds: a numerical and an analytical one. For the sake of clarity, we proceed in a slightly formal manner.

As we have assumed that map T has a nonsingular Jacobian, so the inverse map T^{-1} always exists. Suppose further that we have found a saddle type fixed point (x^*, y^*), the unstable direction $\vec{\xi}$ corresponding to λ_u, and the stable direction $\vec{\eta}$ corresponding to λ_s. Numerically, one can take a tiny segment of a length ϵ on one side of the saddle along the unstable direction, and divide it into, say, $n = 1000$ subsegments. Then the map T is applied to each of the n dividing points $\xi_i, i = 1, 2, \cdots, n$. If λ_u is negative, as that for the fixed point of the Hénon map in the first quadrant, one should apply T^2 instead of T to remain on the continuation of the tiny segment on the same side of the saddle. The n images will outline a part of the unstable invariant manifold $\mathcal{M}^u$. In order to get more points on $\mathcal{M}^u$ one takes the first subsegment closest to the saddle, divides it again with n points and then applies to them the map T (or T^2, if the map reverses direction). This process may be repeated to obtain more and more points. By taking initial points closer and closer to the saddle, we ensure the validity of the linear approximation, and hence the precision of the reconstructed invariant manifold. If the points on $\mathcal{M}^u$ are distributed unevenly, then one has to compensate for the exponential separation of the points under the mapping. It is sufficient to divide the initial segment also in an exponential way, adjusting the numerical factor by trial and error. In order to construct the stable invariant manifold $\mathcal{M}^s$, one starts with the stable direction of the saddle and applies the inverse map T^{-1} (or T^{-2}) to the initial points. It is understandable that one can get a general (usually fairly good) idea regarding the shape of the invariant manifolds, but should not overdo it due to an unavoidable accumulation of numerical errors.

Analytically, a part of the invariant manifolds can be obtained as a series expansion in terms of small deviations from the saddle (Simó [1979]; Franceschini and Russo [1981]). Suppose the invariant curve is given by $y = \phi(x)$. The saddle point (x^*, y^*), any initial point (x_n, y_n) and its image (x_{n+1}, y_{n+1}) should all lie on this curve, i.e.,

$$y^* = \phi(x^*), \quad y_n = \phi(x_n), \quad y_{n+1} = \phi(x_{n+1}). \tag{5.11}$$

Substituting the map into both sides of the last equation of (5.11) and using the second one, we find

$$g(x_n, \phi(x_n)) = \phi(f(x_n, \phi(x_n))). \tag{5.12}$$

By dropping the subscript of x_n, in the case of the Hénon map (5.4), Eq. (5.12) reads

$$x = \phi(1 - \mu x^2 + b\,\phi(x)). \tag{5.13}$$

This is a functional equation for $\phi(x)$. In the vicinity of the saddle one can write

$$\begin{aligned} x &= x^* + \xi, \\ y &= y^* + \eta, \end{aligned}$$

and present ϕ as a series in ξ:

$$y = y^* + \eta = \phi(x^* + \xi) = \phi(x^*) + \sum_{k\leqslant 1} \beta_k \xi^k. \tag{5.14}$$

Substituting (5.14) back into (5.13) and eliminating x^* and y^*, in accordance with the fixed point condition, one gets a set of recursive relations for determining the coefficients β_k (Simó [1979]). Equipped with any algebraic manipulation language, it would not be difficult to get hundreds of terms. Later on, Franceschini and Russo [1981] used 100 terms in a similar expansion, using a parametric representation of the invariant curves.

We have discussed the invariant curves emerging from saddle type fixed points. Similar situations exist for unstable periodic orbits. It is sufficient to consider the fixed points of the pth iterate of the map, i.e., the map T^p, for which the stable and unstable invariant manifolds can be constructed.

The interrelation between stable and unstable manifolds plays a central role in the development of chaotic motion. The unstable manifold of a saddle point, in getting away from the latter, may intersect the stable manifold of the same saddle point. This is called a *homoclinic intersection.* If one homoclinic intersection takes place, then there must be infinitely many such intersections, because the homoclinic point belongs to both invariant manifolds and all its images and pre-images under the mapping necessarily stay in the manifolds. As a rule, a homoclinic intersection first appears as a *homoclinic tangency* when one invariant manifold touches the other tangentially. Then, at a further change of the parameter, the tangent point develops into a *transversal* intersection. The $\mathcal{M}^s$ and $\mathcal{M}^u$ from different unstable fixed points or unstable periodic points may intersect as well, leading to *heteroclinic intersections.*

We cannot do better than cite the first two paragraphs of §397 from the third volume of H. Poincaré's *Les Méthodes Nouvelles de la Mécanique Céleste,* which was first published in 1899[2]:

"Let us attempt to imagine the figure formed by these two curves and their infinite number of intersections, each of which corresponds to a double asymptotic solution. These intersections form a kind of trellis, a fabric, a lattice with an infinitely dense mesh; each of the two curves must never cross itself, but must fold back upon itself in a very complex fashion, in order to cross through the whole infinite number of lattice sites.

"The complexity of this figure, which I shall not even attempt to draw, is truly striking. Nothing else would be more appropriate in giving us an idea of the complications of the three-body problem and, in general, of all the problems in dynamics where there is no uniform integral ..."

When chaotic motion appears as the result of a finite or infinite number of bifurcations, where some periodic regimes loss their stability, the final "chaotic attractor" has a close relation to $\mathcal{M}_p^s$ and $\mathcal{M}_p^u$ of the unstable periods. Geometrically, it is easy to imagine that at least one branch of $\mathcal{M}_p^u$ will approach the attractor, while $\mathcal{M}_p^s$, traced backwards, will outline the boundary of the basin of attraction.

[2] We thank Dr. Ling-An Wu for the English translation from the French original.

5.2.2 Dynamical Foliations of the Phase Plane

Dynamical foliations as manifolds are generalization of the invariant manifolds. They are not invariant in general, but contain the invariant manifolds as subsets. Numerical practice shows that the dynamical foliations in 2D systems of smooth mapping functions seem to be smooth curves. They undergo a gentle change even when the dynamics itself changes quite wildly. At least this is true when one does not care about fine details of the foliations. We will see that a good symbolic dynamics is capable to assign a symbolic name to every foliation, which is segment of general stable or unstable manifolds. It is the tangency between two sets of dynamical foliations that determines the partition line in the two-dimensional phase plane.

Backward contracting foliations

Conceptually, one set of dynamical foliations is constructed as follows. Take an initial point (x_0, y_0) in the phase plane. Iterating n steps backward, using the inverse map of (5.10), we get to the point (x_{-n}, y_{-n}). Draw a small circle around this point and iterate all points on the circle forward n steps, the circle will become an ellipse centered at (x_0, y_0) with its long axis pointing to the most stretching direction. Repeat the construction with greater and greater n. The most stretching direction may change slightly. Finally, when $n \to \infty$, the most stretching direction approaches a limit. Now this most unstable direction is a property of the point (x_0, y_0) and the dynamics. Such most stretching directions of different points in the phase space determine a vector field. The integral curves of this vector field gives unstable manifolds — one family of foliations of the phase plane.

The most unstable directions of forward iterations are the most contracting directions under backward iterations. They form the *backward contracting foliations*, which will be abbreviated as BCFs, and may simply called backward foliations. J. M. Greene [1983] demonstrated the usefulness of the BCF on the example of the conservative standard map, obtained by putting $a = 0$ and $b = 1$ in Eq. (5.128).

Points on one and the same unstable manifold consisting of BCFs will approach each other with the highest speed under backward iterations. Therefore, one may introduce an equivalence relation: points p_1 and p_2 belong to the same manifold if they eventually approach the same destination under backward iterations of the map:

$$p_1 \sim p_2 \quad \text{if} \quad \lim_{n\to\infty} |T^{-n}(p_1) - T^{-n}(p_2)| = 0, \tag{5.15}$$

where $p_i = (x_i, y_i)$ and T is the map (5.10). Any two points in the same manifold have the same past. When the phase plane is partitioned, these points will eventually have the same backward symbolic sequence after a finite step n of backward iterations. It will be seen that, by properly break unstable manifolds into BCFs, all points in a BCF have the same backward symbolic sequence.

Forward contracting foliations

If we draw a small circle around the nth image of (x_0, y_0) and iterate all points on this circle backward n steps using the inverse map T^{-1} of T, we get an ellipse around (x_0, y_0). The long axis of this ellipse points is the most stretching direction of the inverse map. Now by fixing (x_0, y_0) and increasing n, this direction may change slightly. When n goes to infinity, the direction of the long axis approaches a limit. The most unstable direction of the inverse map represents the most stable direction of the map T, i.e., the direction corresponding to the most negative Lyapunov exponent. This is a property of the point (x_0, y_0) and the dynamics. Such most stable directions at various points of the phase plane determine a vector field. The integral curves of this vector field give the other set of foliations for the phase plane called FCFs, or simply forward foliations. They are also called *most stable manifolds* (Gu [1987, 1988]) or *strongly stable manifolds* or just *stable manifolds* in the literature.

Points on one and the same stable manifolds will approach each other with the highest speed under forward iterations of the map. Therefore, one may introduce an equivalence relation: points p_1 and p_2 belong to the same manifold if they eventually approach the same destination under forward iterations of the map:

$$p_1 \sim p_2 \quad \text{if} \quad \lim_{n\to\infty} |T^{n}(p_1) - T^{n}(p_2)| = 0. \tag{5.16}$$

Points in the same stable manifold have the same future. When the phase plane is partitioned, these points will eventually have the same forward symbolic sequence after a finite number n of iterations. By properly breaking a manifold into FCF segments, all points in a FCF have the same forward symbolic sequence. The BCF and FCF are not dynamically invariant in general, as one foliation transforms into other foliations under the map.

In the above discussion "a small circle" has been used for conceptual help. In practice, a small vector from the center to any point on the circle will approach the stretching direction in a few iterations. We have used the stretching directions in both forward and backward iterations, because stretching directions are easier to calculate numerically. For dissipative systems it is usually easier to calculate the FCF than BCF. Due to fast divergence of the backward iterations direct calculation of the BCF may encounter difficulty. In practice, one may obtain an outline of the BCF by calculating the forward orbits. This is because the chaotic attractor is contained in the closure of the unstable manifolds of fixed points and periodic points embedded in the attractor[3], this part of the BCF can be obtained without difficulty.

Symbolic names of invariant manifolds

Since invariant manifolds are subsets of the dynamical foliations, they acquire symbolic names when the partition lines are determined. Take, for example, the "triangle" 2D map (5.1) introduced in Section 5.1 with function f defined by the piecewise linear $f(a,x) = 1 - a\,|x|$. This map has a fixed point in the first quadrant of the phase plane. By the definition of the map, the phase plane is partitioned naturally by the y-axis. Therefore, the forward foliation passing through this fixed point must be named $\bullet R^\infty$, while the backward foliation is $R^\infty\bullet$. An eventually periodic forward foliation $\bullet\Sigma R^\infty$ must belong to the invariant stable manifold $\mathcal{M}^s$ of the fixed point; likewise $R^\infty\Pi\bullet$ belongs to the invariant unstable manifold $\mathcal{M}^u$ of the fixed point. Here Σ or Π denotes a finite string of symbols.

Furthermore, a bi-infinite symbolic sequence $R^\infty\Pi\bullet\Sigma R^\infty$ represents a homoclinic intersection, as it approaches the fixed point in both forward and backward directions. Similarly, $(RL)^\infty\Pi'\bullet\Sigma R^\infty$ and $R^\infty\Pi\bullet\Sigma'(RL)^\infty$ represent two different types of heteroclinic intersections between the fixed point R^∞ and the period 2 points $(RL)^\infty$. We see that in two-dimensional dynamics eventually periodic symbolic sequences are clearly related to the invariant

[3] To our best knowledge this is a general belief, not a completely proved mathematical statement.

manifolds, which is not obviously shown in one-dimensional dynamics. When dealing with piecewise linear 2D maps, we are capable to calculate the dynamical foliations analytically from their symbolic names. This surely enriches our understanding of the corresponding symbolic dynamics.

In Appendix A, a C program to generate dynamical foliations for the Hénon map is listed. It may help the reader to understand how the general principles outlined in this section is implemented numerically.

5.2.3 Summary and Discussion

The notion of dynamical foliations has been in existence under various names for a long time. For example, J. M. Greene [1983] showed that the most stretching directions (the BCFs) converge well in most points of the phase plane. He also gave numerical evidence that the unstable manifolds of unstable periodic points are contained ("coincide" was the word used) in the BCFs. Forward contracting foliations are sometimes called strongly stable manifolds or the most stable manifolds (Gu [1987, 1988]). Gu showed that the FCFs are easy to calculate and the transition to chaos in 2D dissipative systems is caused by tangency of the unstable manifolds of a saddle point with the BCF. The calculated dynamical manifolds are quite smooth although the dynamics itself may vary wildly. In Eckmann and Ruelle [1985] dynamical foliations are simply called stable and unstable manifolds.

The role of tangencies between FCFs and BCFs in determining the partition line and the association of FCFs (BCFs) with forward (backward) symbolic sequences come much later. Grassberger and Kantz [1985] first suggested to use the tangency between the invariant stable and unstable manifolds of the fixed point in the first quadrant to determine the partition line for the Hénon map. They were primarily interested in encoding the unstable periodic orbits for calculating topological entropy. Cvtanović, Gunaratne and Procaccia [1988] used metric representation of symbolic sequences to define a symbolic plane. They introduced the concept of pruning front to check the admissibility of symbolic sequences by explicitly calculated homoclinic tangencies between the invariant manifolds.

From the viewpoint of the phase plane decomposition by dynamical foliations, invariant manifolds of the fixed point, which belong to the FCFs and BCFs as a special subset, are far from complete. In fact, any tangencies

cies between FCFs and BCFs play the same role as the homoclinic tangencies in defolding the dynamics. The tangencies between FCFs and BCFs have been used to define the partition lines (Zheng [1991b, 1992a]; Zhao and Zheng [1993]). This generalization is necessary at least for the following reasons:

1. Conceptually, tangencies between FCFs and BCFs are much more general than homoclinic tangecies. A complete pruning front cannot be determined solely by homolinic tangencies.

2. Without being stuck on homoclinic tangencies, the generalization helps us understand the essence of symbolic dynamics deeper.

3. Operationally speaking, the calculation of dynamical foliations is more convenient, as it may be carried out at any point of the phase plane and there is no need to keep staying within invariant manifolds.

Analytical treatment of the piecewise linear Tél and Lozi maps in subsequent sections will further clarify these remarks. In particular, in a phase plane, in addition to regions where partition lines strictly follow tangencies between FCFs and BCFs, there are regions in which a partition line may be drawn rather freely. Our experience with the two piecewise linear maps will be used to guide the mostly numerical study of the Hénon map in Section 5.5 and the differential equations in Chapter 6.

5.3 The Tél Map

The Tél map is the simplest 2D map in the sense that, owing to the constant Jacobian matrix, it has constant eigenvalues and eigenvectors in the locally linear dynamics at all points. Consequently, the forward and backward foliations have constant slopes in the whole phase plane.

Recall the one-dimensional shift map, introduced in Section 2.8.1,

$$x_{n+1} = ax_n - \operatorname{sgn}(x_n), \tag{5.17}$$

where $\operatorname{sgn}(x) = \pm 1$ is the sign of x, see (5.7). The form of this map has been shown in Fig. 2.20. Now couple the 1D shift map of x to the y direction by a linear term, we get a two-dimensional map first considered by T. Tél in connection with fractal dimension of the strange attractor (Tél [1983]):

$$\begin{aligned} x_{n+1} &= ax_n - \operatorname{sgn}(x_n) + by_n, \\ y_{n+1} &= x_n, \end{aligned} \tag{5.18}$$

where parameters a and b are taken to be positive constants. In fact, we will concentrate on the case $a+b>1$, $b<1$. In addition, we mention that (5.18) is an anti-symmetric map, i.e., it remains the same if we change (x, y) into $(-x,-y)$.

Fixed points

In order to find the fixed points of the map, we equate x_n and x_{n+1} as well as y_n and y_{n+1} in both sides of (5.18). The solution

$$x = y = \frac{\text{sgn}(x)}{a+b-1},$$

yields two fixed points

$$\begin{aligned} \text{H}_+ &: x = y = \frac{1}{a+b-1} > 0, \\ \text{H}_- &: x = y = -\frac{1}{a+b-1} < 0, \end{aligned} \tag{5.19}$$

which are located in the first and third quadrant, respectively.

Period 2 points

The period 2 points of the Tél map may be found as well. They are

$$(x, y) = \left(\frac{\mp 1}{a-b+1}, \frac{\pm 1}{a-b+1}\right). \tag{5.20}$$

This period 2 orbit jumps between the second and fourth quadrants. The symbolic dynamics analysis below will show that when there is a chaotic attractor both fixed points lie outside the attractor if $a < 2\sqrt{1-b}$ and the period 2 points are located on the boundary of the attractor.

The inverse map of (5.18) reads

$$\begin{aligned} x_n &= y_{n+1}, \\ y_n &= [x_{n+1} - a y_{n+1} + \text{sgn}(y_{n+1})]/b. \end{aligned} \tag{5.21}$$

We note that if x and y are interchanged this inverse map is of the same form as the original map (5.18).

5.3.1 Forward and Backward Symbolic Sequences

According to the Tél map (5.18), the y axis divides the phase plane into two halves. The mapping function has a different form in these halves, therefore we denote the left and right halves by the letter L and R, respectively. The y axis may be denoted by $\bullet C$. Starting from an initial point $(x_0, y_0) \equiv (x_0, x_{-1})$ and iterating forward using the map (5.18), we get a forward trajectory

$$x_{-1}\, x_0\, x_1\, x_2\, \cdots\, x_{n-1}\, x_n\, \cdots . \tag{5.22}$$

The reason for only writing the x_i consists in that any consecutive pair in the above trajectory specifies the two coordinates of a point, e.g., $x_{n-1}\, x_n \to (x_n, y_n)$ due to $y_{n+1} = x_n$. Juxtaposing a symbol L or R to each number (except $y_0 = x_{-1}$) in (5.22) according to whether $x_i < 0$ or $x_i > 0$, we get a forward symbolic sequence

$$s_0\, s_1\, s_2\, s_3\, \cdots\, s_{n-1}\, s_n\, \cdots . \tag{5.23}$$

Starting from the same initial point $(x_0, y_0) \equiv (x_0, x_{-1})$, one may iterate backward using the inverse map (5.21) to get a backward trajectory

$$\cdots\, x_{-(m+1)}\, x_{-m}\, \cdots\, x_{-2}\, x_{-1}\, x_0. \tag{5.24}$$

We write a backward trajectory from right to left in order to distinguish it from the forward trajectory. Only coordinates x_i are written for the same reason that a consecutive pair in (5.24) represents the both coordinates of a point. Assigning a letter L or R to each x_i with $i < 0$, we get a backward symbolic sequence from (5.24):

$$\cdots\, s_{\overline{m+1}}\, s_{\bar{m}}\, \cdots\, s_{\bar{2}}\, s_{\bar{1}}, \tag{5.25}$$

where $s_{\bar{m}}$ is the symbol for x_{-m}, as the symbol s_n for x_n appears in (5.23). We always use $s_{\bar{i}} = s_{-i}$ for the symbol of x_{-i}.

The mapping relation $y_{n+1} = x_n$ tells us that the right half-plane is mapped to the upper half-plane, and the left half-plane to the lower half-plane. That is, points in the upper and lower half-planes have their pre-images coded as R and L, respectively. Therefore, we may assign letters L and R respectively to $y_i < 0$ and $y_i > 0$, and regard them as codes according to pre-image.

Obviously, this way of coding according to y_i coincides with the way coding according to x_{i-1}.

The two semi-infinite symbolic sequences may be concatenated to yield a bi-infinite sequence

$$\cdots s_{\bar{m}} \cdots s_{\bar{2}}\, s_{\bar{1}} \bullet s_0\, s_1\, s_2 \cdots s_n \cdots . \tag{5.26}$$

We have put a dot $\bullet$ between $s_{\bar{1}}$ and s_0, i.e., the symbols for $y_0 = x_{-1}$ and x_0, to indicate the present. An iteration forward corresponds to a right shift of dot $\bullet$ by one symbol; an iteration backward to a left shift. As one point in the phase plane is represented by one symbol, the two neighboring symbols of dot $\bullet$ always code the y and x coordinates of a given point. By taking (x_n, y_n) as the "present" point, its bi-infinite sequence is simply

$$\cdots s_{n-2}\, s_{n-1} \bullet s_n\, s_{n+1}\, s_{n+2} \cdots ,$$

where s_n and s_{n-1} indicate the symbols of x_n and $y_n = x_{n-1}$, respectively.

5.3.2 Dynamical Foliations of Phase Space and Their Ordering

The Jacobian matrix of the Tél map

$$\mathbf{J} = \begin{bmatrix} a & b \\ 1 & 0 \end{bmatrix} \tag{5.27}$$

has constant entries, hence a constant determinant and constant eigenvalues everywhere:

$$\det \mathbf{J} = -b, \quad \lambda_{\pm} = \frac{a \pm \sqrt{a^2 + 4b}}{2}. \tag{5.28}$$

This makes the Tél map the simplest 2D map for study. The two eigenvalues correspond to the two eigenvectors

$$\mathbf{v}_{\pm} = (\lambda_{\pm}, 1). \tag{5.29}$$

When $a + b > 1$ the eigenvalue $\lambda_+ > 1$ and $0 > \lambda_- > -1$, so $\mathbf{v}_+$ points to the unstable direction and $\mathbf{v}_-$ to the stable direction. The case of $a + b < 1$ is too simple to be considered.

Generally speaking, symbolic sequences in one-dimensional maps correspond to *points* in the interval, while in two-dimensional maps such semi-infinite sequences correspond to lines in the phase plane. For example, in an

1D unimodal map the letter C indicates the critical point, while in the Tél map the x-axis may be denoted by $C\bullet$ and the y-axis by $\bullet C$.

As $\mathbf{v}_\pm$ are constant vectors, the stable or unstable directions are the same at all points of the phase plane. Owing to the piecewise linear feature of the map, it is possible to find a segment of some straight line so that all the points on the segment may have one and the same backward sequence

$$\cdots s_{n-2}\, s_{n-2}\, s_{n-1} \bullet .$$

Assume the straight line to be

$$x - h_n y = \eta_n. \tag{5.30}$$

Taking a certain point (x_n, y_n) on this line and using the map (5.18), we get

$$\eta_n = x_n - h_n y_n = (a - h_n)x_{n-1} + b y_{n-1} - \epsilon_{n-1}, \tag{5.31}$$

where we have introduced a further shorthand for the sign function $\epsilon_n = \mathrm{sgn}(x_n)$. Regrouping terms in (5.31) to form an $\eta_{n-1} = x_{n-1} - h_{n-1}y_{n-1}$, we are led to

$$\eta_n = -bh_{n-1}^{-1}\eta_{n-1} - \epsilon_{n-1}, \tag{5.32}$$

where h_n and h_{n-1} are related by

$$h_n = a + \frac{b}{h_{n-1}}. \tag{5.33}$$

Using this relation repeatedly, h_n can be expanded into a continued fraction

$$h_n = a + \cfrac{b}{a + \cfrac{b}{a + \cfrac{b}{a + \cdots}}}.$$

A standard shorthand for this continued fraction is

$$h_n = a + \frac{b}{a\,+}\,\frac{b}{a\,+}\,\frac{b}{a\,+}\,\frac{b}{a\,+}\,\cdots. \tag{5.34}$$

We see that h_n does not depend on n, i.e., the location. The infinite continued fraction converges to a value between a and $a+b/a$. In fact, it can be calculated by solving $h = a + b/h$ for h:

$$h_n = h \equiv \frac{a + \sqrt{a^2 + 4b}}{2} = \lambda_+. \tag{5.35}$$

The requirement $h > a$ has picked up λ_+, which indicates that BCFs correspond to the forward stretching direction.

Now introduce a backward operator $\mathcal{B}$ and a forward operator $\mathcal{F}$ defined in the following way:

$$\mathcal{B}u_n = u_{n-1}, \quad \mathcal{F}u_n = u_{n+1}, \tag{5.36}$$

where $\{u_n\}$ is an arbitrary sequence. Using the backward operator $\mathcal{B}$, we write a particular solution of Eq. (5.32) as

$$\begin{aligned}\eta_n &= -(1+bh^{-1}\mathcal{B})^{-1}\epsilon_{n-1} = -\sum_{i=0}^{\infty}(-bh^{-1}\mathcal{B})^i\epsilon_{n-1}\\ &= -\epsilon_{n-1} + bh^{-1}\epsilon_{n-2} - b^2h^{-2}\epsilon_{n-3} + \cdots\\ &= -\epsilon_{n-1} - k\epsilon_{n-2} - k^2\epsilon_{n-3} - k^3\epsilon_{n-4} - \cdots,\end{aligned} \tag{5.37}$$

where $k = -b/h$.

Expression (5.37) shows that η_n is entirely determined by the signatures $\epsilon_{n-1}, \epsilon_{n-2}, \epsilon_{n-3}, \cdots$, i.e., by the backward symbolic sequence. Given a backward symbolic sequence $Q\bullet$ made of R and L, one can determine a straight line $x - hy = \eta$ from (5.37). On a suitable interval of this line all the points have the same backward symbolic sequence, i.e., the given sequence $Q\bullet$. These line intervals, taken together, give one set of foliations of the phase plane, namely, the BCFs. They all have the same slope h given by (5.34), i.e., parallel to $\mathbf{v}_+$.

Let us look at a few specific lines in the BCFs:

1. Through the fixed point H_+ in the first quadrant goes the $R^\infty\bullet$ line. Since all $\epsilon_i = 1$, we get

$$\eta(R^\infty\bullet) = -\frac{1}{1-k}. \tag{5.38}$$

Through the fixed point H_- in the second quadrant goes the $L^\infty\bullet$ line. Due to the anti-symmetry of the map we have

$$\eta(L^\infty\bullet) = \frac{1}{1-k}.$$

2. The two period 2 points contain $(LR)^\infty\bullet$ (the upper line) and $(RL)^\infty\bullet$ (the lower line). Their η values are

$$\eta((LR)^\infty\bullet) = \frac{-1}{1+k}, \quad \eta((RL)^\infty\bullet) = \frac{1}{1+k}. \tag{5.39}$$

3. Symbolic sequences $(LR)^\infty R\bullet$ is a preimage of $(LR)^\infty\bullet$, and $(RL)^\infty L\bullet$ that of $(RL)^\infty\bullet$. Their η values are $\mp(2k+1)/(k+1)$. No sequences are between $C\bullet$ and $(LR)^\infty R\bullet$ or between $C\bullet$ and $(RL)^\infty L\bullet$ (lower one).

These lines are drawn in Fig. 5.4, which will be discussed in more detail at the end of this section. Although the line $x - hy = \eta_n$ goes across the whole phase plane, a foliation with a leading $R\bullet$ exists only in the upper half-plane and that with a leading $L\bullet$ exists only in the lower half-plane. This is why we say that a suitable segment of the line, but not the whole line, belongs to the foliation. A similar remark applies to the forward foliations discussed below.

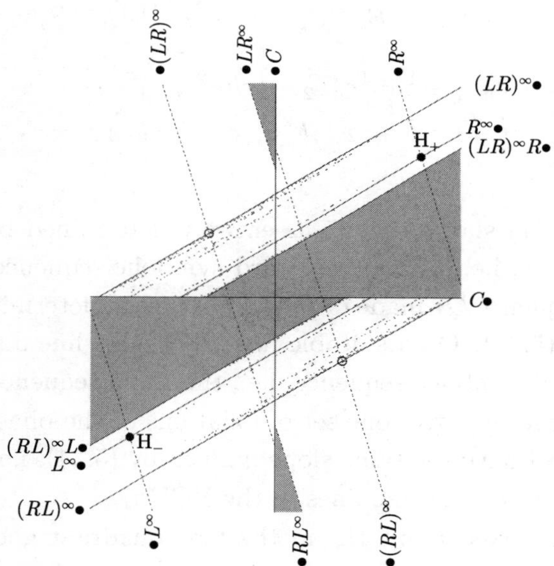

Figure 5.4 Dynamical foliations of the Tél map at $a = 1.35$, $b = 0.5$. BCFs are given by solid lines; FCFs by dotted lines. The dots belong to the strange attractor, which is part of BCFs

Similarly, assuming that some points with one and the same forward symbolic sequence $\bullet P = \bullet s_n\, s_{n+1} \cdots$ are located along a straight line $x - ky = \xi_n$, and taking a point (x_n, y_n) on this line and its image (x_{n+1}, y_{n+1}), we may write

$$\xi_{n+1} \;=\; x_{n+1} - k_{n+1}y_{n+1} = (a - k_{n+1})\xi_n - \epsilon_n, \tag{5.40}$$

where

$$k_n = -\frac{b}{a - k_{n+1}} = -\frac{b}{a\,+}\;\frac{b}{a\,+}\;\frac{b}{a\,+}\;\frac{b}{a\,+}\;\cdots, \tag{5.41}$$

which does not depend on n. In fact, the value of k can be calculated by solving $k = -b/(a-k)$:

$$k = \frac{a - \sqrt{a^2 + 4b}}{2} = \lambda_- < 0. \tag{5.42}$$

This solution shows that a forward contracting foliation relates to the stable direction $\mathbf{v}_-$. From (5.41) it follows that

$$a - k = -\frac{b}{k} = \lambda_+ = h \quad \text{or} \quad h^{-1}b = -k. \tag{5.43}$$

We have used this relation in (5.37).

Using the forward operator $\mathcal{F}$ defined in (5.36), we get a particular solution of (5.40):

$$\begin{aligned} \xi_n &= (1 - h^{-1}\mathcal{F})^{-1}h^{-1}\epsilon_n = h^{-1}\sum_{i=0}^{\infty}(h^{-1}\mathcal{F})^i\epsilon_n \\ &= h^{-1}\epsilon_n + h^{-2}\epsilon_{n+1} + h^{-3}\epsilon_{n+2} + \cdots . \end{aligned} \tag{5.44}$$

Again ξ_n is entirely determined by the signatures ϵ_n, ϵ_{n+1}, $\epsilon_{n+2}, \cdots$, i.e., by the forward symbolic sequences.

We look at a few specific lines among the FCFs:

1. Through the fixed point H_+ in the first quadrant goes the $\bullet R^\infty$ line. Since all $\epsilon_i = 1$, we have

$$\xi(\bullet R^\infty) = \frac{1}{h-1}. \tag{5.45}$$

Through the fixed point H_- in the third quadrant goes the $\bullet L^\infty$ line. Due to the anti-symmetry of the map we have

$$\xi(\bullet L^\infty) = -\frac{1}{h-1}.$$

2. The two period 2 points are contained in $\bullet(RL)^\infty$ (the right line) and $\bullet(LR)^\infty$ (the left line), respectively. Their ξ values are

$$\xi(\bullet(LR)^\infty) = \frac{1}{1+h}, \quad \xi(\bullet(RL)^\infty) = \frac{-1}{1+h}. \tag{5.46}$$

3. Symbolic sequence $\bullet RL^\infty$ is a preimage of $\bullet L^\infty$, and $\bullet LR^\infty$ that of $\bullet R^\infty$. Their ξ values are $\mp(h-2)/[h(h-1)]$. No sequences are between $\bullet C$ and $\bullet RL^\infty$ or between $\bullet C$ and $\bullet LR^\infty$ (left one).

These lines are also drawn in Fig. 5.4. A foliation with a leading $\bullet R$ exists only in the right half-plane and that with a leading $\bullet L$ exists only in the left half-plane.

Now we discuss the ordering of these two sets of foliations according to their symbolic sequences. Suppose that we have two symbolic sequences $Q\bullet P_1$ and $Q \bullet P_2$, i.e., the forward foliations $\bullet P_1$ and $\bullet P_2$ intersect one and the same backward foliation $Q\bullet$. Denote by $(\Delta x)_n$ the x-coordinate difference between the two intersecting points on $Q\bullet$. From the map (5.18), the y-difference between the two points after mapping $(\Delta y)_{n+1} = (\Delta x)_n$. From the form (5.30) for a BCF, $(\Delta x)_n = h_n(\Delta y)_n$. Combining the both, we have $(\Delta x)_{n+1} = h_{n+1}(\Delta x)_n$. Since $h_{n+1} = h_n = h > 1$, the signs of $(\Delta x)_{n+1}$ and $(\Delta x)_n$ will keep the same, being independent of n. Under the convention $\bullet R\cdots > \bullet L\cdots$, we can formulate the ordering rule for forward foliations as

$$\bullet UR\cdots > \bullet UL\cdots , \tag{5.47}$$

where U is an arbitrary finite string made of R and L. According to this ordering rule, the largest symbolic sequence among all forward foliations is $\bullet R^\infty$ and the smallest—$\bullet L^\infty$. They are the forward sequences for the fixed points in the first and third quadrants, respectively. We note that from (5.44) $\xi(\bullet R^\infty)$ is the greatest ξ. We note also that the ordering rule (5.47) is the same as that for the one-dimensional shift map, where both symbols R and L have an even parity, see Section 2.8.6.

Similar reasoning applies to the ordering of backward foliations. Now, for two backward foliations $Q_1\bullet$ and $Q_2\bullet$ intersecting the same forward foliation $\bullet P$, we have $(\Delta x)_n = k(\Delta x)_{n-1}$. Since $k < 0$, each iteration will reverse the sign of (Δx). Under the convention $\cdots R\bullet > \cdots L\bullet$, we formulate the ordering rule for backward foliations as

$$\cdots RE\bullet > \cdots LE\bullet, \quad \cdots RO\bullet < \cdots LO\bullet, \tag{5.48}$$

where the finite strings E and O contain an *even* and *odd* number of symbols R and L, respectively. According to this ordering rule the largest symbolic sequence among all backward sequences is $(LR)^\infty\bullet$, and the smallest sequence —$(RL)^\infty\bullet$. They respectively pass through each of the period 2 points in the second and fourth quadrants. From (5.37) $\eta((RL)^\infty\bullet)$ is the greatest

η. We note that the ordering rule (5.48) is the same as that for the one-dimensional inverse shift map, where both symbols R and L have an odd parity, see Section 2.8.6.

As we shall also see in the subsequent sections, it is often convenient to introduce a metric representation for symbolic sequences. We assign a real number $\alpha(\bullet P)$, to a forward symbolic sequence $\bullet P = \bullet s_1 s_2 \cdots s_n \cdots$ in the following way:

$$\alpha(\bullet P) = \sum_{i=1}^{\infty} \mu_i 2^{-i}, \tag{5.49}$$

where

$$\mu_i = \begin{cases} 0, & \text{when} \quad \epsilon_i = -1, \\ 1, & \text{when} \quad \epsilon_i = 1. \end{cases}$$

The number α may be called the "coordinate" of the symbolic sequence. It takes a value between 0 and 1, with 0 for the minimal and 1 for the maximal sequences.

The coordinate β for a backward symbolic sequence $\cdots s_{\bar{n}} \cdots s_{\bar{2}} s_{\bar{1}} \bullet \equiv Q\bullet$ is defined as

$$\beta(Q\bullet) = \sum_{i=1}^{\infty} \nu_i 2^{-i}, \tag{5.50}$$

where

$$\nu_i = \begin{cases} 0, & \text{when} \quad (-1)^{i-1}\epsilon_{\bar{i}} = -1, \\ 1, & \text{when} \quad (-1)^{i-1}\epsilon_{\bar{i}} = 1. \end{cases}$$

According to (5.49) and (5.50) we have

$$\begin{aligned} &\alpha(\bullet R^\infty) = \beta((LR)^\infty \bullet) = 1, \\ &\alpha(\bullet L^\infty) = \beta((RL)^\infty \bullet) = 0, \\ &\alpha(\bullet RL^\infty) = \alpha(\bullet LR^\infty) = 1/2, \\ &\beta((LR)^\infty R\bullet) = \beta((RL)^\infty L\bullet) = 1/2. \end{aligned} \tag{5.51}$$

In this way any forward or backward sequence is put into accordance with a real number between 0 and 1. A bi-infinite symbolic sequence with the present dot $\bullet$ indicated corresponds to a point in the α-β unit square. This square is called the symbolic plane. Vertical and horizontal lines in the symbolic plane correspond to forward and backward foliations, respectively.

Now we are well prepared to have a closer look at Fig 5.4, which is quite instructive for a deeper understanding of the symbolic dynamics of 2D maps.

This figure shows the phase plane of the Tél map at a typical set of parameters, namely, $a = 1.35$ and $b = 0.5$. All the forward and backward foliations whose η or ξ values were mentioned above have been drawn in the figure.

In Fig. 5.4 we have also superimposed a few hundred points of the attractor. Indeed, they seem to be part of the backward contracting foliations. The linearly layered structure of the attractor is dictated by the piecewise linear nature of the map. One may assign symbolic names to the most clearly seen lines in the attractor with some more effort.

The symbol assignment of the forward iterations (5.18) complies with the sign of x_i, which is the same as in a 1D shift map. Therefore, the maximal symbolic sequence is $\bullet R^\infty$, and the minimal symbolic sequence is $\bullet L^\infty$. These lines go through the upper and lower fixed points $\mathrm{H}_\pm$, respectively. The fact that they are the maximal and minimal sequences implies that to the right of the $\bullet R^\infty$ line all the points in the phase plane have the same forward symbolic sequence $\bullet R^\infty$; to the left of the $\bullet L^\infty$ line all the points in the phase plane have the same forward symbolic sequence $\bullet L^\infty$. For such sequences an area in the phase plane that has one and the same symbolic sequence is represented by just a single line in the symbolic plane. In particular, the areas with $\bullet R^\infty$ and $\bullet L^\infty$ are represented by the vertical sides of the unit square for the symbolic plane.

The symbol assignment of backward iterations complies with the sign of y_n or x_{n-1}. As in a 1D inverse shift map, the maximal and minimal symbolic sequences are given by the backward sequences of the two period 2 points, i.e., $(LR)^\infty\bullet$ and $(RL)^\infty\bullet$. Since these foliations go through the period 2 points, the two lines are part of the unstable manifolds of the periodic points. All the points in the phase plane which are located above the $(LR)^\infty\bullet$ line share the same backward symbolic sequence $(LR)^\infty\bullet$; all the points below the $(RL)^\infty\bullet$ line share the same backward sequence $(RL)^\infty\bullet$.

Symbolically speaking, all the interesting dynamics is then contained in the region demarcated by these four straight lines. In the symbolic plane these lines are represented by the four boundaries of the unit square, which also represent all the four areas outside the four lines in the phase plane. In fact, the actually interesting region of the map is slightly smaller, as it will be discussed in the next section.

The shaded regions in Fig. 5.4 are given by the symbolic sequences which are equivalent to $\bullet C$ or $C\bullet$. In the triangle region between the $\bullet LR^{\infty}$ line and the upper part of $\bullet C$ all the points have the same forward sequence $\bullet LR^{\infty}$, while in the triangle region between the $\bullet RL^{\infty}$ line and the lower part of $\bullet C$ all points have the same sequence $\bullet RL^{\infty}$. The two shaded regions both have metric representation $\alpha = 1/2$. This means that the actual partition line for the forward map may be drawn anywhere within these shaded triangles.

The shaded regions between $(LR)^{\infty}R\bullet$, $C\bullet$, and $(RL)^{\infty}L\bullet$ lines have the same interpretation. The two triangles now touch each other and the whole region has metric representation $\beta = 1/2$. The actual partition line for the backward map may be drawn anywhere in the shaded region.

5.3.3 Forbidden and Allowed Zones in Symbolic Plane

The most important task of applied symbolic dynamics is to describe all symbolic sequences allowed by a given dynamics, i.e., to clarify all possible types of motion in the system. This is conveniently provided by the forbidden and allowed zones in the symbolic plane and by the admissibility conditions embodied in terms of these zones.

In order to accomplish this task we return to the dynamical foliations of the phase plane. Fig. 5.5 is a sketch of a typical case. It differs from Fig. 5.4 in that a few more forward foliations are added. These are the lines labeled by $\bullet RK$ and $\bullet RJ$, and the anti-symmetrically located $\bullet L\bar{K}$ and $\bullet L\bar{J}$. Here we denote by $\bar{P}$ the sequence derived from P by interchanging its R and L. These four FCFs intersect the x-axis at the points where the backward foliations $(RL)^{\infty}\bullet$, $(RL)^{\infty}L\bullet$, $(LR)^{\infty}\bullet$, and $(LR)^{\infty}R\bullet$ lines end.

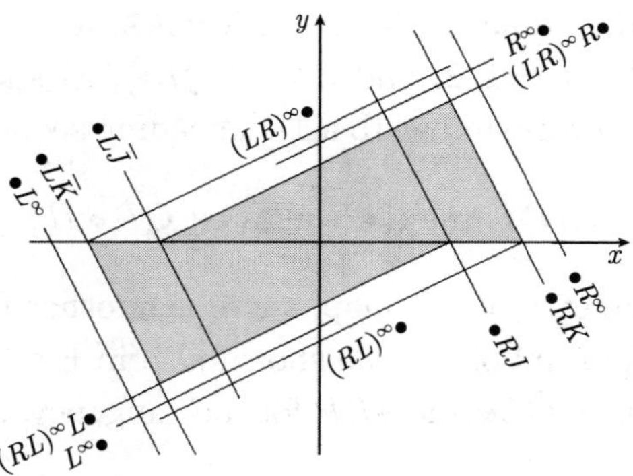

Figure 5.5 A sketch of dynamical foliations for the Tél map

The ordering of dynamical foliations restricts the type of symbolic sequences allowed in a given dynamics. Suppose that in the phase plane a backward foliation $QL\bullet$ intersects the x-axis, i.e., the line $C\bullet$, at a point where a forward foliation $\bullet RP$ goes through, as shown in Fig. 5.6 (a). Such an intersection point is actually a "tangency" between the backward and forward foliations. The term "tangency" has a better graphical justification in other models, e.g., the Lozi or Hénon maps studied later in this chapter.

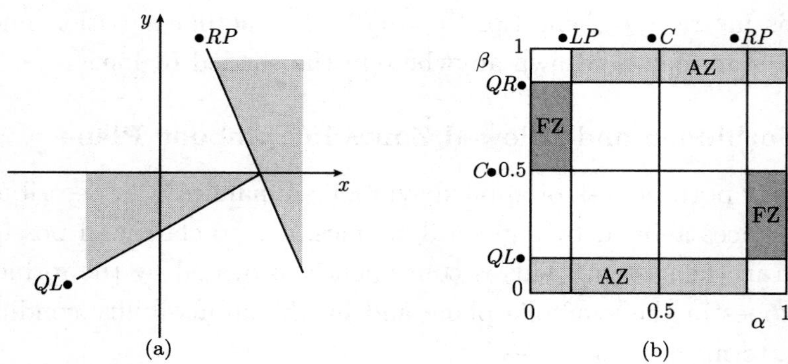

Figure 5.6 A "tangency" between backward foliation $QL\bullet$ and forward foliation $\bullet RP$. (a) The phase plane. (b) The symbolic plane

Whenever such a tangency takes place, any backward foliation which is ordered between $QL\bullet$ and $C\bullet$ cannot touch any forward foliation that is located to the right of $\bullet RP$. Consequently, all backward sequences located between $C\bullet$ and $QL\bullet$ with a leading L are not allowed to have a forward sequence which is larger than $\bullet RP$. In the symbolic plane such a tangency will exclude a rectangle "forbidden zone", whose borderlines are the four straight lines $\alpha = 1$, $\alpha = \alpha(\bullet RP)$, $\beta = 1/2$, and $\beta = \beta(QL\bullet)$, as shown in Figs. 5.6 (b). Sequence $\bullet RP$ plays a role similar to a 1D kneading sequence in the sense that

$$\bullet RP = \max\{\bullet H | \text{allowed } QL \bullet H\}.$$

Due to the anti-symmetry of the map, there is another forbidden zone in the upper left part of the symbolic plane shown also in Fig. 5.6 (b).

When two foliations $QL\bullet$ and $\bullet RP$ form a tangency, a necessary condition is

$$\eta(QL\bullet) = \xi(\bullet RP). \tag{5.52}$$

In particular, for the heteroclinic tangency between invariant manifolds of the period 2 and fixed point, i.e., the tangency between $(RL)^\infty\bullet$ and $\bullet R^\infty$, it follows from (5.39) and (5.45) that

$$\eta((RL)^\infty\bullet) = \frac{1}{1+k} = \frac{1}{h-1} = \xi(\bullet R^\infty). \tag{5.53}$$

By taking into account (5.35) and (5.42), this leads to the relation between the two parameters a and b:

$$a = a_c \equiv 2\sqrt{1-b}. \tag{5.54}$$

If $a < a_c$ the two lines $(RL)^\infty\bullet$ and $\bullet R^\infty$ are separated from each other. This is what happens in Figs. 5.4 and 5.5.

Now a certain line $\bullet RK$, which is smaller than $\bullet R^\infty$, has a tangency with the smallest backward sequence, i.e., $(RL)^\infty\bullet$. This $\bullet RK$ determines the rightmost and the first forbidden zone in the symbolic plane. It means that in all allowed symbolic sequences any subsequence which follows a symbol L cannot be greater than $\bullet RK$. By interchanging $R \leftrightarrow L$ in $\bullet RK$ we get its mirror image $\bullet L\bar{K}$. In all allowed symbolic sequences any subsequence which follows a symbol R cannot be less than $\bullet L\bar{K}$. In the phase plane the region, demarcated by the four lines $\bullet RK$, $\bullet L\bar{K}$, $(LR)^\infty\bullet$, and $(RL)^\infty\bullet$, forms the fundamental trap region for the map. All the points in this region remain in it under iterations.

Similarly, on the line $C\bullet$ there is a point being the rightmost among those whose backward sequences are all $(RL)^\infty L\bullet$, i.e., the largest of the sequences with a leading L. The backward foliation passing through this point has a tangency with some forward foliation, which is denoted by $\bullet RJ$ in Fig. 5.5. This is the last tangency on the right segment of $C\bullet$. There are infinitely many tangencies between the $(RL)^\infty\bullet$ versus $\bullet RK$ and $(RL)^\infty L\bullet$ versus $\bullet RJ$ tangencies, although in practice only a finite number of them may be and need to be determined. Each of these tangencies draws a rectangular forbidden zone in the symbolic plane.

We have discussed the ordering of two families FCFs and BCFs one along the other according to the x-coordinate difference Δx. Due to the one-to-one correspondence between either $\Delta\eta$ and Δx on a FCF or $\Delta\xi$ and Δx on a BCF, the natural order of $-\eta$ and ξ on the x-axis or $C\bullet$ coincides with the ordering rules (5.47) and (5.48) for backward and forward sequences, respectively. This means that, when forward foliations diminish from the first tangency at $\bullet RK$ to the last at $\bullet RJ$ the backward foliations involved in tangencies raise monotonically. As the consequence of this ordering rule the heights of the forbidden rectangles in the symbolic plane diminish from right to left. Taken together, these rectangles form a *fundamental forbidden zone.* The staircase-like outer boundary of the fundamental forbidden zone is called a *pruning front.*[4] Due to the anti-symmetry of the map, there is a similar fundamental forbidden zone and a pruning front in the upper left part of the symbolic plane. Owing to the monotonicity of pruning fronts, as shown in Fig. 5.6(b), corresponding to each forbidden zone there is an *allowed zone* (AZ) which contain no points of any forbidden zones.

An actual symbolic plane is shown in Fig. 5.7 for $a = 1.35$, $b = 0.5$, i.e., at the same set parameters as in Fig. 5.4. Only a finite number of forbidden zones are given. Also drawn in the figure are metric representations of 3000 real orbital points. Indeed, all of these points fall outside the fundamental forbidden zone.

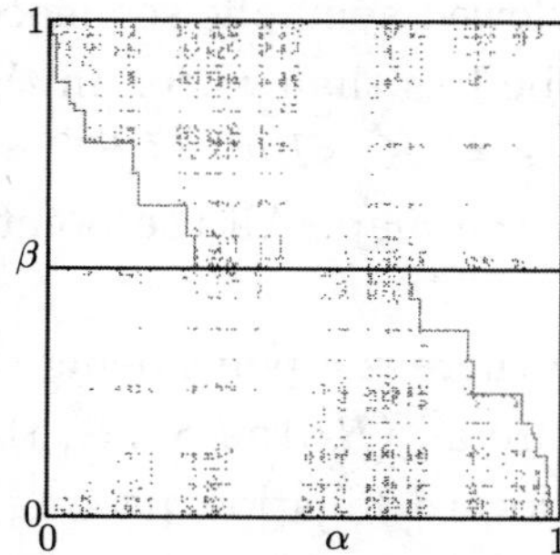

Figure 5.7 Symbolic plane of the Tél map at $a = 1.35$, $b = 0.5$

5.3.4 The Admissibility Conditions

We have discussed restrictions imposed by tangencies between the backward and forward foliations. Given a symbolic sequence with the present dot $\bullet$ indicated, it has a definite representative point in the symbolic plane. Mapping forward or backward corresponds to shift of the present dot rightward

[4] The term pruning front was introduced by Cvitanović, Gunaratne and Procaccia [1988].

or leftward. All the shifts have their representative points in the symbolic plane. Without causing any confusion we can call these points shifts of the original sequence. Obviously, an allowed sequence can not have any shift in the fundamental forbidden zone. In other words, should a shift fall in the fundamental forbidden zone, the original sequence must be forbidden, i.e., the symbolic sequence does not correspond to any real orbit of the map.

No shift falls in the fundamental forbidden zone—this is a necessary condition for a symbolic sequence to be admissible. Now we show that this is also a sufficient condition.

From the definition of $\xi = x - ky$ and $\eta = x - hy$ it follows

$$y = \frac{\xi - \eta}{h - k}. \tag{5.55}$$

Given a symbolic sequence with the present dot indicated, one calculates its ξ and η from the expansions (5.37) and (5.44). Then one gets a y from (5.55). Shifting the present dot, one gets another y. It is easy to check that the y values obtained in this way do satisfy the following difference equation

$$y_{n+1} = ay_n - \epsilon_{n-1} + by_{n-1}, \tag{5.56}$$

which is equivalent to the Tél map (5.18).

The construction of the fundamental forbidden zone tells us that the tangency condition on the pruning front is $\xi = \eta$, see, e.g., (5.52). It follows from the ordering rule that within the lower right fundamental forbidden zone one has $\xi > \eta$, while outside it $\xi < \eta$. In the upper half of the symbolic plane, opposite inequalities hold, i.e., $\xi < \eta$ within the fundamental forbidden zone and $\xi > \eta$ outside. Since $h > 0$ and $k < 0$, $(h - k)$ is a positive quantity. Accordingly, the sign of y coincides with that of $(\xi - \eta)$: $y > 0$ in the upper plane and $\xi > \eta$ in the allowed zone; $y < 0$ in the lower plane and $\xi < \eta$. We see that the y value calculated from an admissible symbolic sequence always agrees with the coding of the symbolic sequence. On the other hand, if any shift of a symbolic sequence falls in the forbidden zone, the calculated y cannot agree with the corresponding symbol. Thus, as long as none of shifts of a symbolic sequence falls in the fundamental forbidden zone, a phase space trajectory can be constructed consistently with the symbolic sequence. This explains the sufficiency of the condition. Therefore, we have a necessary and sufficient admissibility condition: an admissible symbolic sequence has no shifts falling in the fundamental forbidden zone.

In order to determine the fundamental forbidden zone precisely one has to know the infinitely many tangencies along the partition line $C\bullet$ in the phase space. Nonetheless, a finite number of tangencies is enough to tell the admissibility or prohibition of some sequences. To demonstrate this point, we return to the symbolic plane of the $QC\bullet$ and $\bullet CP$ tangency (Fig. 5.6 (b)). We have mentioned before that in the rectangle demarcated by the four straight lines $\alpha = \alpha(\bullet RP)$, $\alpha = \alpha(\bullet R^\infty) = 1$, $\beta = \beta(QL\bullet)$, and $\beta = 1/2$, one has $\xi > \eta$, which contradicts the fact that $y < 0$, hence it is a forbidden zone. However, in the rectangle circumscribed by $\alpha = \alpha(\bullet RP)$, $\alpha = \alpha(\bullet L^\infty) = 0$, $\beta = \beta(QL\bullet)$, and $\beta = \beta((RL)^\infty\bullet) = 0$, $\xi < \eta$ always takes place. Therefore, if *all* shifts of a certain symbolic sequence are located in such rectangles, it is sure that no shifts would fall in any forbidden zones, hence one can construct its phase space trajectory in a self-consistent way.

In summary, the criterion that *none* of shifts falls in any FZ or *all* shifts are in some AZs makes a symbolic sequence admissible. Please note the "none – all" alternative in the statement. All shifts being in a finite number of AZs is a sufficient, but not necessary condition for admissibility. Given a finite number of tangencies, one can construct FZ and AZ for each of them. Clearly, the union of these FZs is contained in the fundamental forbidden zone. Likewise, the union of all these AZs contains no part of the fundamental forbidden zone. If any shift of a sequence falls in the union of FZs, the sequence is forbidden. If all shifts fall in the union of the AZs, it is allowed.

The $b \ll 1$ limit

Now let us look at the particular case of very small b. In the limit $b \to 0$, we have $h \to a$ and $k \to 0$. The map reduces to a 1D shift map. It follows from the expansion (5.37) that $\eta \to 1$ for all backward foliations with a leading L. Therefore, all tangencies reduce to one. In order to construct the fundamental forbidden zone it is enough to calculate the forward foliation with $\xi = 1$. From (5.44) we have

$$1 = h^{-1}\epsilon_n + h^{-2}\epsilon_{n+1} + h^{-3}\epsilon_{n+2} + \cdots . \tag{5.57}$$

This is nothing but the λ-expansion, given in Section 2.8.2. The fundamental forbidden zone is determined by one kneading sequence only in accordance with symbolic dynamics of the one-dimensional map.

Moreover, we take $a = a_c = 2\sqrt{1-b}$ with $b \ll 1$ such that

$$1 - 2^{-l} < \sqrt{1-b} < 1 - 2^{-l-1},$$

where l is a positive integer. For this set of parameters $h = 1 + \sqrt{1-b} < 2$, but $h^2 > 2$. We show that there exist chaotic orbits by inferring the structure of a set of such orbits from symbolic dynamics.

First, the expansion (5.44) yields for large enough m

$$\xi(\bullet R^m L^\infty) = \frac{h^{-1} - 2h^{-m-1}}{1 - h^{-1}} = (1 - 2h^{-m})(1-b)^{-1/2}. \tag{5.58}$$

For the backward foliation $(RL)^\infty L\bullet$, which determines the left-most tangency among all backward foliations with a leading L, the expansion (5.37) leads to

$$\eta((RL)^\infty L\bullet) = \frac{2k+1}{k+1} = 2 - \frac{1}{\sqrt{1-b}}.$$

Therefore, recalling $h < 2$, for $m \leqslant l$ we have

$$\sqrt{1-b}[\eta((RL)^\infty L\bullet) - \xi(\bullet R^m L^\infty)] = 2\sqrt{1-b} + 2h^{-m} - 2 > 2^{-m+1} - 2^{-l} > 0.$$

This means that the forward foliation $\bullet RJ$ which has a tangency with the backward foliation $(RL)^\infty L\bullet$ must be greater than $\bullet R^m L^\infty$. The allowed zone of this tangency is a rectangle formed by the lines $\alpha = 0$, $\alpha = \alpha(\bullet RJ)$, $\beta = 0$, and $\beta = 1/2$.

It is easy to check that any symbolic sequence of the form $\cdots L\bullet R^i \cdots$ with $i < l$ must fall in this AZ. Due to the anti-symmetry of the map, any sequence of the form $\cdots R\bullet L^j \cdots$ with $j < l$ must fall in another AZ in the upper left part of the symbolic plane. Consequently, any sequences made of strings $R^i L^j$ with $i, j < l$ are allowed. In particular, we may form chaotic sequences by concatenating such sequences in a random manner. This proves the existence of chaotic motion in the Tél map for $a = 2\sqrt{1-b}$ and $b \ll 1$. We see that symbolic dynamics allows us to construct some classes of chaotic orbits, although, in general, it is impossible to indicate the structure of all chaotic trajectories.

Furthermore, it is possible to describe explicitly some classes of forbidden symbolic sequences. For example, at the same set of parameters we have

$$\xi(\bullet R^{2t+1} L^\infty) = (1 - 2h^{-2t})(1-b)^{-1/2} > (1 - 2^{-t+1})(1-b)^{-1/2} \tag{5.59}$$

and

$$\eta((LR)^\infty(RL)^s\bullet) = \frac{1-2k^{2s}}{1+k} < (1-2^{-2s(l+1)+1})(1-b)^{-1/2}. \qquad (5.60)$$

(We have made use of $h^2 > 2$.) Therefore, when $t > 2s(l+1)$ all sequences of the form $\cdots R(RL)^s \bullet R^{2t+1} \cdots$ must be forbidden. In other words, symbolic strings $R(RL)^s R^{2t+1}$ never correspond to any real motion in the system. We have, of course, indicated only some, by far not all forbidden orbits.

5.3.5 Summary

The symbolic dynamics of the Tél map may be summarized as follows. The phase space is partitioned by the y-axis ($\bullet C$) into right and left halves according to the present, or equivalently by the x-axis ($C\bullet$) into upper and lower halves according to the pre-image, labeled by R and L, respectively. This leads to a symbolic dynamics of two letters. Three essential issues of the symbolic dynamics are

The ordering rule

Forward symbolic sequences are ordered as

$$\bullet UR\cdots > \bullet UL\cdots, \qquad (5.61)$$

where U is an arbitrary string made of R and L. Backward symbolic sequences are ordered according to

$$\cdots RE\bullet > \cdots LE\bullet, \quad \cdots RO\bullet < \cdots LO\bullet, \qquad (5.62)$$

where E and O are strings, made of an even and odd number of symbols.

The partition line

In sharp contrast to its 1D counterpart where the partition of the phase space is given by a single point, now the partition is made by a line. Correspondingly, instead of a single kneading sequence for the 1D shift map, there are a series of restriction relations given by tangencies on the partition line between backward and forward foliations. The symbolic representation of the partition line in the symbolic plane is the pruning front. As a result of the organization of tangencies, when incrasing one symbolic coordinate along the pruning front the other never increases. This monotonicity describes the coupling among tangencies. The pruning front demarcates the fundamental forbidden zone.

The admissibility condition

Symbolic sequences considered in symbolic dynamics need not always correspond to any real orbits. (In fact, expansions (5.37) and (5.44) are meaningfull only for those sequences of "real" orbits.) The necessary and sufficient condition for a symbolic sequence to be admissible consists in that none of its shifts falls in the fundamental forbidden zone confined by the pruning front. Once the pruning front has been determined, the dynamical behavior of the system may be, in principle, established completely. However, in practice, one has to be content with a finite number of tangencies, which are enough to tell the admissibility or prohibition of some classes of orbits.

5.4 The Lozi Map

The Lozi map

$$\begin{aligned} x_{n+1} &= 1 - a|x_n| + by_n = 1 - a\epsilon_n x_n + by_n, \\ y_{n+1} &= x_n, \end{aligned} \tag{5.63}$$

was first introduced (Lozi [1978]) as a piecewise linear counterpart of the Hénon map, which will be studied in Section 5.5. We restrict the parameters to $a > 0$ and $0 < b < 1$. The ϵ_n in (5.63) is a shorthand for the sign function

$$\epsilon_n \equiv \mathrm{sgn}(x_n) = \mathrm{sgn}(y_{n+1}).$$

We will need the inverse map of (5.63), which reads

$$\begin{aligned} x_n &= y_{n+1}, \\ y_n &= (x_{n+1} + a\,\epsilon_n\,y_{n+1} - 1)/b. \end{aligned} \tag{5.64}$$

The existence of strange attractors in the Lozi map was first proved by Misiurewicz [1980] for a limited range of parameters, then for a much larger range $0 < b < 2/3$ and $1 + b < a < 2 - b/2$ by Liu, Xie, Zhu and Lu [1992]. While the Tél map is a 2D generalization of the 1D shift map, the Lozi map is a 2D generalization of the tent map, which was studied in Section 2.8.1. In spite of the equivalence of the shift and tent maps in one dimension, the Lozi map turns out to be much more complicated in many details than the Tél map, although they do share some essential common features.

Fixed points

In order to find the fixed points of the Lozi map we let $x_{n+1} = x_n = x$ and $y_{n+1} = y_n = y$. The equation $a\epsilon_x x - bx + x = 1$ obtained from (5.63) yields two fixed points:

$$\begin{aligned} \mathrm{H}_+ : x = y = \frac{1}{1+a-b} > 0, \\ \mathrm{H}_- : x = y = \frac{1}{1-a-b} < 0. \end{aligned} \tag{5.65}$$

The existence of the fixed point H_- requires $a + b > 1$.

The Jacobian matrix of (5.63)

$$\mathbf{L}_n = \begin{bmatrix} -a\epsilon_n & b \\ 1 & 0 \end{bmatrix} \tag{5.66}$$

determines the linearized dynamics of the Lozi map in the vicinity of the point (x_n, y_n). The eigenvalues and eigenvectors of $\mathbf{L}_n$ has to be considered separately at the two fixed points. As $\epsilon_n = -1$ at the fixed point H_-, we have

$$\lambda_\pm^{(-)} = \frac{a \pm \sqrt{a^2 + 4b}}{2}, \tag{5.67}$$

and

$$\mathbf{v}_\pm^{(-)} = (\lambda_\pm^{(-)}, 1). \tag{5.68}$$

These are the same as in the Tél map (see (5.28) and (5.29)). At the fixed point H_+ we have $\epsilon_n = 1$ and

$$\lambda_\pm^{(+)} = \frac{-a \mp \sqrt{a^2 + 4b}}{2}, \tag{5.69}$$

and

$$\mathbf{v}_\pm^{(+)} = (\lambda_\pm^{(+)}, 1). \tag{5.70}$$

The eigenvectors of the fixed points $\mathrm{H}_\pm$ are mirror images to each other with respect to the y axis because of $\lambda_\pm^{(-)} = -\lambda_\pm^{(+)}$. When $a > 1 - b$ both fixed points are unstable.

Period 2 points

There is a period 2 orbit in the Lozi map, which jumps between the two points

$F_1 = (x_2, x_1)$ and $F_2 = (x_1, x_2)$, where

$$x_1 = \frac{1+a-b}{a^2+(1-b)^2} > 0, \quad x_2 = \frac{1-a-b}{a^2+(1-b)^2} < 0. \tag{5.71}$$

Therefore, F_1 is located in the second quadrant and F_2 in the fourth when $a > 1-b$.

The stability of the period 2 orbit is determined by the eigenvalues of the matrix

$$\begin{bmatrix} a & b \\ 1 & 0 \end{bmatrix} \begin{bmatrix} -a & b \\ 1 & 0 \end{bmatrix} = \begin{bmatrix} b-a^2 & ab \\ -a & b \end{bmatrix},$$

which are

$$\lambda_{\pm}^{2p} = b - \frac{a^2 \pm a\sqrt{a^2-4b}}{2}. \tag{5.72}$$

With a increasing, this period 2 orbit changes from a focus to a stable node at $a = 2\sqrt{b}$, then it losses stability at $a > 1+b$. In fact, not only the fixed points and the period 2, but also all the periodic points of longer periods become unstable saddles for $a > 1+b$ (Liu, Xie, Zhu and Lu [1992]). We will discuss the details of symbolic dynamics for the Lozi map, confining ourselves to the general parameter range $a > 2\sqrt{b}$ (Zheng [1991b, 1992a], Zheng and Liu [1994]).

5.4.1 Forward and Backward Symbolic Sequences

The assignment of symbols is much similar to what we have done for the Tél map. However, in order to make the discussion of the Lozi map self-content we do it from scratch. The mapping function of the Lozi map (5.63) behaves differently for $x < 0$ and $x > 0$, just as the tent map. Therefore, we denote the left and right halves of the phase plane by the letters L and R, respectively. This assignment carries over to the upper and lower half-planes for y by the second line of (5.63).

Starting from an initial point $(x_0, y_0) \equiv (x_0, x_{-1})$ and iterating forward map (5.63), we get a forward trajectory

$$x_0\, x_1\, x_2\, \cdots\, x_{n-1}\, x_n\, \cdots. \tag{5.73}$$

Since $y_{n+1} = x_n$, it is enough to use only x_i to represent the whole trajectory. Assigning a symbol L or R to each number in (5.73) according to whether

$x_i < 0$ or $x_i > 0$, we get a forward symbolic sequence

$$s_0\, s_1\, s_2\, \cdots\, s_{n-1}\, s_n\, \cdots . \tag{5.74}$$

Starting from the same initial point (x_0, y_0), one may iterate backwards using the inverse map (5.64) to get a backward trajectory

$$\cdots\, x_{-(m+1)}\, x_{-m}\, \cdots\, x_{-2}\, x_{-1}. \tag{5.75}$$

A backward trajectory is written from right to left in order to distinguish it from the forward trajectory. Again, only the x_i coordinates are registered. To keep coding of both the forward and backward trajectories consistent, we use only the symbols of x_i. In this way, we get a backward symbolic sequence from (5.75)

$$\cdots\, s_{\bar{m}}\, \cdots\, s_{\bar{2}}\, s_{\bar{1}}, \tag{5.76}$$

where $s_{\bar{i}} \equiv s_{-i}$ is the symbol for x_{-i}.

The two semi-infinite symbolic sequences are concatenated to form a bi-infinite sequence

$$\cdots\, s_{\bar{m}}\, \cdots\, s_{\bar{2}}\, s_{\bar{1}} \bullet s_0\, s_1\, s_2\, \cdots\, s_n\, \cdots . \tag{5.77}$$

We have inserted a dot $\bullet$ between $s_{\bar{1}}$ and s_0, i.e., the symbols for $y_0 = x_{-1}$ and x_0, to indicate the present time instant. A forward iteration corresponds to a right shift of the dot; a backward iteration corresponds to a left shift. The two neighboring symbols of the dot always code the y and x coordinates of a given point. Since we have assigned letters from one and the same set of symbols to both forward and backward sequences, we may write

$$\cdots\, s_{n-2}\, s_{n-1} \bullet s_n\, s_{n+1}\, s_{n+2}\, \cdots$$

for the n-th shift.

5.4.2 Dynamical Foliations of the Phase Space

Owing to the piecewise linear nature of the Lozi map, most of the construction of dynamical foliations may be carried out analytically. We start from the forward contracting foliations.

Forward contracting foliations

Due to the piecewise linear dynamics in the phase plane there exists a segment of some straight line whose points have the same forward symbolic sequence

$$\bullet s_n s_{n+1} s_{n+2} \cdots . \tag{5.78}$$

Such segments form forward contracting foliations. Assume the line corresponding to sequence (5.78) is $x - k_n y = \xi_n$, where the slope k_n and intercept ξ_n depend on the sequence. Considering a particular point (x_n, y_n) on the segment and its image (x_{n+1}, y_{n+1}) on line $x - k_{n+1} y = \xi_{n+1}$, and making use of the map (5.63), we have

$$\begin{aligned} \xi_{n+1} &= x_{n+1} - k_{n+1} y_{n+1} = -(a\epsilon_n + k_{n+1}) x_n + b y_n + 1, \\ &= -(a\epsilon_n + k_{n+1})(x_n - k_n y_n) + 1 = -b k_n^{-1} \xi_n + 1, \end{aligned} \tag{5.79}$$

where k_n is related to k_{n+1} by

$$k_n = \frac{b}{a\epsilon_n + k_{n+1}}. \tag{5.80}$$

This equation may be expanded into a continued fraction

$$\begin{aligned} k_n &= \cfrac{b}{a\epsilon_n + \cfrac{b}{a\epsilon_{n+1} + \cfrac{b}{a\epsilon_{n+2} + \cdots}}} \\ &\equiv \frac{b}{a\epsilon_n +} \, \frac{b}{a\epsilon_{n+1} +} \, \frac{b}{a\epsilon_{n+2} +} \cdots . \end{aligned} \tag{5.81}$$

Therefore, the slope k is completely determined by $\epsilon_n \, \epsilon_{n+1} \, \epsilon_{n+2} \cdots$, i.e., by the forward symbolic sequence.

From the last equation in (5.79) we get

$$\xi_n = -b^{-1} k_n (\xi_{n+1} - 1) = \frac{k_n}{b}(1 - \mathcal{F}\xi_n),$$

or

$$\left(1 + \frac{k_n}{b}\mathcal{F}\right) \xi_n = \frac{k_n}{b}, \tag{5.82}$$

where $\mathcal{F}$ is the *forward operator*, which, when acting on a member of an arbitrary sequence $\{u_n\}$, increases its subscript by one:

$$\mathcal{F} u_n = u_{n+1}. \tag{5.83}$$

Putting off the problem of convergence to a later time, we can get a particular solution of (5.82) for ξ_n expressed in an infinite series

$$\begin{aligned}\xi_n &= \left(1+\frac{k_n\mathcal{F}}{b}\right)^{-1}\frac{k_n}{b} = \sum_{i=0}^{\infty}\left[\frac{-k_n\mathcal{F}}{b}\right]^i\frac{k_n}{b} \\ &= \frac{k_n}{b} - \frac{k_nk_{n+1}}{b^2} + \frac{k_nk_{n+1}k_{n+2}}{b^3} - \cdots .\end{aligned} \tag{5.84}$$

This result shows that the determination of ξ_n requires the knowledge of k_i for all $i \geqslant n$, i.e., the k_i for all the right shifts of the original sequence. As we shall see in the examples below, this is not a difficult task to accomplish for periodic and eventually periodic sequences.

By now we have given the formulae to determine k_n and ξ_n in the equation $x - k_ny = \xi_n$. Suitable intervals on such lines form the set of forward contracting foliations of the phase plane. In order to have a better feeling of the dynamical foliations we calculate k_i, $i \geqslant n$ and ξ_n for a few important foliations.

1. For the symbolic sequence $\bullet R^\infty$ that goes through the fixed point H_+, we have $\epsilon_i = 1$ for all i. Therefore, (5.81) leads to an equation $u = b/(a+u)$ for $u \equiv k_n$, which yields $u = \lambda_\pm^{(+)}$, i.e., the eigenvalue of the Jacobian at H_+, see (5.69). We should choose $u = \lambda_-^{(+)}$, the one with $|u| < 1$, since from (5.81) $0 < u < b/a$. This is consistent with the fact that forward foliation points in the contracting direction. As $k_i = u$ for all $i \geqslant n$, the series (5.84) sums up easily to $\xi_n = u/(b+u)$. The foliation intercepts the y-axis at $y = -\xi_n/k_n = -1/(b+u)$.

2. Changing the leading symbol in $\bullet R^\infty$ leads to $\bullet LR^\infty$. Now the first $\epsilon_n = -1$ and $\epsilon_i = 1$ for all $i > n$. Therefore, we have $k_n = b/(u-a)$ and $k_i = u$ for all $i > n$ with u defined above. The series (5.84) sums up to give $\xi_n = k_n/(b+u)$. The corresponding foliation intercepts the y-axis at the same $y = -1/(b+u)$ as the $\bullet R^\infty$ foliation.

3. Similarly, the forward foliation $\bullet L^\infty$, which goes through the fixed point H_-, has $k_i = \lambda_-^{(-)} \equiv v$ for all $i \geqslant n$ and $\xi_n = v/(b+v)$. It intercepts the y-axis at $y = -1/(b+v)$. Changing the leading symbol in $\bullet L^\infty$ leads to $\bullet RL^\infty$. Its $k_n = b/(a+v)$, $k_i = v$ for all $i > n$, hence

$$\xi_n = \frac{b}{(a+v)(b+v)}. \tag{5.85}$$

The line $\bullet RL^\infty$ intercepts the y-axis at the same point $y = -1/(b+v)$ as the $\bullet L^\infty$ foliation does. We recollect that in the 1D tent map $\bullet RL^\infty$ is the maximal one and $\bullet L^\infty$ the minimal among all symbolic sequences. We will see in the next section that the ordering rule of the 2D Lozi map yields the same results.

4. Just as in the tent map, no sequence exists between $\bullet RRL^\infty$ and $\bullet LRL^\infty$ except $\bullet CRL^\infty$. For the first sequence $k_n = b(a+v)/[b+a(a+v)]$; for the second $k_n = b(a+v)/[b-a(a+v)]$. They both have the same $k_{n+1} = b/(a+v)$ and $k_i = v$ for all $i \geqslant n+2$. Due to the difference in k_n and

$$\xi_n = \frac{k_n}{b}\left[1 - \frac{b}{(a+v)(b+v)}\right],$$

their ξ_n differ. However, they intercept the y-axis at the same point.

We have drawn these 6 foliations in Fig. 5.8. In fact, all the foliations shown belong to the invariant manifolds of the fixed points marked by filled circles in the figure. In other words, the symbolic sequences of these foliations contain either R^∞ or L^∞. In this figure it is clearly seen that a forward foliation changes its leading symbol upon crossing the y-axis ($\bullet C$). In particular, the two symbolic sequences $\bullet RRL^\infty$ and $\bullet LRL^\infty$, which are limits of $\bullet CRL^\infty$, meet at $\bullet C$ and form a small shaded triangle region. Any point

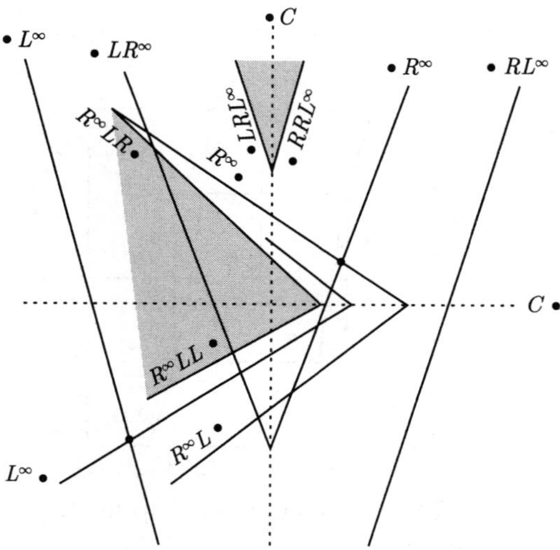

Figure 5.8 Some forward and backward contracting foliations of the Lozi map. Filled circles indicate the fixed points (schematic)

in this shaded region have either $\bullet RRL^\infty$ or $\bullet LRL^\infty$ as its forward symbolic sequence, depending on which side of the y-axis it locates. In the shaded region the partition line may be drawn anywhere. The shaded triangle together with the $\bullet CRL^\infty$ half-line corresponds to a single vertical line in the symbolic plane, see, e.g., Fig. 5.8.

Similarly, all points to the right of the maximal sequence $\bullet RL^\infty$ have the same maximal forward sequence, all points to the left of the minimal sequence $\bullet L^\infty$ have the same minimal forward sequence. Therefore, interesting dynamical behavior takes place only in the corridor between these two extreme foliations and outside the shaded triangle.

There is a 1D map worth studying owing to its relevance to the 2D symbolic dynamics of the Lozi map. In fact, relation (5.80) may be viewed as a map from k_n to k_{n+1}:

$$k_{n+1} = -a\epsilon_n + \frac{b}{k_n}. \tag{5.86}$$

We shall call it the k-map. The mapping function is shown in Fig. 5.9. It has two decreasing branches, corresponding to the two choices $\epsilon_n = 1$ and $\epsilon_n = -1$, i.e., to the symbols $s_n = R$ and $s_n = L$, respectively. We label the two branches by the symbol s_n, then the coding of k_n may adopt the forward symbolic sequence of the point (x_n, y_n). For the time being, ignoring the conditions which the coding of k_n should satisfy, we study the map (5.86) *per se*.

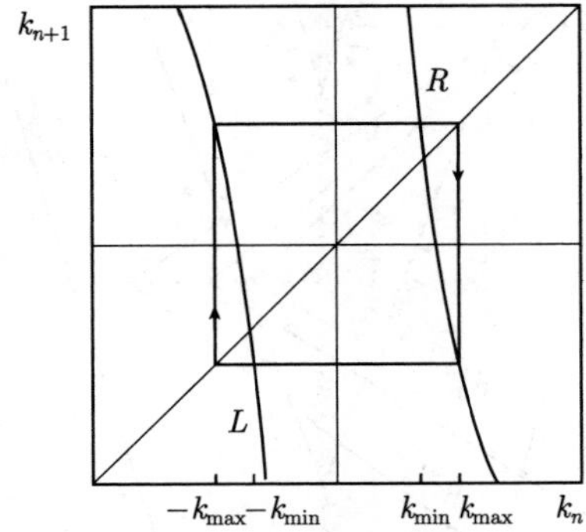

Figure 5.9 The k_{n+1} versus k_n map for the slope

The k-map (5.86) is essentially an inverse shift map. Both symbols R and

L have an odd parity. Therefore, different codings of k are ordered as

$$k(\bullet ER\cdots) > k(\bullet EL\cdots), \quad k(\bullet OR\cdots) < k(\bullet OL\cdots), \tag{5.87}$$

where strings E and O are made of an *even* and *odd* number of symbols R and L. In particular, we have

$$k(\bullet(RL)^\infty) = k_{\max}, \qquad k(\bullet(LR)^\infty) = -k_{\max}. \tag{5.88}$$

When $a > 2\sqrt{b}$ the k-map has a 2-cycle $\{-k_{\max}, k_{\max}\}$, where

$$k_{\max} = \frac{a}{2} - \sqrt{\frac{a^2}{4} - b}. \tag{5.89}$$

Interesting dynamics takes place only on the interval $[-k_{\max}, k_{\max}]$. Therefore, we may take the latter as the mapping interval for the k-map. We see there is a "gap" in the interval between the points labeled as $-k_{\min}$ and $k_{\min}$ in Fig. 5.9. Just as in the expanding tent map with its top lying outside the mapping interval (see Section 2.8.7), points may escape from the interval under iterations. Only a Cantor set of points remains, which forms a strange repeller on the interval $[-k_{\max}, k_{\max}]$. Any sequence made of R and L has its corresponding point on the repeller, and vice versa. The minimal sequence with a leading R, i.e., $\bullet R(RL)^\infty$, determines $k_{\min}$. It can be calculated from (5.81):

$$k_{\min} = \frac{b}{a + k_{\max}} > 0. \tag{5.90}$$

All k values are related to each other as follows:

$$\begin{aligned} k_{\min} &\equiv k(\bullet R(RL)^\infty) \leqslant k(\bullet R\cdots) \leqslant k(\bullet(RL)^\infty) = k_{\max}, \\ -k_{\min} &= k(\bullet L(LR)^\infty) \geqslant k(\bullet L\cdots) \geqslant k(\bullet(LR)^\infty) = -k_{\max}. \end{aligned}$$

These relations also guarantee the convergence of k_n. Throughout this section we require $a > 2\sqrt{b}$ to keep $k_{\max}$ real.

Clearly, k_n now depends on the whole symbolic sequence. However, the sign of k_n is determined by the leading symbol only; as it follows from the above relations a leading R yields a positive k_n and a leading L—a negative k_n. We will see that the sign of k_n affects the ordering rule of the foliations.

Backward contracting foliations

A discussion for the backward contracting foliations may be carried out in full parallel. Skipping the detailed derivation, we mention the essential results.

In the phase plane there is a segment of a certain line $x - h_n y = \eta_n$ whose points have one and the same backward symbolic sequence

$$\cdots s_{n-2}\, s_{n-1}\, \bullet\,.$$

For h_n, or rather for $\bar{h}_n = 1/h_n$, there is a continued fraction expansion

$$\begin{aligned}\bar{h}_n &= \frac{1}{h_n} = \frac{1}{-a\epsilon_{n-1} + b\bar{h}_{n-1}} \\ &= \frac{1}{-a\epsilon_{n-1} +}\,\frac{b}{-a\epsilon_{n-2} +}\cdots.\end{aligned} \tag{5.91}$$

The η_n has an expansion

$$\eta_n = 1 - b\bar{h}_{n-1} + b^2\bar{h}_{n-1}\bar{h}_{n-2} - b^3\bar{h}_{n-1}\bar{h}_{n-2}\bar{h}_{n-3} + \cdots. \tag{5.92}$$

These two formulae determine a straight line from the symbols in the sequence $\cdots s_{n-2}s_{n-1}\bullet$. A suitable segment of this line gives a backward contracting foliation. The collection of all backward contracting foliations makes the other set of dynamical foliations for the Lozi map.

Here are some examples of the most important backward contracting foliations.

1. Through the fixed point H_+ passes the maximal backward foliation $R^\infty\bullet$. Its

$$\bar{h}_n = \frac{a - \sqrt{a^2+4b}}{2b} = \frac{v}{b},$$

where, as in the FCF, $v \equiv \lambda_-^{(-)}$. The definition $\bar{h}_n = h_n^{-1}$ gives $h_n = \lambda_+^{(+)}$, which shows that the BCF corresponds to the forward unstable direction. From (5.92) we then have

$$\eta_n = \frac{1}{1+v}. \tag{5.93}$$

Changing the leading symbol in $R^\infty\bullet$ we get $R^\infty L\bullet$, which is the smallest backward symbolic sequence. It has the same η_n as $R^\infty\bullet$, i.e., it meets the latter at the x-axis. However, it has a different $\bar{h}_n = 1/(a+v)$.

2. Through the fixed point H$_-$ passes the backward foliation $L^\infty\bullet$. Its $\bar{h}_n = -v/b$ and $\eta_n = 1/(1-v)$. Changing the leading symbol we get $L^\infty R\bullet$. It has $\bar{h}_n = -1/(a+v)$.

3. No sequences are between $R^\infty LR\bullet$ and $R^\infty LL\bullet$ except $R^\infty LC\bullet$. They respectively have slopes

$$\bar{h}_n = \frac{1}{-a+b/(a+v)} \quad \text{and} \quad \bar{h}_n = \frac{1}{a+b/(a+v)}.$$

These two lines meet on the x-axis at

$$\eta_n = 1 - \frac{b}{(1+v)(a+v)}. \tag{5.94}$$

4. The period 2 foliations $(LR)^\infty\bullet$ and $(RL)^\infty\bullet$ have $\bar{h}_n = u$, $\eta_n = (1+bu)/(1+b^2u^2)$ and $\bar{h}_n = -u$, $\eta_n = (1-bu)/(1+b^2u^2)$, respectively, where $u \equiv (-a+\sqrt{a^2-4b})/(2b)$. This is to be compared with the period 2 in the $\bar{h}$-map below.

All these backward foliations, except the last two period 2 lines, are drawn in Fig. 5.8. In fact, all foliations shown belong to the invariant manifolds of the fixed points marked by filled circles in the figure. In other words, the symbolic sequences of these foliations contain either R^∞ or L^∞. A backward foliation changes its leading symbol from $\cdots R\bullet$ to $\cdots L\bullet$ and *vice versa* upon crossing the x-axis $C\bullet$. The shaded triangle between $R^\infty LL\bullet$ and $R^\infty LR\bullet$ may be thought as widening of the half-line $R^\infty LC\bullet$. The partition line $C\bullet$ may be continued anywhere into the shaded region. The shaded triangle together with $C\bullet$ corresponds to a single horizontal line $C\bullet$ in the symbolic plane, see Figs. 5.12 and 5.13.

From (5.91) we may also define a map from $\bar{h}_n$ to $\bar{h}_{n-1}$:

$$\bar{h}_{n-1} = \frac{a\epsilon_{n-1}}{b} + \frac{1}{b\bar{h}_n}, \tag{5.95}$$

which may be called a $\bar{h}$-map. Its shape resembles the k-map (5.86) with the labeling of branches interchanged. This is again an inverse shift map, hence the ordering of the $\bar{h}_n$ coding:

$$\bar{h}(\cdots LE\bullet) > \bar{h}(\cdots RE\bullet), \quad \bar{h}(\cdots LO\bullet) < \bar{h}(\cdots RO\bullet), \tag{5.96}$$

where E and O denote strings made of an *even* and *odd* number of symbols R and L. Moreover, we have

$$\begin{aligned}
&\bar{h}_{\max} = \bar{h}((RL)^\infty) = k_{\max}/b, \quad \bar{h}_{\min} = \bar{h}((RL)^L\infty) = k_{\min}/b, \\
&\bar{h}_{\min} \leqslant \bar{h}(\cdots L\bullet) \leqslant \bar{h}_{\max}, \qquad -\bar{h}_{\min} \geqslant \bar{h}(\cdots R\bullet) \geqslant -\bar{h}_{\max}.
\end{aligned}$$

We will not pursue the study of the $\bar{h}$-map as it goes much like the k-map.

Matrix representation of the linearized dynamics

The continued fraction relations for k_n and h_n are closely related to the matrix representation of the linearized dynamics of the Lozi map. In order to reveal this relation we juxtapose a fraction a/b and a vector (b, a):

$$\frac{a}{b} \leftrightarrow \begin{bmatrix} b \\ a \end{bmatrix}.$$

From (5.91) the relation between h_n and h_{n-1} may be represented by

$$\begin{aligned}
\begin{bmatrix} h_n \\ 1 \end{bmatrix} &= \begin{bmatrix} -a\epsilon_{n-1} + b\bar{h}_{n-1} \\ 1 \end{bmatrix} = \bar{h}_{n-1} \begin{bmatrix} -a\epsilon_{n-1}h_{n-1} + b \\ h_{n-1} \end{bmatrix} \\
&= \bar{h}_{n-1} \begin{bmatrix} -a\epsilon_{n-1} & b \\ 0 & 1 \end{bmatrix} \begin{bmatrix} h_{n-1} \\ 1 \end{bmatrix} \equiv \bar{h}_{n-1}\mathbf{L}_{n-1} \begin{bmatrix} h_{n-1} \\ 1 \end{bmatrix}.
\end{aligned}$$

In the last line we have used the Jacobian matrix $\mathbf{L}_n$ of the Lozi map introduced in (5.4.4).

This relation may be iterated to yield:

$$\begin{bmatrix} h_n \\ 1 \end{bmatrix} = \bar{h}_{n-1}\bar{h}_{n-2}\cdots\bar{h}_{n-p}\mathbf{L}_{n-1}\mathbf{L}_{n-2}\cdots\mathbf{L}_{n-p} \begin{bmatrix} h_{n-p} \\ 1 \end{bmatrix}. \tag{5.97}$$

For a period p orbit $h_n = h_{n-p}$, Eq. (5.97) shows that $(h_n, 1)$ is an eigenvector of the period, i.e.,

$$\begin{aligned}
\mathbf{L}_{n-1}\mathbf{L}_{n-2}\cdots\mathbf{L}_{n-p} \begin{bmatrix} h_n \\ 1 \end{bmatrix} &= h_{n-1}h_{n-2}\cdots h_{n-p} \begin{bmatrix} h_n \\ 1 \end{bmatrix} \\
&\equiv \lambda_p \begin{bmatrix} h_n \\ 1 \end{bmatrix}.
\end{aligned} \tag{5.98}$$

Since $|\bar{h}_{n-i}| < \bar{h}_{\max} < 1$ for $a > 1 + b$, the eigenvalue of the period $|\lambda_p| > 1$. Therefore, $(h_n, 1)$ is the unstable direction of the period. We have seen this explicitly on the example of the period 2 foliations $(RL)^\infty\bullet$ and $(LR)^\infty\bullet$.

Similarly, the forward contracting foliations for a period p orbit deal with stable directions. Anticipating that the inverse matrix

$$\mathbf{L}_n^{-1} = \begin{bmatrix} 0 & 1 \\ 1/b & a\epsilon_n/b \end{bmatrix}$$

of the Lozi Jacobian will involve in the matrix representations, we write the k-map (5.80) as

$$\begin{bmatrix} k_n \\ 1 \end{bmatrix} = \begin{bmatrix} 1 \\ a\epsilon_n/b + k_{n+1}/b \end{bmatrix}, \tag{5.99}$$

which can be further transformed into

$$\begin{aligned}\begin{bmatrix} k_n \\ 1 \end{bmatrix} &= k_n\mathbf{L}_n^{-1}\begin{bmatrix} k_{n+1} \\ 1 \end{bmatrix} = k_nk_{n+1}\mathbf{L}_n^{-1}\mathbf{L}_{n+1}^{-1}\begin{bmatrix} k_{n+2} \\ 1 \end{bmatrix} \\ &= k_nk_{n+1}\cdots k_{n+p-1}\mathbf{L}_n^{-1}\mathbf{L}_{n+1}^{-1}\cdots\mathbf{L}_{n+p-1}^{-1}\begin{bmatrix} k_{n+p} \\ 1 \end{bmatrix}.\end{aligned}$$

For a period p orbit $k_{n+p} = k_n$, so $(k_n, 1)$ is an eigenvector of the period along the stable direction for $a > 1 + b$.

5.4.3 Ordering of the Forward and Backward Foliations

Now we determine the ordering of the forward and backward foliations according to their symbolic sequences. Suppose that two forward foliations $\bullet P_1$ and $\bullet P_2$ intersect one and the same backward foliation $Q\bullet$. Denote by $(\Delta x)_n$ the x-coordinate difference between the two intersecting points $Q \bullet P_1$ and $Q \bullet P_2$ on $Q\bullet$. From the map (5.63), the y-difference $(\Delta y)_{n+1} = (\Delta x)_n$. From the form $x - h_ny = \eta_n$ for a BCF, $(\Delta x)_n = h_n(\Delta y)_n$. Combining the both, we have $(\Delta x)_{n+1} = h_{n+1}(\Delta x)_n$. Since $h_n < 0$ for $s_{n-1} = R$ and $h_n > 0$ for $s_{n-1} = L$, a letter $s_n = R$ will reverse the sign of $(\Delta x)_n$ when it is iterated to $(\Delta x)_{n+1}$ while L preserves the order. Under the convention $\bullet R\cdots > \bullet L\cdots$, we can formulate the *ordering rule* for the forward contracting foliations as follows:

$$\bullet E_RR\cdots > \bullet E_RL\cdots, \quad \bullet O_RR\cdots < \bullet O_RL\cdots, \tag{5.100}$$

where strings E_R and O_R made of R and L contain an *even* and *odd* number of symbol R, respectively. We note that this ordering rule is the same as that for the tent or quadratic map.

Similar reasoning applies to the ordering of backward foliations. Now, for two backward foliations $Q_1\bullet$ and $Q_2\bullet$ intersecting the same forward foliation

$\bullet P$, we have $(\Delta x)_n = k_n(\Delta x)_{n-1}$. Since $k_n > 0$ for $s_n = R$ and $k_n < 0$ for $s_n = L$, sign of $(\Delta x)_n$ is kept under a inverse iteration for $s_n = R$ and reversed for L. Under the convention $\cdots R\bullet > \cdots L\bullet$ to make the upper foliation in the plane being greater, we can formulate the *ordering rule* of backward contracting foliations as follows:

$$\cdots RE_L\bullet > \cdots LE_L\bullet, \qquad \cdots RO_L\bullet < \cdots LO_L\bullet, \tag{5.101}$$

where the strings E_L and O_L made of the symbols R and L contain an *even* and *odd* number of symbol L, respectively. We note that this ordering rule is the same as that for the inverse tent map.

As we have done for the tent map and inverse tent map, the ordering rules (5.100) and (5.101) may be embodied by introducing the following metric representation for symbolic sequences. Each forward symbolic sequence $\bullet s_n s_{n+1} s_{n+2} \cdots$ defines a real number α between 0 and 1:

$$\alpha = \sum_{i=n}^{\infty} \mu_i 2^{-i}, \tag{5.102}$$

where

$$\mu_i = \begin{cases} 0, & \text{when} \quad (-1)^{i-n} \prod_{j=n}^{i} \epsilon_j = -1; \\ 1, & \text{when} \quad (-1)^{i-n} \prod_{j=n}^{i} \epsilon_j = 1. \end{cases}$$

This α may be called the *coordinate* of the symbolic sequence. Each backward symbolic sequence $\cdots s_{n-2}s_{n-1}\bullet$ defines a real number β between 0 and 1, i.e., the metric representation of the symbolic sequence:

$$\beta = \sum_{i=n-1}^{\infty} \nu_i 2^{-i}, \tag{5.103}$$

where

$$\nu_i = \begin{cases} 0, & \text{when} \quad \prod_{j=n-1}^{i} \epsilon_j = -1; \\ 1, & \text{when} \quad \prod_{j=n-1}^{i} \epsilon_j = 1. \end{cases}$$

According to these metric representations, the maximal and minimal sequences and two related ones discussed so far acquire the following coordinates:

$$\begin{aligned}
&\alpha(\bullet RL^\infty) = \beta(R^\infty\bullet) = 1,\\
&\alpha(\bullet L^\infty) = \beta(R^\infty L\bullet) = 0,\\
&\alpha(\bullet RRL^\infty) = \alpha(\bullet LRL^\infty) = 1/2,\\
&\beta(R^\infty LR\bullet) = \beta(R^\infty LL\bullet) = 1/2.
\end{aligned} \tag{5.104}$$

In this way, the unit square on the α-β plane constitutes a *symbolic plane.* Vertical and horizontal lines in the symbolic plane correspond to the forward and backward foliations, respectively. A bi-infinite symbolic sequence with the present dot $\bullet$ indicated corresponds to a point in the symbolic plane, namely, the intersection of the vertical and horizontal lines representing its forward and backward symbolic sequence. The symbolic plane is a convenient means to visualize the admissibility conditions for symbolic sequences in the Lozi map.

5.4.4 Allowed and Forbidden Zones in the Symbolic Plane

A schematic phase plane with some forward and backward foliations is shown in Fig. 5.10. In this figure, the unstable manifold of the fixed point H_+ in the first quadrant is given by the foliation with the maximal backward symbolic sequence $R^\infty\bullet$. All points in the whole region above the $R^\infty\bullet$ line have the same backward symbolic sequence $R^\infty\bullet$, as there are no greater sequences at all.

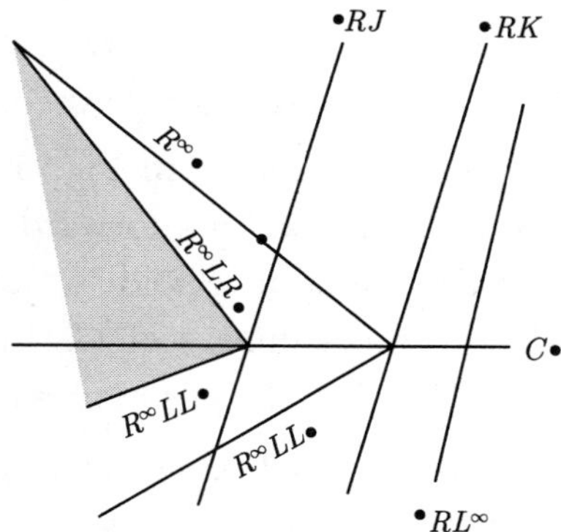

Figure 5.10 A phase plane with some forward and backward foliations (schematic)

The line $R^\infty\bullet$ meets $R^\infty L\bullet$, which is in the lower half-plane and also a part of the unstable manifold of H_+, at a certain point on the x-axis, i.e., $C\bullet$. Through this point passes a forward foliation, which we denote by $\bullet RK$ and which we assume to be smaller than the maximal forward foliation $\bullet RL^\infty$. This is the rightmost tangency one needs to consider in the present situation. Any symbolic sequence greater than $\bullet RK$ cannot become a forward sequence when the associated backward sequence is not the trivial sequence $R^\infty\bullet$ or $R^\infty L\bullet$. In the symbolic plane the two backward foliations correspond to $\beta = 1$ and $\beta = 0$, i.e., the top and bottom of the unit square. The rectangle between the two vertical lines $\alpha = \alpha(\bullet RK)$ and $\alpha = 1$ defines the first forbidden zone: no shifts of any admissible sequence can fall in this zone.

In Fig. 5.10 the shaded triangular region between the backward foliations $R^\infty LR\bullet$ and $R^\infty LL\bullet$, together with the half-line $CR^\infty L\bullet$, corresponds to $\beta = 1/2$ line in the symbolic plane. The tip of this triangle has a tangency with a forward foliation, which we denote by $\bullet RJ$. This is the leftmost tangency one needs to consider in the shown situation. There are areas as pre-images of the above area in the phase plane. Points in such an area share the same backward sequence, corresponding to a single line in the symbolic plane. We will consider only foliations forming the area boundaries.

The ordering rules of foliations impose restrictions on possible types of admissible sequences. Consider a tangency of two backward foliations $QR\bullet$ and $QL\bullet$ with a forward foliation $\bullet P$, as shown in Fig. 5.11 (a). Under the circumstances all the backward foliations included between $QR\bullet$ and $QL\bullet$, i.e., in the shaded triangle, can never touch any forward foliations located to the right of $\bullet P$. In the symbolic plane this tangency excludes a rectangle demarcated by the four lines $\alpha = 1$, $\alpha = \alpha(\bullet P)$, $\beta = \beta(QR\bullet)$, and $\beta = \beta(QL\bullet)$, see Fig. 5.11 (b). This rectangle is a *forbidden zone.*

Now we study how the situation shown in Fig. 5.10 may take place. We have mentioned in Section 5.2.2 that when a forward foliation passes through a fixed point, it belongs to the stable manifold W^s of the fixed point. Thus $\bullet L^\infty$ belongs to W^s of H_-. The pre-images of $\bullet L^\infty$ such as $\bullet RL^\infty$, $\bullet LRL^\infty$ and $\bullet RRL^\infty$, all belong to W^s of H_-. Let us express this fact analytically.

The tangency that takes place between the maximal forward foliation $\bullet RL^\infty$ and the maximal backward foliation $R^\infty\bullet$ is nothing but a heteroclinic tangency between the two fixed points H_+ and H_-. The condition for this tangency to happen can be written down using the expressions for ξ and η of the two foliations, i.e., (5.93) and (5.85):

$$\eta(R^\infty\bullet) = \frac{1}{1+v} = \frac{b}{(a+v)(b+v)} = \xi(\bullet RL^\infty), \tag{5.105}$$

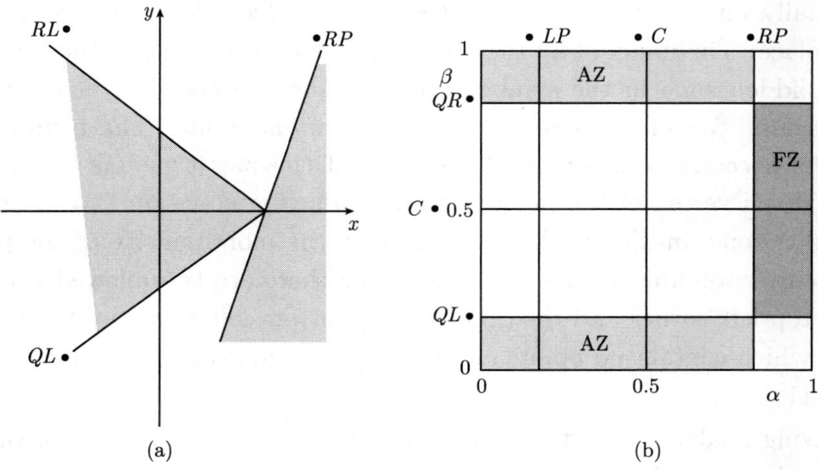

Figure 5.11 A tangency between backward foliations $QR\bullet$ and $QL\bullet$ with a forward foliation $\bullet P$. (a) The Phase plane. (b) The symbolic plane

which yields $b + 2v = 0$, i.e.,

$$a = a_c \equiv 2 - b/2. \tag{5.106}$$

We recollect that $v \equiv \lambda_-^{(-)}$. When $a < a_c$, the foliation $R^\infty\bullet$ cannot reach $\bullet RL^\infty$. Suppose that now the forward foliation $\bullet RK$ smaller than $\bullet RL^\infty$ has a tangency with $R^\infty\bullet$. This tangency defines the rightmost forbidden rectangle in the symbolic plane. Its boundaries are given by the four lines $\alpha = \alpha(\bullet RK)$, $\alpha = 1$, $\beta = 0$, and $\beta = 1$. Restricted by this forbidden zone, any admissible sequence cannot contain a forward sequence which would be greater than $\bullet RK$, unless its associated backward sequence is either $R^\infty\bullet$ or $R^\infty L\bullet$. We shall see that when the 2D Lozi map reduces to a 1D map at $b \to 0$, the sequence $\bullet RK$ becomes the kneading sequence.

Starting from this rightmost tangency and moving leftward along the partition line $C\bullet$, one encounters infinitely many tangencies one by one. These tangencies outline infinitely many forbidden rectangles in the symbolic plane. Due to the one-to-one correspondence between either $\Delta\eta$ or $\Delta\xi$ to Δx, the natural order of $-\eta$ and ξ on the x-axis or $C\bullet$ coincides with the ordering rules for backward and forward sequences. When forward foliations diminish from the first tangency, the backward foliations involved in tangencies diminish or raise monotonically depending on their leading symbol being R or L. Eventually one reaches both $R^\infty LR\bullet$ and $R^\infty LL\bullet$, where the last tangency takes place. The union of all these forbidden zones constitutes the fundamental forbidden zone in the symbolic plane. The staircase-like boarder of the fundamental forbidden zone is called the *pruning front*. The pruning front provides a complete symbolic description of the partition line in the phase space. Each vertex of the pruning front, satisfying equation $\xi = \eta$, defines a forbidden zone, inside which $\xi > \eta$. Due to the monotonicity of the pruning front, corresponding to each forbidden zone there are two allowed zones, one in the top-left corner and the other in the bottom-left corner of the symbolic plane, which contain no points of any forbidden zones and satisfies $\xi < \eta$, see Fig. 5.11.

Having analyzed the structure of the phase plane in terms of forward and backward contracting foliations, we show an actual phase plane of the Lozi map at $a = 1.7$, $b = 0.5$ in Fig. 5.12. We see all the foliations shown before in Fig. 5.8. The $L^\infty R\bullet$ and $R^\infty LR\bullet$ lines are not labeled in this figure, but one easily identifies the former as continuation of $L^\infty\bullet$ above the $C\bullet$ line and the latter as the upper-right boundary of the big shaded triangle. The corresponding symbolic plane with its fundamental forbidden zone is shown in Fig. 5.13. Also given in the figure are 3000 points of real orbits. Indeed, all the points of orbits lie outside the fundamental forbidden zone. This picture may be compared with the fundamental forbidden zone of the Tél map shown in Fig. 5.7. A reflection of the upper left forbidden zone of the latter for the Tél map with respect to the central vertical line would result in a symbolic plane much similar to that of the Lozi map.

With the fundamental forbidden zone defined, the admissibility condition for symbolic sequences in the Lozi map may be formulated as follows: the necessary and sufficient condition for a symbolic sequence to be admissible consists in that all shifts of the sequence do not fall in the fundamental forbidden zone.

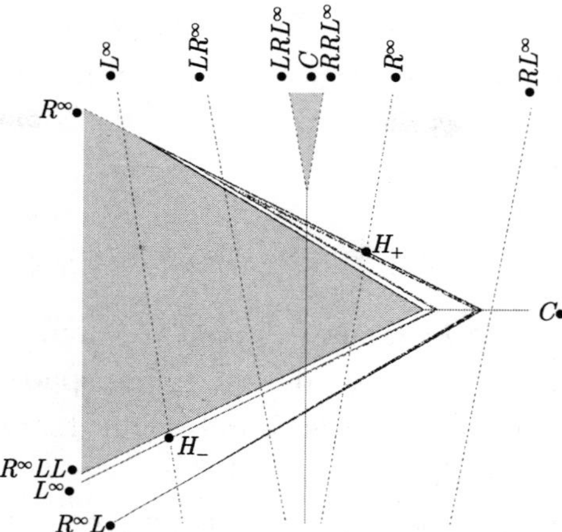

Figure 5.12 Dynamical foliations of the Lozi map at $a = 1.7$, $b = 0.5$. BCFs are given by solid lines; FCFs by dotted lines. The dots belong to the strange attractor, which is part of the BCF

Figure 5.13 Fundamental forbidden zone in the symbolic plane of the Lozi map at $a = 1.7$ and $b = 0.5$

First of all, from the definition of ξ and η one gets

$$y_n = \frac{\xi_n - \eta_n}{h_n - k_n}, \tag{5.107}$$

which is similar to (5.55) in the Tél case. Next, given a bi-infinite symbolic sequence with the present dot • indicated, h and k as well as ξ and η may be calculated from their expansions, thus y_n may be determined from (5.107). By using the recursive relations for ξ_n, η_n, k_n, and h_n, it is straightforward to check that y_{n+2}, y_{n+1}, and y_n defined by (5.107) satisfy the following second order difference equation

$$y_{n+2} = 1 - a\epsilon_n y_{n+1} + b y_n, \tag{5.108}$$

which is equivalent to the Lozi map (5.63). As long as the consistency of the calculated y_n with its coding holds, the phase space trajectory generated from the given symbolic sequence is found. Due to $|\bar{h}| < |\bar{h}_{max}|$, one has $|h| > |h_{\max}| = b/k_{\max}$. Under the condition $a > 2\sqrt{b}$, it follows from (5.89) that $k_{\max}$ remains always positive, and is a decreasing function of a. Therefore, $k_{\max}^2 < b$. Consequently, $|h| > |k_{\max}| > |k|$ and the sign of $h-k$ is determined by the sign of h. Since y and h have opposite signs, the consistency between the symbol and the sign of y is equivalent to $\xi < \eta$. As mentioned above, $\xi > \eta$ inside a FZ, so the occurrence of any shift falling in any FZ breaks the consistency while $\xi < \eta$ in AZs, which guarantees the consistency and justifies the found trajectory.

5.4.5 Discussion of the Admissibility Condition

We start from the $b \to 0$ limit, when the Lozi map reduces to the 1D tent map. It follows from (5.92) that $\eta \to 1$, so there is no dependence on the backward symbolic sequence. Furthermore, one has $\bar{h} \to -\epsilon_{n-1}/a$ and $k \to b\epsilon_n/a$. The expansion (5.84) for ξ reduces to the λ-expansion for the tent map. The number of tangencies, needed for the construction of the fundamental forbidden zone, reduces to one, namely, the one at $\bullet RK$, as shown in Fig. 5.10. The sequence $\bullet RK$ corresponds to the kneading sequence of the tent map and $\xi(\bullet RK) = 1$. The fundamental forbidden zone in the symbolic plane is the rectangle surrounded by $\beta = 0$, $\beta = 1$, $\alpha = \alpha(\bullet RK)$, and $\alpha = 1$. However, there is no structure along the β direction.

We have discussed before that from $\eta(R^\infty\bullet) = \xi(\bullet RL^\infty)$ follows the condition for the heteroclinic tangency between the two fixed points $H_\pm$, namely, $a = a_c = 2 - b/2$. When the forward foliation $\bullet RL^\infty$ has a tangency with the backward foliation $R^\infty LL\bullet$ on the $C\bullet$ line, the area of the fundamental forbidden zone shrinks to zero. Any arbitrary sequence becomes admissible. The condition for this tangency reads:

$$\xi(\bullet RL^\infty) = \frac{b}{(a+v)(b+v)} = 1 - \frac{b}{(a+v)(b+v)} \qquad (5.109)$$
$$= \eta(R^\infty LR\bullet),$$

where $v \equiv \lambda_-^{(-)}$ as before. For small b, to the first order in b we have

$$a = 2 + \frac{b}{2}. \qquad (5.110)$$

Now suppose that a tangency between $R^\infty\bullet$ and $\bullet RLR^\infty$ and another tangency between $(RL)^\infty\bullet$ and $\bullet(RLRRRL)^\infty$ have taken place on the $C\bullet$ line, the latter being a heteroclinic tangency between a period-2 and a period-6 orbit. Let us look at the consequences of these two tangencies, especially, at some forbidden sequences.

The first tangency tells us that all admissible sequences can only be composed of strings of RR and RL types except for the case when the backward sequence is $L^\infty\bullet$, otherwise there would appear forward sequences that are greater than $\bullet RLR^\infty$.

The second tangency shows that for any positive integer n the sequences

$$\cdots RR(RL)^{2n-1}RR\bullet RLRRRR\cdots$$

and

$$\cdots RR(RL)^{2n}\bullet RLRRRR\cdots$$

are forbidden. Therefore, from the latter the string $RR(RL)^{2n+1}R^4$ cannot appear in any admissible sequence.

Generally speaking, to check the admissibility of a given sequence is much more difficult than to show its being forbidden, as all the shifts must be tested. The two tangencies, given above, are not enough for the justification of many admissible sequence. In order to be able to do so, many more tangencies must be invoked.

5.5 The Hénon Map

We have studied two piecewise linear maps in the previous sections with the aim to gain a better understanding of the smooth Hénon map, for which very limited results may be obtained by analytical means.

M. Hénon introduced the following two-dimensional map of the plane (Hénon [1976], Hénon and Pomeau [1977]):

$$\begin{aligned} x_{n+1} &= 1 - ax_n^2 + by_n, \\ y_{n+1} &= x_n. \end{aligned} \tag{5.111}$$

The original motivation for the study of this map was "to find a model problem which is as simple as possible, yet exhibits the same essential properties as the Lorenz system". However, this map does not possess the invariant property with respect to sign change of the variables owned by the Lorenz system, see Eqs. (1.3) in Chapter 1. Nevertheless, the Hénon map has become a subject of research in its own right. Being a canonical form of two-dimensional quadratic maps and simple enough in structure, it has served as a touchstone for many new ideas and numerical procedures. There exists a wide literature on this map (Feit [1978], Curry [1979], Simó [1979], Marotto [1979], Franceschini and Russo [1981], Biham and Wenzel [1990], Gallas [1993], Kaplan [1993], just to name a few).

The map (5.111) transforms the (x, y) plane into itself. The old and new area elements are related by the Jacobian

$$\mathbf{T}(x_n, y_n) = \frac{\partial(x_{n+1}, y_{n+1})}{\partial(x_n, y_n)} = \begin{bmatrix} -2ax_n & b \\ 1 & 0 \end{bmatrix}. \tag{5.112}$$

We will need the inverse map of (5.111)

$$\begin{aligned} x_n &= y_{n+1}, \\ y_n &= (x_{n+1} + ay_{n+1}^2 - 1)/b. \end{aligned} \tag{5.113}$$

Its Jacobian reads

$$\overline{\mathbf{T}}(x_{n+1}, y_{n+1}) = \mathbf{T}^{-1}(x_n, y_n) = \begin{bmatrix} 0 & 1 \\ 1/b & 2ay_{n+1}/b \end{bmatrix}. \tag{5.114}$$

When the determinant $|\mathbf{T}| = b = 1$, it preserves the area and thus imitates a conservative dynamical process. If $|b| < 1$, the contraction of the area may be thought to be due to the presence of dissipation. In the extremely dissipative limit $b = 0$, we recover the logistic map. This is why one-dimensional mappings usually represent simple dissipative dynamics. Nonlinear one-dimensional mappings are often said to be non-invertible, although, in fact, they have multi-valued inverse branches which may be precisely described by means of symbolic dynamics, as we have done in the two preceding chapters. Non-invertibility represents certainly a remarkable distinction of the logistic map from the Hénon map which can be inverted as long as $b \neq 0$, but one should not confuse it with dissipation which is an essentially physical phenomenon.

As it has been done with other maps, we first acquaint ourselves with some analytical properties of the Hénon map before constructing its symbolic dynamics.

5.5.1 Fixed Points and Their Stability

In order to look for the fixed points of the Hénon map (5.111), we solve

$$\begin{aligned} x^* &= 1 - a\,x^{*2} + by^*, \\ y^* &= x^*, \end{aligned}$$

and find

$$x^* = y^* = [-(1-b) \pm \sqrt{(1-b)^2 + 4a}]/(2a). \tag{5.115}$$

We shall follow the notation of Simó [1979] and denote these two fixed points by H_+ and H_- according to the choice of sign in (5.115). In order to remain in the field of real numbers, we must demand that

$$a > a_0 = -(1-b)^2/4. \tag{5.116}$$

As usual, the next question to ask relates to the stability of these fixed points. In the vicinity of the fixed point (x^*, y^*), one writes

$$\begin{aligned} x_n &= x^* + \xi_n, \\ y_n &= y^* + \eta_n, \end{aligned} \tag{5.117}$$

where ξ and η are small deviations from the fixed points. Inserting (5.117) into the map (5.111) and taking into account the fixed point condition, we get in linear approximation

$$\begin{bmatrix} \xi_{n+1} \\ \eta_{n+1} \end{bmatrix} = \begin{bmatrix} -2ax^* & 1 \\ b & 0 \end{bmatrix} \begin{bmatrix} \xi_n \\ \eta_n \end{bmatrix}. \tag{5.118}$$

The 2×2 matrix in (5.118) is the *tangent map* or the linearized map at (x^*, y^*). Eqs. (5.118) realize a linear transformation in the tangent plane. Solving the characteristic equation

$$\begin{vmatrix} -2ax^* - \lambda & 1 \\ b & -\lambda \end{vmatrix} = 0,$$

we get two eigenvalues

$$\lambda_\pm = -ax^* \pm \sqrt{ax^{*2} + b} \tag{5.119}$$

at each fixed point. In the two eigen-directions, small deviations $v^\pm$ change independently according to

$$\begin{aligned} v^+_{n+1} &= \lambda_+ v^+_n, \\ v^-_{n+1} &= \lambda_- v^-_n. \end{aligned} \tag{5.120}$$

The quantities v^+_n and v^-_n are linear combinations of the above ξ_n and η_n, obtainable by applying the similarity transformation that diagonalizes the tangent map, but we skip the simple arithmetic. Several possibilities arise according to the absolute values of the eigenvalues:

1. If both $|\lambda_\pm| < 1$, we have a stable fixed point which is also called a stable node or a sink in the plane.

2. If both $|\lambda_\pm| > 1$, we have an unstable fixed point, also called an unstable node or a source in the plane.

3. If one eigenvalue $|\lambda_>| > 1$, the other $|\lambda_<| < 1$, a small vector in one eigen-direction will be stretched, while in the other direction contracted. This is the interesting case of a saddle point that we shall study in more details in Section 5.5.4.

4. Either or both eigenvalues have an absolute value 1. This happens only as an exceptional case, because it requires that additional conditions are satisfied exactly.

In the first three cases, the system is said to be *hyperbolic*. The last case violates hyperbolicity and occurs when the system undergoes a transition between two different hyperbolic regimes: a *bifurcation* takes place. In general, the eigenvalues may be complex numbers as well, leading to new types of fixed points (stable and unstable foci). We exclude complex eigenvalues for the time being.

The marginal condition that at H^+ one eigenvalue $\lambda = +1$ coincides with $a = a_0$, where a_0 is defined in (5.116). The other requirement for $\lambda_\pm > -1$

leads to a new condition

$$a < a_1 = 3(1-b)^2/4. \tag{5.121}$$

The fixed point H^- is a saddle for all values of a, whereas H^+ remains a sink as long as $a < a_1$. At $a = a_1$, an eigenvalue λ of H^+ reaches -1 and period-doubling bifurcation takes place upon a crossing a_1, just as it happens in the case of a unimodal map. Beyond a_1, a stable 2-cycle comes into existence, but H^+ itself becomes a saddle. Since the ordering of periodic windows along the line $b = 0$ in the a-b parameter plane is nothing but the MSS sequence of the quadratic map, it is natural to look for their extensions into the whole parameter plane. This problem has been studied by El-Hamouly and Mira [1982a and b] numerically. In fact, there exists an algebraic method for the determination of the location of these stable periods, and analytical results can be obtained for orbits up to period 6 (Huang [1985, 1986]). Skipping the detailed derivation, we list the results for a few short periods (see Fig. 5.14). The period 1 orbit appears at the curve

$$a = -(1-b)^2/4,$$

see Eq. (5.116). It destabilizes along the line defined by

$$a = 3(1-b)^2/4,$$

where the stable period 2 appears, see Eq. (5.121). Then the period-4 orbit replaces period 2 at (Simó [1979]):

$$a = (1-b)^2 + (1+b)^2/4.$$

This period-4 orbit remains stable until a reaches the curve

$$a = \frac{3}{4}\{(1+b)^2 + [(1-b)(1+b)^2]^{2/3}\}.$$

The method gives the location of another period-4 orbit (the window of RL^2C at $b = 0$) in the parameter plane. It is remarkable that the boundary of its existence zone reveals an inflection point when approaching the straight line $b = 1$ (see the curve labeled with 4 in Fig. 5.14, where the curve of its destabilizing is not shown). Discovered numerically by Hamouly and Mira, this phenomenon has been confirmed by analytical means. Similar but more complicated results were obtained for all orbits of periods 3 through 6 (Huang [1986]). Curves for periods less than 5 are shown in Fig. 5.14. Note that the quadratic map corresponds to the central horizontal line $b = 0$ in Fig. 5.14.

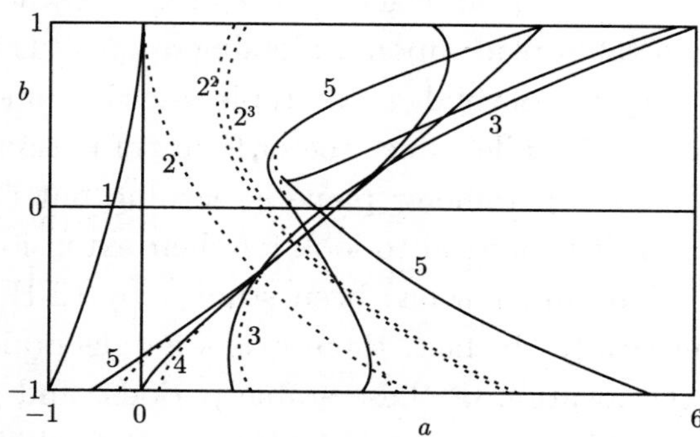

Figure 5.14 The location of stable periodic orbits in the parameter plane of the Hénon map (after Huang [1985, 1986]). Curves are labeled with their corresponding periods. Solid lines are for tangent bifurcation, and dashed lines for period doubling bifurcation

Although there are not many analytical results for the Hénon map, its similarity in dynamical behavior to the Lozi map may serve as a guide in the construction of symbolic dynamics. First of all, we encounter the problem of how to determine the phase space partition lines, as the x-axis is no longer a suitable reference for assigning symbols to resolve different periodic orbits. In what follows we first address the problem of partition lines, then perform a detailed symbolic dynamics analysis at the typical parameter values $a = 1.4$ and $b = 0.3$ (Zhao, Zheng and Gu [1992], Zhao and Zheng [1993]).

5.5.2 Determination of Partition Lines in Phase Plane

Let us take an initial point (x_0, y_0) in the phase space and consider a small circle of radius δ centered at that point. Denote by (ξ_0, η_0) the displacements $(x - x_0, y - y_0)$ of a point (x, y) on the circle. By iterating n steps, the central point arrives at (x_n, y_n) and the displacements (ξ_0, η_0) become (ξ_n, η_n). Due to dynamical instability the initial circle will deform into an ellipse. The equation of this ellipse may be derived as follows:

$$\begin{aligned}\delta &= (\xi_0, \eta_0)\begin{bmatrix}\xi_0\\ \eta_0\end{bmatrix} = (\xi_1, \eta_1)\tilde{\mathbf{T}}_0^{-1}\mathbf{T}_0^{-1}\begin{bmatrix}\xi_1\\ \eta_1\end{bmatrix}\\ &= (\xi_n, \eta_n)\tilde{\mathbf{T}}_{n-1}^{-1}\cdots\tilde{\mathbf{T}}_1^{-1}\tilde{\mathbf{T}}_0^{-1}\mathbf{T}_0^{-1}\mathbf{T}_1^{-1}\cdots\mathbf{T}_{n-1}^{-1}\begin{bmatrix}\xi_n\\ \eta_n\end{bmatrix}.\end{aligned} \tag{5.122}$$

Here $\mathbf{T}_n = \mathbf{T}(x_n, y_n)$ and $\tilde{\mathbf{T}}$ denotes the transpose of $\mathbf{T}$. Suppose that the product matrix

$$\mathbf{M}_n = \mathbf{T}_{n-1}\cdots\mathbf{T}_1\mathbf{T}_0\,\tilde{\mathbf{T}}_0\cdots\tilde{\mathbf{T}}_{n-1}$$

has real eigenvalues λ_1 and λ_2 with $|\lambda_1| > |\lambda_2|$, then the eigenvector of λ_1 points along the major axis of the ellipse. If we fixed the end point (x_n, y_n) of the orbit but keep increasing the length of the trajectory, i.e., let the number of iterations $n \to \infty$, the long axis of the ellipse usually approaches a definite limiting direction. If the end point itself is a hyperbolic periodic point, then the limiting direction is nothing but the unstable eigen-direction of the unstable periodic point and it contracts under backward iterations. If the end point does not correspond to an unstable periodic point, the limit still represents the backward contracting direction of that point. We call this limit the backward contracting direction or simply *backward direction.* In most cases of physical interest it is not difficult to determine the vector field of such directions by numerical means. The family of integral curves of this vector field provides a set of dynamical foliations of the phase space, called the backward contracting foliations or *backward foliations.* A comparison with the backward foliations of the Lozi map discussed in the previous section may help to understand the notion.

Similarly, with respect to the inverse mapping there exists a vector field of forward contracting directions and a family of integral curves of the vector field. The latter forms forward contracting foliations or simply *forward foliations* of the phase space. A forward foliation passing through an unstable periodic point coincides with the stable manifold of that periodic point. Inasmuch as the forward and backward foliations passing through a periodic point of any long period coincide with the stable and unstable manifolds of the periodic point, one can imagine forward and backward foliations in general as the limits of the stable and unstable manifolds of long periods.

The decomposition of the phase space into families of forward and backward foliations enables us to decompose the original two-dimensional map into two one-dimensional maps associated with the forward and backward foliations, respectively. In this way we extend symbolic dynamics of 1D maps to the study of 2D maps. However, the fact that these two 1D dynamics are coupled makes the extension highly non-trivial.

The first step in the construction of a symbolic dynamics consists in partitioning the phase space. Once the partition lines have been determined, one can carry out a coarse-grained description of the phase space trajectories by assigning symbols according to the partition. Based on our experience with the Lozi map it is reasonable to assume that the symbolic dynamics of the Hénon map would require two symbols as well. Thus we need only one partition line. The capacity to resolve as many as possible periodic orbits may be taken as a practical criterion to judge the quality of a partition line: a good partition should assign a unique symbolic name to each and every unstable periodic orbit. (A stable orbit may share name with another periodic orbit in the vicinity of a bifurcation point and acquires its own name at a later stage.) The y-axis does not satisfy this criterion for the Hénon map.

A stronger criterion for partition lines is associated with ordering of symbolic orbits. We have seen that ordering is a key notion in the construction of 1D symbolic dynamics. It plays an essential role in 2D symbolic dynamics as well. A good partition line should guarantee that a set of forward foliations are well ordered along a backward foliation which intersects the forward foliations transversely. Similarly, it should guarantee the well-ordering of backward foliations along a forward foliation which intersects the former transversely.

In order to see that a correct partition line must pass through tangencies between the backward and forward foliations let us look at Fig. 5.15, which essentially reproduces Fig. 5.2. Point D is a tangency between EDF and DH, a backward and a forward foliation. The backward foliation EDF intersects another forward foliation AB at points A and B. Suppose that some image of the partition line (the dashed line) goes through point A, which is not a tangency. Arc ADB then would share the same backward sequence. Consequently, the order of the forward foliations between AB and DH along the segments AD must be opposite to that along DB although both AD and DB are assumed being of the same backward sequence. Only when the preimage of the partition line passes through the tangency D (hence AD and DB are assigned different backward sequences) can the ordering be defined in a consistent way.

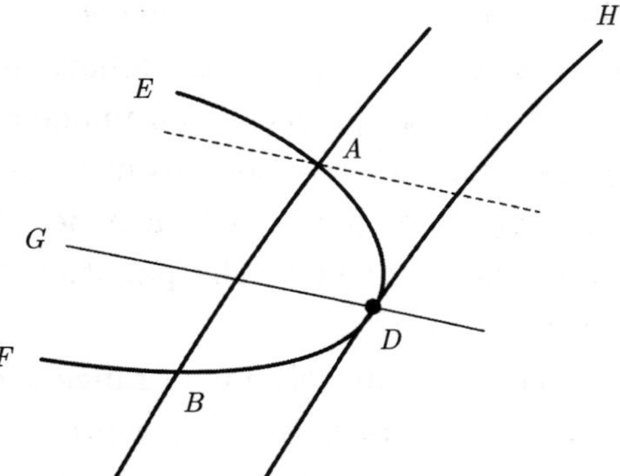

Figure 5.15 Backward foliation EDF is tangent to forward foliation HD at D. A certain image DG of the partition line must pass through the tangent point D. The dashed line corresponds to an image of an improperly chosen partition line, see text

It has been confirmed by numerical studies that the symbolic dynamics of the Lozi map can be well applied to the Hénon map. The two maps both have two fixed points, which may be coded as R^∞ and L^∞. Although the y-axis is not exactly the partition line for the Hénon map the true one can be located near the y-axis by finding tangencies between forward and backward foliations. In the phase plane there are infinitely many such tangencies, but not all play the same role in the construction of a partition line. There is a region $R^\infty\bullet$ in the upper part of the phase plane. Points in the region will quickly move to $+\infty$ under the inverse mapping. The unstable manifold of fixed point R^∞ forms a border of the region. Except the tangency right at the manifold, tangencies in the region play no essential role, so a half-line can be drawn upwards rather freely as a part of the partition line. There are other similar regions inside which tangencies play no role for the construction of the partition line. There are still many tangencies to be determined. Furthermore, images and pre-images of any tangency are also tangencies. Thus, it is far from a trivial task to collect relevant tangencies and finally to figure out the partition line.

If the ordering of symbolic sequences in the Hénon map is similar to that of the Lozi map, the symbolic planes of the two maps then resemble each other. For any finite string U made of R and L the backward symbolic sequence $R^\infty LLU\bullet$ and $R^\infty LRU\bullet$ correspond to one and the same horizontal line in the symbolic plane. The region above the unstable manifold of the fixed point H^+

in the first quadrant has a single backward sequence $R^\infty\bullet$. The region in the phase plane whose borders consist of backward foliations with the above two sequences as backward symbolic sequences is an image of the region of $R^\infty\bullet$. Such a region in the phase space corresponds to just one line in the symbolic plane. In determining the partition line only tangencies on the borders of such a region make sense, and the segment of the partition line between the two tangencies may be chosen arbitrarily.

In the phase plane there are infinitely many tangencies between the backward and forward foliations, some being images or pre-images of the others. In order to determine the partition line $C\bullet$ or $\bullet C$ one must pick up all tangencies belonging to this line. In principle, one may start with the x- or y-axis of the $b = 0$ case and follow the tangencies with b increasing until one reaches the desired parameter. Numerically, these points usually correspond to backward and forward foliations with smaller curvature at the tangency as compared to curvature at other tangencies. Such tangencies along the partition line are thought as the result of the first foldings and called the *primary tangencies.* A few examples of the partition line $\bullet C$ determined numerically are given in Fig. 5.16. Fig. 5.16 (a) corresponds to the case when the fixed point H^+ in the first quadrant is a stable node at $a = 0.3$ and $b = 0.3$. The partition line may pass the region between the two fixed points in the first and third quadrants in a fairly arbitrary manner. No point on the partition line has to be

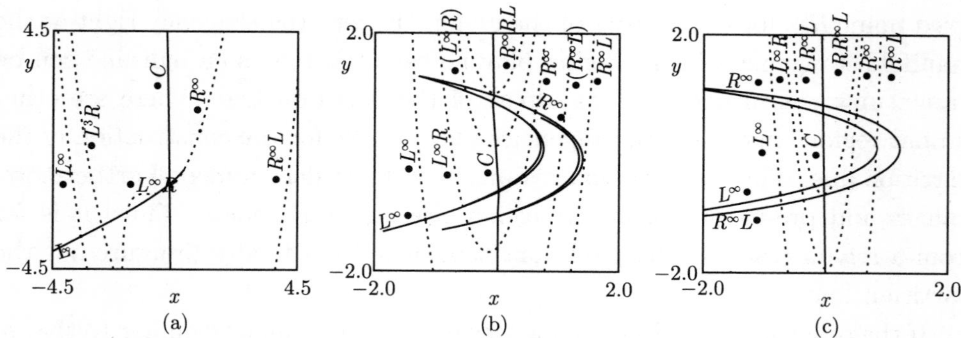

Figure 5.16 Numerically determined partition lines for the Hénon map. (a) $a = 0.3$, $b = 0.3$; (b) $a = 1.4$, $b = 0.3$; (c) $a = 2.85$, $b = 0.3$

[5] These figures are produced by the program listed in Appendix A.

determined uniquely. Fig. 5.16 (b) represents the standard situation of $a = 1.4$ and $b = 0.3$. Now the most important tangency is the one on the backward foliation $R^\infty\bullet$ passing through H^+ with the forward foliation of the greatest forward sequence, which is the counterpart of the kneading sequence at $b = 0$. Fig. 5.16 (c) corresponds to $a = 2.85$ and $b = 0.3$ when the tangency between the backward foliation $R^\infty L\bullet$ and the forward foliation $\bullet RRL^\infty$ contributes to the partition line. In fact, this is the only point on the partition line which has to be determined uniquely.

We conclude the discussion of partition lines by making the following remark. Suppose that we have determined the partition line $\bullet C$, whose image is $C\bullet$. Generally, the crossing point $C \bullet C$ of these two partition lines cannot be a tangency point since it would require too many conditions. Avoided crossing of the tangent point loci then happens except for the "accidental" case of the Lozi map. (One may consider a simple situation: two parabola families $y = x^2/2 + c$ and $x = -y^2/2 + d$. The loci of tangency points between the two curve families are two branches of curve $xy = -1$.) However, the crossing point numerically found is located in a region sharing one and single backward sequence, so does not connect to any tangency. We should emphasize that the only criterion for verifying a partition line is whether it correctly describes the symbolic dynamics.

5.5.3 Hénon-Type Symbolic Dynamics

Reflecting what has been done in the last subsection, we must admit that one has encountered essential difficulty at the first step of introducing coarse-grained description for the Hénon map. A rigorous proof of the symbolic dynamics for the Hénon map depends on non-local dynamical behavior of the system; one cannot avoid the consideration of foliations, and longer and longer orbits. Therefore, instead of establishing the symbolic dynamics precisely for the Hénon map, we choose to construct a prototype symbolic dynamics based on the symbolic dynamics of the Lozi map, postulating the topological similarity of the Lozi and the Hénon maps. We call it a Hénon-type symbolic dynamics and undertake to check the Hénon dynamics against this prototype. The Hénon-type symbolic dynamics deals with two symbols. It consists of the following three essential aspects.

Ordering rule for symbolic sequences

The forward symbolic sequences are ordered as that in the tent map. When comparing two sequences, the R-parity of the common leading string determines whether the order of the first different symbols should be preserved or reversed to get the order of the two sequences. Or put equivalently, the sequence for which the common leading string plus the first different symbol has an odd R-parity is larger. In summary,

$$\bullet E_R R\cdots > \bullet E_R L\cdots, \quad \bullet O_R R\cdots < \bullet O_R L\cdots.$$

The backward symbolic sequences are ordered as that in the inverse tent map. When comparing two sequences, the L-parity of the common leading string determines whether the order of the first different symbols should be preserved or reversed to get the order of the two sequences. Or put equivalently, the sequence for which the common leading string plus the first different symbol has an even L-parity is larger. In summary,

$$\cdots E_L R\bullet > \cdots E_L L\bullet, \quad \cdots O_L R\bullet < \cdots O_L L\bullet.$$

The fundamental forbidden zone

For a given map, the partition lines are determined by the tangencies between the forward contracting and backward contracting foliations. On the partition line $C\bullet$ every point of tangency between a forward and a backward foliation is represented by its corresponding forward and backward symbolic sequences. Each tangency on the partition line, denoted by $QC\bullet P$, confines a rectangular forbidden zone, which is demarcated by lines $\alpha = 1$, $\alpha = \alpha(\bullet P)$, $\beta = \beta(QR\bullet)$ and $\beta = \beta(QL\bullet)$ in the symbolic plane. The union of all forbidden rectangles from all tangency points along the partition line defines the *fundamental forbidden zone* (FFZ). A complete symbolic description of the partition line (the whole set of tangencies $\{QC\bullet P\}$ in the symbolic plane) defines the staircase-like boundary of the FFZ, or the *pruning front*. One profound property of the pruning front is its monotonicity: for any two tangencies $Q_1C\bullet P_1$ and $Q_2C\bullet P_2$, relation $Q_1\bullet \leqslant Q_2\bullet$ implies $\bullet P_1 \leqslant \bullet P_2$ and vice versa. As a consequence of this monotonicity, associated with tangency $QC\bullet P$, there are two allowed zones, which consist of points $V\bullet U$ satisfying $\bullet U \leqslant \bullet P$ and at the same time either $V\bullet \geqslant QL\bullet$ or $V\bullet \leqslant QR\bullet$. This means that any allowed zone contains no points of the fundamental forbidden zone.

Admissibility conditions

The necessary and sufficient condition for a symbolic sequence to be admissible consists in that all its shifts do not fall in the FFZ. Since any allowed zone contains no points of the FFZ a sequence is admissible if all its shifts fall in a finite number of allowed zones. In this case, a finite number of tangencies are enough to verify the admissibility of the sequence.

For the Hénon-type symbolic dynamics, the ordering rules of symbolic sequences are realized only for the true partition line. Any deviation from the true partition line will break the ordering rules somewhere. As the kneading sequences of 1D maps, the pruning front sets limits to allowed forward and backward sequences. A given tangency $QC \bullet P$ on the pruning front implies that, among all allowed sequences VXP with X being either R or L, $Q\bullet$ is the lower limit of V:

$$Q\bullet = \min\{V \bullet | VXP, QCP \text{ are admissible}\}. \tag{5.123}$$

Similarly,

$$\bullet P = \max\{\bullet U | QXU, QCP \text{ are admissible}\}. \tag{5.124}$$

Any violation of (5.123) or (5.124) indicates inconsistency of the partition. Therefore, we may propose a procedure to update an approximate partition line as follows. We generate a very long orbit. Once a partition line is given, we convert the orbit into a sequence. We also take points on the temporary partition line as initial values to generate their forward and backward sequences up to a certain length. We then update the temporary partition line at a fixed $QC\bullet$ with the point $QX \bullet U^*$ where $\bullet U^*$ is the maximal at the fixed Q. This procedure can be iterated. However, the procedure looks only at a local inconsistency of a wrong coding, hence is greedy. Some sophisticated procedure can be designed along this direction. For example, one can gradually increase the sequence length when checking consistency. We accept the partition line once we verify (up to certain sequence length): (1) relation (5.123) or (5.124) is satisfied; (2) Monotonicity of the pruning front is confirmed; (3) no real orbit points fall in the fundamental forbidden zone. A program of simplied greedy version is listed in Appendix A. 3. By the way, it is worth mentioning that for the unimodal map we may justify the use of the above procedure to locate the critical point from sequences.

5.5.4 Symbolic Analysis at Typical Parameter Values

Based on the Hénon-type symbolic dynamics formulated in the last section, we now perform a detailed symbolic analysis of the Hénon map at the most-studied parameter values $a = 1.4$, $b = 0.3$.

Once a partition line in the phase space has been determined, the dynamical behavior of the system is entirely fixed. The dynamical foliations of the Hénon map at $a = 1.4$ and $b = 0.3$ are drawn in Fig. 5.17. Forward foliations are shown by dotted lines, backward foliations—by dashed lines. Solid lines give the attractor, which is part of the backward foliations. The partition line $C\bullet$, determined by the primary tangencies, are shown by a long-dashed curve. We have picked up 94 tangency points along this partition line. The forward and backward symbolic sequences are calculated up to 30 symbols at each point. The fundamental forbidden zone determined by these 94 tangencies is drawn in Fig. 5.18 (a). In order to check the validity of the symbolic dynamics, we have calculated more than 3000 real orbital points. Indeed, none of their metric representations falls in the fundamental forbidden zone, see Fig. 5.18 (b).

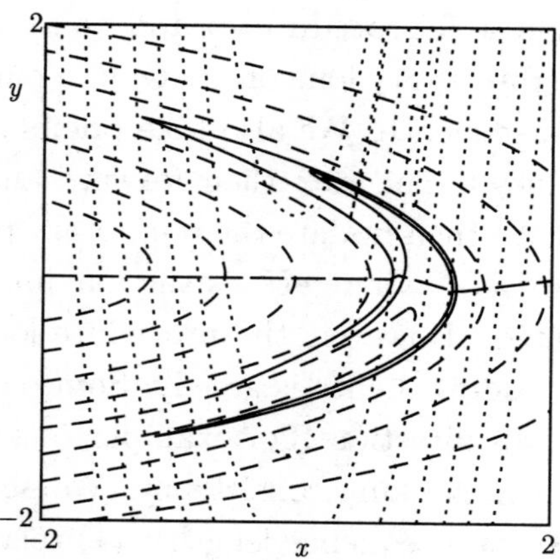

Figure 5.17 Dynamical foliations of the Hénon map at $a = 1.4, b = 0.3$. Dotted lines—forward foliations; dashed lines—backward foliations

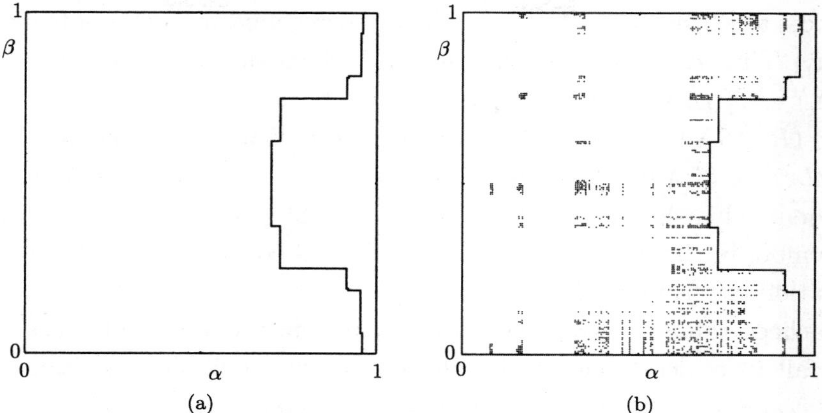

Figure 5.18 Symbolic plane of the Hénon map at $a=1.4, b=0.3$. (a) The fundamental forbidden zone. (b) None of the 3000 real orbital points falls in the fundamental forbidden zone

Although symbolic sequences along the partition line determines the dynamics entirely, usually we are able to collect only a limited set of information about the partition line. Nevertheless, the knowledge of symbolic sequences at a finite number of points along the partition line may yield necessary and sufficient conditions for some admissible sequences. For the standard parameter values studied in this section, the following 5 tangencies may be determined with great confidence:

$$\begin{array}{ll} G: & R^{\infty}LC \bullet R^3LRLRLR^6L\cdots \\ H: & R^{\infty}LRLC \bullet R^3LRLRLR^4L\cdots \\ I: & R^{\infty}L^2C \bullet R^3L^2R^3LRL^3\cdots \\ J: & R^{\infty}LRC \bullet RL^2R^4LRL^3\cdots \\ K: & R^{\infty}C \bullet RL^3R^4LR^2L^2R^5L\cdots \end{array} \tag{5.125}$$

These tangencies are labeled by the letters G through K. In what follows we will use the same letter, say, J to denote the allowed zone in the symbolic plane determined by the tangency J, and $\overline{J}$—the forbidden zone of J. We can draw rather strong conclusions, using these 5 tangencies only.

Periodic orbits

Being the rightmost forbidden rectangle, K excludes any forward sequence that is larger than K. (The only exceptions are sequences whose backward sequences are either $R^\infty\bullet$ or $R^\infty L\bullet$. They correspond to points outside the attractor.) Thus K forbids periodic sequences such as $(RL^3R^2X)^\infty$, $(RL^3X)^\infty$, $(RL^4X)^\infty$, $(RL^5X)^\infty$, etc., where X may be either R or L. Periodic sequence $(RL^3RX)^\infty$ does not have any shift that falls in $\overline{K}$, however, its shifts $(RXRL^3)^\infty\bullet(RXRL^3)^\infty$ falls in the forbidden rectangle $\overline{I}$. Therefore, it is forbidden. This shift is obtained by shifting the present dot "$\bullet$" beyond the 4-th symbol in the original string RL^3RX, so a convenient shorthand for the prohibition criterion is $4\overline{I}$.

To check the admissibility of a given sequence, one should check that its shifts fall in at least one of the allowed zones. The simplest case is X^∞, i.e., R^∞ and L^∞. They have only one shift, which falls in the allowed zone G. Accordingly, the admissibility criterion of X^∞ may be shorthanded as $0G$. The sequence $(RLRR)^\infty$ is also admissible, because its shifts $(RLRR)^\infty\bullet(RLRR)^\infty$, $(LRRR)^\infty\bullet(LRRR)^\infty$, $(RRRL)^\infty\bullet(RRRL)^\infty$, and $(RRLR)^\infty\bullet(RRLR)^\infty$ fall in the allowed zones J, G, J, and G, respectively. Therefore, the shorthand for the admissibility criterion is $0J1G2J3G$. In this way one can check all periodic sequences, say, of length 7 and shorter. The results are given in Table 5.1.

Being reflection of continuity and monotonicity, the ordering rule of symbolic sequences works both in the phase and parameter space. Once the symbolic sequence is given, the actual trajectory may be effectively located in the phase space with the help of the ordering rule. The procedure goes as follows. Take an initial point and generate a symbolic sequence. By comparing the sequence with the given one, it is easy to choose a new point and generate a new symbolic sequence. The ordering of both backward and forward sequence tells us where to move the point. One gets more and more coinciding symbols by using a bisection method, i.e., by interpolation and extrapolation in the correct direction.

As examples, we have determined an orbital point for each of the periodic orbits $(RL^2R^3L)^\infty$ and $(RL^2R^4)^\infty$. They are

$$(x,y)=(1.183\ 226\ 459\cdots,-0.288\ 809\ 258\ 7\cdots)$$

and

$$(x,y)=(1.232\ 250\ 862\cdots,0.128\ 443\ 396\ 6\cdots).$$

Table 5.1 The existence of periodic sequences up to length 7. The letter X denotes R and L. Only non-repeating strings are given

Sequence	Period	Existence	Criterion
X	1	allowed	$0G$
RL	2	allowed	$0J1G$
$RLRR$	4	allowed	$0J1G2J3G$
RLR^3X	6	allowed	$0J1G2G3G4J5G$
RLR^4X	7	allowed	$0J1G2G3G4G5J6G$
RLR^2X	5	forbidden	$3\overline{H}$
RLR^2LRX	7	forbidden	$3\overline{I}$
RLX	3	forbidden	$0\overline{I}$
RL^2RLR	6	forbidden	$0\overline{I}$
RL^2RLRX	7	forbidden	$3\overline{I}$
RL^2RX	5	forbidden	$3\overline{I}$
RL^2R^3X	7	allowed	$0J1G2G3G4G5J6G$
RL^2R^2X	6	forbidden	$4\overline{I}$
RL^2R^2LX	7	forbidden	$4\overline{I}$
RL^2X	4	forbidden	$0\overline{J}$
RL^3RLX	7	forbidden	$0\overline{J}$
RL^3RX	6	forbidden	$4\overline{I}$
RL^3R^2X	7	forbidden	$0\overline{K}$
RL^3X	5	forbidden	$0\overline{K}$
RL^4RX	7	forbidden	$0\overline{K}$
RL^4X	6	forbidden	$0\overline{K}$
RL^5X	7	forbidden	$0\overline{K}$

Chaotic orbits

By using the two tangencies G and J only, it may be verified that all symbolic sequences containing only strings of type R^nL with $n = 1, 3, 5, 6, 7, \cdots$ are admissible. Shifts of such sequences all fall in the allowed zones of G or J. In order to draw this conclusion we first note that the forward sequence of such sequences are smaller than that of J, i.e., $\bullet RL^2R^4LRL^3\cdots$. There are three types of backward sequences to be checked: 1. $\cdots R^2\bullet$, 2. $\cdots RL\bullet$, and 3. $\cdots LR\bullet$.

Replacing C by either R or L in the backward sequence of J, i.e., in $R^\infty LRC\bullet$, we get two backward sequences $R^\infty LR^2\bullet$ and $R^\infty LRL\bullet$. The two allowed zones, cause by J, are located above $R^\infty LRR\bullet$ and below $R^\infty LRL\bullet$. Since $\cdots R^2\bullet > R^\infty LRR\bullet$ and $\cdots RL\bullet < R^\infty LRL\bullet$, the first two types of sequences fall in the allowed zones of the tangency J.

In order to check the third type of backward sequences, one must invoke the allowed zones of G. From the backward sequence $R^\infty LC\bullet$ we derive two backward sequences $R^\infty LR\bullet$ and $R^\infty LL\bullet$, the former being smaller than $\cdots LR\bullet$. When the shifts of the sequence under test starts with $\cdots LR\bullet$, the leading string of its forward sequence must be either L or R^2L or R^4. In any case the forward sequence is smaller than that of G. Therefore, the third type of shifts falls in the allowed zone of G. This completes the proof that a sequence containing only strings of type $R^nL, n = 1, 3, 5, 6, 7, \cdots$ is admissible.

Using strings of R^nL type it is easy to construct a variety of chaotic symbolic sequences. The simplest ones consist of strings RL and R^3L only. The existence of such a set of chaotic orbits implies a positive topological entropy.

It can also be verified that strings of type RL^2R^mL, $m = 5, 6, \cdots$, may be used to construct admissible sequences. If a symbolic sequence contains a string R^2L or R^4L, it may be forbidden. For instance, the sequences

$$\cdots LRLR \bullet RL \cdots$$

and

$$\cdots LRLR \bullet R^3LR^2 \cdots$$

fall in the forbidden zone $\overline{H}$.

5.5.5 Discussion

Once we accept the prerequisites that the symbolic dynamics of the Hénon map is binary and forward (backward) sequences are ordered according to the R- (L-)parity, we may conduct symbolic analysis, including the construction of the partition line, without referring to foliations. Although a rigorous proof of the prerequisites is not available yet, an intuitive explanation can still be given. Take the x-axis as an approximate partition line. The line $y = 0$ maps to a parabola under the forward iteration as shown in Fig. 5.19. The x-axis and the parabola cut the plane into four areas. Each of them has been assigned a code according to the first two symbols common to the backward sequences in the area. It is seen that either the upper or the lower half-planes has its preimage in the whole plane. This explains why the symbolic dynamics is binary. Furthermore, we see that area $RR\bullet$ is on the top of area $LR\bullet$, just as $R\bullet$ on the top of $L\bullet$. However, area $LL\bullet$ is on the top of $RL\bullet$, showing an order reverse. This explains the L-parity for the ordering of backward sequences. We may imagine that when the x-axis is replaced with the true partition line the above qualitative picture will not change. Similarly, the left and right half-planes each maps to the whole plane under a forward iteration. A similar argument can explain the ordering rules for forward sequences.

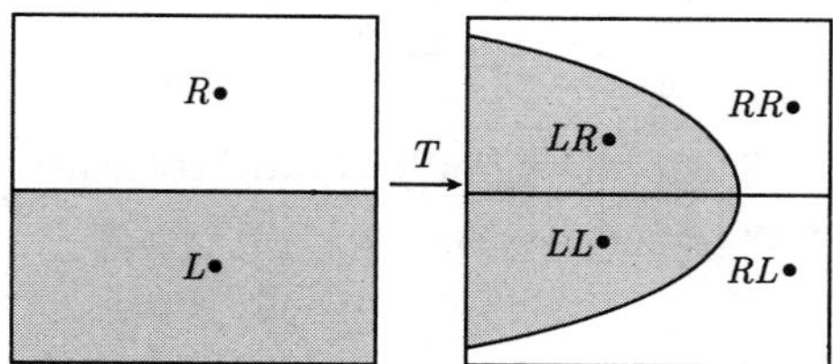

Figure 5.19 The phase plane and the x-axis under the forward mapping

In order to get a better understanding of the ordering rules, let us have a close look at a tangency $Q \bullet CP$ on line $\bullet C$ as shown in Fig. 5.20. Rigorously speaking, we can not write $\bullet C$. Contrary to the 1D unimodal map where C is always followed by the unique kneading sequence, now sequences after or before C are not unique. Anyway, we use $\bullet C$ to mean that the sequences following C are unspecified. The tangency c: $Q \bullet CP$ maps to tangency c': $QC\bullet P$ on line $C\bullet$. Assume that there are two points a: $Q'\bullet LP$ and b: $Q'\bullet RP$; both of them are on the backward foliation $Q'\bullet$. The figure represents the case when the leading symbol of Q is L. The admissibility conditions then require $Q'\bullet > Q\bullet$. The backward sequences on both arcs ac and bc are arranged in descending order. The images of these two points are a': $Q'L \bullet P$ and b': $Q'R \bullet P$. The backward sequences on arc bc and $b'c'$ are still arranged in descending order while the order on $a'c'$ is reversed. This outlines how the ordering rules for backward sequences should be proven.

We have mentioned in the last section how to locate the orbit of a given admissible periodic sequence based on the ordering of foliations. Biham and Wenzel [1990] also proposed a very efficient procedure to locate unstable periodic orbits of the Hénon map. The procedure provides a coding for periodic orbits as its byproduct. In fact, this coding results from the rules of foliation ordering.

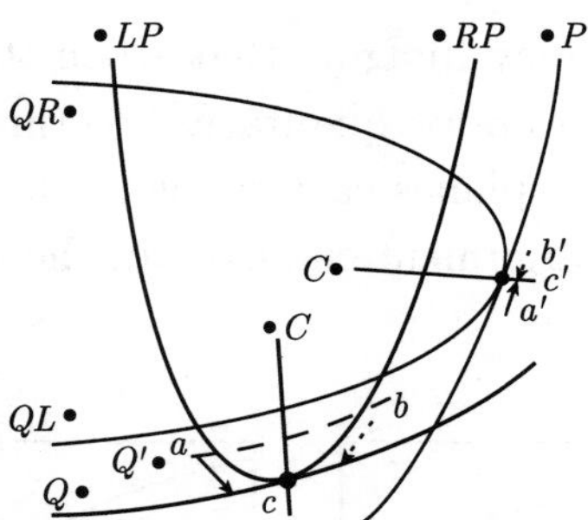

Figure 5.20 The ordering of backward sequences and their images on forward foliations

The Biham-Wenzel Coding

Biham and Wenzel related a periodic orbit of a map to an extremum static configuration of an atomic chain and introduced a type of pseudo-dynamics for the chain. For the Hénon map, an atom at site n experiences a force F_n

$$F_n = \gamma(a - x_n^2 + bx_{n-1} - x_{n+1}), \tag{5.126}$$

where γ is a constant of proportion. An artificial dynamics is then defined by

$$dx_n/dt = \sigma_n F_n, \quad n = 1, \cdots, p, \tag{5.127}$$

where $\sigma_n = \pm 1$. Then Eq. (5.127) is solved under the periodic boundary condition $x_{p+1} = x_1$. With a set of $\{\sigma_n\}$ given, the system is driven towards an extremum static configuration when the forces on all atoms decrease to zero. The resulting configuration gives an exact periodic orbit of order p of the Hénon map. In general, given $\{\sigma_n\}$, the corresponding periodic configuration does not always exist. When the atoms escape to infinity the $\{\sigma_n\}$ corresponds to no orbits. However, the technique was able to find all the existing periodic orbits. Furthermore, Biham and Wenzel observed that there is a good correspondence between the sign of x_n and σ_n as an orbit coding.

We now explain why the procedure works from the viewpoint of symbolic dynamics. It is more convenient to consider discretized time t, and regard Eq. (5.127) as an iteration by setting the time step to 1. Denote by x_n^* the orbit point to which finally x_n converges. The force F_n is essentially the difference $\Delta x_{n+1} = x'_{n+1} - x_{n+1} \approx x'_{n+1} - x^*_{n+1}$, where $x'_{n+1} = a - x_n^2 + bx_{n-1}$ is the image of x_n under the map. We may assume that x'_{n+1} almost falls on the unstable manifold associated with x^*_{n+1}. For the procedure to work, we should push x_n along the unstable manifold of x^*_n to the stable manifold of x^*_n. As known from the symbolic dynamics of the Lozi map, Δx_{n+1} and $\Delta x_n = x_n - x^*_n$ retain the same order when the symbol of x_n is $s_n = L$, while the orders are reversed when $s_n = R$. Thus, assuming that the Hénon map shares the same type of symbolic dynamics with the Lozi map, according to the discussion in Section 5.4.3, to drive x_n in the right direction we should choose $\sigma_n = 1$ for $s_n = R$ and $\sigma_n = -1$ for $s_n = L$.

Similarly, we may construct an inverse map version of the procedure. Now, instead of Δx_{n+1} on the backward foliation of x^*_{n+1}, we consider Δx_{n-1} on the forward foliation of x^*_{n-1}. We have $b\Delta x_{n-1} = bx'_{n-1} - bx_{n-1} = -(a - x_n^2 - x_{n+1}) - bx_{n-1} \approx b(x'_{n-1} - x^*_{n-1})$. According to the symbolic dynamics of the Lozi map, to drive x_n in the right direction to x^*_n along the forward foliation, we should choose $\sigma_n = 1$ for $s_n = L$ and $\sigma_n = -1$ for $s_n = R$. This is consistent with the previous version. Note that the roles of R and L are exchanged, and, correspondingly, the partition line $\bullet C$ is determined instead of $C\bullet$.

There is a main difference between the lozi map and the Hénon map. While the ordering according to the x-coordinate is rigorous for the Lozi map, it could be only approximate for the Hénon map. The ordering is essentially associated with the "dynamical" coordinates, which are the arc lengths along foliations, but not with a "natural" coordinate such as the x-coordinate. We expect that the Biham-Wenzel procedure could fail when x^*_n is close to a tangency point when the dynamical coordinate does not agree with the natural coordinate (Grassberger, Kantz and Moenig [1989]). Such situations often relate to a structure similar to a joint in the Kneading plane of 1D maps.

The structure of parameter space around a joint

For specificity, we consider the "joint" $RLCRC$: at a particular parameter point near $(a, b) = (1.46, 0.16)$ there is a period-5 orbit $(RLCRC)^\infty$, whose two orbit points are located just on the partition line $C\bullet$. The relation among five relevant images and preimages of $C\bullet$ is schematically shown in Fig. 5.21.

Two orbit points $(CRCRL)^\infty \bullet (CRCRL)^\infty$ and $(RLCRC)^\infty \bullet (RLCRC)^\infty$ are marked with a filled cirlce in the figure.

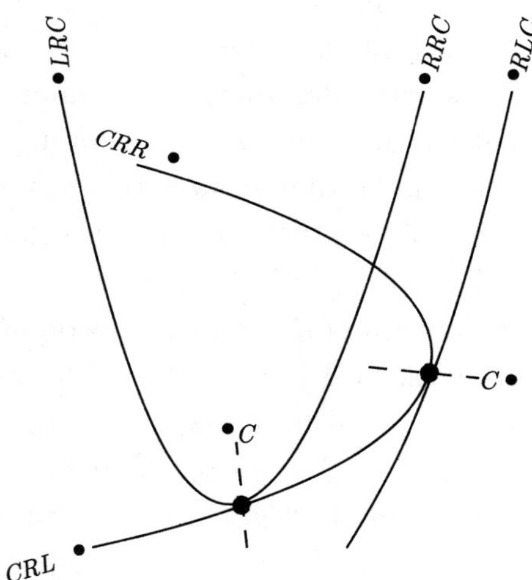

Figure 5.21 The relation among five images and preimages of $C\bullet$ relevant to the period-5 $(RLCRC)^\infty$. The two points marked by a filled circle belong to the orbit

When the parameters change around the joint, four "leg" regions characterized by symbolic sequences $RLRRC$, $RLLRC$, $RRRLC$ and $RLRLC$ can be detected as shown in Fig. 5.22. While the left figure is drawn to reminisce the kneading plane of 1D maps, the right figure is drawn using the given parameters a and b. (Note that the map has essentially an infinite number of parameters. Other orbits may co-exist with different mechanisms and basins.) We sketch the changes in the phase space seen when the parameters vary along certain directions in Figs. 5.23 and 5.24. In the figures curves have been simplified as straight lines, so only their relative positions are of meaning.

Before closing this section we point out that, as it should be, the eigenvalues of the two fixed points tell us the ordering rules since the local ordering near fixed points must be consistent with the global ordering. For the fixed point H^+ of R^∞, the positive sign of its stable eigenvalue indicates the R-preserval for the order of backward foliations while the negative sign of its unstable eigenvalue indicates the R-reversal for the order of forward foliations. Similarly, for the fixed point H^- of L^∞, the negative sign of its stable eigenvalue gives the

L-reversal for the order of backward foliations while the positive sign of its unstable eigenvalue gives the L-preserval for the order of forward foliations.

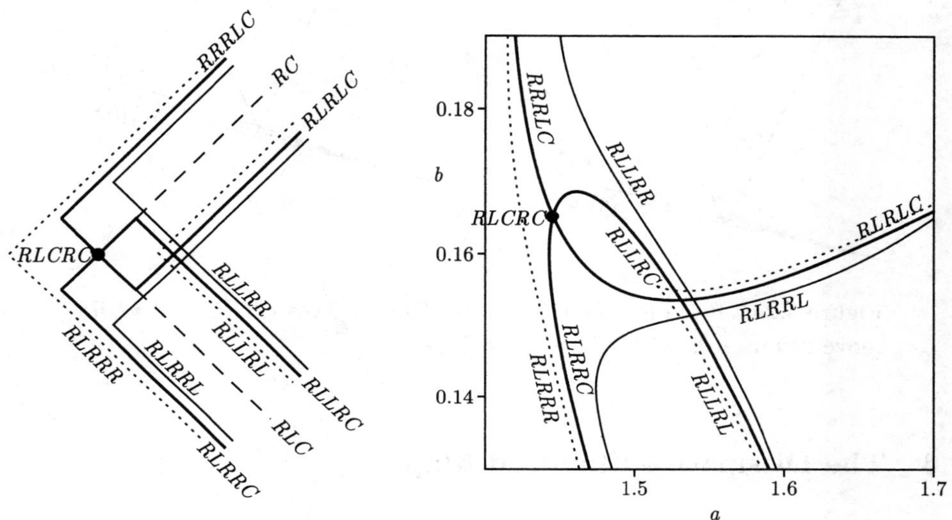

Figure 5.22 Four "legs" characterized by symbolic sequences $RLRRC$, $RLLRC$, $RRRLC$ and $RLRLC$ "grow" from the joint $RLCRC$. The left figure is drawn to reminisce the kneading plane of 1D maps while the right is drawn using the given parameters a and b. The two dotted lines represent the curves of the tangent bifurcation while the two thin ones represent those of the period-doubling bifurcation

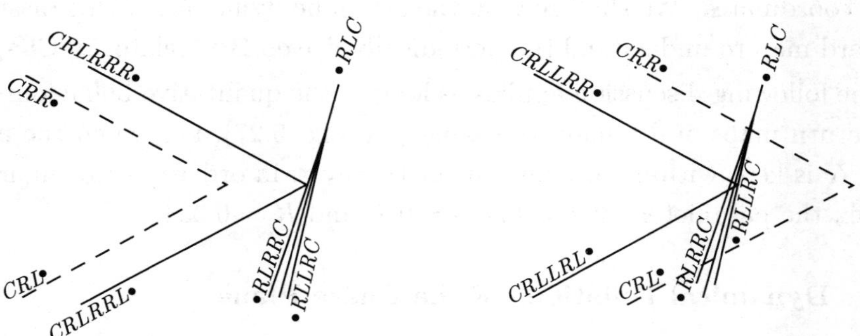

Figure 5.23 From leg $RLRRC$ to $RLLRC$, curves $CRR\bullet$ and $CRL\bullet$ move rightwards with respect to $\bullet RLC$

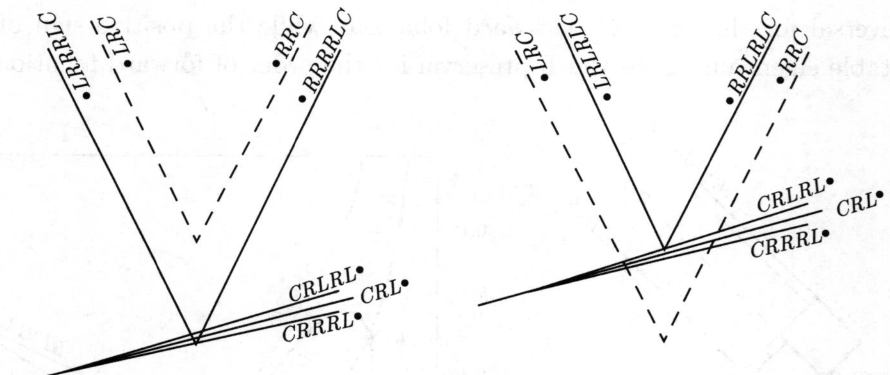

Figure 5.24 From leg $RRRLC$ to $RLRLC$, curves $\bullet RRC$ and $\bullet LRC$ move downwards with respect to $CRL\bullet$

5.6 The Dissipative Standard Map

In this section we study a two-dimensional generalization of the one-dimensional circle map—the dissipative standard map

$$\begin{aligned} r_{n+1} &= a + br_n + K \sin 2\pi\theta_n, \\ \theta_{n+1} &= \theta_n + r_{n+1} \pmod 1). \end{aligned} \tag{5.128}$$

Geometrically, this dynamics takes place on a 2D annulus described by the r and θ coordinates. We shall rely on the symbolic dynamics of this dissipative standard map to understand the periodically driven Brusselator in Chapter 6.

The following discussion applies as long as the qualitative behaviour of the first return maps of θ remain the same (see Fig. 5.27), i.e., when the nonlinearity K is kept within a certain range. However, in order to draw figures, we shall fix the parameters at $a = 0.4$, $b = 0.2$, and $K = 0.334$.

5.6.1 Dynamical Foliations of the Phase Plane

The Jacobian of the map (5.128) is

$$\mathbf{J}_n = \begin{bmatrix} 1 + 2\pi K \cos 2\pi\theta_n & b \\ 2\pi K \cos 2\pi\theta_n & b \end{bmatrix}. \tag{5.129}$$

Let us consider a point (θ_n, r_n), which has been reached from (θ_{n-j}, r_{n-j}) by iterating (5.128) j times, and introduce the backward product matrix

$$\mathbf{M}_n^b \equiv \mathbf{J}_{n-1}\mathbf{J}_{n-2}\cdots\mathbf{J}_{n-j}\,\tilde{\mathbf{J}}_{n-j}\tilde{\mathbf{J}}_{n-j-1}\cdots\tilde{\mathbf{J}}_{n-1}. \tag{5.130}$$

The eigenvector corresponding to the larger eigenvalue of the matrix $\mathbf{M}_n^b$ converges for increasing j (Greene [1983], Gu [1987]). In (5.130) $\tilde{\mathbf{J}}$ denotes the transpose of $\mathbf{J}$. In the limit $j \to \infty$ the eigenvector defines the *unstable direction* at the point (θ_n, r_n). Integral curves of the field of such directions determine the *backward contracting foliations* (BCFs) or simply the *backward foliations* of the phase plane.

Similarly, the *stable direction* at (θ_n, r_n) can be found as the $k \to \infty$ limit of the eigenvector corresponding to the smaller eigenvalue of the forward product matrix

$$\mathbf{M}_n^f \equiv \tilde{\mathbf{J}}_{n+1}\tilde{\mathbf{J}}_{n+2}\cdots\tilde{\mathbf{J}}_{n+k}\,\mathbf{J}_{n+k}\mathbf{J}_{n+k-1}\cdots\mathbf{J}_{n+1}. \tag{5.131}$$

The integral curves of such directions determine the *forward contracting foliations* (FCFs) or simply the *forward foliations* of the phase plane.

According to Grassberger and Kantz [1985] a partition line can be determined from primary tangencies of stable and unstable manifolds of the unstable fixed point. It is natural to extend this procedure to the tangencies between the two classes of foliations (Zheng [1991b, 1992a]; Zhao and Zheng [1993]). This generalization extends the partition line into the phase plane beyond the attractor.

In Fig. 5.25 we show the attractor (solid curves) and two primary partition lines ($\bullet S$ and $\bullet G$) on the background of the forward foliations (dash curves). The line marked with $\bullet D$ is the pre-image of $\bullet S$. The notations S, G, and D are mnemonic for smallest, greatest, and discontinuity, as we have used in Fig. 4.12 for the circle map of Section 4.5, and as it will be seen in the corresponding first return map (Fig. 5.27). The areas between these lines are labeled by $\bullet R$, $\bullet L$, and $\bullet N$.

Since the dynamics takes place on an annulus, we may cut the annulus along the line $\bullet S$. By dropping (mod 1) in (5.128) we get the *lifted* (θ, r) plane. In the lifted plane one may set the *fundamental strip* to be the strip whose border is the line $\bullet S$ and the line obtained by shifting $\bullet S$ to the right by one unit. Thus, the area to the left of the $\bullet S$ line in Fig. 5.25, also marked with $\bullet N$, belongs to the left neighbor of the fundamental strip. In order to avoid any mis-understanding we note that the partition lines $\bullet S$ and $\bullet G$ are not parallel to the r-axis, as they may seem to be.

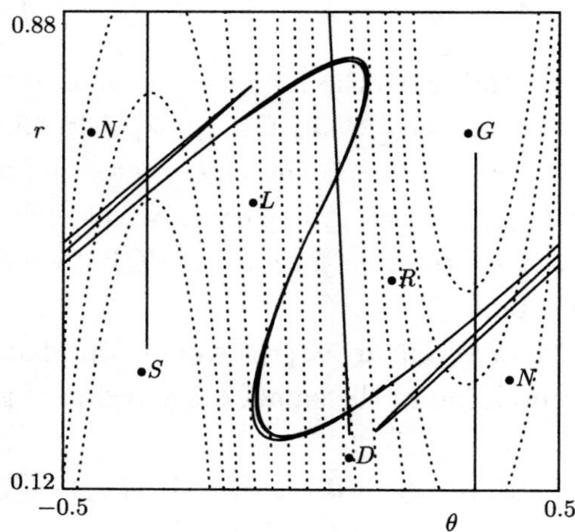

Figure 5.25 Two primary partition lines $\bullet S$ and $\bullet G$ of the dissipative standard map. The fundamental strip of the unwrapped annulus is the area between the line $\bullet S$ and its parallel shift to the right by one unit

Any orbit in the phase plane are coded as a bi-infinite symbolic sequence

$$\cdots s_{-2}s_{-1} \bullet s_0s_1s_2 \cdots \tag{5.132}$$

where s_i stands for the letter L, R or N, depending on which area the i-th point of the orbit falls in, and the heavy dot $\bullet$ indicates the "present" point. The sequence

$$\bullet s_0s_1s_2 \cdots$$

is the forward symbolic sequence with respect to the present dot, and the backward symbolic sequence is

$$\cdots s_{-2}s_{-1} \bullet .$$

Another way to partition the phase space is based on pre-images, as shown in Fig. 5.26. Under the inverse map the area marked with $L\bullet$ in the figure will map to the area $\bullet L$ of Fig. 5.25, while the areas marked with $R\bullet$ and $N\bullet$ map to areas $\bullet R$ and $\bullet N$.

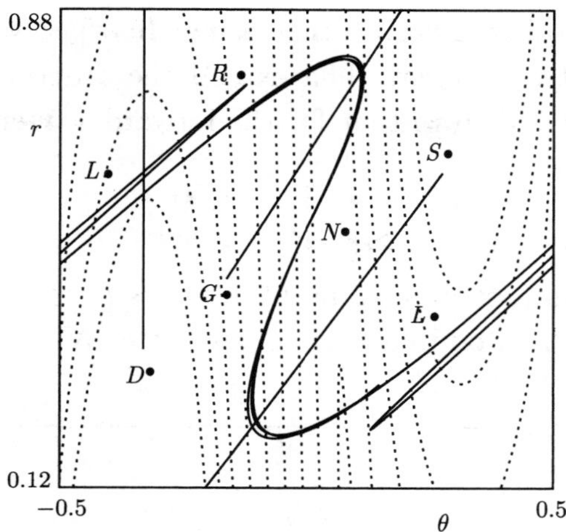

Figure 5.26 The partition of the annulus according to pre-images

5.6.2 Ordering of Symbolic Sequences

We may introduce an ordering in the fundamental strip. According to Fig. 5.25 we may take the "natural" order

$$\bullet S < \bullet L < \bullet R < \bullet G < \bullet N, \tag{5.133}$$

and from Fig. 5.26 we may take the order

$$R\bullet < G\bullet < N\bullet < S\bullet < L\bullet\,. \tag{5.134}$$

The return map θ_{n+1}-θ_n, constructed from Fig. 5.25, is shown in Fig. 5.27. This map clearly shows two-dimensional feature, since it is multi-valued near the discontinuity as well as near the maximum and minimum. However, this does not prevent us from treating it as a one-dimensional map for the following reason. In Fig. 5.25 all points on one and the same forward foliation (dash line) have the same future, and, after appropriate coarse-graining, share the same forward symbolic sequence. Therefore, if we are only interested in forward symbolic sequences, we can shrink different pieces of the attractor along forward foliations to make the attractor approximately a one-dimensional object.

If ignoring the layered structure, we may treat the return map Fig. 5.27 as a circle map. Thus in the $\bullet N$ area, variable θ_{n+1} as a function of θ_n is decreasing, while in other areas it is increasing. In other words, only the letter N has an odd parity. We may then extend the ordering rule for symbolic sequences of the circle map to that for the forward sequences of map (5.128) as

$$\begin{aligned} &\bullet EL\cdots < \bullet ER\cdots < \bullet EN\cdots , \\ &\bullet OL\cdots > \bullet OR\cdots > \bullet ON\cdots , \end{aligned} \tag{5.135}$$

where E and O are finite strings made of the letters L, R and N, and containing respectively an even and odd numbers of the letter N.

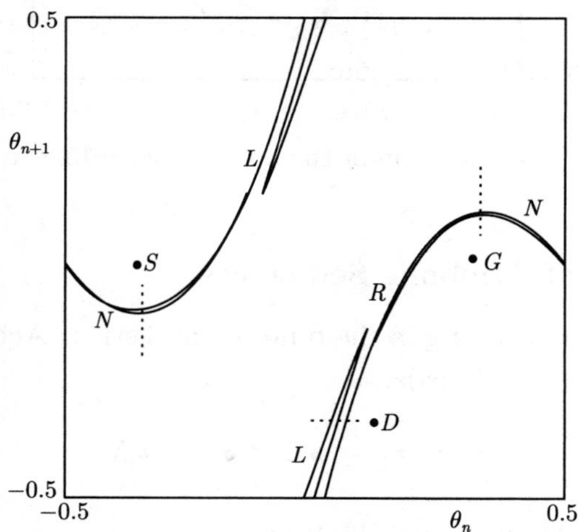

Figure 5.27 The θ_{n+1}-θ_n first return map

Similarly, all points on one and the same backward foliation (not shown in Fig. 5.25 except the attractor, which is part of the backward foliations,) have the same past, and, after appropriate coarse-graining, share the same backward symbolic sequence. Therefore, based on the ordering (5.134), backward symbolic sequences are ordered as

$$\begin{aligned} &\cdots RE\bullet < \cdots NE\bullet < \cdots LE\bullet , \\ &\cdots RO\bullet > \cdots NO\bullet > \cdots LO\bullet , \end{aligned} \tag{5.136}$$

where E and O have the same meaning as in (5.135).

We can summarize the ordering of symbolic sequences in two dimensional maps such as (5.128) as follows. A family of forward foliations are ordered according to their transverse intersections with a backward foliation; a family of backward foliations are ordered according to their transverse intersections with a forward foliation. The order is well-defined as long as there are no tangencies between the two foliations. The occurrence of tangencies is associated with foldings of one family of foliations and with the necessity of introducing a partition line through the tangencies. The ordering reverses on crossing the partition line. In practice, the ordering may be inferred from the local order near fixed points, i.e., from the signs of their eigenvalues.

5.6.3 Symbolic Plane and Admissibility of Symbolic Sequences

Once the ordering of foliations has been established, a tangency on a partition puts a restriction on allowed symbolic sequences. For example, when the backward foliations $QR\bullet$ and $QN\bullet$ are tangent to the forward foliation $\bullet P$ on the partition line $G\bullet$, forming a tangency $QG \bullet P$, any forward sequence that is greater than $\bullet P$ (geometrically located to the right of the foliation $\bullet P$) cannot intersect with backward foliations between $QR\bullet$ and $QN\bullet$, i.e., those which are smaller than $QN\bullet$ and greater than $QR\bullet$. Thus sequences of the type Q_+RP_+ or Q_+NP_+ are forbidden by the tangency $QG \bullet P$, where $Q_+\bullet > Q\bullet$ and $\bullet P_+ > \bullet P$. Similarly, sequences of the type U_-NV_- or U_-LV_- are forbidden by a tangency $US \bullet V$ on the partition line $S\bullet$, where $U_-\bullet < U$ and $\bullet V_- < \bullet V$.

What has been said is best represented in a *symbolic plane* (Cvitanović, Gunaratne and Procaccia [1988]). In order to construct the symbolic plane we introduce a metric representation of symbolic sequences, which embodies the ordering rules (5.135) and (5.136). We first define an integer ϵ_i and for every symbol s_i:

$$\epsilon_i = \begin{cases} -1, & \text{if} \quad s_i = N, \\ 1, & \text{otherwise.} \end{cases} \tag{5.137}$$

We then assign to each forward sequence $\bullet s_1 s_2 \cdots s_i \cdots$ a real number $\alpha \in [0, 1]$:

$$\alpha = \sum_{i=1}^{\infty} \mu_i 3^{-i}, \tag{5.138}$$

where $\mu_i \in \{0, 1, 2\}$ is defined by

$$\mu_i = \begin{cases} 1, & \text{if} \quad s_i = R, \\ |\epsilon_1 \epsilon_2 \cdots \epsilon_{i-1} - \epsilon_i|, & \text{otherwise.} \end{cases} \tag{5.139}$$

It is easy to see that all forward sequences are ordered according to their α values.

Similarly, we define another integer σ_i for a symbol s_i

$$\sigma_i = \begin{cases} -1, & \text{if} \quad s_i = L, \\ 1, & \text{otherwise.} \end{cases} \tag{5.140}$$

and assign to each backward sequence $\cdots s_{-i} \cdots s_{-2}s_{-1}\bullet$ a real number $\beta \in [0, 1]$:

$$\beta = \sum_{i=1}^{\infty} \nu_i 3^{-i}, \tag{5.141}$$

where

$$\nu_i = \begin{cases} 1, & \text{if} \quad s_{-i} = N, \\ |\epsilon_{-1}\epsilon_{-2}\cdots\epsilon_{-i+1} - \sigma_{-i}|, & \text{otherwise.} \end{cases} \tag{5.142}$$

All backward symbolic sequences are ordered according to their β values. The unit square $\alpha \in [0,\ 1] \times \beta \in [0,\ 1]$ forms the symbolic plane. Horizontal and vertical lines in the symbolic plane correspond to backward and forward foliations, identified by their symbolic sequences respectively. A point in the unit square represents a bi-infinite symbolic sequence with a given present dot. It can be verified, e.g., that in the metric representation we have

$$\begin{aligned} &\alpha(\bullet NL^\infty) = \beta(L^\infty\bullet) = 1, && \alpha(\bullet L^\infty) = \beta(R^\infty\bullet) = 0, \\ &\alpha(\bullet NNL^\infty) = \alpha(\bullet RNL^\infty) = 2/3, && \beta(R^\infty L\bullet) = \beta(R^\infty N\bullet) = 2/3, \\ &\alpha(\bullet RL^\infty) = \alpha(\bullet LNL^\infty) = 1/3, && \beta(L^\infty N\bullet) = \beta(L^\infty R\bullet) = 1/3. \end{aligned}$$

The two tangencies $QG \bullet P$ and $US \bullet V$ discussed above demarcate two forbidden zones in the symbolic plane. Along a partition line an infinite number of tangencies may be found in principle. However, when one is interested in sequences not exceeding a fixed length, a finite number of tangencies is enough. Such a case is shown in Fig. 5.28, where 20 tangencies are used to outline the forbidden zones. In the figure 60 000 points representing real orbits are also drawn. All of them are located outside the forbidden zones. Although the image of a forbidden zone is also forbidden, it suffices to consider only the two unions of forbidden zones associated with $G\bullet$ and $S\bullet$. They are called the *fundamental forbidden zones.* The boundary of a FFZ defines a "pruning front" in the symbolic plane. Therefore, we may say that the partition lines $S\bullet$ and $G\bullet$ in the phase plane transform into the "pruning fronts" in the symbolic plane.

In Fig. 5.28 the area enclosed by the pruning front of $S\bullet$ and the line $\alpha = 0$ and the area enclosed by the pruning front of $G\bullet$ and the line $\alpha = 1$ form the fundamental forbidden zones. Any point within the FFZs corresponds to a forbidden sequence.

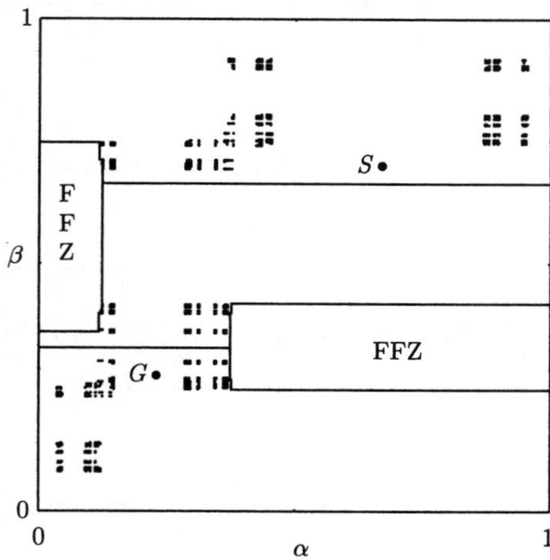

Figure 5.28 Symbolic plane of the dissipative standard map. Together with the FFZ, 60 000 points of real orbits are also shown. None of them falls inside the FFZ

Furthermore, from the construction of the FFZs we see that for the tangency $QS\bullet P$ on the line $S\bullet$, the square enclosed by $\beta = 1$, $\beta = \beta(QL\bullet)$, $\alpha = 1$ and $\alpha = \alpha(\bullet P)$ does not overlap with the FFZs, hence forms an allowed zone.

If a symbolic sequence has no shifts falling in the FFZs, it is admissible. Therefore, in order to formulate the admissibility conditions for all symbolic sequences an infinite number of tangencies are needed. As long as the role of a single tangency is concerned, it determines its own forbidden zone and allowed zones in the symbolic plane. One can only reach the following conclusions:

1. Even if just one shift of a sequence falls in the forbidden zone, the sequence is forbidden.

2. If a sequence has all its shifts in the allowed zones of the given tangency, it is admissible.

In Fig. 5.29 we sketch the allowed and forbidden zones formed by four tangencies. A tangency may be compared with a kneading sequence in one-dimensional maps. While a one-dimensional map with a finite number of critical points only possesses a finite number of kneading sequences, a two-dimensional map has infinite many tangencies. However, when we are interested in symbolic orbits not exceeding a finite length, a finite number of tangencies suffice for the job.

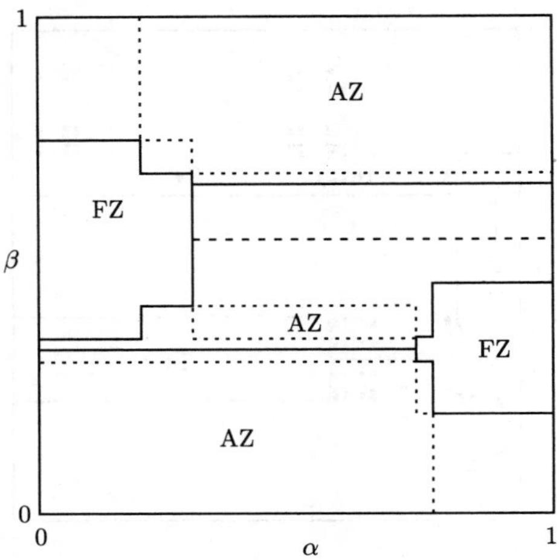

Figure 5.29 A sketch of forbidden and allowed zones when two tangencies are used both on $S\bullet$ and $G\bullet$ for the construction. Notice that the preimage of $S\bullet$ is $D\bullet$, and $DS\bullet$ is perturbed to $RL\bullet$ and $LN\bullet$. The upper-left FZ is then centered at $\beta = 5/9$

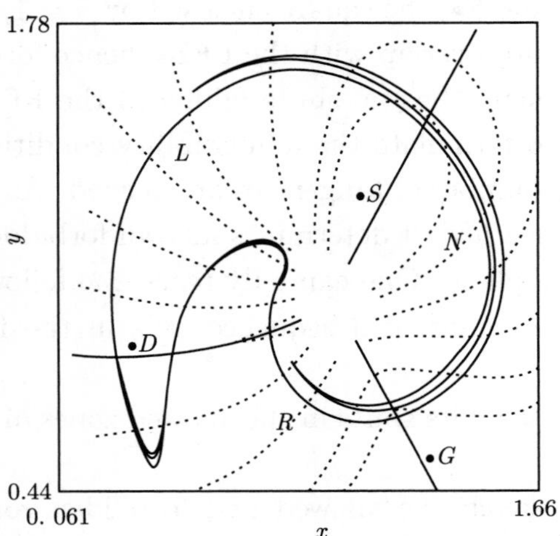

Figure 5.30 Attractor in the dissipative standard map, shown in the x-y plane instead of the r-θ plane (cf. Fig. 5.25)

For later comparison with the Poincaré map of the forced Brusselator it is convenient to map the θ-r phase plane into the x-y plane by making a transformation

$$(\theta,\ r) \to (x,\ y) = (x_0 - r\cos 2\pi\theta,\ y_0 - r\sin 2\pi\theta). \tag{5.143}$$

Take $x_0 = y_0 = 1$, the r-θ plane in Fig. 5.25 becomes x-y plane in Fig. 5.30.

5.7 The Stadium Billiard Problem

The maps discussed so far are all dissipative. In this section we shall apply symbolic dynamics to two-dimensional billiards, which are conservative systems. A billiard consists of a point mass freely moving inside a connected domain enclosed by a continuous boundary. The particle elastically reflects from the boundary upon collision according to the rule of specular reflection. A main difference between smooth mechanical systems and billiards consists in that the Hamiltonian or Lagrangian function is insufficient to uniquely define the dynamics of billiards since the infinite forces at the wall cause a discontinuity. Therefore, a discrete map specifying collisions with the boundary is more appropriate to describe the motion than differential equations.

Rectangular and circular billiards, belonging to integrable cases, exhibit regular motions. Another such example is the elliptic billiard, for which the nontrivial integral of motion is

$$F = l^2 - d^2\sin^2\theta,$$

where l is the distance from the ellipse center to a chord, i.e., a straight segment of a trajectory, d is the half distance between the foci, θ the angle between the chord and the major axis. Note that the velocity has been scaled to be 1. For a general billiard no global invariant function like F exists, so the dynamics is no longer integrable. The KAM theory extended to billiards is Lazutkin's theorem: For a convex billiard with sufficiently smooth boundary, in the billiard mapping there exists a set of invariant curves whose rotation numbers are irrational and form a set with a positive Lebesgue measure (Lazutkin [1973]). The theorem requires a sufficiently smooth deformation of the elliptic billiard in order to preserve the invariant curves with sufficient irrational rotation numbers. The radius of curvature for boundary of the deformed billiard must be bounded from above and below.

Any flat enough segments in the boundary turns a billiard chaotic. Bunimovich studied a general class of billiards composed of straight segments and circle arcs and found the conditions under which the dynamics of billiards is completely chaotic. The Bunimovich stadium is a well-known example of complete chaos (Bunimovich [1974, 1979]). The boundary of the billiard consists of two semi-circles of radius $R = 1$ joined by two straight segments of length $2a$, see Fig. 5.31.

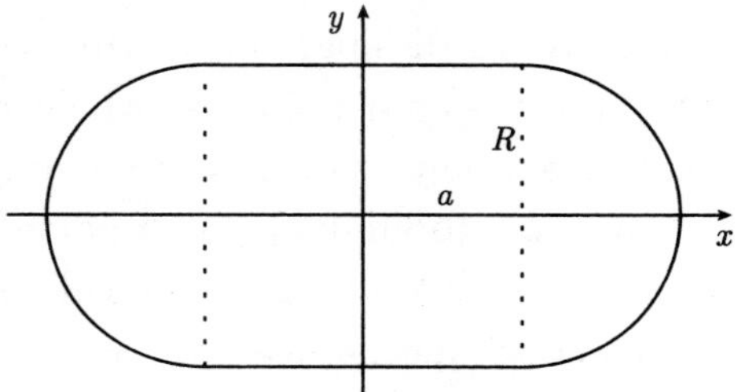

Figure 5.31 The stadium billiard

Biham and Kvale [1992] introduced a symbolic description for the stadium which associates each orbit with a unique symbolic sequence. Later, Hansen and Cvitanović [1994] developed a more compact, desymmetrized covering symbolic dynamics. Meiss [1992] explored ordered periodic orbits and cantori classified with rotation numbers. By means of the arc length s along the boundary and the tangent component v of the momentum, i.e. the Birkhoff coordinates, a map of the annulus is generated from the stadium. However, the information on the rotation number of an orbit is not explicitly contained in neither the Biham-Kvale coding nor the Hansen-Cvitanović. In this section we propose another coding based on lifting, since the dynamics of the map can be better understood by first lifting and then wrapping it. In this way we are able to study the orientation and rotation numbers of orbits. The proposed coding will be compared with other codings and used to construct symbolic dynamics for the billiard system (Zheng [1997a]).

5.7.1 A Coding Based on Lifting

We measure the arc length of the boundary along the counterclockwise direction from the lower end of the left semi-circle, i.e., at the joint point $s = 0$. We further denote by L the half length $\pi + 2a$ of the boundary. The annulus corresponds to the fundamental domain (FD) of $s \in [0, 2L)$ and $v \in [-1, 1]$. The image and preimage of this domain in the lifted space are shown in Figs. 5.32 and 5.33, respectively. In Fig. 5.33 the area hij of the FD (to the right of the curve connecting points h and j) maps to the area bcd right of the FD shown in Fig. 5.32 under the forward map, while under the backward map the area fbe of Fig. 5.32 (to the left of the lower curve connecting points b and e) maps to the area glk of Fig. 5.33 left of the FD. It is known that a bounce off a straight wall reverses the ordering of two neighboring orbits, while a bounce off a semi-circle preserves the ordering. Thus, the four strips separated by lines $s = 2a$, $s = L$ and $s = L + 2a$ in the FD must be assigned different symbols. To reflect the different behavior in the lifted map, we further divide the FD with the pre-image curve hj of the line $s = 0$. In this way we partition the FD into seven pieces labeled with symbols M, L, L_1, N, N_1, R and R_1 as shown in Fig. 5.33. This 7-letter alphabet will be called the full-stadium code later on. By means of this partition we may code an orbit with a bi-infinite sequence

$$\cdots s_{-2}s_{-1} \bullet s_0 s_1 \cdots ,$$

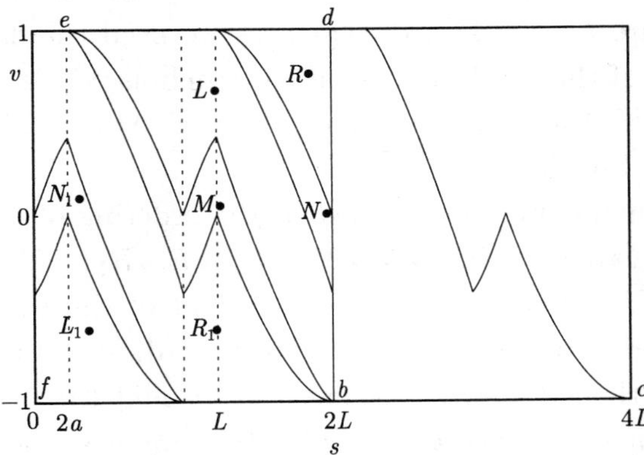

Figure 5.32 The image $bcde$ of the fundamental domain $s \in [0, 2L)$ and $v \in [-1, 1]$ and the partition of the domain according to pre-images of orbital points for $a = 0.5$ (cf. Fig. 5.33)

where • indicates the present. The corresponding partition according to pre-images is shown in Fig. 5.32, where each piece is the image of its counterpart in Fig. 5.33. For example, $L_1\bullet$ is the image of $\bullet L_1$, where we have used • to distinguish the two different partitions. Since the image and pre-image advance in opposite directions $L_1\bullet$ is an area left of the FD in the lifted space.

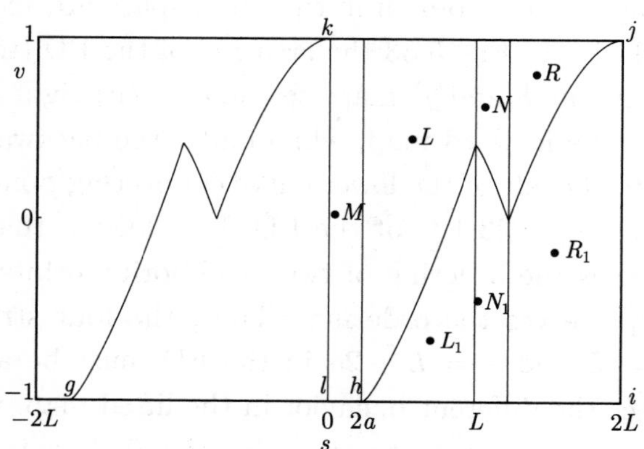

Figure 5.33 The pre-image $ghjk$ of the fundamental domain and the partition of the domain according to the present point for $a = 0.5$ (cf. Fig. 5.32)

We adopt the natural order of the lifted phase space to write

$$L_1\bullet < N_1\bullet < R_1\bullet < M\bullet < L\bullet < N\bullet < R\bullet, \tag{5.144}$$

since their pre-images are arranged successively from left to right in the lifted space. To compare two backward sequences we need the definition of parity of a finite string. If the total number of the symbols M, N and N_1 contained in the string is odd, the string has an odd parity. Any odd leading string will reverse the above ordering.

The same principle may be applied to the ordering of forward sequences. We have $\bullet M < \{\bullet L, \bullet L_1\} < \{\bullet N, \bullet N_1\} < \{\bullet R, \bullet R_1\}$. While the image of $\bullet L$ stays in the FD, that of $\bullet L_1$ is located right of the FD. Since L and L_1 preserve the ordering of neighboring orbits, according to their image we have $\bullet L < \bullet L_1$. Similarly, $\bullet R < \bullet R_1$. Taking into account the fact that N and N_1 reverse the ordering, we have $\bullet N_1 < \bullet N$. In summary, the ordering rule for forward sequences is

$$\bullet M < \bullet L < \bullet L_1 < \bullet N_1 < \bullet N < \bullet R < \bullet R_1. \tag{5.145}$$

Similarly, an odd leading string will also reverse this ordering.

Based on the ordering rules metric representations for both forward and backward sequences may be introduced to construct the symbolic plane. Every forward or backward sequence then corresponds to a number between 0 and 1. An orbit point corresponds to a point (α, β) in the unit square, where α and β are associated with the forward and backward sequences, respectively. In the symbolic plane forbidden sequences are pruned by the so-called primary pruning front which consists of the points in the symbolic plane representing all the points on the partition lines ($s = 0$, $s = 2a$, $s = L$ and $s = L + 2a$). As an example, we show the symbolic plane for $a = 0.5$ in Fig. 5.34 (a) where 15 000 points of several real orbits are drawn. The corresponding primary pruning front is shown in Fig. 5.34 (b).

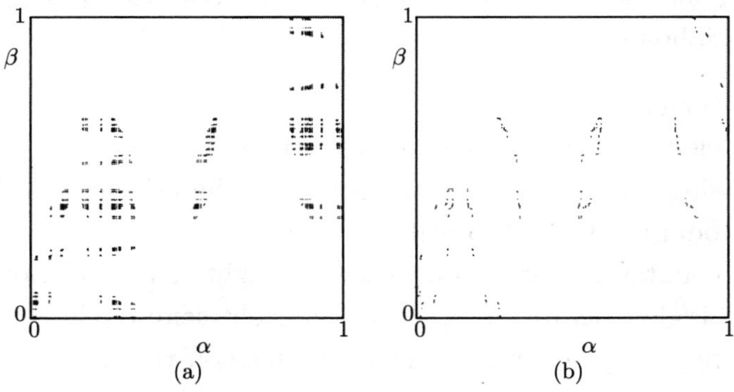

Figure 5.34 (a) The symbolic plane for $a = 0.5$. Approximately 15 000 real orbital points are drawn. (b) The primary pruning front for $a = 0.5$. It forms the border of the region within which orbit points are restricted to fall

The above coding, describing the dynamics of lifting and wrapping, directly gives the rotation number of a periodic orbit as the total number of symbols with subscript 1 in the non-repeating string of its symbolic sequence divided by the length of the string. For an orientation preserving circle map symbols with subscript 1 must be greater than any symbols without a subscript 1. According to the relation (5.145), this is possible only when

A. $\bullet M < \bullet L < \bullet N < \bullet R < \bullet R_1$;

B. $\bullet M < \bullet L < \bullet L_1 < \bullet N_1 < \bullet R_1$.

Consequently, an orientation preserving orbit consists of at most five letters from one of the two categories. In this way the ordering for backward and forward sequences coincides. In both cases M, N or N_1 appears at most once except for periodic orbits. This implies the non-existence of invariant circles for $a \neq 0$. Furthermore, when two different symbols t and t' are both with or without a subscript 1 and satisfy $t > t'$, any subsequence following t can never be smaller than any sequence following t'. Any subsequence following a letter with subscript 1 is never greater than any subsequence following a letter without subscript 1. For example, the two (1,3) orbits shown in Fig. 4 of Meiss [1992] are $(LLLRRR_1)^\infty$ and $(MLLNRR_1)^\infty$. Two examples of orientation preserving infinite sequences are $(LRR_1L)^\infty(RR_1LL)^\infty$ and $(LRR_1L)^\infty NR_1M(LRR_1L)^\infty$.

5.7.2 Relation to Other Codings

The Biham-Kvale code associates a bounce with a 6-letter alphabet $\{0, 1, 2, 3, 4, 5\}$ defined as follows:

0	a bounce off the bottom wall
1	a clockwise bounce off the left semi-circle; or a single counterclockwise bounce off the left semi-circle
2	a bounce off the top wall
3	a counterclockwise bounce off the right semi-circle; or a single clockwise bounce off the right semi-circle
4	a non-single counterclockwise bounce off the left semi-circle
5	a non-single clockwise bounce off the right semi-circle

In the full-stadium code the arc length is allowed to increase continuously past $2L$. In this way a forward orbit always advances forward in the lifted space. Thus, a point should be coded by a symbol with subscript 1 if its image has an s smaller than itself in the FD. The correspondence from the Biham-Kvale code to the full-stadium code is then as follows:

0	M
2({1,4})	N
2({0,3,5})	N_1
3({2,3,1,4}), 5({2,1,4})	L
3(0), 5({0,5})	L_1
4(4)	R
1, 4({0,2,3,5})	R_1

where we have used the fact that 14, 41, 35 and 53 are forbidden (Biham and Kvale [1992]). For example, $(0321)^\infty$ corresponds to $(MLNR_1)^\infty$, $(3212)^\infty$ to $(LNR_1N_1)^\infty$, and $(055211)^\infty$ to $(ML_1LNR_1R_1)^\infty$.

The definition of a single bounce depends on the preceding or following bounce while the full-stadium code depends only on the present bounce. It can be verified that the correspondence from the full-stadium code to the Biham-Kvale code is as follows:

$\bullet M$	0
$\bullet N$, $\bullet N_1$	2
$(L_1)\bullet\{L, L_1\}$, $\bullet L_1(\{L, L_1\})$	5
$\bullet L_1(M)$, $(\{N_1, R_1, M, L\})\bullet L$	3
$(R)\bullet R_1$, $\bullet R$	4
$(\{R_1, M, L, N\})\bullet R_1$	1

where we have indicated a preceding or following bounce with a parenthesis. For example, $(LLLRRR_1)^\infty$ corresponds to $(333444)^\infty$, and $(MLLNRR_1)^\infty$ to $(033244)^\infty$.

Superimposing the FD in Figs. 5.32 and 5.33, one sees that the allowed pairs of symbols are

$$M\{L, L_1, N, N_1, R, R_1\},\quad L\{L, L_1, N, N_1, R, R_1\},\quad L_1\{M, L, L_1\},$$
$$N\{R, R_1\},\quad N_1\{M, L, L_1\},\quad R\{R, R_1\},\quad R_1\{M, L, L_1, N_1, R_1\},$$

or, written in an equivalent way,

$$\{L_1, N_1, R_1\}M,\quad \{M, L, L_1, N_1, R_1\}L,\quad \{M, L, L_1, N_1, R_1\}L_1,$$
$$\{M, L\}N,\quad \{M, L, R_1\}N_1,\quad \{M, L, N, R\}R,\quad \{M, L, N, R, R_1\}R_1.$$

The Hansen-Cvitanović code is a symmetry-reduced 5-letter alphabet, introduced by defining

$\bar{0}$	a first bounce off either semi-circle
$\bar{1}$	a clockwise non-first bounce off a given semi-circle
$\bar{2}$	a counterclockwise non-first bounce off a given semi-circle
$\bar{3}$	a bounce off a flat wall with positive v
$\bar{4}$	a bounce off a flat wall with negative v

It can be verified that the correspondence between this code and the full-stadium code is

$\bar{0}$	$\{M, N_1, R_1\} \bullet L,\ \{M, N_1, R_1\} \bullet L_1,\ \{M, L, N\} \bullet R,\ \{M, L, N\} \bullet R_1$
$\bar{1}$	$L_1 \bullet L,\ L_1 \bullet L_1,\ R_1 \bullet R_1$
$\bar{2}$	$L \bullet L,\ L \bullet L_1,\ R \bullet R,\ R \bullet R_1$
$\bar{3}$	$\{R_1, N_1^+\} \bullet M,\ \{M, L\} \bullet N,\ \{L, M^+\} \bullet N_1$
$\bar{4}$	$\{L_1, N_1^-\} \bullet M,\ \{M^-, R_1\} \bullet N_1$

where by M^+ (M^-) we mean a backward sequence greater (smaller) than $(N_1M)^\infty\bullet$, and by N_1^+ (N_1^-) a sequence greater (smaller) than $(MN_1)^\infty\bullet$. In fact, this 5-letter code follows from the half stadium which we shall discuss in the next subsection.

5.7.3 The Half-Stadium

Being invariant under $(s, v) \to (s - L, v)$, the stadium map has a translation symmetry. This symmetry may be used to reduce the motion to the half domain (Meiss [1992]). We may partition the phase space of this half stadium in a way similar to that of the full stadium. The partition is shown in Fig. 5.35 (a). The phase space is divided into $\bullet M$, $\bullet M_1$, $\bullet L$, $\bullet L_1$ and $\bullet L_2$ by the line $s = 2a$ and the two-piece pre-image of the line $s = 0$. The corresponding partition according to the past $M\bullet$, $M_1\bullet$, $L\bullet$, $L_1\bullet$ and $L_2\bullet$ is shown in Fig. 5.35 (b). This 5-letter alphabet will be referred to as the half-stadium code. It can be verified that the correspondence between the half-stadium and full-stadium codes is as follows:

5-letter	7-letter code
$\bullet M$	$\bullet M\{L, L_1\},\ \bullet N$
$\bullet M_1$	$\bullet M\{N, N_1, R, R_1\},\ \bullet N_1$
$\bullet L$	$\bullet L\{L, L_1\},\ \bullet R\{R, R_1\}$
$\bullet L_1$	$\bullet L\{N, N_1, R, R_1\},\ \bullet R_1\{M, L, L_1\}$
$\bullet L_2$	$\bullet L_1\{M, L, L_1\},\ \bullet R_1\{N_1, R_1\}$

Using the same argument that yields the ordering of the seven symbols in the full stadium, we have the following ordering for the half stadium:

$$L_2\bullet < M_1\bullet < L_1\bullet < M\bullet < L\bullet, \tag{5.146}$$

$$\bullet M_1 < \bullet M < \bullet L < \bullet L_1 < \bullet L_2. \tag{5.147}$$

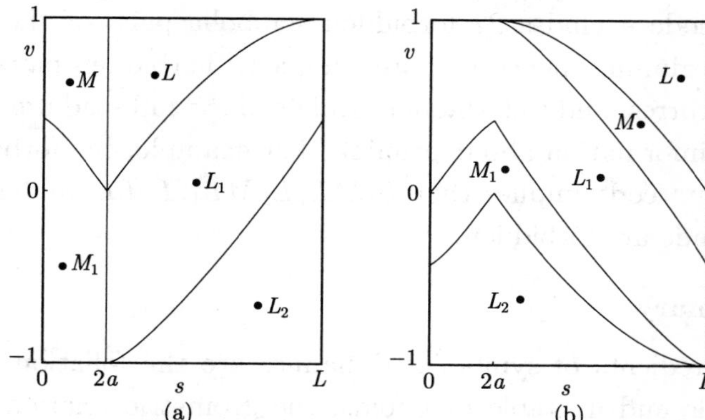

Figure 5.35 The partition of the phase space for the half stadium with a=0.5. (a) according to the present point. (b) according to the pre-image of the present point

Similarly, a symbol M or M_1 will reverse the order of forward or backward sequences, while L, L_1 and L_2 preserve the order. We can also construct the symbolic plane for the half-stadium. The symbolic plane for $a = 5$ is shown in Fig. 5.36 (a) where 15 000 points from a number of real orbits are drawn. The corresponding primary pruning front is shown in Fig. 5.36 (b).

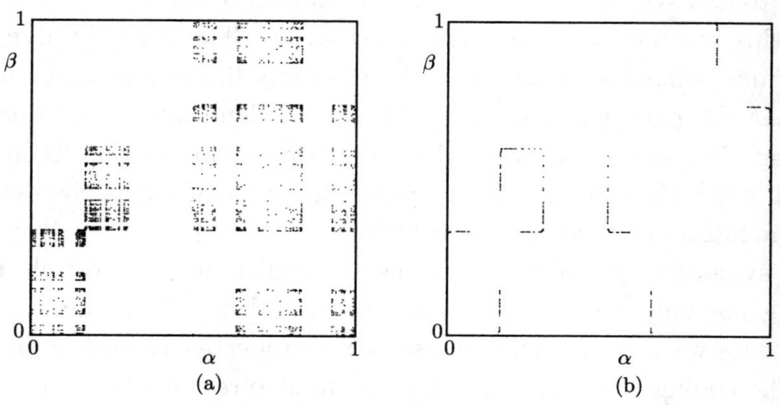

Figure 5.36 The half stadium at $a = 5$. (a) the symbolic plane; approximately 15 000 real orbit points are drawn. (b) the primary pruning front that forms the border of the region within which orbit points are bound to fall

Either from Fig. 5.35 (a) and (b), we see that in the half-stadium code the 6 symbolic pairs LM_1, LM, MM_1, MM, L_2M and L_2L are forbidden. The half-stadium code excludes 22 forbidden symbolic pairs of the full-stadium code at the beginning, hence is more compact. Forbidden pairs of the half-stadium code correspond to forbidden triplets of the full-stadium code, so they contain some information about pruning. For example, the forbidden L_2L of the half-stadium code implies that L_1ML, L_1ML_1, L_1LL, and L_1LL_1 of the full-stadium code are forbidden.

5.7.4 Summary

The main ingredients of symbolics dynamics are the foliation of the phase space by stable and unstable manifolds, the geometric relation between the two kinds of manifolds, partition of the phase space and monotonic ordering in the symbolic coding for both kinds of manifolds. In the construction of the primary pruning front every point on a partition line maps to a point in the symbolic plane, which forms a vertex of a forbidden rectangle in the plane. Generally, such a pruning front consists of an infinite number of vertices. However, in some limiting cases, e.g. in the limit of infinite a, the number of vertices may appear to be finite. These vertices give a complete description of the geometric pruning, which hold for all values of the parameter a.

The arguments may be applied to other systems such as the n-disk pinballs. In Hamiltonian systems we often deal with angle variables. A lift of the phase space helps to understand the dynamics better. The symbolic dynamics of the stadium billiard is as simple as the piecewise linear standard map in the sense that the partition lines are given by the definition of the map at the beginning. For smooth systems like a particle moving in the 2D hyperbolic potential x^2y^2, the partition of the phase space usually must be determined from dynamical manifolds (Zheng [1997b]).

The symbolic dynamics enables us to predict all the possible motions, including not only periodic, but also chaotic orbits. By means of the new simple codes we may further discuss metric properties as well as topological ones. The coding based on the lifting contains direct information about the rotation number and orientation. Orbits with different orientation preserving properties might play different roles. In this case the above full-stadium and half-stadium codes would be more useful.

6
Application to Ordinary Differential Equations

Many interesting nonlinear models in physical sciences and engineering are well described by systems of ordinary differential equations (ODEs). While bifurcation and chaos in discrete maps can be studied more or less thoroughly by both analytical and numerical means, similar task for differential equations may encounter great difficulty. Analytical tools can be of some help only in limited cases, e.g., when the criterion of Melnikov (see, e.g., Chapter 4 of Guckenheimer and Holmes [B1983]) or Shilnikov (see, e.g., Gaspard and Nicolis [1983]) is applicable. These methods usually only provide a criterion for the existence of chaotic motion and give an estimate for the threshold of transition to chaos, but cannot help with the exploration of the global structure of the parameter as well as phase space. However, what is important is just the global understanding of the bifurcation and chaos "spectrum": the systematics of periodic orbits, stable as well as unstable ones, at fixed and varying parameters, the type of chaotic attractors which usually occur as limits of sequences of periodic regimes, etc. This is by far not a simple job to accomplish neither by analytical means nor by numerical study alone.

In analytical aspect, just recollect the problem of the number of limit cycles in *planar* systems of ODEs. Despite of the great effort of mathematicians, it has not been completely solved yet.[1] As chaotic behavior may appear only in systems of more than three autonomous ODEs or non-autonomous systems with more than two variables, it naturally leads to problems that are much

[1] For a review on this problem see, e.g., Y.C. Yeh, C.Y. Lo, and S.L. Cai, *Theory of Limit Cycles*, Translation of Mathematical Monographs vol. 66, American Mathematical Society, Providence, 1986.

more difficult than counting the number of limit cycles in planar systems.

Numerically speaking, one can never be confident that all stable periodic orbits up to a certain length have been found in a given parameter range or no short unstable periodic orbits in the phase space have been missed at a fixed parameter set, not to mention that it is extremely difficult to draw global conclusions from numerical data alone.

On the other hand, it is known from numerical observations that chaotic attractors of many higher-dimensional dissipative systems with one positive Lyapunov exponent usually show one-dimension-like structures in some Poincaré sections. Some years ago it was suggested to associate the systematics of numerically found periodic orbits in ODEs with symbolic dynamics of 1D maps (Hao [1986]). While this approach has had some success (see the introductory part of Sections 6.2 and 6.3 below), many more questions have been raised, for example,

1. The number of periodic orbits found in ODEs is usually less than that allowed by the admissibility conditions of the corresponding one-dimensional symbolic dynamics. Within the one-dimensional framework it is hard to tell whether this is caused by insufficient numerical search or is forbidden by the dynamics.

2. In the Poincaré sections of ODEs, at a closer examination, the attractors often reveal two-dimensional structures such as layers. On one hand, one has to explain the success of 1D symbolic description, which sometimes even seems too good to be expected. On the other hand, the limitation of 1D approach has also to be analyzed, as the Poincaré maps are necessarily two-dimensional.

3. Early efforts of applying symbolic dynamics to ODEs were more or less concentrated on stable orbits, while unstable orbits play a fundamental role in organizing chaotic motion. One needs to develop symbolic dynamics capable to treat stable and unstable periodic orbits alike up to a certain length. In addition, it is desirable to indicate the structure of some, if not all, chaotic orbits at a given parameter set.

The elucidation of these problems has to await a significant progress of symbolic dynamics of two-dimensional mappings. In this Chapter we will show that numerical study under the guidance of symbolic dynamics is capable to yield global and qualitative results for ODEs which can hardly be obtained by purely numerical methods or by analytical tools only. One can even predict and locate short periodic orbits in an exhaustive way. With a good symbolic dynamics established, one can gain much knowledge on the nature of chaotic motion in the system. Although so far this has been a largely experimental approach, with further automation of the technique and accumulation of experience on more systems this method will surely find wider applications.

This chapter is organized as follows. Section 6.1 starts with a general discussion on the type of ODEs, the numerical methods involved, and the strategy of research. Then two detailed case studies follow.

Section 6.2 is devoted to a non-autonomous system, namely, the periodically forced Brusselator. Stroboscopic sampling method at the forcing frequency and its subharmonics simplifies the calculation of Poincaré maps. Symbolic dynamics of both the 2D dissipative standard map and the 1D unimodal map is invoked to interpret the results. In particular, the transition from annular to interval type dynamics calls for comparison with different types of symbolic dynamics.

Section 6.3 treats an autonomous system—the Lorenz model. The discrete symmetry of this system plays an essential role in constructing the symbolic dynamics in both the 1D and 2D approaches. A by-product of the symbolic dynamics study is now there is an absolute nomenclature of periodic orbits in the Lorenz model, i.e. we are able to tell the period of each periodic orbit despite of the lacking of a reference period in an autonomous system. Section 6.4 finishes the chapter with a brief summary of the symbolic dynamics of other ODE systems and a discussion on the prospect of the method.

6.1 General Discussion

Before undertaking the construction of symbolic dynamics for several systems of ordinary differential equations we touch a few general questions. These include the classification of ODEs, the numerical integration of ODEs and the calculation of Poincaré maps.

6.1.1 Three Types of ODEs

Three classes of ordinary differential equations are often encountered in various physical models of natural processes.

1. Autonomous differential equations, i.e., equations without explicit time

dependence on their right-hand sides, when one requires at least three independent variables, in order to observe period-doubling or chaotic transitions. A classical and much-studied example of autonomous equations is the Lorenz model (Lorenz [1963]) obtained from a three-mode truncation of the thermal convection problem.

2. Non-autonomous systems with at least two independent variables. It is well known that by addition of one or more independent variables a non-autonomous system can be transformed into an autonomous one. The most important class of non-autonomous systems are those driven by periodic external force. In numerical studies, the availability of a control frequency opens the possibility of reaching very high frequency resolution. In the next Section we will study the periodically forced Brusselator as a typical example of driven systems.

3. Time-delayed differential equations. Formally, it is enough to have a single independent variable, in order to display complicated bifurcating and chaotic behavior. However, time-delayed equations are, in essence, functional equations with an infinite number of degrees of freedom. This can be seen easily by rewriting the time delay as a sum of high order derivatives:

$$f(t-T) = \exp\left(-T\frac{d}{dt}\right)f(t) = \sum_{i=0}^{\infty}(-T)^i\frac{d^i}{dt^i}f(t), \tag{6.1}$$

or by considering the dependence of the solution on an initial function instead of on an initial point. Time-delayed equations have been used, for example, in physiological models (Mackey and Glass [1977]) and in the description of optical bistable devices, one of the simplest cases being that described by Li and Hao [1989]:

$$\tau\dot{x}(t) + x(t) = 1 - \mu[x(t-T)]^2. \tag{6.2}$$

Since time-delayed differential equations will not be studied further in this book, we make a few remarks on the case (6.2).

Firstly, when $\tau \ll 1$, away from rapid transition regions, measuring time in units of T, one can transform (6.2) into a 1D map. In fact, the sine-square map studied in Section 3.7 was obtained in this way from a time-delayed system. We emphasize, however, no matter how detailed is the knowledge which one possesses for this map, the results for the 1D map cannot be extrapolated into the τ-μ parameter plane for small τ, because there is an infinite number of "linear" modes which are excitable near $\tau = 0$. In other words, we have a singular perturbation case, since the vanishing small parameter removes the only derivative and changes the nature of the equation drastically.

Secondly, in the opposite small delay limit, although one may use a truncated expansion (6.1) to transform (6.2) into a system of finite order ordinary differential equations, the resulting system may resemble the original equation only for small enough t. Therefore it is not of much help in the study of the asymptotic $t \to \infty$ behavior. In fact, this is another extreme of singular perturbation, since the truncation always means neglect of higher order derivatives.

Thirdly, the solution of (6.2) depends on the choice of the initial function. One might try to use a one-parameter family of functions and look for the dependence on that parameter. It has been shown that sensitive dependence on the parameter appears when the asymptotic regime is chaotic. For example, in some region of the parameter space there are two coexisting chaotic attractors, distinguishable by different characteristic frequencies in the power spectra. The system jumps at random between these two attractors while the parameter in the initial function varies continuously (Li and Hao [1989]). Recently, this phenomenon has been studied in great details (Losson, Mackey and Longtin [1993]).

6.1.2 On Numerical Integration of Differential Equations

A prerequisite for the application of any method to study ODEs is that a good algorithm for the integration of the equations. Gone has the time when a scientist must write his own mathematical routines, since so many tested programs have accumulated that surpass any amateur's handiwork in stability, precision and efficiency. All one has to do is to choose an appropriate subroutine from one of the existing libraries (e.g., IMSL, NAG, CACM). Therefore, we shall confine ourselves to a few comments.

In order to fix the notation, we write a general autonomous system of nonlinear ordinary differential equations in the standard vector form

$$\frac{d\mathbf{x}}{dt} = \mathbf{F}(\mathbf{x}). \tag{6.3}$$

It should be supplemented with the initial condition

$$\mathbf{x}(t = 0) = \mathbf{x}_0.$$

The solution of this initial value problem is an integral curve passing through $\mathbf{x}_0$. All possible integral curves taken together constitute a *flow* $\Phi_t(\mathbf{x})$ in the phase space and the solution of the above problem picks up a particular curve

$$\mathbf{x}(t) = \Phi_t(\mathbf{x}_0), \tag{6.4}$$

which satisfies, of course, the condition

$$\mathbf{x}_0 = \Phi_0(\mathbf{x}_0).$$

The system (6.3) can be linearized at a point on a flow curve, say $\mathbf{x}_1$, by letting

$$\mathbf{x} = \mathbf{x}_1 + \mathbf{W},$$

where $d\mathbf{x}_1/dt = \mathbf{F}(\mathbf{x}_1)$, and $\mathbf{W}$, being a small vector, satisfies the linearized equation

$$\frac{d\mathbf{W}(t)}{dt} = \mathbf{J}(\mathbf{x}_1)\mathbf{W}(t), \tag{6.5}$$

with

$$\mathbf{J}(\mathbf{x}_1) = \left.\frac{\partial \mathbf{F}(\mathbf{x})}{\partial \mathbf{x}}\right|_{\mathbf{x}=\mathbf{x}_1}$$

being an $n \times n$ matrix.

Since it is a linear system, the solutions of (6.5) may be expressed by means of a linear evolution operator $\mathbf{U}(t) \equiv \mathbf{U}(t, 0)$:

$$\mathbf{W}(t) = \mathbf{U}(t)\mathbf{W}(0).$$

It is readily seen that $\mathbf{U}(t)$ satisfies the same equation (6.5)

$$\frac{d\mathbf{U}(t)}{dt} = \mathbf{J}(\mathbf{x}_1)\mathbf{U}(t)$$

with the initial conditions

$$\mathbf{U}(t) \mid_{t=0} = \mathbf{I},$$

where $\mathbf{I}$ is the unit matrix.

Numerical algorithms for the integration of ordinary differential equations can be subdivided into two classes. The first class is based on Taylor expansion and step by step progress from the initial point. The second class employs some numerical quadrature method over a small interval. As a rule, algorithms in the first class only require the result of the previous step and are called one-step or single-step methods. The second class leads to multi-step algorithms; in particular, it is quite useful for the treatment of time-delayed problems, when the initial function is known over an interval, for example, we have used the fourth order Adam's method to study (6.2) (Li and Hao [1989]). As a rule, multi-step methods require less arithmetic operations per step than single-step algorithms, but at the start they demand initial values at several points. On the other hand, multi-step schemes are usually derived for equidistant partitions of the integration interval, whereas single-step methods are flexible enough to allow for varying step lengths.

A very frequently used single-step method is the Runge-Kutta algorithm. A fourth order Runge-Kutta scheme has a local truncation error of the order h^5 with h being the integration step. It measures the difference between the discrete iteration equations and the original continuous differential equations at one single step, hence the adjective "local". However, a small truncation error at each step does not guarantee a good global approximation of the difference scheme to the system of differential equations. In addition, our concern with a chaotic regime implies a sensitive dependence on initial values as well as on local truncation errors. Nevertheless, a number of global characteristics of the motion such as the Lyapunov exponents, dimensions and entropies, remain quite insensitive to the algorithm or local truncation, provided small enough integration steps are used.

In connection with the Runge-Kutta method we would like to point out that when the differential equations represent a conservative system, the resulting difference scheme may become dissipative. This can be checked on the example of a simple linear oscillator

$$\begin{aligned}\dot{x} &= \omega y,\\ \dot{y} &= -\omega x.\end{aligned}$$

In this case, the corresponding Runge-Kutta difference scheme may be written down explicitly and there appears a dumping factor as well as a shift of frequency. It is much better to use the so-called symplectic difference schemes in dealing with numerical integration of Hamiltonian systems (Feng [1985], Feng *et al.* [1989, 1991, 1994]).

A last, but not least important point to be mentioned concerns so-called stiff equations, when the eigenvalues of the matrix $\mathbf{J}$ in (6.5) differ in orders of magnitude. One should consult the literature, e.g., the book of Lambert[2], to avoid numerical quirks.

6.1.3 Numerical Calculation of the Poincaré Maps

In a study of bifurcation and chaos in ordinary differential equations, one does not deal with a single system, but has to treat a family of equations with varying control parameters. Furthermore, using a computer with finite word length and finite run time, it is impossible to distinguish, say, a very long periodic orbit from a quasi-periodic or chaotic trajectory, if one observes the trajectory in the phase (configuration) space $\mathcal{R}^n$ only. However, by invocation of the tangent space together with the phase space one can calculate with confidence such quantitative characteristics of the motion as the Lyapunov exponents which, in turn, enable us to distinguish between chaotic and non-chaotic behavior. In addition, many systems display multi-stable solutions for one and the same set of parameter values, i.e., different attractors, trivial and strange ones, may coexist. There appears a basin dependence: the destiny of the motion depends on the chosen initial values.

In summary, we see that a full-scale numerical study of a system of ordinary differential equations would require a scanning of the product space $\mathcal{R}^n \otimes \mathcal{R}^n \otimes \mathcal{R}^n \otimes \mathcal{R}^m$, where n is the dimension of the phase (hence the initial value and tangent) space, and m is the dimension of the control parameter space. For the simplest autonomous system, one should take at least $n = 3$, $m = 2$ (a codimension 2 study). This would be a job which exceeds the capability of many present-day computers. One has to be satisfied with the knowledge of a few sections of this huge product space. Fortunately, this happens to be sufficient in many cases, and the Poincaré sections are the most important ones to study.

The Poincaré section is a low-dimensional (usually two-dimensional, but not necessarily so) intersection of the phase space, chosen in such a way that the trajectories intersect it transversally, i.e., do not touch it tangentially. The choice of a Poincaré section must be preceded at least by a linear stability analysis of the system (6.3), in order to ensure that all qualitatively interesting trajectories actually intersect it. Once the section has been chosen, we focus

[2] J. D. Lambert, *Computational Methods in Ordinary Differential Equations*, Wiley, 1973.

on consecutive intersecting points in the Poincaré section and view the motion as a point-to-point mapping in the section itself. Only in a local sense can this map sometimes be written down approximately. In general, one has to resort to numerical calculations. Anyway, adoption of the Poincaré sections reduces the description of the dynamics significantly and it still reflects the essential features of the motion. For example, a simple periodic orbit would become a single fixed point in the map; a periodic orbit with two commensurable frequency components would give rise to a finite number of points; a quasi-periodic trajectory would draw a closed curve in the Poincaré section, and chaotic motion would show off as erratically distributed points, etc.

Autonomous systems: Hénon's method

A simple-minded approach to the calculation of a Poincaré map would involve step by step integration of the equations and a test of the sign changes of a certain component, say z, when the $z = 0$ plane is used as the Poincaré section. However, in order to locate the intersection point with a sufficiently precision to match that of the integration algorithm, one must use a high order interpolation scheme, which requires saving and updating of several consecutive points at each integration step. M. Hénon [1982] described a clever method for the determination of the intersection point at one shot. We explain the idea by means of the following example:

$$\begin{aligned}\dot{x} &= f_1(x, y, z),\\ \dot{y} &= f_2(x, y, z),\\ \dot{z} &= f_3(x, y, z),\end{aligned} \tag{6.6}$$

with $z = 0$ being the Poincaré section. Suppose we have found that at the nth step

$$t_n,\ x_n,\ y_n,\ z_n < 0,$$

and at the next step ($t_{n+1} = t_n + \Delta t$)

$$t_{n+1},\ x_{n+1},\ y_{n+1},\ z_{n+1} > 0.$$

The intersection must occur between these two steps. Now, interchange the role of z and t by dividing the first two equations by the third and inverting the third equation:

$$\begin{aligned}\frac{dx}{dz} &= \frac{f_1(x,y,z)}{f_3(x,y,z)},\\ \frac{dy}{dz} &= \frac{f_2(x,y,z)}{f_3(x,y,z)},\\ \frac{dt}{dz} &= \frac{1}{f_3(x,y,z)}.\end{aligned} \tag{6.7}$$

Integrating these equations in the new independent variable z backward by one step $\Delta z = -z_{n+1}$ from z_{n+1}, using

$$x_{n+1}, \quad y_{n+1}, \quad t_{n+1}$$

as initial values, or, integrating one step $\Delta z = z_n$ forward from the $z_n < 0$ point, one reaches exactly the $z = 0$ plane.

Hénon's method can be used to locate the intersection point of an orbit with a general surface

$$S(x, y, z) = \text{const.}. \tag{6.8}$$

It is sufficient to introduce an additional function

$$u = S(x, y, z) - \text{const.},$$

and to derive the differential equation for u using the original system (6.6):

$$\frac{du}{dt} = f_1\frac{\partial S}{\partial x} + f_2\frac{\partial S}{\partial y} + f_3\frac{\partial S}{\partial z}.$$

Thus, we return to the old problem of looking for the intersection with the $u = 0$ plane for a system with one more equation.

In practice, due to the presence of transients, which may be very long when one approaches a bifurcation point ("critical slowing down", see Hao [1981]), it is a time-consuming task to calculate a Poincaré map with high precision, even with the help of Hénon's procedure. However, in the case of periodic solutions, there exist efficient ways to locate the exact periodic orbit starting from an approximate one.

Non-autonomous systems: subharmonic stroboscopic sampling

We shall only consider the case of periodically driven systems, where the external frequency ω provides us with a reference period $T = 2\pi/\omega$. The extended phase space, i.e., the x-y plane plus time t, can be considered as closed in the t-direction and thus to form a torus or a truncated cylinder of height T. The calculation of Poincaré maps now reduces to a sampling of x, y values at multiples of T (up to an unessential shift of the starting time). This procedure corresponds exactly to the stroboscopic sampling techniques used by experimentalists. We should emphasize that the widespread opinion on the uselessness of stroboscopic methods in studies of ordinary differential equations applies only to autonomous systems when there is no constant characteristic period at hand. For periodically driven systems, however, the stroboscopic technique becomes a most powerful method for the exploration of subtle dynamic details.

The subharmonic stroboscopic sampling method is a very simple, yet quite effective extension (Hao and Zhang [1982a, 1983]) of the stroboscopic sampling idea mentioned above. Besides sampling at the driving period T, one also samples at multiples of the fundamental period, i.e., at pT, where p is a correctly chosen integer. When properly used, this method may provide very high frequency resolution at the cost of longer computing time. Period-doubling cascades up to the 8192nd subharmonic have been resolved in this way and the hierarchy of chaotic bands as well as the systematics of the periodic windows, according to the two-letter symbolic dynamics of unimodal maps have been observed in the forced Brusselator (Hao and Zhang [1982a]; Hao, Wang and Zhang [1983]). Similar resolution could usually be attained only for one-dimensional maps, not for ordinary differential equations.

When sampling at pT, p being an integer, one has the freedom of shifting the starting time by qT, $q = 0, 1, \cdots, p-1$, thereby picking up one of the p components of the attractor. However, as a method of discrete sampling, the subharmonic stroboscopic sampling technique has the same demerit as the discrete Fourier transform, namely, non-uniqueness in its interpretation and the impossibility of resolving frequency components higher than the sampling frequency. Suppose the actual period T^* is related to the sampling period T by

$$T^* = \frac{n}{m} T,$$

where n and m are coprime integers, then sampling at multiples of T will always yield n points (or clusters) in the map, for all $m \geqslant 1$. If n is the product of two integers k and l, then one can change the sampling interval to lT or kT, increasing the frequency resolution correspondingly. However, if one has misused a k which was not a factor of n, then there would appear a spurious factor k in the measured period. For the sake of safety, one must go gradually from the fundamental frequency to subharmonics, comparing the results with power spectrum analysis whenever available.

6.2 The Periodically Forced Brusselator

The term Brusselator was coined by J. J. Tyson [1973] to denote a set of two ODEs introduced originally by Ilia Prigogine's group in Brussels to describe a model of tri-molecular chemical kinetics. The Brusselator with a diffusion term exhibits a rich variety of transitions and spatial patterns. A simpler model with complicated temporal dynamics, the periodically forced Brusselator (Tomita and Kai [1978]; Kai and Tomita [1979]), is obtained by adding a periodic force to the Brusselator:

$$\begin{aligned} \dot{x} &= A - (B+1)x + x^2y + \alpha\cos(\omega t), \\ \dot{y} &= Bx - x^2y, \end{aligned} \tag{6.9}$$

where x and y are concentrations of intermediate products, A and B are concentrations of some chemicals under control, α and ω are the forcing amplitude and frequency.

The periodically forced Brusselator (6.9) displays rich bifurcation and chaos behavior. Tomita and Kai discovered a small chaotic region and many periodic "bubbles" in the α-ω parameter plane. Hao and Zhang [1982a and b], using subharmonic stroboscopic sampling method, explored in great detail the hierarchical structure of chaotic bands and periodic orbits embedded in these bands. This led to the discovery of the first "universal" ordering of kneading sequences in a system of ODEs (Hao, Wang and Zhang [1983]). It turns out to be the U-sequence of Metropolis, Stein, and Stein, described in Section 2.4.3 on unimodal maps, up to period 6 except for one missing word in their study of periodic sequences in unimodal maps. Other findings include intermittent transitions to chaos (Wang, Chen and Hao [1983]) and transition from quasi-periodic regime to chaos (Wang and Hao [1984]).

Fig. 6.1, taken from Hao, Wang and Zhang [1983], shows an A-ω section of the parameter space for fixed $B = 1.2$ and $\alpha = 0.05$. The solid lines denote boundaries between periodic regimes, ticked-solid lines—boundaries to period-doubling cascades. The numbers in the figure indicate the periods and Q stands for quasi-periodic regime. The shaded regions in Fig. 6.1 are chaotic with many embedded periodic strips, of which only a few are indicated in the figure. In fact, a method was devised to assign symbolic names to all stable periods which were found numerically. These words happen to be ordered just as that in the unimodal map. It is along a slanting straight line, say, $A = 0.46 - 0.2\omega$ in this plane, where the above-mentioned U-sequence of MSS was discovered. All but one stable periodic orbits up to period 6, corresponding to those in the unimodal map, exist and are ordered in the same way as in the U-sequence. A large amount of direct numerical search for the only missing period 6 RL^3RC has ended in vain. In Section 6.2.3 we will show how the absence of this and many other words follows from the admissibility condition of the corresponding two-dimensional symbolic dynamics.

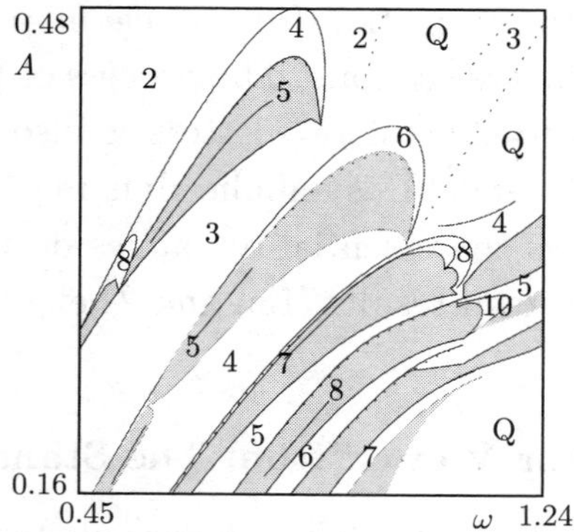

Figrue 6.1 An A-ω section in the parameter space with fixed $B = 1.2$ and $\alpha = 0.05$ (for details see the text)

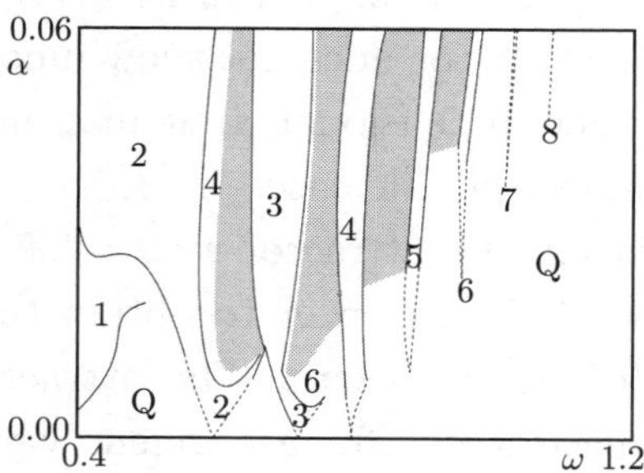

Figure 6.2 An α-ω section in the parameter space with $A = 0.48 - 0.2\omega$ and $B = 1.2$

If we cut Fig. 6.1 along the straight line $A = 0.48 - 0.2\omega$ and take a perpendicular plane, we get an α-ω section, which is shown in Fig. 6.2. This is the section where the transition from quasi-periodicity to chaos was first studied in the forced Brusselator (Wang and Hao [1984]). In Fig. 6.2, solid lines denote the boundaries between periodic regimes, dashed lines—boundaries between periodic and quasi-periodic regions. The numbers indicate periods. The letter Q stands for quasi-periodicity and shaded regions are chaotic. We are going to show that behavior of the forced Brusselator in the lower part of this parameter plane is described first by symbolic dynamics of a 1D circle map, then by a 2D symbolic dynamics similar to that of the dissipative standard map. With increasing α the symbolic dynamics changes first to that of the 2D Hénon map (with a positive Jacobian), then to 1D unimodal map.

In Section 6.2.1 we begin the discussion of the forced Brusselator by associating it with the dissipative standard map studied in Section 5.6. Section 6.2.2 is specially devoted to the transition from circle-map type to unimodal-map type behavior in the Poincaré sections. The process of how symbolic dynamics of three letters reduces to that of two letters is also elucidated there. Section 6.2.3 constructs 1D and 2D symbolic dynamics when an unimodal or Hénon-type map captures the essential dynamics on the attractor. The presentation is based on our own results (Liu and Zheng [1995b]; Liu, Zheng and Hao [1996]).

6.2.1 The Brusselator Viewed from The Standard Map

The lower part of Fig. 6.2 looks much like the parameter plane of a typical circle map. Indeed, quasi-periodic motion and the transition from quasi-periodicity to chaos has been discovered in this region. In order to construct symbolic dynamics we first draw the attractor in the Poincaré section and determine the partition lines. If the attractor does not show much 2D feature, reduction to symbolic dynamics of one-dimensional circle map may capture much of the essentials. We start from this simple case.

Fig. 6.3 (a) shows the chaotic attractor at $\omega = 0.775$, $\alpha = 0.0124$, $B = 1.2$ and $A = 0.48 - 0.2\omega = 0.325$. In order to obtain the stroboscopic portrait an initial phase $t_0 = 0.7\pi/\omega$ is taken. The attractor resembles that of a one-dimensional circle map except for a segment where two sheets are just perceptible. From the tangencies between the forward and backward foliations two primary partition lines $\bullet G$ and $\bullet S$ are determined.

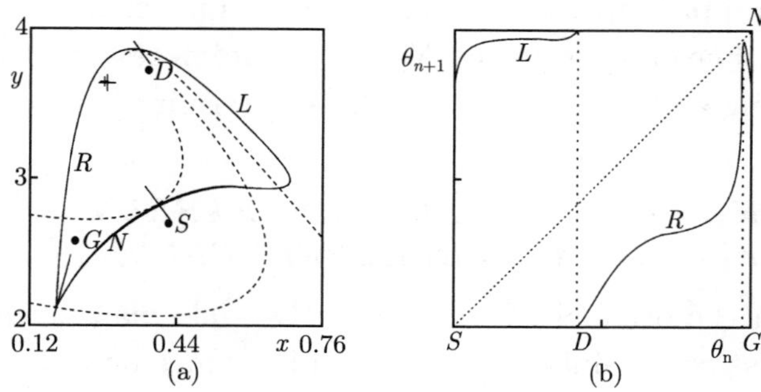

Figure 6.3 (a) A Poincaré section at $\omega = 0.775$, $A = 0.325$, $B = 1.2$, and $\alpha = 0.0124$. The partition lines $\bullet S$, $\bullet G$, and the pre-image $\bullet D$ of $\bullet S$ divide the attractor into three parts, denoted by the letters L, R, and N. (b) A first return map obtained by mapping the attractor in the left figure to the θ_n-θ_{n+1} plane

Taking the point $(x_r,\ y_r) = (0.275,\ 3.64)$ inside the circle as a reference point, which is indicated by a cross in the figure, we define an angle

$$\theta = \frac{1}{2\pi} \arctan\left(\frac{y - y_r}{x - x_r}\right) + \phi_0$$

for any point $(x,\ y)$ on the attractor. The phase offset $\phi_0 = 0.225\ 826$ is chosen in such a way as to put $\bullet S$ at $\theta \approx 0$. A first return map $\theta_n \mapsto \theta_{n+1}$ calculated from Fig. 6.3 (a) is shown in Fig. 6.3 (b). The three monotone segments in it may be assigned the letters L, R, and N in accordance with the 2D partitions in Fig. 6.3 (b).

Many tangent points may be identified on the partition line $S\bullet$. It turns out that all the forward symbolic sequences have at least 11 leading letters in common. For example, two points are

$$T_1 : \cdots LLLLRLRRLRRLRRNRDS \bullet RLNRR\,LRRLN\,RLNRR\,LRRLN \cdots,$$
$$T_2 : \cdots RRRRRRRRRRRRRRRRDS \bullet RLNRR\,LRRLN\,RRLRR\,RLRRL \cdots.$$

On the line $G\bullet$ the common leading string of all forward sequences are determined to be at least 16 letters long. Two of these tangencies are

$$T_3 : \cdots RLRLRRLRLRRLRLRRLG \bullet RRLRR\,RLRRR\,LRRRL\,RLNRR \cdots,$$
$$T_4 : \cdots RRRRRRRRRRRRRRRRLG \bullet RRLRR\,RLRRR\,LRRRL\,RRLNR \cdots.$$

In reducing the two-dimensional attractor to 1D return map, we pick up the smallest forward sequence among all tangencies at $S\bullet$ to be the kneading sequence K_S, and pick up the greatest forward sequence among all tangencies at $G\bullet$ to be the kneading sequence K_G, which are necessary, but need not be sufficient. Among all the tangencies we have determined these are T_1 and T_4, respectively.

$$\begin{aligned} K_S &= RLNRR\,LRRLN\,RLNRR\,LRRLN \cdots , \\ K_G &= RRLRR\,RLRRR\,LRRRL\,RRLNR \cdots . \end{aligned} \tag{6.10}$$

Compared with the original 2D map, the 1D circle map given by these K_S and K_G puts less constraints on allowed orbits. Since on either partition line the common leading string of forward sequences is quite long (over 11 or 16 letters) no difference between the 1D and 2D maps can be recognized if only periodic orbits with periods shorter than 11 are concerned.

The knowledge of the two kneading sequences (6.10) determines everything in the symbolic dynamics of the circle map, see Section 4.6. For example, one may define a *rotation number* ρ for a symbolic sequence by counting the weight of letters R and N, i.e., those on the right branch of the first return map, in the total number n of all letters:

$$\rho = \lim_{n\to\infty} \frac{1}{n}(\text{number of } R \text{ and } N). \tag{6.11}$$

Chaotic regime is associated with the existence of a rotation interval, a closed interval in the parameter plane (Ito [1981]). Within a rotation interval there must be well-ordered orbits. We can construct some of these well-ordered sequences explicitly, knowing the kneading sequences K_S and K_G.

In our case it can be verified that the ordered periodic orbits $(RRL)^\infty$ and $[(R^3L)^3R^2L]^\infty$ are admissible. These two sequences have rotation numbers 2/3 and 11/15, so the rotation interval of the circle map contains [2/3, 11/15], inside which there are rational rotation numbers 5/7, 7/10, 8/11, and 9/13 with denominators less than 15. Their corresponding ordered orbits are $(R^3LR^2L)^\infty$, $[R^3L(R^2L)^2]^\infty$, $[(R^3L)^2R^2L]^\infty$, and $[R^3L(R^2L)^3]^\infty$.

We can further construct ill-ordered sequences from well-ordered ones by the following transformation. One notes that the left limit of the point D is the greatest point on the subinterval L, while the right limit of D is the smallest R. When the point D is crossed by a continuous change of initial points the corresponding symbolic sequences must change as follows:

$$\text{greatest } LN \cdots \rightleftharpoons \text{smallest } RL \cdots .$$

Similarly, on crossing the critical point G another change of symbols takes place:

$$\text{greatest } R \rightleftharpoons \text{smallest } N.$$

Neither change has any effect on rotation numbers. As an example, starting with the ordered period 7 orbit $(R^3LR^2L)^\infty$ we get

$$LRRLRRR \to LRRLNRR \to LNRLNRR \to LNRLNNR$$

as candidates for the fundamental strings in ill-ordered sequences of period 7. Among the four sequences, the latter two are forbidden by K_G.

In this way we have determined all periodic sequences allowed by the two kneading sequences (6.10) up to period 15. The result is summarized in Table 6.1. We have examined the admissibility of all these sequences by using the four tangencies T_1 through T_4. They are all allowed. In fact, we have numerically found all these orbits in the Brusselator.

Table 6.1 Allowed periodic sequences up to period 15 for the circle map corresponding to $\omega = 0.775$ and $\alpha = 0.0124$. Only non-repeating strings of the sequences are given. P denotes the period and ρ the rotation number

P	ρ	Sequences
3	2/3	RLR RLN
6	2/3	$RRLRLN$
7	5/7	$RRLRRLR$, $RRLRRLN$
9	2/3	$RRLRLNRLN$, $RRLRRLRLN$
10	7/10	$RRLRRLRRLR$, $RRLRRLRRLN$
11	8/11	$RRLRRRLRRLR$, $RRLRRRLRRLN$
12	2/3	$RRLRRLRLNRLN$, $RRLRRLRRLRLN$, $RRLRLNRLNRLN$
13	9/13	$RRLRRLRRLRRLR$, $RRLRRLRRLRRLN$, $RRLRRLRRRLRLN$, $RRLRRLRRLNRLN$
14	5/7	$RRLRRRLRRLRRLR$, $RRLRRRLRRLRRLN$, $RRLRRLNRRLRRLR$
15	11/15	$RRLRRRLRRRLRRLR$, $RRLRRRLRRRLRRLN$

A more interesting case is encountered at $\omega = 0.66$ and $\alpha = 0.0145$. The attractor and the primary partition lines are shown in Fig. 6.4, which manifestly shows two-dimensional features. From the tangencies along the $S\bullet$ and $G\bullet$ lines we list the following seven:

$$T_1: \cdots RRRRRRRRRRRRRRRLRDS \bullet LNRLR\, LRLLN\, RLRRL\, RRLLN \cdots,$$
$$T_2: \cdots RRRRRRRRRRRLRLRLNDS \bullet LNRLR\, LRLRL\, RLRRL\, RRLRR \cdots,$$
$$T_3: \cdots RRRRRRRRRRRRRRRRLDS \bullet LNRLR\, LRRLR\, RLRLR\, RLRRL \cdots,$$

$T_4: \cdots RRRLRLRLRLRLRLLNRLDS \bullet LNRLR\,RLRLR\,LRRLR\,RLRRL \cdots,$
$T_5: \cdots RRRLRLRLRLRLRLRLDS \bullet LNRLR\,RLRRL\,RLNRL\,LNRLR \cdots,$
$T_6: \cdots RRLRLRLRLRLRLRLLLG \bullet RLRRL\,RRLRR\,LLNLN\,RLRLR \cdots,$
$T_7: \cdots RRRRRRRRRRRRRRRRLG \bullet RLRRL\,RRLRR\,LRLLN\,LNLNL \cdots.$

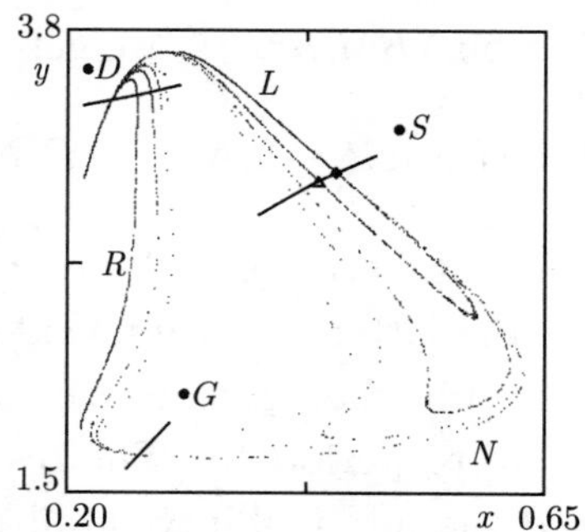

Figure 6.4 A Poincaré section at $\omega = 0.66$, $A = 0.348$, $B = 1.2$, and $\alpha = 0.0145$

Picking up the smallest forward sequences along $S\bullet$ at T_5 and the greatest one along $G\bullet$ at T_7, we get

$$K_S = LNRLR\,RLRRL\,RLNRL\,LNRLR \cdots,$$
$$K_G = RLRRL\,RRLRR\,LRLLN\,LNLNL \cdots.$$

For the 1D circle map, we have determined all allowed periodic sequences up to period 11, which are listed in Table 6.2. Some of these cycles are now forbidden by certain tangencies of the 2D Poincaré map. For example, among the four sequences of period 7 the cycle $(NRLRLNL)^\infty$, denoted by an asterisk in Table 6.2, is forbidden by the tangency T_2 since its shift $(LNRLRLN)^\infty \bullet (LNRLRLN)^\infty$ is in the forbidden zone of T_2. There are four forbidden sequences of period 9. All other cycles up to period 9 are allowed. Their admissibility has been examined and all of these periodic orbits have been located numerically. The symbolic plane of the Brusselator at this parameter set is shown in Fig. 6.5. Its similarity to the symbolic plane Fig. 5.28 of the dissipative standard map is clear.

So far we have discussed only periodic orbits. In principle, the admissibility of any given orbit can be examined. Although it is impossible to construct all admissible chaotic sequences, we are able to tell the structure of some chaotic orbits. For instance, a chaotic sequence may be obtained by randomly concatenating the segments LR and $LLRR$, as it can be verified that all shifts of such a sequence fall in the allowed zone of the tangency T_1 or T_6. Furthermore, from the tangencies T_1 and T_6 it follows that any sequence consisting of only the segments RL and LN are also admissible.

Table 6.2 Allowed periodic sequences up to period 11 for the 1D circle map corresponding to $\omega = 0.66$ and $\alpha = 0.0145$. An asterisk denotes those forbidden by 2D tangencies

P	ρ	Sequences
2	1/2	$RL\ NL$
4	1/2	$NRLL$
5	3/5	$RRLRL, NRLRL$
6	1/2	$NRLRLL, NLNRLL$
7	4/7	$RRLRLRL, NRLRLRL, RRLLNRL, NRLRLNL^*$
8	1/2	$NRLRLRLL, NLNRLRLL, NRLLNLNL$
	5/8	$RRLRRLRL, NRLRRLRL$
9	5/9	$RRLRLRLRL, NRLRLRLRL, RRLLNRLRL^*, NLNRLRLNL^*, NRLRLRLNL^*, NRLRRLLNL, NRLRRLRLL, NRLRLNRLL^*$
10	1/2	$NRLLNLNLNL, NRLLNLNRLL, NRLRLLNLNL, NRLLNRLRLL, NRLRLRLRLL, NRLRLRLLNL$
11	6/11	$RRLRLRLRLRL, NRLRLRLRLRL, NLNRLRLRLRL, NRLRLRLRRLL, NRLLNRLRLRL, NLNLNRLRLRL, NRLRLRRLRLL, NRLRLLNRLRL, NRLRRLRLRLL, NRLLNLNRLRL, NLNLNLNRLRL, NLNRLLNRLRL, NRLLNRLRRLL$
	7/11	$RRLRRLRRLRL, NRLRRLRRLRL$

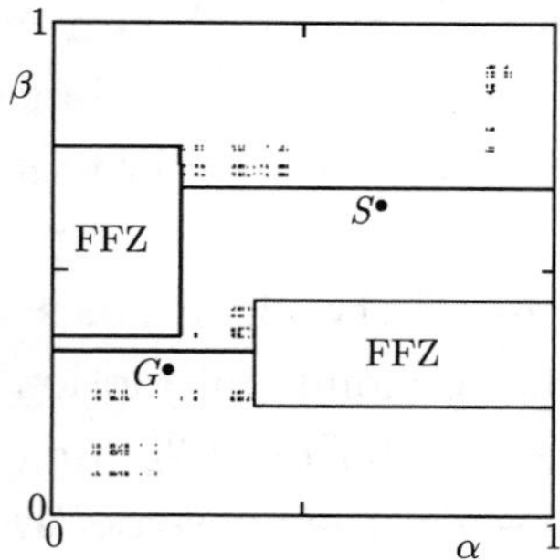

Figure 6.5 The symbolic plane of the attractor in Fig. 6.4. Together with the FFZ, 70 000 points of real orbits are shown, none of which falls inside the FFZ

6.2.2 Transition from Annular to Interval Dynamics

When the nonlinear coupling α in the forced Brusselator (6.9) increases, the dynamics undergoes a transition from annular type to that of an interval. It is interesting to trace the change of the corresponding symbolic dynamics from circle type to unimodal type, in particular, to watch how the number of symbols reduces from three to two. In the free Brusselator there is a Hopf bifurcation, where the stable fixed point at $(x, y) = (A, B/A)$ loses stability and a limit cycle comes into life. All the rich dynamical behavior of the forced Brusselator appears as the interaction between the limit cycle and the linear oscillator $\cos(\omega t)$ changes. Therefore, it is normal to expect that the nature of the unstable fixed point plays an essential role in the transition under study.

Roughly speaking, the transition undergoes the following stages:

1. For α small enough, the fixed point is an unstable focus, i.e., both eigenvalues are complex with module bigger than 1. The phase portrait is an 1D closed curve and the return map θ_{n+1}-θ_n is a sub-critical circle map without any decreasing branch. The symbolic dynamics is that of rigid rotation, i.e., with two letters of even parity.

2. At a first critical α_{c1} the circle map undergoes a transition from sub-critical to supercritical regime. While the fixed point remains an unstable focus, the backward and forward foliations begin to show tangencies. This signals the appearance of a decreasing branch in the return map, requiring a third letter with odd parity to construct the symbolic dynamics.

3. Upon further increase of α the fixed point becomes an unstable node, i.e., the two eigenvalues both become real with module greater than 1. The phase portrait and first return map show two-dimensional feature clearly, e.g., multi-layered structure. However, a standard-map type symbolic dynamics still works well.

When the module of one of the eigenvalues gets less than 1, the fixed point becomes a saddle. The predominant motion changes from rotational to vibrational. The "return plot" in terms of θ_{n+1}-θ_n can no longer be treated as a one-dimensional map. However, the phase portrait and the y_{n+1}-y_n or x_{n+1}-x_n return map may be analyzed by using a Hénon type symbolic dynamics.

4. Further decrease of the smaller eigenvalue makes the attractor and the y_{n+1}-y_n return map even more close to one-dimensional. The dynamics fits well into that of a unimodal map.

These stages are demonstrated in Fig. 6.6, where we have collected the phase portraits of the attractor, the θ_{n+1}-θ_n and x_{n+1}-x_n return maps at five different α values along the vertical line $\omega = 0.775$ in Fig. 6.2.

The $\alpha = 0.0124$ case (top row in Fig. 6.6 has been discussed in detail in Section 6.1, see Fig. 6.3. In fact, it is representative for a wide range of α, say, from 0.006 to 0.0154. The θ_{n+1}-θ_n return map is essentially a one-dimensional circle map.

At $\alpha = 0.027$ the fixed point is an unstable node with eigenvalues $\lambda_1 = -1.442\,634$ and $\lambda_2 = -1.262\,673\,4$ (2nd row in Fig. 6.6). The two-dimensional feature of the attractor and the θ_{n+1}-θ_n return map calls for symbolic dynamics analysis, similar to that of the dissipative standard map. However, after shrinking along the forward foliations, symbolic dynamics of a one-dimensional circle map still captures the essentials.

Once the fixed point has become a saddle, e.g., at $\alpha = 0.05$, $\lambda_1 = -2.522\,606\,7$ and $\lambda_2 = -0.495\,022\,4$ (3rd row in Fig. 6.6), the dynamics is no longer annular type and Hénon type symbolic dynamics must be developed, as we shall do in Section 6.2.3.

Due to the change of the dynamics to interval type, the θ_{n+1}-θ_n return maps for greater values of α are of no use. Now the x_{n+1}-x_n or y_{n+1}-y_n return maps should be used instead. While the $\alpha = 0.08$ case (4-th row in Fig. 6.6) still needs two-dimensional consideration, the $\alpha = 0.2$ case (last row in Fig. 6.6) turns out to be one-dimensional to high precision. This explains the early success in applying purely one-dimensional symbolic dynamics to the study of the forced Brusselator.

An interesting, but open question is the connection between the three letters used in annular type symbolic dynamics and the two letters used in Hénon type maps. It is a prerequisite for the clarification of the relation between the Farey sequence in circle map and the U-sequence in unimodal map and requires further investigation.

6.2.3 Symbolic Analysis of Interval Dynamics

Now we turn to symbolic dynamics analysis of the forced Brusselator when the interval dynamics predominates. This happens, e.g., at the $\alpha = 0.05$ level in Fig. 6.2, i.e., along the $A = 0.48 - 0.2\omega$ or $A = 0.46 - 0.2\omega$ slanting lines in Fig. 6.1.

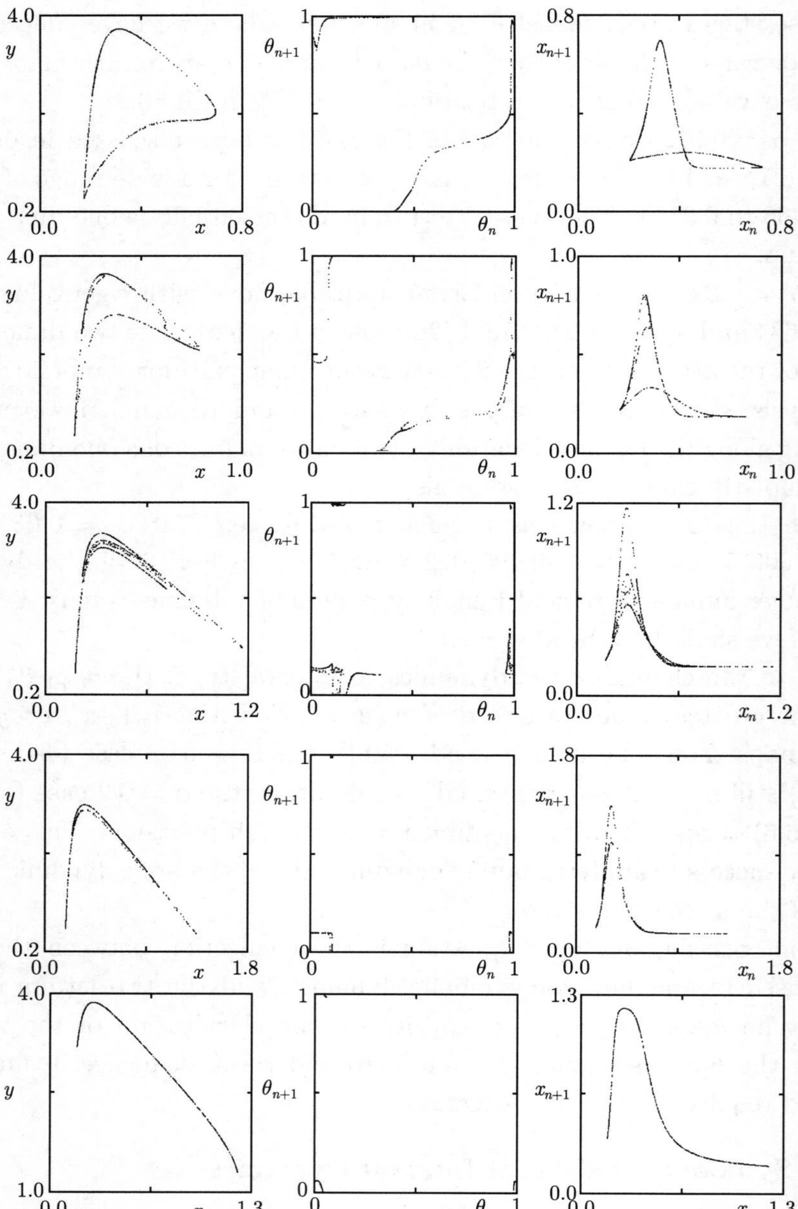

Figure 6.6 Transition from annular to interval dynamics. Phase portraits of the attractor, θ_{n+1}-θ_n return map, and x_{n+1}-x_n return maps are juxtaposed at five different α values for $\omega = 0.775$ and $A = 0.48 - 0.2\omega = 0.325$. From top to bottom: $\alpha = 0.0124$, 0.027, 0.05, 0.08 and 0.2

Only when the dynamics is controlled by a saddle-type unstable fixed point, the symbolic dynamics is close to that of a mapping on an interval. This is the case, e.g., at $\omega = 0.705$ and its vicinity, within the second chaotic band shown in Fig. 6.2.

Fig. 6.7 shows the chaotic attractor in the background of the FCF (dash lines) and the BCF (dotted lines). The attractor, of course, is part of the BCF. The partition line $C\bullet$ is determined from tangencies between FCF and BCF. From the tangencies along $C\bullet$ we choose the following six, which are enough to examine the admissibility of some orbits:

$T_1 : L^\infty RC \bullet RL^2RLRL^2RLR^2L^2RLRLRL^2RL \cdots (0.2897031163, 3.5304029647),$
$T_2 : L^\infty R^5C \bullet RL^2RLRL^2RLR^2L^2RL^2RLRL^2 \cdots (0.2800612211, 3.5251072434),$
$T_3 : L^\infty R^4C \bullet RL^2RLRL^2RLR^2LR^5L^2RLRL^2 \cdots (0.2781676187, 3.5238786963),$
$T_4 : L^\infty R^2LR^2C \bullet RL^2RLRL^2RLR^2LRL^2RL \cdots (0.2766910281, 3.5229044584),$
$T_5 : L^\infty R^2C \bullet RL^2RLRL^2RLR^3L^2RLRLRL^2 \cdots (0.2617705248, 3.5132942637),$
$T_6 : L^\infty C \bullet RL^2RLRL^2RLR^3L^2RLRLR^2L^2R \cdots (0.2613067470, 3.5124905195).$

The letter C stands for either R or L. We have given the precise locations of the tangencies in the last column.

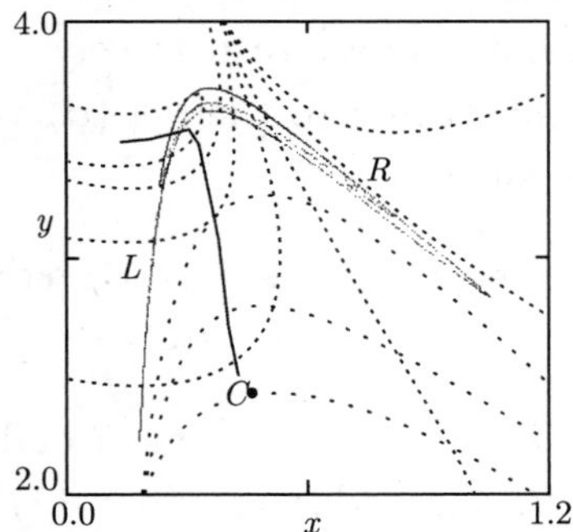

Figure 6.7 The chaotic attractor, FCF (dashed lines), BCF (dotted lines), and the partition line $C\bullet$ at $\omega = 0.705$

Once a partition line has been determined, each orbit may be encoded with a doubly infinite symbolic sequence. However, now it consists of only two letters, R and L. Accordingly, the metric representation of symbolic sequences is somewhat simpler as defined below.

We assign an integer $\epsilon_i = 1$ or -1 to a symbol s_i when it is the letter L or R. To each forward sequence $\bullet s_1 s_2 \cdots s_n \cdots$ we assign a real number α:

$$\alpha = \sum_{i=1}^{\infty} \mu_i 2^{-i}, \tag{6.12}$$

where

$$\mu_i = \begin{cases} 0, \\ 1, \end{cases} \text{if} \quad \prod_{j=1}^{i} \epsilon_j = \begin{cases} 1, \\ -1. \end{cases} \tag{6.13}$$

Similarly, to each backward sequence $\cdots s_{\bar{m}} \cdots s_{\bar{2}} s_{\bar{1}} \bullet$ we assign a real number β:

$$\beta = \sum_{i=1}^{\infty} \nu_{\bar{i}} 2^{-i}, \tag{6.14}$$

where

$$\nu_{\bar{i}} = \begin{cases} 0, \\ 1, \end{cases} \text{if} \quad \prod_{j=1}^{i} \epsilon_{\bar{j}} = \begin{cases} 1, \\ -1. \end{cases} \tag{6.15}$$

According to these definitions we have

$$\begin{aligned} &\alpha(\bullet RL^{\infty}) = \beta(L^{\infty}R\bullet) = 1, \quad \alpha(\bullet L^{\infty}) = \beta(L^{\infty}\bullet) = 0, \\ &\alpha(\bullet RRL^{\infty}) = \alpha(\bullet LRL^{\infty}) = \beta(L^{\infty}RR\bullet) = \beta(L^{\infty}RL\bullet) = 1/2. \end{aligned} \tag{6.16}$$

In the symbolic plane spanned by α and β, forward and backward foliations become vertical and horizontal lines, respectively. All forward (backward) sequences may be ordered according to their α (β) values. The ordering rule may also be formulated as follows:

$$\begin{aligned} &\bullet ER \cdots > \bullet EL \cdots, \quad \bullet OR \cdots < \bullet OL \cdots, \\ &\cdots RE\bullet > \cdots LE\bullet, \quad \cdots RO\bullet < \cdots LO\bullet, \end{aligned} \tag{6.17}$$

where the finite string E (O) consists of letters R and L and contains an *even* (*odd*) number of the letter R. The ordering rule (6.17) turns out to be the same as that of the Hénon map with a *positive* Jacobian. For the Hénon map with a positive Jacobian the relations (6.17) follow from the "local" ordering caused by the fact that both eigenvalues of the fixed point R^{∞} (L^{∞}) are negative (positive).

When foliations are well ordered, the location of a tangency places a restriction on allowed symbolic sequences. A point on the partition line $C\bullet$ may symbolically be represented as $QC\bullet P$. The rectangle enclosed by the lines $QR\bullet$, $QL\bullet$, $\bullet P$, and $\bullet RL^{\infty}$ forms a forbidden zone in the symbolic plane. Therefore, a symbolic sequence UV with $U\bullet$ between $QR\bullet$ and $QL\bullet$, and, at the same time, $\bullet V > \bullet P$ must be forbidden by the tangency $QC\bullet P$. In the symbolic plane the sequence $U \bullet V$ corresponds to a point inside the forbidden zone of $QC \bullet P$. Each tangency point on the partition line rules out a rectangle in the symbolic plane. The union of the forbidden rectangles forms the fundamental forbidden zone (FFZ), the staircase-like boundary of which is the pruning front.

Take a finite set of tangencies $\{Q_i C \bullet P_i\}$. If the shift of a sequence $\cdots s_{k-2}s_{k-1} \bullet s_k s_{k+1} \cdots$ satisfies the condition that the backward sequence $\cdots s_{k-2}s_{k-1}\bullet$ is not between $Q_i R\bullet$ and $Q_i L\bullet$, and, at the same time, $\bullet P_i > \bullet s_k s_{k+1} \cdots$ for some i, then this shift is not forbidden by any tangencies, owing to the property of well-ordering of foliations. Thus, we may say that the shift is allowed according to that tangency. A necessary and sufficient condition for a sequence to be allowed is that all of its shifts are allowed according to the set of tangencies.

In order to check the admissibility condition, we draw 100 000 points representing real sequences generated from the Poincaré map together with the FFZ in the symbolic plane in Fig. 6.8. Indeed, the FFZ contains no point of allowed sequences. At first glance, the pruning front seems to be a straight line. A blow-up in the α direction shows the "structure", displayed in the right part of Fig. 6.8. Note that the α range in the blow-up is $[0.8902, 0.9032]$ and the structure shows off in even narrower range. The width of the FFZ steps in α may be taken as an indicator to judge how good a one-dimensional symbolic dynamics will capture the dynamics of the higher-dimensional system.

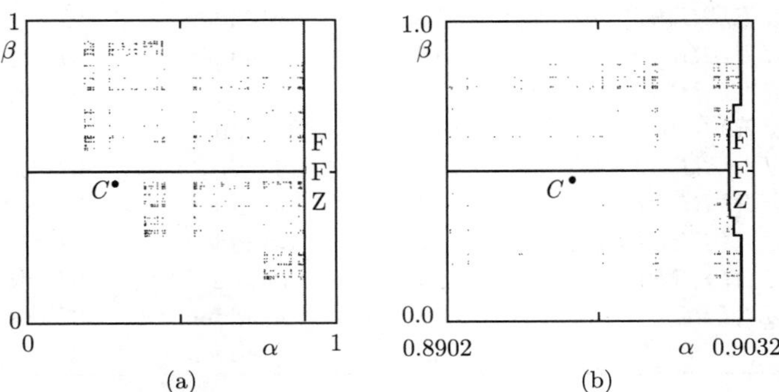

Figure 6.8 The symbolic plane at $\omega = 0.705$. (a) A total of 100 000 points representing real orbits are drawn together with the FFZ. No point falls in the FFZ. (b) A blow-up of the symbolic plane in the interval $\alpha = [0.8902, 0.9032]$

In the one-dimensional unimodal map there are 802 admissible periodic sequences from C to $(RLLRC)^\infty$ with periods less than or equal to 17, as it may be verified by direct generation of these words. Using the 6 tangencies, listed in the beginning of this Section, one may check their admissibility in the forced Brusselator at $\omega = 0.705$. The results are partially listed in Table 6.3. A short-hand notation is used in the table. If the k-th shift of a periodic sequence $P^\infty \bullet P^\infty$ is allowed or forbidden by a tangency T, we write the criterion as kT or $k\overline{T}$, respectively. In addition, $(j:k)T = jT(j+1)T \cdots kT$.

Table 6.3 Admissibility of some periodic sequences not greater than $(RLLRC)^\infty$ at $\omega = 0.705$. A letter C stands for either L or R. Only non-repeating strings of the sequences are given. For the short-hand notation in the "Criterion" column see text

Sequence	Period	Admissibility	Criterion
$RLLRLRLLRLC$	11	allowed	$(0{:}\ 10)T_1$
$RLLRLRLLRLRRLLRLC$	17	allowed	$0T_3(1{:}\ 16)T_1$
$RLLRLRLLRLRRLC$	14	allowed	$0T_3(1{:}\ 13)T_1$
$RLLRLRLLRLRRLRRLC$	17	allowed	$0T_3(1{:}\ 16)T_1$
$RLLRLRLLRLRRLRRC$	16	allowed	$0T_3(1{:}\ 15)T_1$
$RLLRLRLLRLRRLRRRC$	17	forbidden	$0\overline{T}_2$
$RLLRLRLLRLRRLRC$	15	forbidden	$0\overline{T}_2$
$RLLRLRLLRLRRLRLRC$	17	forbidden	$0\overline{T}_2$
$RLLRLRLLRLRRLRLC$	16	allowed	$0T_5(1{:}\ 15)T_1$
$RLLRLRLLRLRRC$	13	forbidden	$0\overline{T}_4$
$RLLRLRLLRLRRRRLC$	16	allowed	$0T_5(1{:}\ 15)T_1$
$RLLRLRLLRLRRRRLRC$	17	forbidden	$0\overline{T}_4$
$RLLRLRLLRLRRRRC$	15	forbidden	$0\overline{T}_4$
$RLLRLRLLRLRRRRRRC$	17	forbidden	$0\overline{T}_4$
$RLLRLRLLRLRRRRRC$	16	forbidden	$0\overline{T}_4$
$RLLRLRLLRLRRRRRLC$	17	allowed	$0T_5(1{:}\ 16)T_1$
$RLLRLRLLRLRRRC$	14	forbidden	$0\overline{T}_4$
$RLLRLRLLRLRRRLRLC$	17	allowed	$0T_5(1{:}\ 16)T_1$
$RLLRLRLLRLRRRLRC$	16	forbidden	$0\overline{T}_4$
$RLLRLRLLRLRRRLRRC$	17	forbidden	$0\overline{T}_4$
$RLLRLRLLRLRRRLC$	15	allowed	$0T_5(1{:}\ 14)T_1$
$RLLRLRLLRLRC$	12	forbidden	$0\overline{T}_6$
$RLLRC$	5	forbidden	$0\overline{T}_6$

First, all sequences from C to RL^2RLRL^2RLC are allowed by T_1. They correspond to the horizontal $C\bullet$ in Fig. 6.8. We have shown only the last period 11 in the first row of Table 6.3. Second, all sequences greater than the period 12 in the next to last row of Table 6.3 are ruled out by T_6, whose forward sequence plays the role of a kneading sequence in a one-dimensional map. Only in between these two limits two-dimensional symbolic dynamics is essential in telling the admissibility of symbolic sequences. A case which exhibits two-dimensional features more clearly was studied in Liu and Zheng [1995b].

Now we are in a position to treat problems like the missing 6P orbit $(RL^3RC)^\infty$ mentioned in the beginning of this section. It was the only missing member in the first U-sequence ever reported in the driven Brusselator when numerical study was carried out up to period 6 orbits (Hao, Wang and Zhang [1983]). At $\omega = 0.705$ the $\bullet(RL^3RC)^\infty$ sequence is bigger than the forward sequence in the tangency T_6, which lies at the border of the attractor as its backward sequence $L^\infty C\bullet$ suggests. The same happens at $\omega = 0.8086$ when

$$T_6 : L^\infty C \bullet RL^3(RLR)^2L^2R^2L^3RLR^2L^2R^2\cdots \quad (0.2227039132, 3.8946519193).$$

Therefore, in order to check the existence of $(RL^3RC)^\infty$ we have to increase ω further. At $\omega = 0.813$ we have found the following tangencies, among others:

$$\begin{array}{ll}
T_1 : L^\infty RC \bullet R^2LR^2L^3R^4L^2RLR^2LR\cdots & (0.306\,175\,384\,6, 3.751\,890\,561\,9),\\
T_2 : L^\infty RL^2RC \bullet R^2LR^2L^3R^4LR^2L^3R^2\cdots & (0.304\,627\,269\,6, 3.752\,397\,421\,4),\\
T_3 : L^\infty R^2LRC \bullet R^2LR^2LR^2L^2RL^3RL^3R^2\cdots & (0.303\,475\,017\,5, 3.755\,752\,594\,0),\\
T_4 : L^\infty R^2LR^2C \bullet RL^3RLR^2L^2R^2L^3R^4L\cdots & (0.279\,680\,026\,4, 3.903\,668\,015\,6),\\
T_5 : L^\infty R^2C \bullet RL^3RLR^2LR^2L^3R^4L^2\cdots & (0.279\,230\,515\,2, 3.905\,714\,160\,3),\\
T_6 : L^\infty C \bullet RL^3R^2LRL^3RLR^2L^2RL^2\cdots & (0.228\,000\,592\,9, 3.907\,657\,139\,1).
\end{array}$$

(We use the same notations T_i. This will not cause confusion as ω is indicated.) Now T_6 alone cannot either exclude or justify the existence of $(RL^3RC)^\infty$, but T_5 does forbid it. At $\omega = 0.821$ the relevant tangencies are:

$$\begin{array}{ll}
T_1 : L^\infty RC \bullet R^2L^2R^2L^3R^2L^2R^2L^3R^2\cdots & (0.298\,921\,874\,8, 3.765\,433\,393\,8),\\
T_2 : L^\infty RL^2RC \bullet R^2L^2R^2L^3R^2L^2R^2L^3RL\cdots & (0.298\,891\,102\,1, 3.765\,511\,029\,2),\\
T_3 : L^\infty RL^2R^2LRC \bullet R^2LR^2L^3R^2L^2R^2L^3R^2L\cdots & (0.293\,730\,077\,8, 3.778\,573\,024\,8),\\
T_4 : L^\infty R^2LR^2C \bullet RL^3R^2LRL^3R^4L^2R^2L\cdots & (0.270\,273\,520\,9, 3.940\,527\,446\,6),\\
T_5 : L^\infty C \bullet RL^3R^2L^3RLR^2L^3RLR^2\cdots & (0.218\,389\,565\,3, 3.947\,673\,904\,6).
\end{array}$$

Now both T_3 and T_4 exclude the $(RL^3RC)^\infty$ sequence, but T_5 says nothing. The situation becomes clearer when we draw the foliations going through the two tangencies. In Fig. 6.9 the diamond indicates the tangency T_3 and the square—T_4. Dash lines show the forward foliations through T_3 and T_4, the solid line represents the backward foliation of T_3.

According to the ordering rule (6.17), the forward foliations associated with $\bullet(RL^3RR)^\infty$ or $\bullet(RL^3RL)^\infty$ must locate in the region between the two forward foliations of T_3 and T_4. At the same time, the backward foliation associated with the sequences $(RRL^3R)^\infty C\bullet$ and $(LRL^3R)^\infty C\bullet$ necessarily lie in the region below the backward foliation of T_3. Consequently, they are separated by the foliations of T_3. This leads to the non-existence of $(RL^3RC)^\infty$.

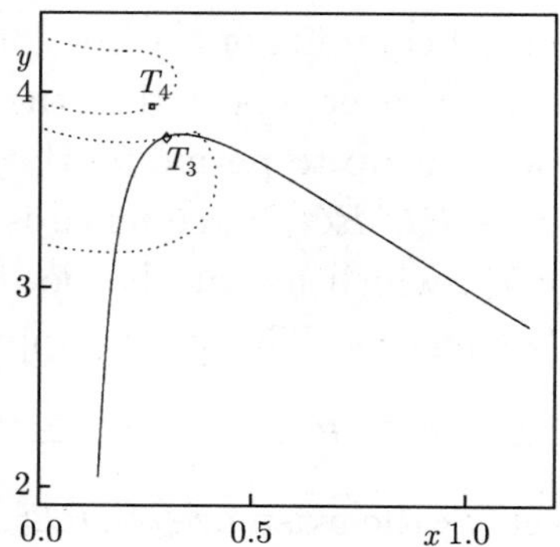

Figure 6.9 The forward foliations (dash lines) of the tangencies T_3 (diamond) and T_4 (square) and the backward foliation (solid line) of T_3 at $\omega = 0.821$

The foregoing discussion clearly shows that $(RL^3RR)^\infty$ or $(RL^3RL)^\infty$ is always forbidden as ω varies, i.e. the "missing" period six RL^3RC actually does not exist.

In this way we have checked the U-sequence in the periodically forced Brusselator up to period 7. The results are listed in Table 6.4. This is to be compared with the early numerical results of Hao, Wang and Zhang [1983]. Now, instead of "missing", we are able to say firmly which orbits are forbidden. This kind of results can never be obtained neither by numerical work alone nor by purely analytical means. It also tells the necessity of invoking two-dimensional symbolic dynamics in exploring global behaviour of ODEs.

Table 6.4 Periodic windows along the $A = 0.46 - 0.2\omega$ line in Fig. 6.1. The words with an asterisk are added as compared to the U-sequence published in Hao, Wang and Zhang [1983]

Word	Period	Range in ω	Width of the Window
C^*	1	0 – 0.368 522 4	0.368 522 4
RC	2	0.368 6 – 0.555 489	0.186 889
$RLRC$	4	0.555 5 – 0.577 7	0.022 2
$RLRRRC$	6	0.582 49 – 0.582 51	0.000 02
$RLRRRRC^*$	7	0.582 989 983 4 – 0.582 991 029 838	0.000 001 046 438

Continued

Word	Period	Range in ω	Width of the Window
RLRRC	5	0.584 5 – 0.584 8	0.000 3
*RLRRLRC**	7	0.588 405 805 31 – 0.588 458 188 7	0.000 052 383 39
RLC	3	0.594 7 – 0.691 5	0.096 8
RLLRLC	6	0.691 5 – 0.702 5	0.011 0
*RLLRLRC**	7	0.702 993 8 – 0.703 028 1	0.000 034 3
RLLRC	5	0.706 8 – 0.711 5	0.004 7
*RLLRRRC**	7	0.714 386 619 36 – 0.714 445 3	0.000 058 680 64
RLLRRC	6	0.718 – 0.718 5	0.000 5
*RLLRRLC**	7	0.721 559 986 71 – 0.721 727 07	0.000 167 083 29
RLLC	4	0.732 5 – 0.792	0.059 5
RLLLRLC	7	0.803 5 – 0.805 6	0.002 1
RLLLRC	6	forbidden	
RLLLRRC	7	0.819 381 023 7 – 0.819 659 13	0.000 278 106 3
RLLLC	5	0.825 9 – 0.867 5	0.038
*RLLLLRC**	7	forbidden	
RLLLLC	6	0.901 5 – 0.923	0.021 5
RLLLLLC	7	0.959 – 0.974	0.015

As the periodically forced Brusselator is concerned, the transition from annular to interval dynamics has enabled us to trace how the symbolic dynamics evolves from that of 1D circle map to 1D unimodal map via 2D dissipative standard map and 2D Hénon map. Although some deeper questions, e.g., the analysis of the transition from U-sequence in 1D unimodal map to Farey sequence in the circle map, remain to be tackled, it has furnished a framework for the study of other periodically driven systems.

6.3 The Lorenz Equations

The Lorenz equations were first derived by mode truncating for a set of partial differential equations describing thermal convection between two infinite plates (Saltzman [1962]; E. N. Lorenz [1963]). It has been shown that several models of hydrodynamic, mechanical, dynamo and laser problems may also lead to this set of ODEs, see, e.g., the monograph by C. Sparrow [B1982].

The Lorenz system consists of three ordinary differential equations:

$$\begin{aligned}\dot{x} &= \sigma(y - x),\\ \dot{y} &= rx - xz - y,\\ \dot{z} &= xy - bz,\end{aligned} \tag{6.18}$$

where σ, r, b are positive parameters, corresponding to the Prandtl number, the Rayleigh number, and a geometric ratio, respectively. So far, most of the numerical and analytical studies on the Lorenz equations have been restricted to a one-parameter sub-family, defined by the straight line $\sigma = 10$, $b = 8/3$, and $r > 0$ in the parameter space. Another sub-family, along the line $\sigma = 16$, $b = 4$, and $r > 0$, was investigated mainly by Japanese authors (Shimizu and Morioka [1978]; Shimada and Nagashima [1978, 1979]; Tomita and Tsuda [1980]), but did not lead to qualitatively new picture. In this Chapter we will study this system in a wide range $r > 28$ at fixed $\sigma = 10$ and $b = 8/3$ by constructing 1D and 2D symbolic dynamics from its Poincaré maps.

Being one of the first examples exhibiting a strange attractor in dynamical systems, this system has become a touchstone for many new ideas in chaotic dynamics. A number of interesting phenomena such as pre-turbulence (Kaplan and Yorke [1979]), noisy-periodicity (Lorenz [1980]) and intermittency (Manneville and Pomeau [1979]), have been first observed and interpreted in these equations. However, we will concentrate on the symbolic dynamics study of this model. In fact, this is the first set of autonomous differential equations whose qualitative behavior has been investigated in great details by using the method of symbolic dynamics (Ding and Hao [1988]; Fang and Hao [1996]; Zheng [1997c]; Hao, Liu and Zheng [1998]).

6.3.1 Summary of Known Properties

We summarize a few properties of the model which readily follow from a simple analysis of the equations (6.18).

Anti-symmetry of the Lorenz equations

First of all, the equations (6.18) are invariant under the transform

$$x \leftrightarrow -x, \quad y \leftrightarrow -y, \quad z \leftrightarrow z. \tag{6.19}$$

If one wishes to compare the systematics of periodic orbits in the Lorenz equations with that in low-dimensional maps, this anti-symmetry must be taken into account. Indeed, the 1D antisymmetric cubic map, with or without a gap in the middle, and their two-dimensional extensions, are capable to capture much of the qualitative behavior of the Lorenz model (Ding and

Hao [1988]; Fang and Hao [1996]). The Lorenz-Sparrow map, introduced in Section 3.8 and to be used in what follows, yields even some quantitatively meaningful results (Zheng and Liu [1997]). In contrast to these 1D and 2D maps, the two-dimensional Hénon map, originally devised to mimic the Lorenz attractor (Hénon [1976]), does not share this symmetry. Consequently, it turns out to be of little help to the understanding of periodic and chaotic orbits in the Lorenz model.

The z-axis

The dynamical behavior along the z-axis is obtained by setting $x = y = 0$ in Eqs. (6.18), yielding $\dot{x} = \dot{y} = 0$ and

$$\dot{z} = -bz.$$

Therefore, starting from any point on the z-axis one returns to the origin $(0, 0, 0)$ with exponential rate as long as $b > 0$. This means that the z-axis belongs to the stable manifold of the fixed point $(0, 0, 0)$. Whenever the unstable manifold of the origin approaches the z-axis a homoclinic situation takes place and there appears an invariant set which includes a countably infinite number of unstable periodic orbits, an uncountable number of aperiodic orbits, and an uncountable number of initial values which are eventually attracted to the origin. Sparrow [B1982] called this phenomenon a homoclinic explosion. Due to the intricate bending and folding of the unstable manifold homoclinic explosions happen infinitely many times when the parameters are varied.

Contraction of phase volume

The three equations (6.18) determine the time evolution of three components of a vector field. How a volume element $dV = dx\,dy\,dz$ varies with the dynamics may be seen by differentiating it with respect to time:

$$d\dot{V} = d\dot{x}\,dy\,dz + dx\,d\dot{y}\,dz + dx\,dy\,d\dot{z} = \left(\frac{d\dot{x}}{dx} + \frac{d\dot{y}}{dy} + \frac{d\dot{z}}{dz}\right) dV.$$

Using the right-hand side of Eqs. (6.18), we get

$$\frac{d\dot{V}}{dV} = \text{div}\,\mathbf{f} = -(\sigma + 1 + b), \tag{6.20}$$

where **f** is a vector with the right-hand side functions of (6.18) as components. The constant negative value of div **f** implies phase volume contraction that explains the success of approximating the dynamics on the attractor by using 1D maps.

Fixed points and their stability

In order to locate the fixed points of the Lorenz system we put the first derivatives in (6.18) to zero. The system of algebraic equations thus obtained has three sets of solutions:

1. The origin $(0,0,0)$.
2. A pair of solutions which exist only when $r > 1$:

$$C_{\pm} = (\pm\sqrt{b(r-1)}, \pm\sqrt{b(r-1)}, r-1). \tag{6.21}$$

To analyze the stability of the system in the vicinity of a fixed point, the system is linearized around that point. The origin $(x, y, z) = (0, 0, 0)$ is globally attracting for $r < 1$. At $r = 1$ the origin loses stability and the fixed points $C_{\pm}$ emerge from the origin. Their eigenvalues undergo some qualitative changes at $r = 1.345\,617$. For $1 < r < r_0 = 13.926$ every trajectory approaches one of the two fixed points. At $r_0 = 13.926$ a strange invariant set (not an attractor yet) comes into life. It is said that the first homoclinic explosion takes place. For $r > r_0$ the invariant set born at the homoclinic explosion remains qualitatively unchanged.

For $r = r_A = 24.06$ another transition takes place. The invariant set becomes a strange attractor. In addition, an infinite sequence of homoclinic explosions begins at this r value. For $r_A < r < r_H = 24.74$ there exist a stable chaotic attractor and a pair $q^{\pm}$ of stable fixed points. At $r_H = 24.74$, both $q^{\pm}$ become unstable. For $r > r_H$ no stable fixed points exist any more. For $r > r_1 = 30.1$ stable periodic orbits may appear. For larger r there are infinitely many periodic windows and chaotic bands. It has been proven that for r large enough there are only simply periodic motion in the system (Robbins [1979]). We will concentrate on the dynamics of the Lorenz model for a wide range of r from $r_H = 24.74$ to practically infinity.

Before undertaking the symbolic dynamics analysis we summarize briefly what has been done on the Lorenz system from the viewpoint of symbolic dynamics. Guckenheimer and Williams introduced the geometric Lorenz model

(Guckenheimer and Williams [1979], Williams [1979], Guckenheimer and Holmes [B1983]) for the vicinity of $r = 28$ which leads to symbolic dynamics of two letters, proving the existence of chaos in the geometric model. However, as Smale [1991] pointed out, it remains an unsolved problem as whether the geometric Lorenz model means to the real Lorenz system. Though not using symbolic dynamics at all, the paper by Tomita and Tsuda [1980] studying the Lorenz equations at a different set of parameters $\sigma = 16$ and $b = 4$ is worth mentioning. They noticed that the quasi-1D chaotic attractor in the $z = r - 1$ Poincaré section outlined by the upward intersections of the trajectories may be directly parameterized by the x coordinates. A 1D map was devised to numerically mimic the global bifurcation structure of the Lorenz model. C. Sparrow [B1982] used two symbols x and y to encode orbits without explicitly constructing symbolic dynamics. In Appendix J of Sparrow [B1982] a family of 1D maps was described as "an obvious choice if we wish to try and model the behavior of the Lorenz equations in the parameter range $\sigma = 10$, $b = 8/3$ and $r > 24.06$". In this book this family of maps has been called the *Lorenz-Sparrow map* and studied in Section 3.8. The work of Tomita and Tsuda [1980] and Sparrow [B1982] have been quite instrumental for the construction of symbolic dynamics in what follows. In fact, the 1D maps to be obtained from the 2D upward Poincaré maps of the Lorenz equations after some manipulations belong precisely to the family suggested by Sparrow. Ding and Hao [1988] associated the systematics of stable periodic orbits in the Lorenz equations with that of a 1D anti-symmetric cubic map. The choice of an anti-symmetric map was dictated by the invariance of the Lorenz equations under the transformation (6.19). It was shown that most of the periods known then are ordered in a "cubic" way. However, many short periods present in the 1D map have not been found in the Lorenz equations. It was realized later (Fang and Hao [1996]) that a cubic map with a discontinuity in the center may better reflect the ODEs and many of the missing periods are excluded by the 2D nature of the Poincaré map. Instead of devising model maps one should generate all related 1D or 2D maps directly from the Lorenz equations and construct the corresponding symbolic dynamics. This has been done by Hao, Liu and Zheng [1998] and makes the main body of the present section.

6.3.2 Construction of Poincaré and Return Maps

The Poincaré map in the $z = r - 1$ plane captures most of the interesting dynamics as it contains both fixed points $C_{\pm}$. The z-axis is contained in the stable manifold $\mathcal{W}^s$ of the origin $(0,0,0)$. All orbits reaching the z-axis will be attracted to the origin, thus most of the homoclinic behavior may be tracked in this plane. In principle, either downward or upward intersections of trajectories with the $z = r - 1$ plane may be used to generate the Poincaré map. However, upward intersections with $dz/dt > 0$ have the practical merit to yield 1D-like objects which may be parameterized by simply using the x coordinates.

Fig. 6.10 (a) shows a Poincaré section at $r = 118.15$. The dashed curves and diamonds represent one of the FCFs and its tangent points with the BCF. The 1D-like structure of the attractor is apparent. Only the thickening in some part of the attractor hints on its 2D nature. Ignoring the thickening for the time being, the 1D attractor may be parameterized by the x coordinates only. Collecting successive x_i, we construct a first return map $x_n \mapsto x_{n+1}$ as shown in Fig. 6.10 (b). It consists of four symmetrically located pieces with gaps on the mapping interval. For a first return map a gap belonging to both $\{x_n\}$ and $\{x_{n+1}\}$ plays no role in the dynamics. If necessary, we can use this specificity of return maps to squeeze some gaps in x. Furthermore, we can interchange the left subinterval with the right one by defining, e.g.,

$$x' = x - 36 \quad \text{for} \quad x > 0; \quad x' = x + 36 \quad \text{for} \quad x < 0. \tag{6.22}$$

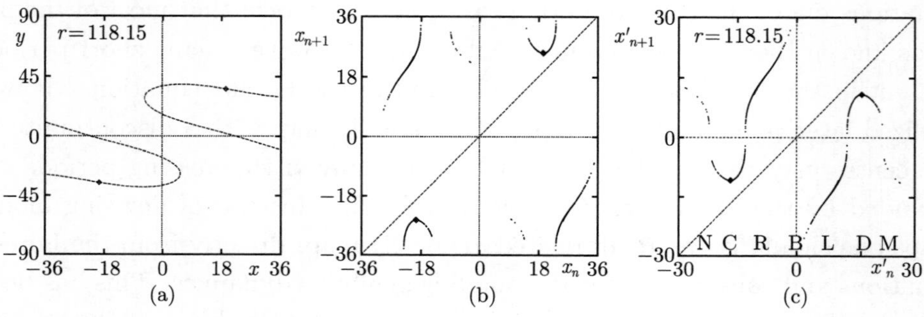

Figure 6.10 (a) The Poincaré map at $r = 118.15$. (b) The first return map constructed from the Poincaré map using the x coordinates only. (c) The swapped return map x'_n-x'_{n+1} constructed from the middle figure

The precise value of the numerical constant here is not essential; it may be estimated from the upper bound of $\{|x_i|\}$ and is so chosen as to make the final figure look nicer. The swapped first return map, as we call it, is shown in Fig. 6.10 (c). The corresponding tangent points between FCF and BCF (the diamonds) are also drawn on these return maps for later use.

It is crucial that the parameterization and swapping do keep the 2D features present in the Poincaré map. This is important when it comes to take into account the 2D nature of the Poincaré maps.

In Fig. 6.11 Poincaré maps at 9 different values from $r = 28$ to 203 are shown. The corresponding swapped return maps are shown in Fig. 6.12. Generally speaking, as r varies from small to greater values, these maps undergo

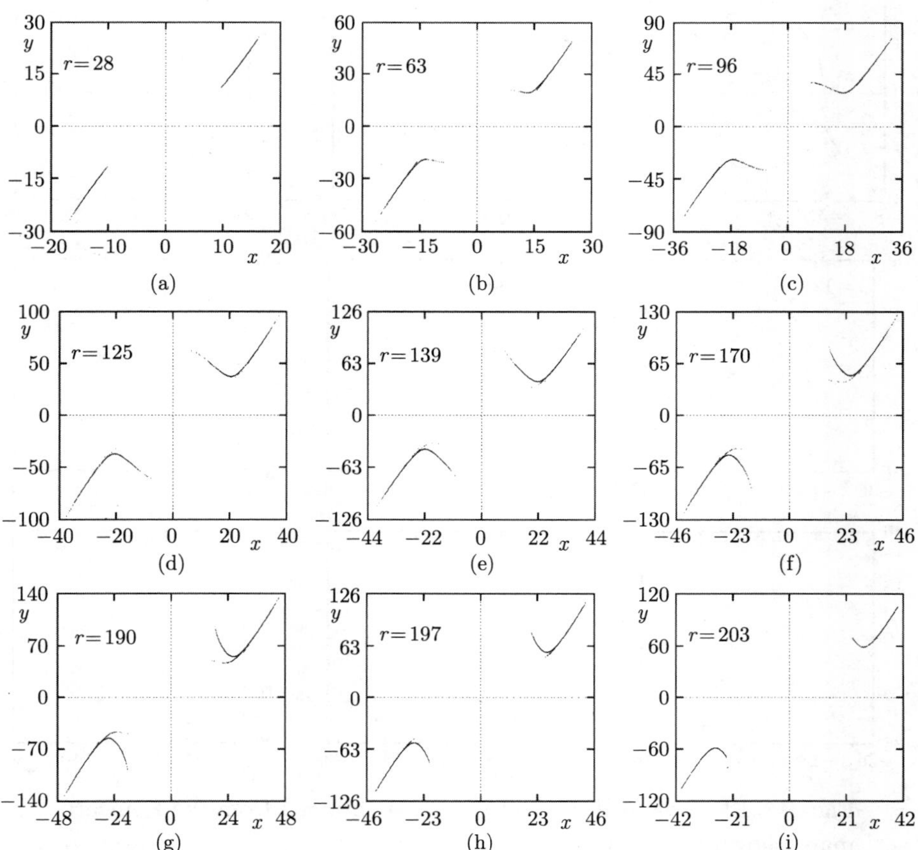

Figure 6.11 Upward Poincaré maps at 9 different r-values

transitions from 1D-like to 2D-like, and then to 1D-like again. Even in the 2D-like range the 1D backbones still dominate. This partly explains the early success (Ding and Hao [1988]; Fang and Hao [1996]) in applying purely 1D symbolic dynamics to the Lorenz model. We will see how to judge this success later on. Ignoring the 2D features for the time being, all the return maps shown in Fig. 6.12 fit into the family of Lorenz-Sparrow map studied in Section 3.8. Even the qualitative changes seen in Fig. 6.12 at varying r may be associated with the limiting cases discussed in Section 3.8. We note also that the return map at $r = 28$ complies with what follows from the geometric Lorenz model. The symbolic dynamics of this Lorenz-like map has been completely constructed in Section 3.5.

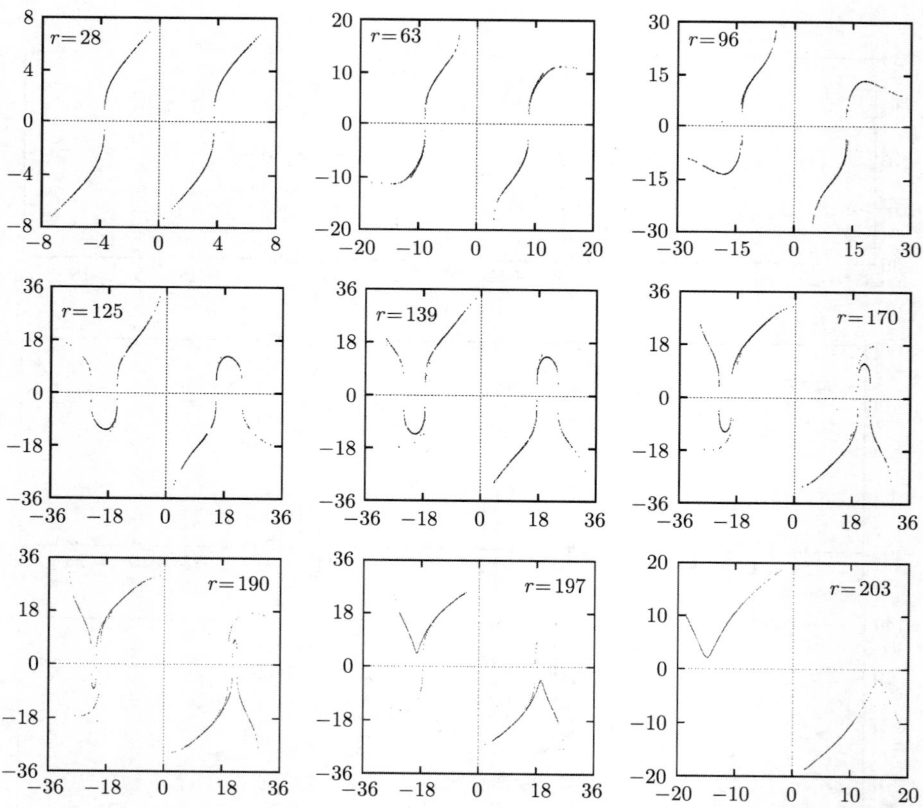

Figure 6.12 Swapped first return maps obtained from the Poincaré maps shown in Fig. 6.11

6.3.3 One-Dimensional Symbolic Dynamics Analysis

Equipped with the symbolic dynamics constructed in Section 3.8 we are now well prepared to carry out a 1D symbolic dynamics analysis of the Lorenz equations using the swapped return maps shown in Fig. 6.12. Take $r = 118.15$ as an working example, the rightmost point in $\{x_i\}$ and the minimum at C determine the two kneading sequences:

$$H = MRLNRLRLRLRLNRL\cdots,$$
$$K = RLRLRLRRLRLRLRR\cdots.$$

Indeed, they satisfy (3.115) and form a compatible kneading pair. Using the propositions formulated in Section 3.8.2, all admissible periodic sequences up to period 6 are generated. They are LC, $LNLR$, $LNLRLC$, $RMRLR$, $RMLNLC$, $RMLN$, $RLLC$, $RLLNLC$, $RLRMLC$, and $RLRLLC$. Here the letter C is used to denote both N and R. Therefore, there are altogether 17 unstable periodic orbits with period equal to or less than 6. Relying on the ordering of symbolic sequences and using a bisection method, these unstable periodic orbits may be quickly located in the phase plane.

It should be emphasized that we are dealing with unstable periodic orbits at a fixed parameter set. The existence of LN and LR does not necessarily imply the existence of LC.

Similar analysis may be carried out for other r. In Table 6.5 we collect some kneading sequences at different r-values. Their corresponding metric representations are also included. We first note that they do satisfy the admissibility conditions (3.115), i.e., K and H at each r make a compatible kneading pair. An instructive way of presenting the data consists in drawing the plane of metric representation for both $\alpha(K)$ and $\alpha(H)$, see Fig. 6.13. The compatibility conditions require, in particular, $K \leqslant H$, therefore only the upper left triangular region is accessible.

As we have indicated at the end of the last Section, the Lorenz-Sparrow map has two one-parameter limits. The first limit (3.118) takes place somewhere at $r < 36$, maybe around $r < 30.1$, as estimated in Sparrow [B1982] in a different context. In Table 6.5 there is only one kneading pair $K = R^\infty$ and $H = L^\infty$ satisfying $H = L\overline{K}$. In terms of the metric representations the condition (3.118) defines a straight line (line a in Fig. 6.13)

$$\alpha(H) = (3 - \alpha(K))/4.$$

Table 6.5 Kneading pairs (K, H) and their metric representations at different r-values

r	K	$\alpha(K)$	H	$\alpha(H)$
203.0	$LNLRLRLRLRLRLRL$	0.525 000	$MRLRLRLRLRLRLRL$	0.900 000
201.0	$LNLRLRLRMRLRLRM$	0.524 995	$MRLRLRLNLRLRLNL$	0.900 018
199.04	$LNLRLRMRLRLNLRL$	0.524 927	$MRLRLNLRLRMRLRL$	0.900 293
198.50	$LNLRMRMRLRMRMRL$	0.523 901	$MRLNLNLRLNLNLRL$	0.904 396
197.70	$LNLRMRLRLRLNLRL$	0.523 828	$MRLNLRLRLRMRLRL$	0.904 688
197.65	$LNLRMRLRLNLRLRM$	0.523 827	$MRLNLRLRMRLRLNL$	0.904 692
197.58	$LNLRMRRMRMRLRMR$	0.523 796	$MRRMRMRLNLNLRLR$	0.908 110
196.20	$LNLLNLRLNLRLRLN$	0.523 042	$MRRLRLRLRLRLNLR$	0.912 500
191.0	$RMRLNLNLLRLRLRM$	0.476 096	$MNRLRLRMRLRMLNL$	0.962 518
181.8	$RMRLLNLRRMRLLNL$	0.475 410	$MNRLRLRLRLRLRLN$	0.962 500
166.2	$RMLNRMLNRMLNRML$	0.466 667	$MNRLNRMRLRLLNRM$	0.961 450
139.4	$RLRMRLRRLLNLRRL$	0.404 699	$MNRRLLNRMLRRLLR$	0.959 505
136.5	$RLRMLRRLRMRLLRL$	0.403 954	$MNRRLLNLRRLRMLR$	0.959 498
125.0	$RLRLLNLRLRRLRLL$	0.400 488	$MRRLRMLRLRRLRLL$	0.912 256
120.0	$RLRLRLRLRLRLRML$	0.400 000	$MRRMLRLRLRLRLRR$	0.908 594
118.15	$RLRLRLRRLRLRLRR$	0.399 988	$MRLNRLRLRLRLNRL$	0.903 906
117.7	$RLRLRLNRLLRLRLR$	0.399 938	$MRLRRLRLRLNRLLR$	0.900 781
114.02	$RLRLRRLRLRRLRLR$	0.399 804	$MRLRLNRLRLRRLRL$	0.900 244
107.7	$RLRRLLRRLRRLLRR$	0.397 058	$MRLLRRLLRLLNRLR$	0.897 058
104.2	$RLRRLRLRRLRRLRL$	0.396 872	$MRLLRLLRLRRLLNR$	0.896 826
101.5	$RLRRLRRLRRLRLNR$	0.396 825	$MLNRMLRLLRLLRLL$	0.866 598
99.0	$RLNRLRRLRRLRRLR$	0.384 425	$MLNRLRRLNRLRRLN$	0.865 572
93.4	$RRMLLNRRMLLNRLL$	0.365 079	$MLRRLLRRRMLRLLR$	0.852 944
83.5	$RRLLRRRLLRRRLLR$	0.352 884	$MLRLLRLRRRLLLRR$	0.849 233
71.7	$RRLRRRLRRRLRRLR$	0.349 020	$MLLRLRRLLRRLLLR$	0.837 546
69.9	$RRRMLLLNRRRMLLR$	0.341 176	$MLLRLRLLLRLLRLR$	0.837 485
69.65	$RRRLLLLNRRRMLLR$	0.338 511	$MLLRLLNRRLRRLLR$	0.837 363
65.0	$RRRLLRRRRLLRLLR$	0.338 217	$MLLLRRRLRRRLLRL$	0.834 620
62.2	$RRRLRLRRRLRRRRM$	0.337 485	$MLLLRRLLLLRLRLL$	0.834 554
59.4	$RRRLRRRRLRRRLLL$	0.337 243	$MLLLRLLRLRRLRLL$	0.834 326
55.9	$RRRRLLRRRRLRLLR$	0.334 554	$MLLLLRRLRRLLLLR$	0.833 643
52.6	$RRRRLRRRRRLLRRR$	0.334 310	$MLLLLRLLLLRRRRR$	0.833 578
50.5	$RRRRRLLRRRLRLRL$	0.333 639	$MLLLLLRRLLRRLRR$	0.833 410
48.3	$RRRRRLRRRRRLLLL$	0.333 578	$MLLLLLRLLLLRLLR$	0.833 394
48.05	$RRRRRRLLLLLRLLL$	0.333 415	$MLLLLLRLLLLLLLR$	0.833 394
46.0	$RRRRRRLRLRRRRLL$	0.333 398	$MLLLLLLRLRLLLLN$	0.833 350
44.0	$RRRRRRRLRLLLLLL$	0.333 350	$MLLLLLLLRLRRRRR$	0.833 337
36.0	$RRRRRRRRRRRRRLR$	0.333 333	$MLLLLLLLLLLLLLR$	0.833 333
28.0	$RRRRRRRRRRRRRRR$	0.333 333	$LLLLLLLLLLLLLLL$	0.666 667

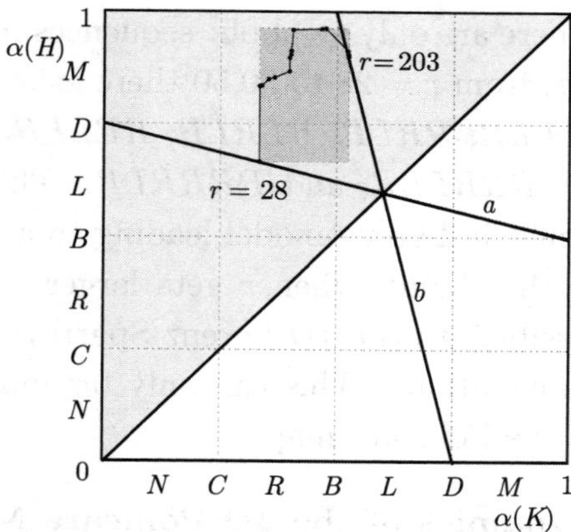

Figure 6.13 The $\alpha(H)$ versus $\alpha(K)$ plane shows kneading pairs (solid circles) corresponding to the Lorenz equations from $r = 28$ to 203. The two straight lines a and b represent the two one-parameter limits of the Lorenz-Sparrow map

(We have used (3.117).) The point $r = 28$ drops down to this line almost vertically from the $r = 36$ point. This is the region where "fully developed chaos" has been observed in the Lorenz model and perhaps it outlines the region where the geometric Lorenz model may apply.

The other limit (3.120) happens at $r > 197.6$. In Table 6.5 all 6 kneading pairs in this range satisfy $K = L\overline{H}$. They fall on another straight line (line b in Fig. 6.13)

$$\alpha(K) = (3 - \alpha(H))/4,$$

but can hardly be resolved. The value $r = 197.6$ manifests itself as the point where the chaotic attractor stops to cross the stable manifold $\mathcal{W}^s$ of the origin in the 2D Poincaré map.

We note that the kneading pair at $r = 203$ is very close to a limiting pair $K = LN(LR)^\infty$ with a precise value $\alpha = 21/40 = 0.525$ and $H = M(RL)^\infty$ with exact $\alpha = 0.9$. For any kneading pair in Table 6.5 one can generate all admissible periods up to length 6 inclusively. For example, at $r = 125$ although the swapped return map shown in Fig. 6.12 exhibits some 2D feature as throwing a few points off the 1D attractor, the 1D Lorenz-Sparrow map still works well. Besides the 17 orbits listed above for $r = 118.15$, five new periods appear: LLN, $LNLRMR$, $LNLLC$, and $RMRLN$. All these 22 unstable periodic orbits have been located with high precision in the Lorenz equations. Moreover, if we confine ourselves to short periods not exceeding period 6, then from $r = 28$ to 59.40 there are only symbolic sequences made of the two letters R and L. In particular, from $r = 28$ to 50.50 there exist the same 12 unstable periods: LR, RLR, $RLLR$, $RRLR$, $RLRLR$, $RRLLR$, $RRRLR$, $RLRLLR$, $RRLLLR$, $RRLRLR$, $RRRLLR$, and $RRRRLR$. This may partly explain the success of the geometric Lorenz model leading to a symbolic dynamics of two letters. On the other hand, when r gets larger, e.g., $r = 136.5$, many periodic orbits "admissible" to the 1D Lorenz-Sparrow map cannot be found in the original Lorenz equations. This can only be analyzed by invoking 2D symbolic dynamics of the Poincaré map.

6.3.4 Symbolic Dynamics of the 2D Poincaré Maps

As the Poincaré maps of the Lorenz equations are by definition two-dimensional, knowledge of symbolic dynamics of 2D maps is a prerequisite for the understanding of the dynamics in the Lorenz model. Referring to Chapter 5 for the basics of 2D symbolic dynamics, we immediately undertake the job for this system of ODEs.

Partitioning of the Poincaré section

Lorenz equations at $r = 136.5$ provide a typical situation where 2D symbolic dynamics must be invoked. Fig. 6.14 shows an upward $z = r - 1$ Poincaré section of the chaotic attractor. The dashed lines indicate the contour of the FCFs. The two symmetrically located families of FCFs are demarcated by the intersection of the stable manifold $\mathcal{W}^s$ of the $(0, 0, 0)$ fixed point with the $z = r - 1$ plane. The actual intersection located between the dense dashed lines is not shown. The BCFs are not shown either except for the attractor itself, which is a part of the BCFs.

The 1D symbolic dynamics analysis performed in Section 6.3.3 deals with forward symbolic sequences only. However, the partition of the 1D interval shown, e.g., in Fig. 6.10, may be traced back to the 2D Poincaré section to indicate the partition for assigning symbols to the forward symbolic sequences. Two segments of the partition lines are shown in the left figure of Fig. 6.14 as dotted lines. The labels $\bullet C$ and $\bullet D$ correspond to C and D in the Lorenz-Sparrow map, see Fig. 6.10. The ordering rule (3.112) should now be understood as

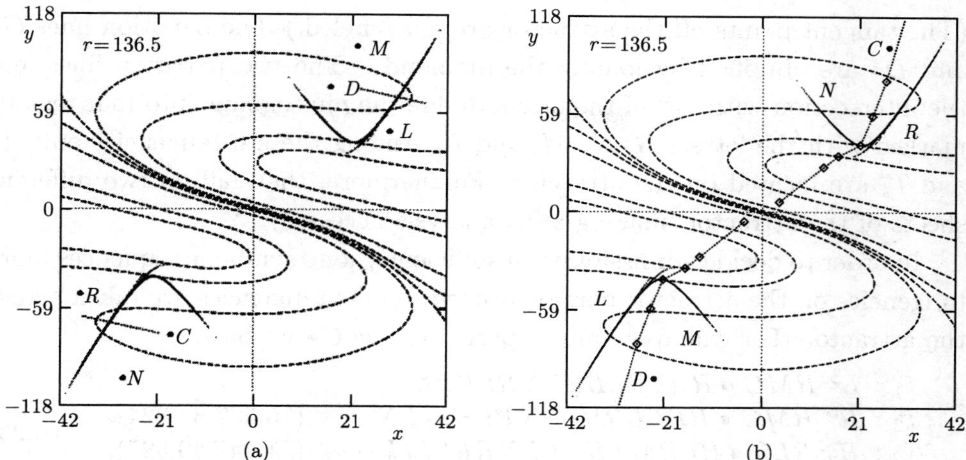

Figure 6.14 An upward Poincaré section at $r = 136.5$ showing the chaotic attractor and a few FCFs (dashed lines). Segments of the partition lines for forward symbols (dotted lines) are shown in (a). Twelve tangent points (diamonds) and the partition lines for backward symbols (dotted lines) are shown in (b)

$$
\begin{aligned}
&\bullet EN \cdots < \bullet EC \cdots < \bullet ER \cdots < \bullet EB \cdots \\
&\qquad < \bullet EL \cdots < \bullet ED \cdots < \bullet EM \cdots , \\
&\bullet ON \cdots > \bullet OC \cdots > \bullet OR \cdots > \bullet OB \cdot \\
&\qquad > \bullet OL \cdots > \bullet OD \cdots > \bullet OM \cdots ,
\end{aligned}
$$

with E and O being *even* and *odd* strings of M, L, R, and N. In fact, from Fig. 6.10 one could only determine the intersection point of the partition line with the 1D-like attractor. To determine the partition line in a larger region of the phase plane one has to locate more tangencies between the FCFs and the BCFs. Two sets of tangent points can equally used to determine the partition lines, one set of which is the image of the other. Six tangent points and their mirror images are located and indicated as diamonds in the right figure of Fig. 6.14 for the partition lines defining backward symbols. The tangencies in the first quadrant are

$$
\begin{aligned}
&(3.833\,630\,661\,151, 5.915\,245\,399\,002), \\
&(13.347\,217\,142\,10, 27.069\,324\,409\,06), \\
T_1 :\ &(16.501\,306\,048\,50, 33.814\,256\,215\,18), \\
T_4 :\ &(21.240\,128\,507\,67, 40.568\,508\,427\,96), \\
&(23.867\,574\,249\,70, 58.009\,259\,119\,37), \\
&(26.738\,296\,763\,87, 79.375\,838\,379\,12).
\end{aligned}
$$

(The tangent points off the attractor are not labeled.) The partition lines $C\bullet$ and $D\bullet$ are obtained by joining the diamonds. The two partition lines and the intersection with $\mathcal{W}^s$ of the origin divide the phase plane into four regions marked with the letters R, N, M, and L. Among these 6 tangencies only T_1 and T_4 are located on the attractor. Furthermore, they fall on two different sheets of the attractor, making a 2D analysis necessary.

In order to decide admissibility of sufficiently long symbolic sequences more tangencies on the attractor may be needed. These tangencies are taken across the attractor. For example, on the partition line $C\bullet$ we have

$$\begin{array}{lll} T_1: L^\infty RMC \bullet RRLRLLRLNRLLRLN\cdots, & & \\ T_2: R^\infty RMC \bullet RRLLRRLLNRLRMLN\cdots & (16.567,34.824), & \\ T_3: R^\infty RLC \bullet RLRMLRRLLNRLRML\cdots & (21.247,40.525), & \\ T_4: L^\infty LLC \bullet RLRMLRRLRMRLLRL\cdots. & & \end{array} \tag{6.23}$$

Due to insufficient numerical resolution in Fig. 6.14 the diamond on the main sheet of the attractor represents T_3 and T_4, while the diamond on the secondary sheet represents T_1 and T_2. The mirror images of these tangencies are located on the $D\bullet$ partition line:

$$\begin{array}{l} \overline{T}_1: R^\infty LND\bullet LLRLRRLRMLRRLRM\cdots, \\ \overline{T}_2: L^\infty LND\bullet LLRRLLRRMLRLNRM\cdots, \\ \overline{T}_3: L^\infty LRD\bullet LRLNRLLRRMLRLNR\cdots, \\ \overline{T}_4: R^\infty RRD\bullet LRLNRLLRLNLRRLR\cdots. \end{array}$$

We denote the symbolic sequence of the tangency T_i as $Q_iC\bullet K_i$, keeping the same letter K as the kneading sequence K in the 1D Lorenz-Sparrow map, because K_i complies with the definition of a kneading sequence as the next iterate of C. If one is interested in forward sequences alone, only these K_i will matter. Moreover, one may press together different sheets seen in Fig. 6.14 along the FCFs, as points on one and the same FCF have the same forward symbolic sequence. Here lies a deep reason for the success of 1D symbolic dynamics at least when only short periodic orbits are concerned with. Therefore, before turning to the construction of 2D symbolic dynamics let us first see what a 1D analysis would yield.

1D symbolic analysis at $r = 136.5$

Fig. 6.15 is a swapped return map obtained from the first return map by letting the numerical constant be 41 in (6.22). The 2D feature manifests itself as layers near C and D. The four tangencies are plotted as two diamonds in the figure, since T_1 is very close to T_2 and T_3 to T_4. As no layers can be seen away from the turning points one could only get one H. Now there are 4 kneading sequences K_i ordered as

$$K_1 < K_2 < K_3 < K_4$$

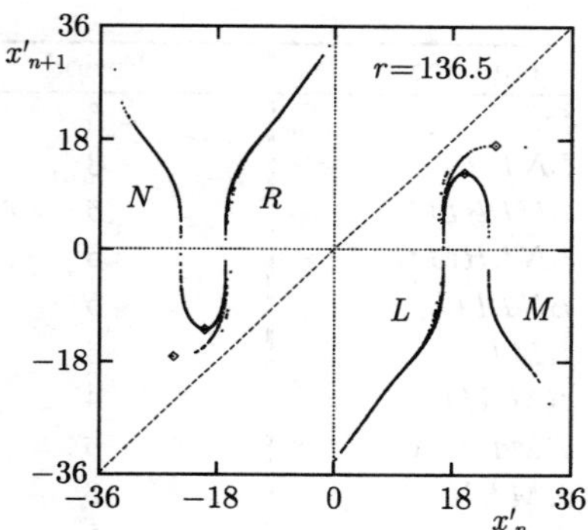

Figure 6.15 The swapped return map at $r = 136.5$

according to ordering rules (3.112). From the admissibility conditions (3.115) it follows that if two $K_- < K_+$ are both compatible with H, then any symbolic sequence admissible under (K_+, H) remains admissible under (K_-, H) but not the other way around. In our case, (K_4, H) puts the most severe restriction on admissibility while (K_1, H) provides the weakest condition. We start with the compatible kneading pair

$$\begin{aligned} K_1 &= RRLRLLRLNRLLRLN\cdots, \\ H &= MNRRLLNLRRLRMLR\cdots. \end{aligned} \tag{6.24}$$

We produce all periodic symbolic sequences admissible under (K_1, H) up to period 6 using the procedure described in Section 3.8.2. The results are listed in Table 6.6. Only shift-minimal sequences with respect to N and R are given. Their mirror images, i.e., shift-maximal sequences ending with M or L, are also admissible. There are in total 46 periods in Table 6.6, where a C stands for both N and R. The kneading pair (K_2, H) forbids 2 from the 46 periods. The two pairs (K_3, H) and (K_4, H) lying on the main sheet of the attractor have the same effect on short periodic orbits. They reduce the allowed periods to 20, keeping those from LC to $RLRMLC$ in Table 6.6. The actual number of unstable periodic orbits up to period 6 may be less than 46, but more than 20. A genuine 2D symbolic dynamics analysis is needed to clarify the situation.

Table 6.6 Admissible periodic sequences up to period 6 under the kneading pair (K_1, H) at $r = 136.5$. Only the non-repeating shift-minimal strings with respect to N or R are given. An asterisk marks those forbidden by 2D tangencies (see text and Table 6.7)

Period	Sequence	Period	Sequence
2	LC	6	$RLRRMR$
4	$LNLR$	3	RLC^*
6	$LNLRLC$	6	$RLNRLR^*$
6	$LNLRMR$	6	$RLNRMR$
5	$LNLLC$	5	$RLNLC^*$
3	RMR	6	$RLNLLC^*$
5	$RMRLC$	4	$RRMR$
6	$RMLNLC$	6	$RRMRLC^*$
4	$RMLN$	6	$RRMRMR$
4	$RLLC$	5	$RRMLC^*$
6	$RLLNLC$	6	$RRMLLN^*$
6	$RLRMLC$	6	$RRLLLC^*$
6	$RLRLLC^*$	5	$RRLLC^*$
5	$RLRLC^*$	6	$RRLRMC$

2D symbolic dynamics analysis at $r = 136.5$

In order to visualize the admissibility conditions imposed by a tangency between FCF and BCF in the 2D phase plane we need metric representations both for the forward and backward symbolic sequences. The metric representation for the forward sequences remains the same as defined by (3.116).

The partition of phase plane shown in Fig. 6.14 leads to a different ordering rule for the backward symbolic sequences. Namely, we have

$$L\bullet < D\bullet < M\bullet < B\bullet < N\bullet < C\bullet < R\bullet,$$

with the parity of symbols unchanged. (The unchanged parity is related to the positiveness of the Jacobian for the flow.) The ordering rule for backward sequences may be written as

$$\cdots LE\bullet < \cdots ME\bullet < \cdots NE\bullet < \cdots RE\bullet,$$
$$\cdots LO\bullet > \cdots MO\bullet > \cdots NO\bullet > \cdots RO\bullet,$$

where E (O) is a finite string containing an *even* (*odd*) number of M and N. From the ordering rule it follows that the maximal sequence is $R^\infty\bullet$ and the minimal is $L^\infty\bullet$.

The metric representation of forward symbolic sequences of the Lorenz-Sparrow map has been given in Section 3.8.4, see Eq. (3.116). In order to introduce a metric representation for backward symbolic sequences, we associate each backward sequence $\cdots s_{\bar{m}} \cdots s_{\bar{2}} s_{\bar{1}}\bullet$ with a real number β:

$$\beta = \sum_{i=1}^{\infty} \nu_i 4^{-i},$$

where

$$\nu_i = \begin{cases} 0, \\ 1, \\ 2, \\ 3, \end{cases} \text{for} \quad s_{\bar{i}} = \begin{cases} L, \\ M, \\ N, \\ R \end{cases} \text{and} \prod_{j=1}^{i-1} \epsilon_{\bar{j}} = 1,$$

and

$$\nu_i = \begin{cases} 3, \\ 2, \\ 1, \\ 0 \end{cases} \text{for} \quad s_{\bar{i}} = \begin{cases} L, \\ M, \\ N, \\ R \end{cases} \text{and} \prod_{j=1}^{i-1} \epsilon_{\bar{j}} = -1.$$

According to the definition we have

$$\beta(L^\infty\bullet) = 0, \quad \beta(D_\pm\bullet) = 1/4, \quad \beta(B_\pm\bullet) = 1/2,$$
$$\beta(C_\pm\bullet) = 3/4, \quad \beta(R^\infty\bullet) = 1.$$

In terms of the two metric representations a bi-infinite symbolic sequence with the present dot specified corresponds to a point in the unit square spanned by α of the forward sequence and β of the backward sequence. This unit square is the symbolic plane, where forward and backward foliations become vertical and horizontal lines, respectively. The symbolic plane is an image of the whole phase plane under the given dynamics. Regions in the phase plane that have one and the same forward or backward sequence map into a vertical or horizontal line in the symbolic plane. The symbolic plane should not be confused with the $\alpha(H)$-$\alpha(K)$ plane (Fig. 6.13), which is the metric representation of the kneading plane, i.e., the parameter plane of a 1D map.

As long as foliations are well ordered, a tangency on a partition line puts a restriction on allowed symbolic sequences. Suppose that there is a tangency

$QC{\bullet}K$ on the partition line $C\bullet$. The rectangle enclosed by the lines $\beta(QR\bullet)$, $\beta(QN\bullet)$, $\alpha(\bullet K)$, and $\alpha((MN)^\infty) = 0$ forms a forbidden zone (FZ) in the symbolic plane. In the symbolic plane a forbidden sequence corresponds to a point inside the FZ of $QC \bullet K$. A tangency may define some allowed zones as well. However, in order to confirm the admissibility of a sequence all of its shifts must fall in the allowed zones, while one point in the FZ is enough to exclude a sequence. This "all or none" alternative tells us that it is easier to exclude a sequence by a single tangency than to confirm it. Similarly, a tangency $\overline{Q}D \bullet \overline{K}$ on the partition line $D\bullet$ determines another FZ, symmetrically located to the FZ mentioned above. Due to the anti-symmetry of the map one may confine oneself to the first FZ and to shift-minimal sequences ending with N and R only when dealing with finite periodic sequences. The union of FZs from all possible tangencies forms a fundamental forbidden zone (FFZ) in the α-β symbolic plane. A necessary and sufficient condition for a sequence to be allowed consists in that all of its shifts do not fall in the FFZ. Usually, a finite number of tangencies may produce a fairly good contour of the FFZ for checking the admissibility of finite sequences. In Fig. 6.16 we have drawn a symbolic plane with 60 000 points representing real orbits generated from the Poincaré map at $r = 136.5$ together with a FFZ, outlined by the four tangencies (6.23). The other kneading sequence H in the 1D Lorenz-Sparrow map bounds the range of the 1D attractor. In the 2D Poincaré map the sequence H corresponds to the forward foliation which intersects with the attractor and bounds the subsequences following an R. In the symbolic plane the rectangle formed by $\alpha = \alpha(H)$, $\alpha = 1$, $\beta = 0.5$, and $\beta = 1$ determines the forbidden zone caused by H. It is shown in Fig. 6.16 by dashed lines. Indeed, the FFZ contains no point of allowed sequences.

In order to check the admissibility of a period n sequence one calculates n points in the symbolic plane by taking the cyclic shifts of the non-repeating string. All symbolic sequences listed in Table 6.6 have been checked in this way and 20 out of 46 words are forbidden, in fact, by T_3. This means that among the 26 sequences forbidden by K_4 in a 1D analysis actually 6 are allowed in 2D. We list all admissible periodic sequences of length 6 and less in Table 6.7. The 6 words at the bottom of the table are those forbidden by 1D but allowed in 2D. All the unstable periodic orbits listed in Table 6.7 have been located with high precision in the Lorenz equations. The knowledge of symbolic names and the ordering rule significantly facilitates the numerical work. The coordinates

(x, y) of the first symbol of each sequence are also given in Table 6.7.

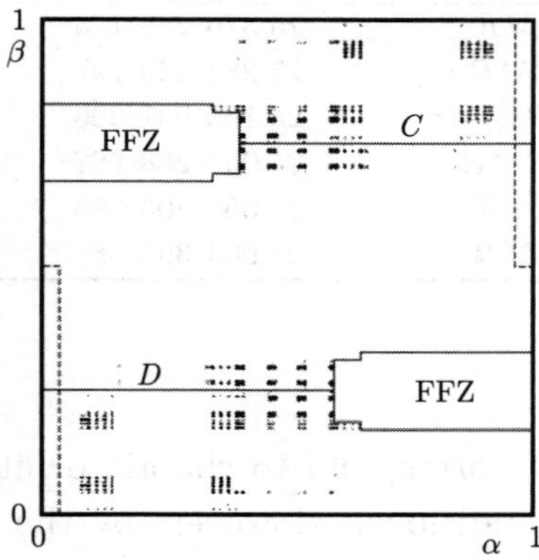

Firure 6.16 The symbolic plane at $r = 136.5$. A total of 60 000 points representing real orbits are drawn together with the FFZ outlined by the 4 tangencies and the forbidden zone caused by H

Table 6.7 Location of admissible periodic orbits left from Table 6.6 by 2D analysis. The coordinates (x, y) are that of the first symbol in a sequence

Period	Sequence	x	y
2	*LR*	−26.789 945 953	−51.732 394 996
2	*LN*	−33.741 639 204	−79.398 248 620
4	*LNLR*	−34.969 308 137	−84.807 257 714
6	*LNLRLN*	−34.995 509 382	−84.923 968 314
6	*LNLRLR*	−35.378 366 481	−86.639 695 512
6	*LNLRMR*	−36.614 469 777	−92.269 654 794
5	*LNLLN*	−36.694 480 374	−92.638 862 207
5	*LNLLR*	−37.362 562 975	−95.744 924 312
3	*RMR*	36.628 892 834	92.335 783 415
5	*RMRLN*	36.548 092 868	91.963 380 870
5	*RMRLR*	35.927 763 416	89.123 769 188
6	*RMLNLR*	33.541 019 900	78.514 904 719
6	*RMLNLN*	33.465 475 168	78.187 866 239
4	*RMLN*	33.432 729 468	78.045 902 429
4	*RLLR*	29.500 017 415	61.545 331 390
4	*RLLN*	28.800 493 901	58.709 002 527
6	*RLLNLN*	28.566 126 025	57.759 480 656
6	*RLLNLR*	28.548 686 604	57.682 625 650
6	*RLRMLN*	28.310 187 611	56.723 089 485
6	*RLRMLR*	28.299 181 162	56.676 646 246

Continued

Period	Sequence	x	y
6	$RLRRMR$	26.376 239 173	48.942 140 325
6	$RLNRMR$	25.282 520 197	44.566 367 330
4	$RRMR$	25.163 031 306	44.088 645 613
6	$RRMRMR$	25.047 268 287	43.625 877 472
6	$RRLRMN$	24.055 406 683	39.708 285 078
6	$RRLRMR$	24.064 390 385	39.704 320 864

Chaotic orbits

Symbolic sequences that correspond to chaotic orbits also obey the ordering rule and admissibility conditions. However, by the very definition these sequences cannot be exhaustively enumerated. Nevertheless, it is possible to show the existence of some chaotic symbolic sequences in a constructive way.

We first state a proposition similar to the one mentioned in the paragraph before Eq. (6.24). If (K, H_-) and (K, H_+) are two compatible kneading pairs with $H_- < H_+$, then all admissible sequences under (K, H_-) remain so under (K, H_+), but not the other way around. It is seen from Table 6.5 that from $r = 120.0$ to 191.0 all K starts with R while the minimal H starts with MRR. Let $K = R\cdots$ and $H = MRR\cdots$. It is easy to check that any sequence made of the two segments LR and $LNLR$ satisfies the admissibility conditions (3.115). Therefore, a random combination of these segments is an admissible sequence in the 1D Lorenz-Sparrow map. A similar analysis can be carried out in 2D using the above tangencies. For example, at $r = 136.5$ any combination of the two segments remains an admissible sequence in 2D based on the tangency T_4. Therefore, we have indicated the structure of a class of chaotic orbits.

6.3.5 Stable Periodic Orbits

So far we have only considered unstable periodic orbits at fixed r. A good symbolic dynamics should be capable to deal with stable orbits as well. One can generate all compatible kneading pairs of the Lorenz-Sparrow map up to a certain length by using the method described in Section 3.8.2. Although there is no way to tell the precise parameter where a given periodic orbit will become stable, the symbolic sequence does obey the ordering rule and may be located on the r-axis by using a bisection method. Another way of finding a stable period is to follow the unstable orbit of the same name at varying parameters by using a periodic orbit tracking program. Anyway, many periodic windows have been known before or encountered during the present study. We collect them in Tables 6.8 and 6.9. Before making further remarks on these tables, we indicate how to find symbolic sequences for stable periods.

Table 6.8 Some stable periodic orbits in the Lorenz equations associated with the main sheet of the dynamical foliations (to be continued on next page)

Period	Sequence	r
2	DC	315 – 10 000
2	LN	229.42 – 314
4	$LNLC$	218.3 – 229.42
8	$LNLRLNLC$	216.0 – 218.3
16	$LNLRLNLNLNLRLNLC$	215.5 – 216.0
24	$LNLRLNLNLNLRLNLRLNLRLNLC$	215.07 – 215.08
12	$LNLRLNLNLNLC$	213.99 – 214.06
6	$LNLRLC$	209.06 – 209.45
12	$LNLRLRLNLRLC$	208.98
10	$LNLRLRLNLC$	207.106 – 207.12
8	$LNLRLRLC$	205.486 – 206.528
10	$LNLRLRLRLC$	204.116 – 204.123
12	$LNLRLRLRLRLC$	203.537
14	$LNLRLRLRLRLRLC$	203.273 5
16	$LNLRLRLRLRLRLRLC$	203.151 1
18	$LNLRLRLRLRLRLRLRLC$	203.093 332
30	$LNLRLRLRLRLRLRLRLRLRLRLRLRLRLC$	203.041 203 679 65
14	$LNLRLRDRMRLRLC$	200.638 – 200.665
10	$LNLRDRMRLC$	198.97 – 198.99
5	$LNLLC$	195.576
10	$LNLLRLNLLC$	195.564
20	$LNLLRLNLLNLNLLRLNLLC$	195.561
5	$RMRLC$	190.80 – 190.81
10	$RMRLRRMRLC$	190.79
20	$RMRLRRMRLNRMRLRRMRLC$	190.785
7	$RMRLRLC$	189.559 – 189.561
9	$RMRLRLRLC$	188.863 – 188.865
16	$RMRLRLRDLNLRLRLC$	187.248 – 187.25
12	$RMRLRDLNLRLC$	185.74 – 185.80
8	$RMRDLNLC$	181.12 – 181.65
10	$RMRMRMLNLC$	178.074 5
12	$RMRMRDLNLNLC$	177.78 – 177.81
6	$RMLNLC$	172.758 – 172.797

Continued

Period	Sequence	r
12	*RMLNLNRMLNLC*	172.74
16	*RMLNRMRDLNRMLNLC*	169.902
10	*RMLNRMLNLC*	168.58
4	*RDLC*	162.1 – 166.07
4	*RLLC*	154.4 – 162.0
4	*RLLN*	148.2 – 154.4
8	*RLLNRLLC*	147.4 – 147.8
16	*RLLNRLLRRLLNRLLC*	147
12	*RLLNRLLRRLLC*	145.94 – 146
20	*RLLNRLLRRDLRRMLRRLLC*	144.35 – 144.38
12	*RLLNRDLRRMLC*	143.322 – 143.442
6	*RLLNLC*	141.247 – 141.249
12	*RLLNLRRLLNLC*	141.23
6	*RLRMLC*	136.800 – 136.819
12	*RLRMLRRLRMLC*	136.788 – 136.799
10	*RLRMLRRLLC*	136.210 – 136.211 2
16	*RLRMLRRDLRLNRLLC*	135.465 – 135.485
8	*RLRDLRLC*	132.06 – 133.2
16	*RLRLLRRDLRLRRLLC*	129.127 – 129.148
6	*RLRLLC*	126.455 – 126.52
12	*RLRLLNRLRLLC*	126.415 – 126.454
24	*RLRLLNRLRLLRRLRLLNRLRLLC*	126.41
12	*RLRLRDLRLRLC*	123.56 – 123.63
8	*RLRLRLLC*	121.687 – 121.689
7	*RLRLRLC*	118.128 – 118.134
14	*RLRLRRDLRLRLLC*	116.91 – 116.925
5	*RLRLC*	113.916 – 114.01
10	*RLRLNRLRLC*	113.861 – 113.915
10	*RLRRDLRLLC*	110.57 – 110.70
9	*RLRRLLRLC*	108.977 8
7	*RLRRLLC*	107.618 – 107.625
14	*RLRRLRDLRLLRLC*	106.746 – 106.757
8	*RLRRLRLC*	104.185
16	*RLRRLRRDLRLLRLLC*	103.632 – 103.636
3	*RLC*	99.79 – 100.795
6	*RLNRLC*	99.629 – 99.78
12	*RLNRLRRLNRLC*	99.57
9	*RLNRLRRLC*	99.275 – 99.285
12	*RRMLRDLLNRLC*	94.542 – 94.554
6	*RRDLLC*	92.51 – 93.20
6	*RRLLLC*	92.155 – 92.5
12	*RRLLLNRRLLLC*	92.066 – 92.154

Continued

Period	Sequence	r
12	*RRLLRDLLRRLC*	90.163 – 90.20
8	*RRLLRLLC*	88.368
7	*RRLLRLC*	86.402
14	*RRLLRRDLLRRLLC*	85.986 – 85.987
8	*RRLLRRLC*	84.336 5
5	*RRLLC*	83.36 – 83.39
10	*RRLLNRRLLC*	83.35
10	*RRLRDLLRLC*	82.040 – 82.095
8	*RRLRLLLC*	81.317
6	*RRLRLC*	76.818 – 76.822
12	*RRLRLNRRLRLC*	76.815
12	*RRLRRDLLRLLC*	76.310 – 76.713
8	*RRLRRLLC*	75.140 5
7	*RRLRRLC*	73.712
14	*RRLRRRDLLRLLLC*	73.457
4	*RRLC*	71.452 – 71.52
8	*RRLNRRLC*	71.410 – 71.451
8	*RRRDLLLC*	69.724 – 69.839
8	*RRRLLRLC*	66.204 6
9	*RRRLLRRLC*	65.502 5
6	*RRRLLC*	64.895 – 64.898
12	*RRRLLNRRRLLC*	64.894 6
24	*RRRLLNRRRLLRRRRLLNRRRLLC*	64.893
12	*RRRLRDLLLRLC*	64.572 – 64.574
7	*RRRLRLC*	62.069
14	*RRRLRRDLLLRLLC*	61.928
9	*RRRLRRLLC*	61.314 97
8	*RRRLRRLC*	60.654
5	*RRRLC*	59.242 – 59.255
10	*RRRLNRRRLC*	59.24
10	*RRRRDLLLLC*	58.700 – 58.715
9	*RRRRLLLLC*	58.076 3
9	*RRRRLLRLC*	56.533 15
7	*RRRRLLC*	55.787
14	*RRRRLRDLLLLRLC*	55.675
6	*RRRRLC*	52.457 – 52.459
12	*RRRRLNRRRRLC*	52.455 – 52.456
12	*RRRRRDLLLLLC*	52.245 – 52.248
8	*RRRRRLLC*	50.303 8 – 50.324 0
7	*RRRRRLC*	48.118 8 – 48.119 4
14	*RRRRRLNRRRRRLC*	48.118 4 – 48.118 7
14	*RRRRRRDLLLLLLC*	48.027

Table 6.9 Some stable periodic orbits in the Lorenz equations associated with secondary sheets of the dynamical foliations

Period	Sequence	r
3	RMN	328.083 8
3	RMC	327.58 – 327.88
6	$RMRRMC$	327.3 – 327.5
12	$RMRRMNRMRRMC$	327.26
24	$RMRRMNRMRRMRRMRRMNRMRRMC$	327.2
10	$RMRRDLNLLC$	191.982 – 191.985
20	$RMRRMLNLLRRMRRMLNLLC$	191.979 5
6	$RLRRMC$	183.043 5
12	$RLRRMRRLRRMC$	183.043 4
24	$RLRRMRRLRRMNRLRRMRRLRRMC$	183.043 38
6	$RLNRMC$	168.249 2
12	$RLNRMNRLNRMC$	168.249 189
4	$RRMR$	162.138 1
8	$RRMRRRMC$	162.138 06
16	$RRMRRRMNRRMRRRMC$	162.138 04
6	$RRMRMC$	157.671 066
12	$RRMRMNRRMRMC$	157.671 065 6
24	$RRMRMNRRMRMRRRMRMNRRMRMC$	157.671 065 4
6	$RRLRMC$	139.923 843 3
12	$RRLRMRRRLRMC$	139.923 843 0
24	$RRLRMRRRLRMNRRLRMRRRLRMC$	139.923 842 8

When there exists a periodic window in some parameter range, one cannot extract a return map of the interval from a small number of orbital points so there may be ambiguity in assigning symbols to numerically determined orbital points. Nonetheless, there are at least two ways to circumvent the difficulty. First, one can take a nearby parameter where the system exhibits chaotic behavior and superimpose the periodic points on the chaotic attractor. In most cases the (K, H) pair calculated from the chaotic attractor may be used to generate unstable periods coexisting with the stable period. Second, one can start with a set of initial points and keep as many as possible transient points before the motion settles down to the final stable periodic regime. (A few points near the randomly chosen initial points have to be dropped anyway.) From the set of transient points one can construct return maps as before. Both methods work well for short enough periods, especially in narrow windows.

Fig. 6.17 shows a stable period 6 orbit $RLRRMR$ at $r = 183.0435$ as diamonds. The background figure looks much like a chaotic attractor, but it is actually a collection of its own transient points. The last symbol X in $RLRRMX$ corresponds to a point $(20.945\ 669, 45.391\ 029)$ lying to the right of a tangency at $(20.935\ 971, 45.393\ 162)$. Therefore, it acquires the symbol R, not N. This example shows once more how the x-parameterization helps in accurate assignment of symbols.

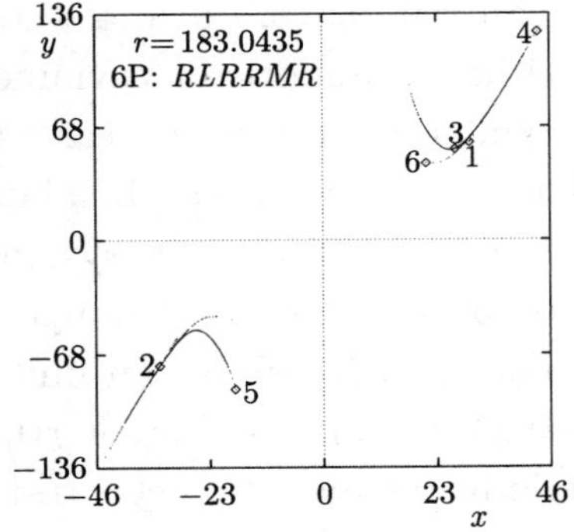

Figure 6.17 A stable period 6 orbit at $r = 183.0435$ on the background of its own transient points

Absolute nomenclature of periodic orbits

In a periodically driven system such as the forced Brusselator the period of the external force serves as a unit to measure the periods in the system. This is not the case in autonomous systems like the Lorenz equations, since the fundamental frequency, if any, drifts with the varying parameter. No wonder in the several hundred papers on the Lorenz model no authors had ever described a period as, say, period 5 until a calibration curve of the fundamental frequency was obtained by extensive Fourier analysis of the Lorenz system at various r values (Ding, Hao and Hao [1985]). It is remarkable that the absolute periods obtained from the power spectrum analysis coincide with those determined from symbolic dynamics. As a consequence, we know now that the window first studied by Manneville and Pomeau [1979] starts with a period 4, the period-doubling cascade first discovered by Franceschini [1980] lives in a period 3 window, etc. Moreover, we know their symbolic names and their location in the overall systematics of all stable periods. In Tables 6.8 and 6.9 many period-doubling cascades are indicated. Guided by ordering rule and admissibility conditions of symbolic dynamics it is easy to locate more periods if the necessity arises.

Symmetry breakings and restorations

In a dynamical system with discrete symmetry the phenomenon of symmetry breaking and symmetry restoration comes into play. In the Lorenz equations periodic orbits are either symmetric or asymmetric with respect to the transformation (6.19); asymmetric orbits appear in symmetrically located pairs. Some essential features of symmetry breaking and restoration have been known. For example, in many dissipative systems when viewed in properly chosen direction of the paramter space symmetry breakings usually precede period-doubling—no symmetric orbits can undergo period-doubling directly without the symmetry being broken first. Furthermore, while symmetry breakings take place in periodic regime, symmetry restorations occur in chaotic regime. As we have seen in Section 3.3 all these features may be explained by using symbolic dynamics. Although the analysis performed in Section 3.3 was based on the anti-symmetric cubic map, it is applicable to the Lorenz equations via the Lorenz-Sparrow map.

A doubly "super-stable" symmetric orbit must be of the form $\Sigma D\overline{\Sigma}C$, therefore its period is even and only even periods of this special form may undergo symmetry breaking. The shortest such orbit is DC. To keep the symmetry when extending this super-stable period into a window, one must change D and C in a symmetric fashion, i.e., either replacing D by M and C by N or replacing D by L and C by R at the same time, see (3.111) and (3.113). Thus we get a window (MN, DC, LR). (MN does not appear in the Lorenz equations while LR persists to very large r.) This is indeed a symmetric window, as the transformation (3.113) brings it back after cyclic permutations. Moreover, this window has a signature $(+, 0, +)$ according to the parity of the symbols. (We assign a null parity to C and D.) It cannot undergo period-doubling as the latter requires a $(+, 0, -)$ signature. By continuity LR extends to an asymmetric window (LR, LC, LN) with signature $(+, 0, -)$ allowing for period-doubling. It is an asymmetric window as its mirror image (RL, RD, RM) is different. They represent the two symmetrically located asymmetric period 2 orbits. The word $(LR)^\infty$ describes both the second half of the symmetric window and the first half of the asymmetric window. The precise symmetry breaking point, however, depends on the mapping function and cannot be told by symbolic dynamics.

In general, a word λ^∞ representing the second half of the symmetric window continues to become the first half of the asymmetric window $(\lambda, \tau C, \rho)$. The latter develops into a period-doubling cascade described by the general rule of symbolic dynamics. The cascade accumulates and turns into a period-halving cascade of chaotic bands. The whole structure is asymmetric. Finally, the chaotic attractor collides with the symmetric unstable periodic orbit λ^∞ and takes back the symmetry to become a symmetric chaotic attractor. This is a symmetry restoration crisis, taking place at the limit described by the eventually periodic kneading sequence $\rho\lambda^\infty$. In our period 2 example this happens at $LN(LR)^\infty$. In Table 6.8 this limit has been traced by $LN(LR)^{n-2}LC$ up to $n = 15$. The only period 30 sequence in Table 6.8 indicates closely the location of the symmetry restoration point corresponding to the asymmetric period 2^n cascade. All other symmetric orbits in Table 6.8 are put conditionally in the form $\Sigma D\overline{\Sigma}C$ as the parameters given can hardly match a doubly super-stable orbit. For example, the three consecutive period 4 from $r = 148.2$ to 166.07 in Table 6.8 actually mean:

$$(RMLN, RDLC, RLLR) \to (RLLR, RLLC, RLLN),$$
$$(+, 0, +) \to (+, 0, -),$$

followed by an asymmetric period-doubling cascade. The symmetry restoration takes place at $RLLN(RLLR)^\infty$ whose parameter may easily be estimated.

Symbolic dynamics also yields the number of periodic orbits that are capable to undergo symmetry breaking. In the parameter range of the Lorenz equations there are one period-2, one period-4, and two period-6 of such orbits, all listed in Table 6.8.

"2D" orbits and co-existing attractors

Now we return to Table 6.8 which is a list of stable periods associated with the main sheet of the dynamical foliations. When there is an attracting stable period these sheets are not readily seen. They still resemble the main sheets shown in Fig. 6.12 or 6.15. In fact, one may merge all the kneading sequences

K_i listed in Table 6.5 with Table 6.8 according to their r-values. They all fit well into the overall ordering. The ordered list of stable periods plus that of kneading sequence K_i determined from the main sheets of the chaotic attractor makes an analogue to the MSS-sequence in the symbolic dynamics of unimodal maps, cf. Section 2.4.3.

It is an amazing fact the 1D Lorenz-Sparrow map captures so much of the real Lorenz equations. Are there any manifestly 2D features after all? As long as stable periods are concerned, some orbits showing 2D features are collected in Table 6.9. As a rule, these are very narrow windows living on some secondary sheets of the dynamical foliations. It is remarkable that they may be named according to the same rule of the Lorenz-Sparrow map; they form a different ordered list as compared with Table 6.8. Among them there are a few orbits co-existing with a periodic orbit from the main sheet especially when the latter forms a wide window. For instance, RMN and LR coexist in the vicinity of $r = 328.0838$, see Fig. 6.18 (a). This period 3 orbit develops a period-doubling cascade, traced to period 24 in Table 6.9. The period 2 orbit $(LR)^\infty$ may even be seen co-existing with a tiny chaotic attractor from the same 3^n cascade at $r = 327.167\ 55$, see Fig. 6.18 (b). Other cases given in Table 6.9 include $RRMR$ and $RRMRMC$ as well as their period-doubled regimes, both coexisting with the symmetric period 4 orbit $RLLR$ below $r = 162.1381$ and $157.671\ 066$, respectively. In addition, there are orbits involving both sheets. We attribute all these orbits to the manifestation of 2D features.

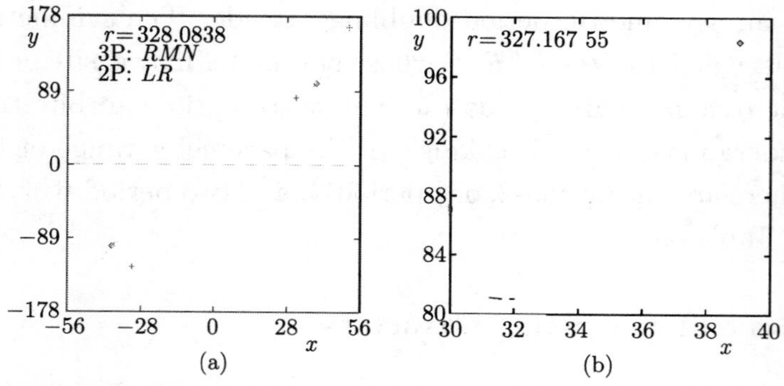

Figure 6.18 Attractors coexisting with the period 2 LR orbit (diamonds). (a) A period-3 (crosses) orbit at $r = 328.0838$. (b) A chaotic attractor at $r = 327.167\ 55$ (only half of the symmetric trajectories are shown)

6.3.6 Concluding Remarks

We have seen that fairly detailed global knowledge of the Lorenz equations in the phase space as well as in the parameter space may be obtained by numerical work under the guidance of symbolic dynamics. Two-dimensional symbolic dynamics of the Poincaré map may provide, in principle, a complete list of stable and unstable periodic orbits up to a given length and a partial description of some chaotic orbits. However, 1D symbolic dynamics extracted from the 2D Poincaré map is simpler and instructive. The 2D features seen in the Poincaré and first return maps may safely be circumvented by shrinking along the FCFs in a 1D study which deals with forward symbolic sequences only. Whether 1D or 2D symbolic dynamics is needed and how many tangencies to keep in a 2D study is a matter of precision. Even in a seemingly "pure" one-dimensional situation 2D features may need to be taken into account when it comes to cope with very long symbolic sequences. This has to be decided in practice. Therefore, what has been described so far remains a physicist's approach for the time being. However, it may provide food for thought to mathematicians.

To this end we would like to make a few more remarks. S. Smale has summarized ten great and unsolved problems in dynamical system theory (Smale [1991]), the first problem being: is the dynamics of the Lorenz equations (6.18), with $\sigma = 10$, $r = 28$, and $b = 8/3$, described by the "geometric Lorenz attractor" of Williams, Guckenheimer and Yorke? The results of the preceding sections push the problem into much wider, obviously non-hyperbolic, range of parameter r. It seems that there is a good hope to apply, rigorously or approximately, the method of symbolic dynamics to some non-hyperbolic systems and to add this method to the arsenal of practitioners in nonlinear dynamics.

Guckenheimer and Holmes [B1983] indicated in their book that non-hyperbolic limit sets are often encountered in examples of practical importance: "One would like to study their symbolic dynamics, but this has not been done in a satisfactory or systematic manner, except in the special case of mappings defined on the line". The results on the Lorenz equations, though not being claimed to be satisfactory, do present a more or less systematic attempt in that direction.

6.4 Summary of Other ODE Systems

Numerical investigation of many ordinary differential equations (Rössler [1976]; Hao and Zhang [1982]; Wang, Hao and Zhang [1983]; Ueda [1991]) as well as some partial differential equations (Knobloch, Moore, Toomre, and Weiss [1986]; Proctor and Weiss [1990]; Moore and Weiss [1990]), has revealed the similarity of the bifurcation structure with that of low-dimensional mappings. This has stimulated the application of symbolic dynamics of one-dimensional maps to the study of differential equations. The physical reason for low-dimensional symbolic dynamics working so well is of course the presence of dissipation, which causes the shrinking of phase space volume. The symbolic dynamics analysis of the Lorenz equations also shows the importance of symmetry in constructing symbolic dynamics for a physical system. Another apparent demerit of the symbolic dynamics approach for the time being is its restriction to systems with one stretching direction, i.e., one positive Lyapunov exponent.

In addition, there are some technical subtleties in carrying out the program that we could not touch upon due to limited space, but there is also a good hope to automate the process and to apply it to more systems of physical importance. We summarize briefly the results of symbolic dynamics analysis of some other systems of ordinary differential equations.

6.4.1 The Driven Two-Well Duffing Equation

The driven two-well Duffing equation reads

$$\frac{d^2x}{dt^2} + k\frac{dx}{dt} + \alpha x + x^3 = f\cos t, \tag{6.25}$$

where k, α and f are the damping coefficient, potential parameter and forcing amplitude, respectively. The equation exhibits rich bifurcation behavior with respect to varying parameters (Sato, Sano and Sawada [1983]; Solari and Gilmore [1988b]; Hu and Hao [1990], Ueda [1991]). The potential function of Eq. (6.25)

$$V(x) = \frac{x^4}{4} + \frac{\alpha x^2}{2}$$

consists of two wells when α is negative, otherwise it has a single well. Denoting $y = dx/dt$, Eq. (6.25) can be rewritten in the standard form of a system of first order ODEs:

$$\frac{dx}{dt} = y,$$

$$\frac{dy}{dt} = -kx - \alpha x - x^3 + f \cos t.$$

The equation possesses an anti-symmetry with respect to a time shift in half of the driving period which has been set to 2π. Namely, by defining a transformation $\mathcal{T}$:

$$\mathcal{T}(x, y, t) = (-x, -y, t + \pi) \tag{6.26}$$

with the obvious property $\mathcal{T}^2 = 1$, we see that if $X = \{x(t), y(t)\}$ is a solution, so does $\mathcal{T}X = \{-x(t+\pi), -y(t+\pi)\}$. The transformation $\mathcal{T}$ generates an anti-symmetry between any two Poincaré sections with a phase shift of a half period π. This makes its symbolic dynamics closer to that of the autonomous Lorenz system. In fact, it was associated with a symbolic dynamics of four letters with an interchange transformation both in 1D and 2D maps obtained from the Poincaré sections (Xie, Zheng and Hao [1995]). A subtle point in constructing the symbolic dynamics is one has to superimpose two Poincaré sections with a phase difference of π in order to determine the partition lines in accordance with the anti-symmetry.

In weak forcing regime the Duffing equation has a closed bifurcation region in the parameter plane inside which there are many different subharmonic periodic windows. One-dimensional symbolic dynamics derived from the Poincaré section is capable to label and order all the 45 different periodic windows observed along the $f = 1$ line of the parameter plane. Further 2D analysis by using only two tangent points between the dynamical foliations yields both a set of admissible periodic orbits and a set of forbidden orbits. Referring to the original work (Xie, Zheng and Hao [1995]) we skip the technical details.

6.4.2 The NMR-Laser Model

The chaotic behavior in an NMR-laser has been described by an extended Bloch-type laser (EBL) model using a system of ODEs (Flepp *et al.* [1991]; Finardi *et al.* [1992]):

$$\dot{x} = \sigma \left[y - \frac{x}{1 + A\cos(\omega t)} \right],$$

$$\dot{y}= -y(1+ay)+rx-xz,$$
$$\dot{z}= -bz+xy. \tag{6.27}$$

The meaning of the parameters σ, A, ω, a, r, and b may be found in the original papers. A binary generating partition was approximately obtained from a number of unstable periodic orbits up to period 9. The guideline to draw the partition line was uniqueness in assigning symbolic names to different unstable periods. However, no ordering rule and admissibility condition may be formulated in this way.

A full symbolic dynamics analysis of the NMR-laser model was carried out by determining the partition line from 11 tangencies of backward and forward foliations in the Poincaré maps (Liu, Wu and Zheng [1996]). Both the forward and backward foliations are ordered as those in a Hénon map with a positive Jacobian. Based on the analogue with the Hénon type symbolic dynamics the admissibility conditions of symbolic sequences in the NMR-laser model may be formulated and used to produce a list of admissible periodic orbits up to a certain length. It turns out that some orbits previously considered as forbidden are in fact admissible. All the admissible periods are found numerically using the ordering of orbits. The structure of some chaotic orbits is also indicated. Skipping the details of this analysis, we mention only that in some parameter range the symbolic dynamics of the EBL model is of the same type as that of the periodically forced Brusselator (Liu and Zheng [1995a and b]).

7

Counting the Number of Periodic Orbits

In this Chapter we present the complete solution of a counting problem, namely, the number of different periodic orbits of a given length in any continuous map of the interval with multiple critical points. Partial results for maps with discontinuity will also be given. This counting problem has close relation with symbolic dynamics but it goes beyond the scope of the latter. Therefore, we will make the presentation more or less independent of the previous chapters.

In Section 7.1 we will first discuss the relevance of periodic orbits to chaotic dynamics, then summarize the general results and introduce a few notations to be used in the sequel.

A crucial point in understanding the general counting problem consists in solving the problem for a special class of one-parameter maps. This makes the content of Sections 7.2 and 7.3. Section 7.2 extends several counting methods from the unimodal case to continuous maps with multiple critical points. We emphasize that there are a few independent problems that lead to the same counting results. Some aspects of the counting problem tackled in Section 7.3 have not been extended to maps with multiple critical points yet.

Section 7.4 is devoted to the derivation of the general counting formula. It shows how to get the number of periodic orbits in any continuous map from that of the special class of one-parameter maps. This will be carried out in a simple graphical way.

Section 7.5 treats the counting problem for two maps with discontinuity, the gap map and the Lorenz-like map. We sum up the results in Section 7.6. The chapter ends with a brief discussion on the cycle expansion for topological entropy in Section 7.7.

7.1 Periodic versus Chaotic Regimes

In nonlinear dynamical systems which exhibit chaotic behavior, periodic orbits are in close relation with chaotic regime. On one hand, when a parameter is varied, the system may undergo a series of periodic events, usually stable subharmonics of a fundamental frequency, before arriving at a chaotic state. This is called a route or a scenario to chaos. By looking at the periodic events alone, one can anticipate much about the nature of the limiting chaotic regime; this is particularly true when the corresponding symbolic dynamics may be inferred from the periodic orbits.

On the other hand, chaotic orbits are "organized" around unstable periodic orbits (Gunaratne and Procaccia [1987]). Once in the chaotic regime, the attractor contains an infinite number of unstable periods. A chaotic orbit may be thought of as a random walk among various unstable periods for longer or shorter time intervals. In principle, these unstable periods may be extracted from experimental data, provided the sampled series is long enough.

Furthermore, the characterization of chaotic attractors may proceed at two levels. At the "macroscopic" level, global characteristics, such as dimensions, entropies, and Lyapunov exponents, are calculated from time average over a long enough trajectory or from ensemble average over the attractor. These characteristics, usually a few numbers, cannot tell much about the difference or similarity of various attractors.

A more definite description may be reached at the "microscopic" level, when the number of different types of unstable periods is extracted from the data. These numbers are topological invariants, and they do not change under continuous coordinate transformation. Sometimes, local expansion and contraction rates, i.e., eigenvalues of various unstable periods may also be obtained. These are metric invariants and may furnish a more detailed characterization of the attractor. P. Cvitanović [1988] advocates the idea that an experimental attractor should be represented by a list of periodic types and their eigenvalues.

Topologically, higher-dimensional dissipative chaotic systems may have the same kind of attractors as low-dimensional ones. Therefore, if we are able to count the number of periodic orbits in one-dimensional mappings, then we may use them as templates to compare with high-dimensional systems. In addition, the counting problem itself has many facets and is interesting on its own. It may be formulated and solved in many independent ways.

7.1.1 Stable Versus Unstable Periods in 1D Maps

In one-dimensional mappings, stable and unstable orbits are closely related. Sometimes, by counting the number of stable periodic orbits one can get the number of the unstable ones. Perhaps, the best way to explain the relation between stable and unstable periods in one-dimensional mappings is to recall the Sarkovskii ordering (Sarkovskii [1964]; Stefan [1977]) and the U-sequence of Metropolis, Stein and Stein [1973].

Sarkovskii first introduced a precedence relation "$\prec$" among various periods in one-dimensional mappings. If the existence of period p implies the existence of period q, then p is said to precede q: $p \prec q$. Then Sarkovskii re-ordered all integers as follows:

$$\begin{aligned} 3 \prec 5 \prec 7 \prec \cdots \prec 3 \times 2 \prec 5 \times 2 \prec 7 \times 2 \prec \cdots \\ \prec 3 \times 2^2 \prec 5 \times 2^2 \prec 7 \times 2^2 \prec \cdots \prec 2^3 \prec 2^2 \prec 2 \prec 1 \end{aligned} \tag{7.1}$$

(the Sarkovskii ordering), and proved that if in a certain continuous map of the interval there exists a period p, then all periods that follow p in the ordering (7.1) also exist. In particular, from the existence of a period 3 it follows that there must be orbits of all possible periods, since 3 precedes all numbers in the Sarkovskii ordering. This is partially what the Li-Yorke theorem says, although the theorem contains more (Li and Yorke [1975]).

The Sarkovskii ordering deals with periods in a continuous map at fixed parameter values; it is a statement on the existence of periods as numbers without indication on their location in the phase space. Neither does it care about the stability of the orbits. In fact, almost all the periods the Sarkovskii ordering deals with are unstable ones.

In 1973, Metropolis, Stein and Stein [1973] studied the systematics of periods in unimodal maps in terms of symbolic dynamics of two letters. They gave the rule to construct and to order all admissible superstable periodic words and called this ordering a U-sequence (U for universal), see our discussion of the MSS Table in Section 2.4.3. For example, there is only one period 2 word RC; one period 3 word RLC; two period 4 words $RLRC$, which is the period-doubled regime of RC, and $RLLC$; and three period 5 words $RLLRC$, $RLRRC$, $RLLLC$, etc., as listed in Table 2.2.

The U-sequence is best visualized in a bifurcation diagram, i.e., a phase space versus parameter space diagram in which the orbits are plotted against the parameter. A bifurcation diagram of the quadratic map has been shown in Fig. 2.5. As we have learned from Chapter 2, every periodic "window", seen in the diagram, is described by a word in the U-sequence. As MSS pointed out, the number of different words of the same period equals to the number of periodic sequences invariant under the discrete group $C_n \otimes S_2$, where C_n is the cyclic group of order n and S_2—the symmetric group of order 2. The U-sequence gives not only the number of stable periods, but also their ordering along the parameter axis. When the parameter is fixed, symbolic dynamics also gives the ordering of orbits (stable and unstable, finite and infinite) in the phase space, i.e., on the interval. In this sense, the U-sequence contains more detailed information on the periods, but it is proved for unimodal maps only, while the Sarkovskii ordering applies to general continuous mappings.

The relation between the Sarkovskii ordering and the U-sequence (or the generalization of U-sequence to maps with multiple critical points) is revealed if one plots not only the stable periods, but also the unstable ones in the bifurcation diagram. The Sarkovskii theorem works in the phase space; the U-sequence appears along the parameter axis. They are, so to speak, "orthogonal" and interrelated to each other. When one is interested in the number of unstable periods in an chaotic attractor, it is helpful to know the total number of stable periods which exist at least once in the parameter space. This is the problem we study in what follows.

7.1.2 Notations and Summary of Results

Consider a continuous map f of an interval I into itself:

$$f: I \to I, \tag{7.2}$$

or

$$x_{n+1} = f(\mu, x_n), \quad x \in I, \tag{7.3}$$

where μ stands for one or more parameters. The mapping function f may have m monotone branches, also called *laps* by Milnor and Thurston [1977, 1988]. The $m = 2$ case corresponds to the well-studied quadratic map. A general map with $m = 6$ is shown in Fig. 7.1 (a). For convenience we take the interval I to be $(0, 1)$ and to fix f at the two ends of the interval:

$$f(0) = 0, f(1) = \begin{cases} 0, & m \text{ even}, \\ 1, & m \text{ odd}. \end{cases} \tag{7.4}$$

A continuous map with m laps has $m - 1$ critical points, among which there are $[m/2]$ maxima and $[(m-1)/2]$ minima, where [ ] means taking the integer part. If necessary, we will denote the maximal points by C_i, $i = 1$ to $[m/2]$, and the minimal points—by D_i, $i = 1$ to $[(m-1)/2]$.

In the language of symbolic dynamics, the first iterate of each critical point leads to a *kneading sequence*, see Sections 2.3.3 and 3.1.1. A kneading sequence corresponds to a real number, calculable from a given map, for example,

$$K_{C_i} = f(C_i) = \sigma_0\sigma_1\sigma_3\cdots, \tag{7.5}$$

where σ_i are letters used in the corresponding symbolic dynamics, see Fig. 7.2 below. Kneading sequences as numbers are the most natural parameters of a map. Put in other words, a continuous map is best parameterized by its kneading sequences (Zheng and Hao [1990]; Xie and Hao [1993]).

There are many degenerated cases of a map. For instance, one or more kneading sequences may be fixed at their maximal or minimal value, as shown in Figs. 7.1 (b), (d), (e), (f), (h), and (i). Furthermore, two or more kneading sequences may change in unison, corresponding to only one parameter. In Figs. 7.1 (b), (c), (d), (g), (h), and (i), those critical points which change simultaneously are indicated by a dash line.

The degenerated case shown in Fig. 7.1 (b) is a one-parameter map, which turns out to be the key to the counting problem. We denote by $N_m(n)$ the total number of period n orbits in the one-parameter m-lap map of this type. The number $N_m(n)$ may be calculated or enumerated in many different ways by straightforward extension of results for $N_2(n)$. This is discussed in Section 7.2 below. The numbers $N_m(n)$ for $m = 2$ to 7 are listed in Table 7.1. The number of periods in all other cases may be expressed as combinations of $N_k(n)$ with $k \leqslant m$. Using various cases of the map with $m = 6$ shown in Fig. 7.1, we list some results:

1. The number of periods of the one-parameter map (e) (as well as (f) and other similar cases) is given by $N_6(n) - N_4(n)$. In a m-lap map it is given by $N_m(n) - N_{m-2}(n)$.

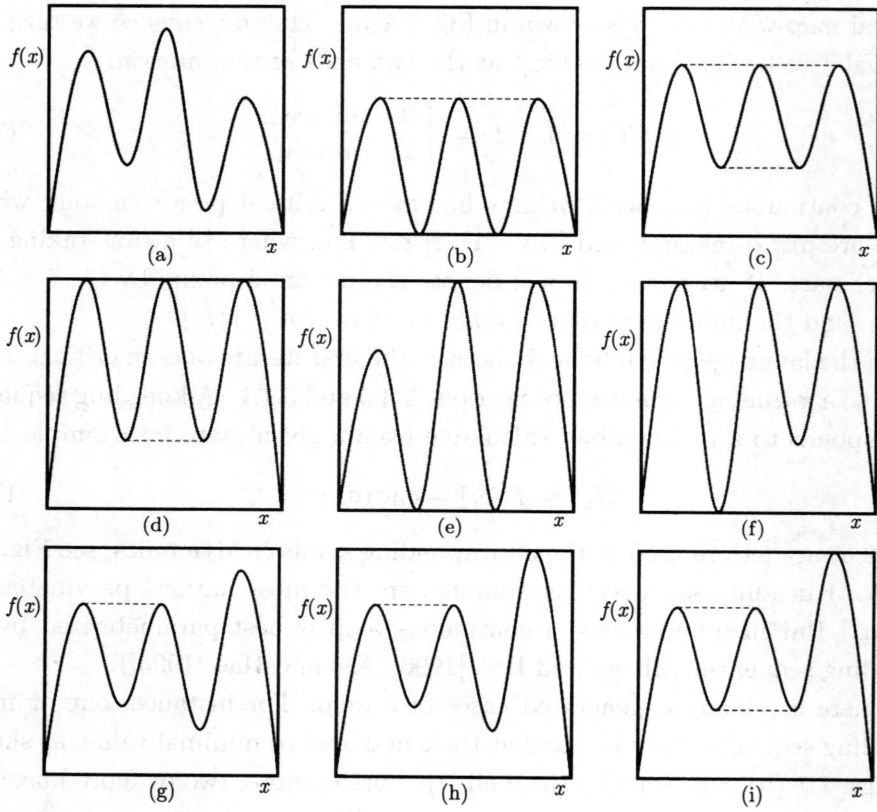

Figure 7.1 Various cases of an $m = 6$ map

2. The number of periods of the map (d) is given by $N_6(n) - N_2(n)$. A similar case in general would have $N_m(n) - N_{m-2\times 2}(n)$ periods.

3. The number of periods of the map (i) is twice of (d), i.e., $2(N_6(n) - N_2(n))$.

4. The number of periods of the map (c) is the sum of (b) and (d), i.e., $2N_6(n) - N_2(n)$.

5. The number of periods of the map (h) is the sum of (d) and twice of (e), i.e., $3N_6(n) - N_2(n) - 2N_4(n)$.

6. The number of periods of the map (g) is the sum of (d) and thrice of (e), i.e., $4N_6(n) - N_2(n) - 3N_4(n))$.

7. The number of periods in the general $m = 6$ map (a) is $5(N_6(n) - N_4(n))$. It is $(m-1)(N_m(n) - N_{m-2}(n))$ for a general m-lap map.

If a m-lap map has k_1 critical points which change independently, k_2 pairs of maximal (or minimal) critical points which change simultaneously, k_3 triples of maximal (or minimal) critical points which change simultaneously, and so on, then the total number of period n superstable orbits is given by

$$N = \sum_{i=1} k_i[N_m(n) - N_{m-2i}(n)], \tag{7.6}$$

where

$$k_0 + k_1 \times 1 + k_2 \times 2 + k_3 \times 3 + \cdots = m - 1,$$

where k_0 is the number of extremes fixed at the top or bottom of the unit square. This general counting formula will be explained in Section 7.3.1.

Table 7.1 Number of period n orbits $N_m(n)$ for maps with m laps

n \ m	2	3	4	5	6	7
1	1	1	2	2	3	3
2	1	2	4	6	9	12
3	1	4	10	20	35	56
4	2	10	32	78	162	300
5	3	24	102	312	777	1 680
6	5	60	340	1 300	3 885	9 800
7	9	156	1 170	5 580	19 995	58 824
8	16	410	4 096	24 414	104 976	360 300
9	28	1 092	14 560	108 500	559 860	2 241 848
10	51	2 952	52 428	488 280	3 023 307	14 123 760
11	93	8 052	190 650	2 219 460		
12	170	22 140	699 040	10 172 500		
13	315	61 320	2 581 110			
14	585	170 820	9 586 980			
15	1 091	478 288				
16	2 048	1 345 210				
17	3 855	3 798 240				
18	7 280	10 761 660				
19	13 797	30 585 828				
20	26 214	87 169 608				

Many authors have worked on the counting problem from various viewpoints. Results for $m = 2$ and $m = 3$ are scattered in the literature. For example, in his 1976 paper (May [1976]), R. May referred to six groups of authors who had studied the $m = 2$ case. In order to make this chapter self-content, we will summarize some of these results in our own narration while presenting some new results and generalizations.

7.1.3 A Few Number Theory Notations and Functions

Any integer-valued function defined on integers is called a number-theory function. The Möbius function $\mu(n)$ and Euler function $\varphi(n)$ are two frequently used such functions. As we shall encounter these functions in what follows, we list their definition and properties for quick reference.

The Möbius function is defined as follows:

$$\mu(n) = \begin{cases} 1, & \text{if} \quad n = 1, \\ (-1)^r, & \text{if} \quad n \text{ is a product of } r \text{ distinct primes,} \\ 0, & \text{otherwise,} \end{cases} \tag{7.7}$$

We denote the largest common divisor of two integers n and m by (n, m). If $(n, m) = 1$ these two numbers are co-prime to each other. The number of co-primes to n among $1, 2, \cdots, n-1$, defines the Euler function $\varphi(n)$. For example, $\varphi(p) = p - 1$ for any prime p. The first $\mu(n)$ and $\varphi(n)$ are listed in Table 7.2.

Table 7.2 Values of the Möbius and Euler functions

n	1	2	3	4	5	6	7	8	9	10	11	12
$\mu(n)$	1	−1	−1	0	−1	1	−1	0	0	1	−1	0
$\varphi(n)$	1	1	2	2	4	2	6	4	6	4	10	4

A real function f defined on the positive integers is said to be *multiplicative* if $f(m)f(n) = f(mn)$ for all m, n with $(m, n) = 1$. Both $\mu(n)$ and $\varphi(n)$ are multiplicative. We list a few useful relations for $\mu(n)$ and $\varphi(n)$:

1. We denote by $d|n$ that d is a factor of n, or, in other words, d divides n. When a sum runs over all $\{d : d|n, 1 \leqslant d \leqslant n\}$, we simply write $\sum_{d|n}$. The largest common divisor (n, j) for $j \leqslant n$ is certainly one of d. In fact, a given factor d of n appears $\varphi(n/d)$ times among all (n, j) when j runs from 1 to n. In particular, there is a useful relation

$$\sum_{s=1}^{n} q^{(n,s)} = \sum_{d|n} \varphi\left(\frac{n}{d}\right) q^d = \sum_{d|n} \varphi(d)\, q^{n/d}, \tag{7.8}$$

where q is any quantity that does not depend on s.

2. The Möbius functions for given n sums up to the identity number-theory function $I(n)$:

$$\sum_{d|n} \mu(d) = I(n), \tag{7.9}$$

which is defined as

$$I(n) = \left[\frac{1}{n}\right] = \begin{cases} 1, & n = 1, \\ 0, & n > 1. \end{cases} \tag{7.10}$$

3. The Euler functions for a given n sums up to another number-theory function $N(n) \equiv n$, i.e.,

$$\sum_{d|n} \varphi(d) = N(n). \tag{7.11}$$

This is the special case of (7.8) at $q = 1$.

4. If a number-theory function $g(n)$ is given by a sum over another function $h(n)$:

$$g(n) = \sum_{d|n} h(d), \tag{7.12}$$

then $h(n)$ may be expressed through $g(n)$, using the inverse Möbius transform:

$$h(n) = \sum_{d|n} \mu\left(\frac{n}{d}\right) g(d) = \sum_{d|n} \mu(d) g\left(\frac{n}{d}\right). \tag{7.13}$$

(There is an analogue of Möbius inversion for functions defined over the reals, namely if $g(x) = \sum_{n\leqslant x} f(x/n)$ then $f(x) = \sum_{n\leqslant x} \mu(n) g(x/n)$.) If g is multiplicative, so is h.

7.2 Number of Periodic Orbits in a Class of One-Parameter Maps

Now we turn to the key quantity $N_m(n)$ and look at the many interesting aspects of the counting problem. If the various facets of the counting problem are specific to unimodal maps only, the topic would be much less interesting. Fortunately, most of these aspects, if not all, may be extended to maps with more critical points, even to maps with discontinuities. In each subsection below, we will first treat the unimodal, $m = 2$, case and then, where possible, extend the results to larger m.

The quantity $N_m(n)$ applies only to a quite degenerated form of polynomial maps of order m; namely, when all the maxima yield the same kneading sequence and all the minima produce an ever-minimal kneading sequence so there is no need to check for shift-minimality. However, the counting problem of more general mappings decomposes into that of various degenerated cases.

At least, one may pose and solve independently the counting problems for the nine cases listed in Fig. 7.1. All the answers are related in one or another way to the key quantity $N_m(n)$.

7.2.1 Number of Admissible Words in Symbolic Dynamics

We partition the phase space of the one-dimensional dynamics, i.e., the interval I, in the standard way as discussed in Chapter 3 on maps with multiple critical points: divide I by maximal and minimal points of the map f and label each subinterval, on which f behaves monotonically, by a symbol. This is shown schematically in Fig. 7.2, in which the monotone branches of f near three critical points C_i, D_k, and C_j are labeled by letters I_i, I_{i+1}, I_k, I_{k+1}, I_j, and I_{j+1}, respectively. We shall use these notations in subsequent discussions.

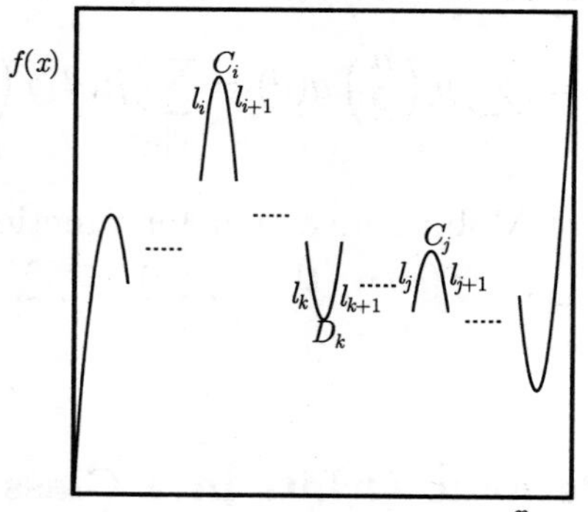

Figure 7.2 A general continuous map

Generally speaking, not every symbolic sequence may correspond to a trajectory in the dynamics. In order to become an allowed sequence at some parameter value a symbolic sequence Σ must satisfy the following *admissibility conditions*, based on ordering rule for symbolic sequences:

$$\begin{aligned} &\mathcal{I}_i(\Sigma), \mathcal{I}_{i+1}(\Sigma) \leqslant K_{C_i}, \\ &K_{D_k} \leqslant \mathcal{I}_k(\Sigma), \mathcal{I}_{k+1}(\Sigma), \end{aligned} \tag{7.14}$$

plus possible equalities between some kneading sequences if they are bound to vary simultaneously. In (7.14) $\mathcal{I}(\Sigma)$ denotes the set of all subsequences in Σ that follow one or another symbol I in Σ. When the kneading sequence associated with a maximal (minimal) point C (D) is the largest (smallest) among all possible ones, there is no need to check the conditions with this K_C (K_D).

The ordering rule and admissibility conditions may easily be programmed to generate and count all periodic words of a given length. Anyway, this "brute-force" counting is not restricted to continuous maps only and it provides numbers with which other methods are to be compared. In fact, all the results of other counting formulae described in what follows have been checked with this direct counting. A C program PERIOD.C designed for this purpose is listed in Appendix A.

7.2.2 Number of Tangent and Period-Doubling Bifurcations

There are several sets of recursive relations for the number of periods. In one dimensional maps a period n orbit may be generated either at a tangent (saddle node) bifurcation or at a period-doubling bifurcation. We denote the number of orbits that come from saddle node bifurcations by $M(n)$, and the number of orbits that come from period-doubling bifurcations by $P(n)$. Since no odd period can be born at period-doubling bifurcation, we have

$$P(2k+1) = 0 \quad \text{for all } k > 0. \tag{7.15}$$

Any period k orbit can undergo period-doubling, no matter how itself comes about. Therefore,

$$P(2k) = P(k) + M(k) = N(k) \quad \text{for all } k \geqslant 1. \tag{7.16}$$

where $N(k)$ is the total number of period-k. Since at a saddle node bifurcation a pair of stable and unstable periodic orbits are always born simultaneously, when counting the numbers of roots of $x = f^{(n)}(\mu, x)$, where f is a quadratic function describing the unimodal map, we should add a factor 2 to M:

$$2^n = \sum_{d|n} d[2M(d) + P(d)]. \tag{7.17}$$

To simplify the notation, we denote

$$C(d) = d[2M(d) + P(d)], \tag{7.18}$$

and write Eq. (7.17) as

$$2^n = \sum_{d|n} C(d). \tag{7.19}$$

Eqs. (7.15) to (7.19) give us a recursion scheme to calculate $P(n)$, $M(n)$, and the total number of periods

$$N(n) = M(n) + P(n). \tag{7.20}$$

In order to generalize the recursive relation to polynomial mappings of order m, one has to distinguish even m from odd m. The case of even m is straightforward: one simply replaces 2 by m in the left-hand side of Eq. (7.19) to get

$$m^n = \sum_{d|n} C_m(d), \quad m \text{ even}, \tag{7.21}$$

with Eqs. (7.15) and (7.16) unchanged.

Under the normalization (7.4), mapping function with odd m always has a fixed point at $x = 1$, which leads to a trivial periodic orbit 1^n for all n. Therefore, the general formula reads

$$m^n = \sum_{d|n} C_m(d) + 1, \quad m \text{ odd}. \tag{7.22}$$

Eqs. (7.21) and (7.22) may be combined by writing, say,

$$m^n = \sum_{d|n} C_m(d) + s(m), \quad \forall\, m \geqslant 2, \tag{7.23}$$

where

$$s(m) = m \pmod 2 = \begin{cases} 0, & m \text{ even}, \\ 1, & m \text{ odd}. \end{cases}$$

We can solve (7.23) by using the inverse Möbius transform (7.13):

$$C_m(n) = \sum_{d|n} [\mu(d) m^{\frac{n}{d}} - s(m)] = \sum_{d|n} \mu(d) m^{\frac{n}{d}} - s(m) I(n), \tag{7.24}$$

where $I(n)$ is the identity function (7.9). We see that $s(m)$ only affects $C_m(1)$ and only when m is odd.

We list here the first $C_m(n)$ for easy reference:

$$\begin{aligned} C_m(1) &= m - s(m), \\ C_m(2) &= m^2 - m, \\ C_m(3) &= m^3 - m, \\ C_m(4) &= m^4 - m^2, \\ C_m(5) &= m^5 - m, \\ C_m(6) &= m^6 - m^3 - m^2 + m, \\ C_m(7) &= m^7 - m, \\ C_m(8) &= m^8 - m^4, \\ C_m(9) &= m^9 - m^3, \\ C_m(10) &= m^{10} - m^5 - m^2 + m, \\ C_m(11) &= m^{11} - m, \\ C_m(12) &= m^{12} - m^6 - m^4 + m^2. \end{aligned} \tag{7.25}$$

Recursive arguments have been used by many authors to determine the number of periods in unimodal, i.e., $m = 2$, maps, e.g., May [1976], Piña [1983], Chavoya-Aceves, Angulo-Brown and Piña [1985], Zeng [1985, 1987], Hao and Zeng [1987], Brucks [1987], Doi [1993].

7.2.3 Recursion Formula for the Total Number of Periods

A simple recursion formula for the total number of periods $N_m(n)$ is readily derived from Eq. (7.23). We add and subtract a $dP_m(d)$ term in the sum to get

$$m^n - s(m) = 2\sum_{d|n} dN_m(d) - \sum_{d|n} dP_m(d). \tag{7.26}$$

Suppose the period n is decomposed as

$$n = 2^k n', \quad k \geqslant 0, \ n' \text{ odd}, \tag{7.27}$$

then any $2^j d'$, where $0 \leqslant j \leqslant k$ and $d'|n'$, is a factor of n. Therefore, a sum over $d|n$ may be written as

$$\sum_{d|n} \cdots = \sum_{j=0}^{k} \sum_{d|n'} \cdots .$$

Consequently, Eq. (7.26) becomes

$$m^n - s(m) = 2\sum_{j=0}^{k}\sum_{d|n'} 2^j dN_m(2^j d) - \sum_{j=1}^{k}\sum_{d|n'} 2^j dP_m(2^j d).$$

In the second sum the $j = 0$ term drops out due to (7.15). Using (7.16) and (7.20), we get

$$m^n - s(m) = 2\sum_{j=0}^{k}\sum_{d|n'} 2^j dN_m(2^j d) - 2\sum_{j=1}^{k}\sum_{d|n'} 2^{j-1} dN_m(2^{j-1}d).$$

It is easy to see that all terms cancel out except the $j = k$ term in the first sum. Finally, we are led to

$$m^n - s(m) = 2\sum_{d|n'} 2^k dN_m(2^k d), \tag{7.28}$$

with n' defined in (7.27). Again using the inverse Möbius transform (7.13) and (7.9), we get an explicit expression for $N(n)$:

$$N_m(n) = \frac{1}{2n}\sum_{d|n'} \mu(d) m^{\frac{n}{d}} - s(m)I(n'). \tag{7.29}$$

When using this formula, one must always pay attention to the difference between n and n' given by (7.27). In particular, for odd m, now $s(m)$ affects all periods which are powers of 2, i.e., $n = 2^k$, $k = 0, 1, \cdots$. This is in contrast to (7.24), where only the $n = 1$ term is affected.

The counting formula (7.29) for $m = 2$ appeared in Lutzky [1988]. It was also derived by W. M. Zheng for $m = 2, 3$, see page 30 in Hao [B1990].

7.2.4 Symmetry Types of Periodic Sequences

The number of periods in a unimodal $m = 2$ map is related to a classical combinatorial problem: how many different necklaces may be made from n pieces of stones, each stone having a choice from q colors. The answer naturally depends on which necklaces are considered different. Of course, one should allow for cyclic permutations of stones and even allow for permutation of colors. If so, the problem turns out to be a particular case of a group-theoretical problem on the number of periodic sequences invariant under a certain discrete group. In this particular case the group is $C_n \otimes S_q$ and the answer has been known since late 1950s (Fine [1958]; Gilbert and Riordan [1961]).

Again one distinguishes the number $F_q(n)$ of primitive period n, which is not a multiple of a shorter period, and the total number $F_q^*(n)$, which are related to $F_q(n)$ by the Möbius transformation:

$$
\begin{aligned}
F_q^*(n) &= \sum_{d|n} F_q(d), \\
F_q(n) &= \sum_{d|n} \mu\left(\frac{n}{d}\right) F_q^*(d).
\end{aligned}
\tag{7.30}
$$

The $F_q^*(n)$'s are given explicitly by Gilbert and Riodan [1961]:

$$
F_q^*(n) = \frac{1}{q!n} \sum_{d|n,\,\{k_i\}} \varphi(d) N(k_1, k_2, \cdots, k_q)[m(d)]^{n/d}, \tag{7.31}
$$

where

$$
\begin{aligned}
& k_1 + 2k_2 + \cdots + qk_q = q, \\
& N(k_1, k_2, \cdots, k_q) = \frac{q!}{k_1!k_2!\cdots k_q!2^{k_2}\cdots q^{k_q}}, \\
& m(d) = \sum_{c|d} ck_c,
\end{aligned}
\tag{7.32}
$$

and $\varphi(d)$ is the Euler function. The $F_q(n)$'s were tabulated in Gilbert and Riordan [1961]. Metropolis, Stein and Stein [1973] realized that the number of periods in the unimodal map is the same as $F_2(n)$. The values of $F_2(n)$ for the group $C_n \otimes S_2$ indeed agree with $N_2(n)$.

The fact that it is incorrect to use the group $C_n \otimes S_3$ for the cubic map has been understood by several authors (Piña [1984]; Zeng [1985, 1987]). For the antisymmetric cubic map, Zeng pointed out that the correct group is $C_n \otimes S_2 \otimes S_2$ and she changed condition (7.32) to cope with the case. How about $m \geqslant 3$ maps in general?

It turns out that the counting problem no longer has simple relation with the necklace problem. However, a formula may be derived along the line of Gilbert and Riodan [1961], by making use of Pólya's theorem, which is the cornerstone of enumerative combinatorics. We formulate this theorem, see, e.g., Fine [1958], Gilbert and Riordan [1961], and show its use on our problem.

Pólya-Burnside Theorem If G is a finite group of order g, whose elements act on a finite set of objects, and if two objects are equivalent when one is transformed into the other by an element of G, then the number of inequivalent objects is

$$
T = \frac{1}{g} \sum_{t \in G} I(t), \tag{7.33}
$$

where $I(t)$ is the number of objects left invariant by $t \in G$ and the sum runs over all elements of G.

Consider a degenerated map of the type shown in Fig. 7.1 (b) with m laps labeled each by a symbol. Thus, m different symbols are available. The number m may be even or odd. Our objects are periodic symbolic sequences of length n. Periodicity implies invariance under a cyclic group C_n of order n. This particular type of map has only one kneading sequence determined by all the maxima, which are bound to move together. Therefore, the admissibility condition reduces to the requirement of shift-maximality. Among all the cyclic shifts of a given periodic sequence, there must be a shift-maximal sequence. Due to the periodic window theorem (see Section 2.4.1), one can interchange two symbols which are nearest neighbors of a maximal point without violating admissibility of the sequence.

For example, when $m = 6$ we might have the following symbols:

$$L < C_1 < M < D_1 < N < C_2 < P < D_2 < Q < C_3 < R, \tag{7.34}$$

while for $m = 7$ we might have

$$\begin{aligned} L < C_1 < M < D_1 < N < C_2 < P \\ < D_2 < Q < C_3 < S < D_3 < R. \end{aligned} \tag{7.35}$$

In the first case, the transformation s which brings a periodic sequence into an equivalent one is

$$s : \begin{cases} L \leftrightarrow M, \\ N \leftrightarrow P, \\ Q \leftrightarrow R, \end{cases} \tag{7.36}$$

while in the second case the transformation s is

$$s : \begin{cases} L \leftrightarrow M, \\ N \leftrightarrow P, \\ Q \leftrightarrow S, \\ R \text{ unchanged}. \end{cases} \tag{7.37}$$

Clearly, $s^2 = e$ is an identity element in a group of order 2, which we will call S_2 (all groups of order 2 are isomorphic). In other words, the group S_2 has two elements $\{e, s\}$. The cyclic group C_n has elements $\{E, p, p^2, \cdots, p^{n-1}\}$, where E is the identity and p denotes the cycle $(123\cdots n)$: $p^n = E$. Therefore, the group under study is $G = C_n \otimes S_2$ with elements t chosen from the set $\{Ee, Es, \cdots, p^{n-1}e, p^{n-1}s\}$.

In order to derive the counting formula let us start from the simplest case $n = 1$. The group G has only two elements: Ee and Es. For an object made of $n = 1$ symbol, there are m choices. They are all invariant under the identity Ee. For even m none of these objects remains invariant under Es, whereas for odd m there is always one object, namely, R, which is not affected by s due to the last line in (7.37). Therefore,

$$I(Ee) = m, \quad I(Es) = \begin{cases} 0, & m \text{ even}, \\ 1, & m \text{ odd}. \end{cases} \tag{7.38}$$

The Pólya theorem (7.33) gives

$$F_m^*(1) = \begin{cases} \dfrac{m}{2}, & m \text{ even}, \\ \dfrac{m+1}{2}, & m \text{ odd}. \end{cases} \tag{7.39}$$

When $n = 2$ the group G has 4 elements $\{Ee, Es, pe, ps\}$. For the objects made of two symbols, there are m^2 choices. It is easy to see that

$$I(Ee) = m^2, \quad I(Es) = \begin{cases} 0, & m \text{ even}, \\ 1, & m \text{ odd}, \end{cases} \quad I(pe) = I(ps) = m. \tag{7.40}$$

The Pólya theorem yields

$$F_m^*(2) = \begin{cases} \dfrac{m^2+2m}{4}, & m \text{ even}, \\ \dfrac{m^2+2m+1}{4}, & m \text{ odd}. \end{cases} \tag{7.41}$$

In order to see the regularity we list $I(t)$ for $n = 3, 4$ in Table 7.3.

Table 7.3 Number of invariant objects under $C_n \otimes S_2$ for $n = 3, 4$

		$I(Ee)$	$I(Es)$	$I(pe)$	$I(ps)$	$I(p^2e)$	$I(p^2s)$	$I(p^3e)$	$I(p^3s)$
$n=3$	m even	m^3	0	m	0	m	0		
	m odd	m^3	1	m	1	m	1		
$n=4$	m even	m^4	0	m	m	m^2	m^2	m	m
	m odd	m^4	1	m	m	m^2	m^2	m	m

Now we can split the sum in Pólya's theorem into two terms:

$$\sum_{t\in C_n\otimes S_2} I(t)=\sum_{j=1}^{n} I(p^j e)+\sum_{j=1}^{n} I(p^j s). \tag{7.42}$$

The first sum is nothing but the number of periodic sequences of length n made of m symbols and invariant under cyclic group C_n, which has been given in Gilbert and Riordan [1961]:

$$Z_m(n)=\sum_{j=1}^{n} I(p^j)=\sum_{j=1}^{n} m^{(n,j)}=\sum_{d|n}\varphi(d)m^{\frac{n}{d}}, \tag{7.43}$$

see also Eq. (7.8). We have introduced a notation $Z_m(n)$ for this quantity for later use in Section 7.5. The tabulated values of $Z_2(n)$ will be given there in Table 7.5.

The second sum may be transformed in a similar way to yield

$$\sum_{j=1}^{n} I(p^j s)=\sum_{j=1}^{n} \overline{m}^{(n,j)}=\sum_{d|n}\varphi(d)\overline{m}^{\frac{n}{d}},$$

where

$$\overline{m}=\begin{cases} m, & m \text{ odd},\\ 0, & \text{both } d \text{ and } m \text{ even},\\ 1, & d \text{ odd but } m \text{ even}.\end{cases}$$

The group $C_n\otimes S_2$ is of order $2n$. Therefore, we get the final result:

$$F_m^*(n)=\frac{1}{2n}\sum_{d|n}\varphi(d)(m^{\frac{n}{d}}+\overline{m}^{\frac{n}{d}}). \tag{7.44}$$

We remind the reader that $F_m^*(n)$ contains periods which are factors of n. In order to calculate the number of primitive periods, one has to use the second equation in (7.30) to get $F_m(n)$, which coincides with $N_m(n)$ for $n>1$. For odd m, $F_m(1)=N_m(1)+1$, because R is invariant under (7.37), but does not correspond to a superstable period. In fact, R^n is present at all n, but it contributes only to $n=2^k$, as we have learned from Eq. (7.29).

For odd m the map may possess an anti-symmetry, i.e., invariant under inversion $x\to -x$. Periodic orbits in this case are subdivided into symmetric and asymmetric ones. The number of asymmetric orbits is half of what given by (7.44). The result is the same as taking the group to be $C_n \otimes S_2 \otimes S_2$. Moreover, symmetric orbits, which must have an even period $2k$, may undergo symmetry-breaking in periodic regime and restore symmetry in chaotic regime. The number of such orbits equals that of asymmetric orbits of period k, i.e., $N_m(k)$. This problem has been analyzed in Hao [B1989], by using symbolic dynamics for $m = 3$ map, but the essential result applies also to all maps with odd m.

To avoid any confusion, we emphasize that while the group remains to be $C_n \otimes S_2$, it should be interpreted according to Eqs. (7.36) and (7.37).

7.2.5 Explicit Solutions to the Recurrence Relations

Explicit solutions of various recurrence relations introduced in the previous subsections usually contain number-theory functions such as the Möbius function $\mu(n)$ or the Euler function $\varphi(n)$, which one can always avoid. As a matter of fact, explicit expressions, which are equivalent to the solutions of the aforementioned recursive relations, were given by Gumowski and Mira [1980] based on conservation of the number of roots of algebraic equations for the case of $m = 2$ and 3. Recently, Huang [1994] observed that the counting formula of Gumowski and Mira also works well for higher values of m.

$$H_m(n) = \frac{1}{n}\left[m^n - \sum_{i=1}^{r} m^{e_i} + \sum_{j=2}^{r}\sum_{i=1}^{j-1} m^{e_{ij}} + \cdots + (-1)^r m^{e_{12\cdots r}}\right], \qquad (7.45)$$

where r and $e_{ij\cdots}$ are determined from decomposition of n into prime factors:

$$n = p_1^{k_1} p_2^{k_2} \cdots p_r^{k_r},$$

k_i being positive integers and

$$e_i = \frac{n}{p_i}, \quad e_{ij} = \frac{n}{p_i p_j} \ (i \neq j), \quad \cdots, \quad e_{12\cdots r} = n \Big/ \prod_{i=1}^{r} p_i.$$

In order to calculate $M(n)$ and $P(n)$, there are different formulae for n odd or even; when n is even one distinguishes the case of n being a power of 2 or not (Huang [1994]). In practice, there are many cancellations of contributions from different terms in Eq. (7.45). The number-theory functions in the explicit expressions of Sections 7.2.2 and 7.2.3 take into account these cancellations automatically. All this makes many terms of Eq. (7.45) obsolete.

7.2.6 Finite Lambda Auto-Expansion of Real Numbers

By looking at a piecewise linear tent map, Derrida, Gervois and Pomeau [1978] showed that the number of periods in the quadratic map may be obtained from an extension of the finite λ-auto-expansion of real numbers. Take a real number $1 < \lambda < 2$ and expand it in its own inverse powers:

$$\lambda = \sum_{i=0}^{\infty} \frac{a_i}{\lambda^i}, \tag{7.46}$$

where $a_0 = a_1 = 1$, $a_i = 0, \pm 1$ for $i > 0$; if $a_j = 0$, then $a_k = 0$ for all $k \geqslant j$. In the latter case, one has a finite expansion. It turns out that a finite expansion

$$\lambda = \sum_{i=0}^{n-1} \frac{A_i}{\lambda^i}$$

with A_i defined below in (7.50) has a close relation to the number of periods in the quadratic map. Namely, the number of different λ in the interval $(1, 2)$ which leads to a finite expansion of length n is exactly the number of period n orbits in a $m = 2$ map.

This λ-auto-expansion extends easily to an m-lap map. It is enough to look at a m-piece linear function which maps the interval $[0, m]$ into itself, see Fig. 7.3. We parameterize the straight lines in Fig. 7.3 in the following way:

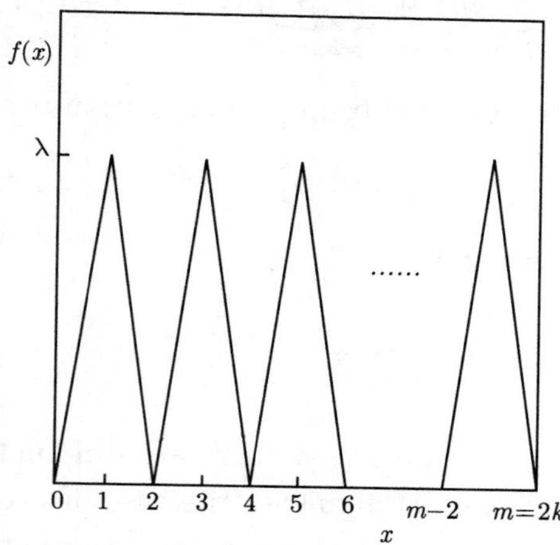

Figure 7.3 A piecewise linear map with m laps

$$f(x) = \lambda \alpha_i x - \lambda \beta_i, \quad \lambda \in (1, m], \tag{7.47}$$

where

$$\begin{aligned} &\alpha_i = 1, \quad \beta_i = 2i, \quad \text{for } 2i \leqslant x \leqslant 2i+1, \; i = 0, 1, \cdots, [(m-1)/2]; \\ &\alpha_i = -1, \quad \beta_i = -2i, \quad \text{for } 2i-1 \leqslant x \leqslant 2i, \; i = 1, 2, \cdots, [m/2]. \end{aligned} \tag{7.48}$$

Starting from an initial point $x_0 \in [0, m]$, we get the iterations

$$\begin{aligned} x_1 &= f(x_0) = \lambda\alpha_0 x_0 - \lambda\beta_0, \\ x_2 &= f(x_1) = \lambda^2\alpha_1\alpha_0 x_0 - \lambda^2\alpha_1\beta_0 - \lambda\beta_1, \\ &\cdots \\ x_n &= f(x_{n-1}) \\ &= \lambda^n \alpha_{n-1}\alpha_{n-2}\cdots\alpha_1\alpha_0 x_0 - \sum_{i=0}^{n-1} \lambda^{n-i}\alpha_{n-1}\alpha_{n-2}\cdots\alpha_{i+1}\beta_i. \end{aligned}$$

For a period n orbit $x_n = x_0$ and we take $x_0 = f(1) = \lambda$ (corresponding to a "superstable" orbit). Multiplying all terms in the last equality by

$$\alpha_{n-1}\alpha_{n-2}\cdots\alpha_1\alpha_0$$

and noticing that

$$\alpha_i^2 = 1, \quad \forall\, i,$$

we get

$$\lambda = \frac{(\beta_{n-1}+1)\alpha_{n-1}\alpha_{n-2}\cdots\alpha_1\alpha_0}{\lambda^{n-1}} + \sum_{i=0}^{n-2} \frac{\beta_i\alpha_i\alpha_{i-1}\cdots\alpha_0}{\lambda^i}. \tag{7.49}$$

In deriving Eq. (7.49) we have divided both sides by λ^n and regrouped terms. Now, introducing new notations

$$\begin{aligned} A_i &= 2i\alpha_{i-1}\cdots\alpha_0, \quad i = 0, 1, \cdots, n-2, \\ A_{n-1} &= [2(n-1) + \alpha_{n-1}]\alpha_{n-2}\cdots\alpha_1\alpha_0, \end{aligned} \tag{7.50}$$

we arrive at a finite auto-expansion of $\lambda \in (1, m]$ in terms of its own inverse powers:

$$\lambda = \sum_{i=0}^{n-1} \frac{A_i}{\lambda^i}. \tag{7.51}$$

When writing down (7.50) we put $\beta_i = 2i\alpha_i$ according to (7.48). Clearly, an arbitrarily chosen set $\{A_i\}$ may not lead to a finite auto-expansion of a real number $\lambda \in (1, m]$. The admissibility condition for being so have been discussed by Derrida, Gervois and Pomeau [1978] (for $m = 2$) and Zeng [1987] (for $m = 3$). It consists in the fulfillment of the following inequalities

$$\pm(A_i, A_{i+1}, A_{i+2}, \cdots) \leqslant (A_0, A_1, A_2, \cdots), \quad i = 1, 2, \cdots , \tag{7.52}$$

where the comparison is made on a term-by-term basis. The $\leqslant$ relation holds for two sequences of coefficients

$$(c_1, c_2, \cdots , c_{k-1}, c_k, \cdots) \leqslant (d_1, d_2, \cdots , d_{k-1}, d_k, \cdots),$$

as far as $c_i = d_i$ for $i = 1, 2, \cdots , k-1$ and $c_k \leqslant d_k$ in the sense of the two real numbers c_k and d_k. Put in other words, the first occurrence of an unequal relation determines the whole inequality.

In practice, one may write a program to generate α_i and β_i for $i = 1, 2, \cdots , m$ according to (7.48), form all possible products A_i, $i = 1, 2, \cdots , n-1$, according to (7.50), then check them against the admissibility condition (7.52). The numbers obtained in this way coincide with $N_m(n)$ calculated by other methods in the previous subsections.

7.3 Other Aspects of the Counting Problem

There are several other aspects of the counting problem which are interesting themselves. Some of these problems are formulated and solved only in the unimodal case. So far we do not know how to generalize to maps with more critical points. We describe these aspects of the counting problem for completeness.

7.3.1 The Number of Roots of the "Dark Line" Equation

We have discussed the "dark lines" in the bifurcation diagram of 1D maps in Section 2.2.3. We repeat the essential point here. For the unimodal map, using the same nonlinear function $f(\mu, x)$ in Eq. (7.3), one recursively defines a set of functions $\{P_n(\mu)\}$ of the parameter:

$$\begin{aligned} P_0(\mu) &= C, \\ P_n(\mu) &= f(\mu, P_{n-1}(\mu)) = f^n(\mu, C), \quad n = 1, 2, \cdots , \end{aligned} \tag{7.53}$$

where C is the critical point of the map. These functions describe all the dark lines and sharp boundaries of chaotic bands seen in the bifurcation diagram

(Zeng, Hao, Wang and Chen [1984]). The real roots of the following equation:

$$P_n(\mu) = C, \tag{7.54}$$

located in the suitable parameter range of the map, give the number of period-n orbits. This number is $N_2(n)$.

For maps with multiple critical points, we may extend Eq. (7.53) to define a set of "dark line" functions for each independently varying critical point:

$$\begin{aligned} P_0^{(i)}(\mu) &= C_i, \\ P_{n+1}^{(i)}(\mu) &= f(\mu, P_n^{(i)}(\mu)), \end{aligned} \tag{7.55}$$

where C_i is one of the critical points, including D_i, and μ stands for the set of parameters, preferably expressed through the kneading sequences. When one fix all but one kneading sequence, the real roots of the equation

$$P_n^{(i)}(\mu) = C_i \tag{7.56}$$

determine the superstable period n orbits along the parameter axis. Their number is $N_m(n) - N_{m-2}(n)$, see Section 7.4 below.

7.3.2 Number of Saddle Nodes in Forming Smale Horseshoe

Yorke and Alligood [1985] extended the Smale construction by allowing not to put the folded strip completely back to the original region. They describe this transformation by an additional topological parameter $0 < \mu \leqslant 1$, the Smale case being the "surjective" limit $\mu = 1$. They tabulated the number $S(n)$ of saddle-node bifurcations which may appear when varying the parameter μ from 0 to 1. It turns out that $S(n)$ is just $M(n)$ defined earlier. Intuitively, it should be possible to carry over this construction to maps with multiple critical points by invoking deformations with more bends.

7.3.3 Number of Solutions of Renormalization Group Equations

We consider first the unimodal case with $m = 2$. The Cvitanović-Feigenbaum renormalization group fixed point equation (2.40) in the space of unimodal functions (Feigenbaum [1978])

$$\alpha^{-1} g(\alpha x) = g^{(2)}(x), \tag{7.57}$$

is the first one of an infinite set of similar equations. Eq. (7.57) determines the scaling property of period-doubling cascades in both the phase space and the parameter space, e.g., the Feigenbaum universal exponents δ and α.

In principle, one can take any superstable periodic word $W = \Sigma C$ and extend it into a window

$$((\Sigma C)_-, \Sigma C, (\Sigma C)_+),$$

where $(\Sigma C)_-$ and $(\Sigma C)_+)$ are the smaller and larger sequence in ΣR and ΣL. They are called the lower and upper sequence of ΣC, respectively. We use these sequences to define a *generalized composition rule*

$$\begin{aligned} R &\to (\Sigma C)_+, \\ L &\to (\Sigma C)_-, \end{aligned} \tag{7.58}$$

which is equivalent to the $*$-composition of Derrida, Gervois and Pomeau [1978] in this simple case (Zheng and Hao [1990]). When (7.58) are applied to the word W itself an infinite number of times, it leads to a fixed sequence in the space of all possible symbolic sequences. This corresponds to the limiting set of one of the so-called period-n-tupling sequences. It leads to a renormalization group equation

$$\alpha^{-1} g(\alpha x) = \underbrace{g \circ g \circ \cdots \circ g}_{n \text{ times}}(x) \tag{7.59}$$

with boundary conditions $g(0) = 1$ and $g'(0) = 0$. To solve it one has to fix the shape of the function $g(x)$ near its maximum, i.e., to indicate z in the expansion:

$$g(x) = 1 + Ax^z + Bx^{2z} + \cdots .$$

When $n = 2$, these conditions guarantee a unique solution which determines, together with the universal function $g(x)$, the Feigenbaum universal constants.

However, when $n \geqslant 5$, there are more than one solutions. Each set of solution determines its own universal constants δ_W, α_W, etc. The Feigenbaum constants arise for the simplest choice of $W = RC$. For $n \geqslant 5$ a good estimate of the initial value α_0 is required in order to settle down to a particular solution (Zeng, Hao, Wang and Chen [1984]). The number of different solutions of (7.59) for a fixed n is $M(n)$.

In maps with multiple critical points it is always possible to pick up some subsets of symbolic sequences, which are analogous with the period-n-tupling sequences in unimodal maps. They need not lead to any new scaling property. Therefore, the problem of renormalization group equations and the number of their solutions remains open.

7.4 Counting Formulae for General Continuous Maps

Having calculated the quantities $N_m(n)$, we are in a position to express the number of superstable periodic kneading sequences for any continuous map.

1. In a general continuous map, the total number of kneading sequences associated with a given critical point is the same as that associated with any other critical point. To prove this statement, let us look at Fig. 7.4. For symbolic sequences ending with C_i to be a kneading sequence, it is enough to check only one of the admissibility conditions, namely,

$$\mathcal{I}_i(K_{C_i}), \mathcal{I}_{i+1}(K_{C_i}) \leqslant K_{C_i}, \tag{7.60}$$

all others being conditions for compatibility, i.e., conditions for K_{C_i} to coexist with kneading sequences associated with other critical points. In (7.60), as in (7.14), $\mathcal{I}(K)$ denotes a set of all subsequences in K which follow one or another symbol I in K. Now take the maximal point C_j and introduce a transformation between symbols:

$$\begin{aligned} &I_i \leftrightarrow I_j, \\ &I_{i+1} \leftrightarrow I_{j+1}, \\ &C_i \leftrightarrow C_j, \end{aligned} \tag{7.61}$$

keeping other I_k unchanged. Applying the transformation (7.61) to the condition (7.60), we get

$$\mathcal{I}_j(K_{C_j}), \mathcal{I}_{j+1}(K_{C_j}) \leqslant K_{C_j}, \tag{7.62}$$

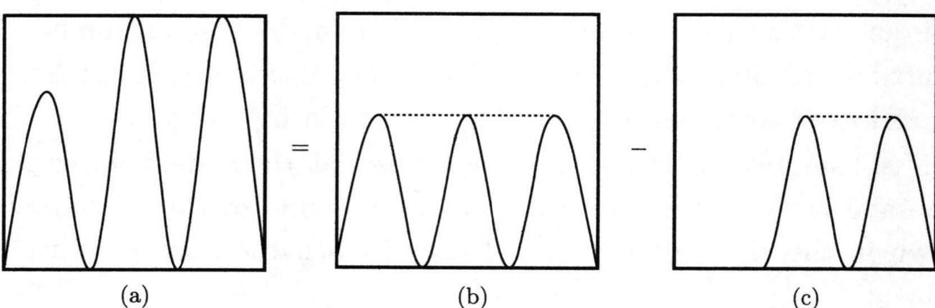

Figure 7.4 Derivation of Eq. (7.64)

which is nothing but the admissibility condition for K_{C_j}. The transformation (7.61) may be applied to any kneading sequence, ending with C_j, to get another kneading sequence ending with C_i. Therefore, their numbers must be the same.

When comparing K_{C_i} with kneading sequences ending with a minimal point D_k, we have to modify the transformation (7.61) slightly to

$$I_i \leftrightarrow I_{k+1}, \quad I_{i+1} \leftrightarrow I_k, \quad C_i \leftrightarrow D_k, \tag{7.63}$$

with other I unchanged. The hint is to exchange symbols with the same parity. When C_i and D_k are close neighbors, the symbol between them remains unchanged, i.e., I_{i+1} and I_k may be the same letter.

2. In a map with m laps, if all but one critical points are located at their maximal or minimal position and the only kneading sequence, associated with the "free" critical point, is allowed to vary, then the number of periods is given by

$$N = N_m(n) - N_{m-2}(n). \tag{7.64}$$

Examples of this case are shown in Figs. 7.1 (e) and (f).

Instead of proving this statement by carefully analyzing the admissibility conditions, we look at Fig. 7.4, which is a schematic equation between the numbers of periods in the three maps. The two maxima to the right of the "free" critical point in Fig. 7.4 (a) touch the top of the unit square, leading to a maximal kneading sequence for the map. Therefore, there is no need to check admissibility of this kneading sequence. Moreover, this kneading sequence associated with the first maximum is compatible with the other two as long as the two other maxima do not become lower than the first one. Therefore, the $N_6(n)$ periods of the map in Fig. 7.4 (b) contain all the periods present in map (see Fig. 7.4 (a)). However, in Fig. 7.4 (b) there are periods which do not contain at all the two letters associated with the first maximum. These are contributions from Fig. 7.4 (c). There are $N_4(n)$ of them and they are subtracted from contribution of see Fig. 7.4 (b) in Eq. (7.64).

3. If all but two critical points are located at their maximal or minimal position and the two "free" critical points are bound to vary simultaneously, as shown in the right most map of Fig. 7.5, then the number of periods is given by

$$N = N_m(n) - N_{m-4}(n). \tag{7.65}$$

Fig. 7.5 is a schematic equation between the maps in the numbers of periods. The interpretation is similar to Fig. 7.4, so we skip it.

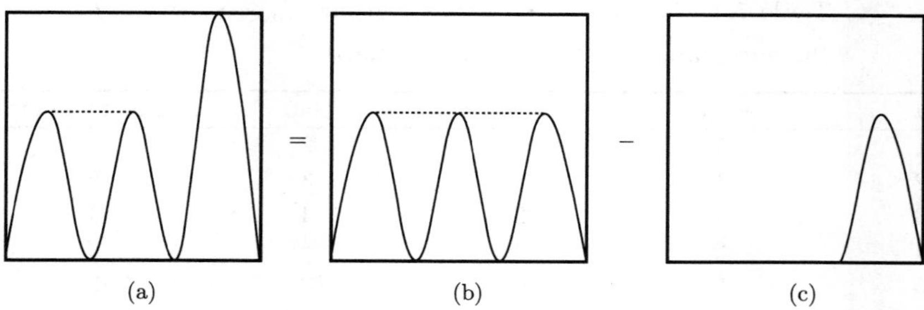

Figure 7.5 Derivation of Eq. (7.65)

4. In general, if i maximal (or minimal) critical points are bound to vary together while all others are located at their maximal or minimal possible position, the number of periods is given by

$$N = N_m(n) - N_{m-2i}(n). \tag{7.66}$$

The general formula (7.6) follows directly from (7.66).

7.5 Number of Periods in Maps With Discontinuity

The counting procedure described in the previous Sections can be extended to some maps with a finite number of discontinuities, when the left and right limits, $f(d_-)$ and $f(d_+)$, of a discontinuity at $x = d$ are allowed to touch 0 and 1 correspondingly. We show this on the examples of the gap map (see Fig. 3.4 in Section 3.4) and the Lorenz-like map (see Fig. 3.1 in Section 3.5), when there is only one discontinuity at $x = C$ (Xie and Hao [1995]).

We recall two quantities introduced in the previous sections. The first is $N_2(n)$, the number of period n orbits in the unimodal map at $m = 2$. The second is $Z_2(n)$, the number of periodic sequences of length n, invariant under cyclic group C_n of order 2, see (7.43) for its definition. These quantities are related by

$$\begin{aligned} Z_2(n) &= 2N_2(n), && n \text{ odd}, \\ Z_2(n) &= 2N_2(n) - N_2(n/2), && n \text{ even}. \end{aligned}$$

We are going to show that the number of periods of length n in the gap map is given by $4N_2(n)$ for $n > 2$, while that in the Lorenz-like map—by

$2Z_2(n)$ for $n > 1$. The quantities $N_2(n)$ and $Z_2(n)$ as well as the number of periods in the two maps are listed in Table 7.4 for $n \leqslant 20$.

Table 7.4 Number of periods for group $C_n \otimes S_2$, group C_n, the gap map, and the Lorenz-like map

n	$N_2(n)$	$Z(n)$	Gap	Lorenz Map
1	1	2	2	2
2	1	1	3	2
3	1	2	4	4
4	2	3	8	6
5	3	6	12	12
6	5	9	20	18
7	9	18	36	36
8	16	30	64	60
9	28	56	112	112
10	51	99	204	198
11	93	186	372	372
12	170	335	680	670
13	315	630	1260	1260
14	585	1161	2340	2322
15	1091	2182	4364	4364
16	2048	4080	8192	8160
17	3855	7710	15420	15420
18	7280	14532	29120	29064
19	13797	27594	55188	55188
20	26214	52377	104856	104754

7.5.1 Number of Periods in the Gap Map

The gap map has been discussed in Section 3.4, see Fig. 3.4 for its shape. The proof that the number of period n in the gap map is $4N_2(n)$ follows from an analysis of the admissibility conditions for the unimodal and the gap maps. The admissibility condition for a kneading sequence K in the unimodal map has been given in Section 2.3.4, see (2.51). It reads

$$\mathcal{L}(K), \mathcal{R}(K) \leqslant K. \tag{7.67}$$

The admissibility condition for the gap map has been given in (3.46) in Section 3.4. Since we are concerned with the number of periods in the whole parameter range, it is enough to check the eligibility of a sequence to be a kneading sequence K_+ or K_- without taking into account the compatibility condition (3.49). In other words, for sequences ending with C_- we have to check only the condition

$$\mathcal{L}(K_-) \leqslant K_-, \tag{7.68}$$

while for sequences ending with C_+ only to check the condition

$$\mathcal{R}(K_+) \leqslant K_+. \tag{7.69}$$

Now suppose that for the unimodal map there is a kneading sequence $K = \Sigma C$. The Periodic Window Theorem (see Section 2.6.1) says that both $(\Sigma L)^\infty$ and $(\Sigma R)^\infty$ are admissible, i.e., shift-maximal sequences. Thus we have

$$\mathcal{L}((\Sigma L)^\infty) \leqslant (\Sigma L)^\infty, \tag{7.70}$$
$$\mathcal{R}((\Sigma L)^\infty) \leqslant (\Sigma L)^\infty, \tag{7.71}$$
$$\mathcal{L}((\Sigma R)^\infty) \leqslant (\Sigma R)^\infty, \tag{7.72}$$
$$\mathcal{R}((\Sigma R)^\infty) \leqslant (\Sigma R)^\infty. \tag{7.73}$$

By continuity we can replace the last L in (7.70) by C_- and replace the last R in (7.73) by C_+. According to (7.68) and (7.69), ΣC_- may be taken as a K_- and ΣC_+—a K_+ in the gap map. We have dealt with $\mathcal{L}((\Sigma L)^\infty)$ and $\mathcal{R}((\Sigma R)^\infty)$.

In order to proceed further we note that an L-shift of a periodic symbolic sequence always ends with an L and an R-shift—with an R. Therefore, $(\Sigma R)^\infty$ does not belong to $\mathcal{L}((\Sigma R)^\infty)$ and $(\Sigma L)^\infty$ does not belong to $\mathcal{R}((\Sigma L)^\infty)$.

Among the L-shifts $\mathcal{L}((\Sigma R)^\infty)$ there is a maximal one, let us denote it by $(UL)^\infty$:

$$(UL)^\infty = \max\{\mathcal{L}((\Sigma R)^\infty)\}.$$

By definition all shifts of UL cannot be larger than UL, i.e.,

$$\mathcal{L}((UL)^\infty) \leqslant (UL)^\infty.$$

Replacing the last L by C_-, UC_- meets the condition (7.68) for K_-. Being a member of $\mathcal{L}((\Sigma R)^\infty)$, UL is definitely different from ΣL. Therefore, UC_- is a new kneading sequence K_- of the gap map, derived from ΣC of the unimodal map.

The same argument, applied to

$$(VR)^\infty = \max\{\mathcal{R}((\Sigma L)^\infty)\},$$

shows that VC_+ is a new kneading sequence K_+ of the gap map.

To summarize, from one and the same kneading sequence ΣC of the unimodal map we have derived four periodic kneading sequences for the gap map:

1. $K_- = \Sigma C_-$.
2. $K_+ = \Sigma C_+$.
3. $K_- = \max\{\mathcal{L}((\Sigma R)^\infty)\}|_{C_-}$.
4. $K_+ = \max\{\mathcal{R}((\Sigma L)^\infty)\}|_{C_+}$.

The notation $\Sigma|_C$ means replacing the last symbol in the string Σ by the symbol C. We have used this result in Section 3.4.

7.5.2 Number of Periods in the Lorenz-Like Map

The Lorenz-like map shown in Fig. 7.6 is slightly different from Fig. 3.11 due to our normalization (7.4). It has two kneading sequences $K_+ = f(C_+)$ and $K_- = f(C_-)$, where C_- is the rightmost point of L and C_+ — the leftmost point of R. All symbolic sequences are made of the two symbols L and R, both of even parity. The admissibility condition for kneading sequences without considering their compatibility follows from (3.71) and reads:

$$\begin{aligned}\mathcal{L}(K_-) &\leqslant K_-,\\ \mathcal{R}(K_+) &\geqslant K_+.\end{aligned} \tag{7.74}$$

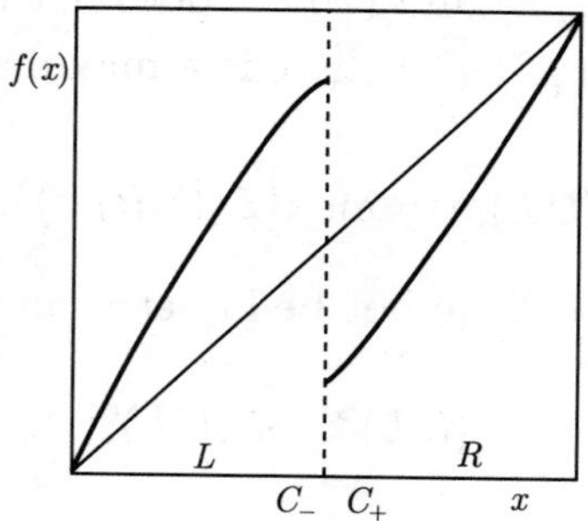

Figure 7.6 The Lorenz-like map

Due to the presence of discontinuity one has to consider $(XL)^\infty$ type and $(XR)^\infty$ type sequences separately. The number of such periodic sequences, as we have known in (7.43), is just $Z_2(n)$. Any periodic symbolic sequence of length n may be brought into shift-maximal form by cyclic permutations. All cyclic permutations are equivalent—this is just what "invariant under cyclic group C_n" means. In what follows we always take the shift-maximal form of a periodic sequence. Recollect that shift-maximality of a sequence Σ means

$$\begin{aligned} \mathcal{L}(\Sigma) &\leqslant \Sigma, \\ \mathcal{R}(\Sigma) &\leqslant \Sigma. \end{aligned} \tag{7.75}$$

Now let us first discuss K_-, i.e., kneading sequences ending with C_-. Take any one of the $Z_2(n)$ sequences. If it is of $(XL)^\infty$ type, the shift-maximality condition says

$$\mathcal{L}((XL)^\infty) \leqslant (XL)^\infty.$$

Replacing the last L by C_- by continuity, XC_- meets the admissibility condition for K_- in (7.74).

If the sequence is of $(XR)^\infty$ type, all members in the L-shifts $\mathcal{L}((XR)^\infty)$ must end with an L. Take the maximal one of them and denote it by $(UL)^\infty$. Being maximal in $\mathcal{L}((XR)^\infty)$ implies

$$\mathcal{L}((UL)^\infty) \leqslant (UL)^\infty.$$

Again by continuity we can replace the last L in UL by C_- and get a kneading sequence $K_- = UC_-$, which satisfies the first line in (7.74). Thus from each of $Z_2(n)$ sequences we get one K_-.

Repeating the above discussion with interchange $R \leftrightarrow L$ and $C_- \leftrightarrow C_+$, we get one K_+ from each of $Z_2(n)$ periodic sequences. Therefore, the total number of periodic kneading sequences of length n in the Lorenz-like map is $2Z_2(n)$.

However, the above argument does not work for $n = 1$ due to emptiness of the shifts. Direct inspection shows that we get C_- from L^∞ and C_+ from R^∞. Therefore, we have two kneading sequences $K_- = C_-$ and $K_+ = C_+$, i.e., $Z_2(1) = 2$.

7.6 Summary of the Counting Problem

Straightforward extension of various counting methods for the unimodal map of $m = 2$ to maps with larger m leads to the quantity $N_m(n)$, which is the number of periods in a degenerated one-parameter map whose maxima are bound to vary simultaneously and whose minima are located at $f = 0$. Consequently, the admissibility condition for symbolic sequences remains to be shift-maximality. This is crucial for the success of the generalization. Nevertheless, combinations of $N_k(n)$ with $k \leqslant m$ give the number of periods in all possible variations of a continuous m-lap map.

We mention in passing that the neat form of counting formulae does depend on the perfect normalization (7.4) of the mapping function f. Should one somehow deform the end points of the map, there might be changes in the number of periods, which cannot always be counted by a single term in the corresponding formulae. Anyway, "brute-force" counting by checking the admissibility conditions in symbolic dynamics always works.

When the end point $x = 1$ of the interval I is a critical point, it requires some minor modification of our results to get the number of periods (a critical point at $x = 0$ does not matter). This includes the sine-square map studied by Xie and Hao [1993]. Circle maps, e.g., the $m = 4$ degree 0 map and $m = 7$ degree 1 map studied by Zeng and Glass [1989], also fit well into the framework.

Retrospectively speaking, there are two counting problems and we have solved the easier one, namely, counting the total number of periodic orbits which may appear within the whole range of variation of the kneading sequences. The other, more difficult, problem consists in telling how these periods appear with the variation of one or more kneading sequences. In one-dimensional mappings, even when there are more than one independently varying parameters, all kneading sequences may be ordered in a unique way. Therefore, it makes sense to ask, given a kneading sequence K, how many superstable periods, which are smaller than K, have appeared. Naturally, the question may be answered on a computer, by using the kind of program we mentioned in Section 7.2.1. However, analytical approach so far has made only limited progress along this line. For example, Smale and Williams [1976] proved that, in the unimodal map, when the parameter reaches the period 3 window, i.e., $K = (RLR)^\infty$, the number of period n orbits is given by $F_{n+2} + F_{n-2} + 1$, where F_n is the nth Fibonacci number. Piña [1983] made some further study of this problem.

7.7 Cycle Expansion for Topological Entropy

In a chaotic attractor there are infinitely many unstable periodic orbits. An approximate picture of chaotic motion is random jumping of the system among various unstable periods. If all the periodic orbits, including their "local" behavior, i.e., the eigenvalues and eigenvectors of the periodic points, up to a certain length are known, the overall motion on the attractor may be recovered to a large extent. The time a chaotic orbit spends to follow an unstable period is inversely proportional to the absolute value of the eigenvalue. The motion following the unstable periodic orbit may be well represented by the linearized approximation to that period. This period expansion or "cycle expansion" has become an effective way to calculate various characteristics of chaotic systems (Cvitanović [1988]; Artuso, Aurell, and Cvitanović [1990]). This approach originates from the thermodynamic formalism of dynamical systems, which is beyond the scope of this book. However, we would like to stop at the simplest case of cycle expansion in order to show its close relation to symbolic dynamics.

The simplest case of cycle expansion is counting the number of periodic points. It has nothing to do with the metric property of a system, e.g., the eigenvalues of a periodic orbit. The number N_n of orbital points of a period n orbit is given by

$$N_n = \sum_{\{x \mid f^n(x)=x\}} 1. \tag{7.76}$$

The sum runs over all periodic points from 1 to n, each point having a weight 1. We further introduce a generating function $\Omega(z)$ for N_n:

$$\begin{aligned}\Omega(z) = \sum_{n=1}^{\infty} N_n z^n &= \sum_{\{n_p|n\}} n_p \sum_r (z^{n_p})^r \\ &= \sum_{\{n_p|n\}} \frac{n_p z^{n_p}}{1-z^{n_p}} = -z\frac{d}{dz}\ln\prod_p (1-z^{n_p}).\end{aligned} \tag{7.77}$$

Here p is the index for periodic orbits. The second sum in (7.77) runs over all primitive period n_p which is a factor of the period n. In other words, $\{n_p|n\}$ denotes all n_p that divides n. We have mentioned before that the growth rate of the number of orbital points is measured by the topological entropy h, i.e., $N_n \propto e^{hn}$. Then from the first equality in (7.77) it follows that

$$\Omega(z) \propto (1-ze^h)^{-1}. \tag{7.78}$$

Therefore, topological entropy is determined by the pole $z^* \sim e^{-h}$ of $\Omega(z)$. Similar to the Riemann ζ function, we can define a topological ζ function

$$\frac{1}{\zeta} = \prod_p (1 - z^{n_p}). \tag{7.79}$$

Then from the last equality of (7.77) we have

$$\Omega(z) = z\frac{d}{dz}\ln\zeta(z), \tag{7.80}$$

thus connecting the topological entropy to the zeros of the $1/\zeta$ function.

Given N letters, the total number of strings of length n made of these letters is N^n. Suppose that n has a factor d and the number of primitive periodic strings made of these letters is M_d. If no additional restrictions are imposed, the maximal possible M_d and N^n are related as

$$N^n = \sum_{\{d|n\}} dM_d. \tag{7.81}$$

Therefore, we can write down the number of primitive periods by using the inverse Möbius transform (7.13):

$$M_n = \frac{1}{n}\sum_{\{d|n\}} \mu(d)N^{n/d}, \tag{7.82}$$

where the Möbius function $\mu(n)$ has been defined in (7.7). Usually some primitive periods are forbidden by the dynamics, so M_n is less than what is given by (7.82).

All the M_n primitive periods have the same contribution to ζ. Therefore, (7.79) may be written as

$$\frac{1}{\zeta(z)} = \prod_{n=1}^{\infty}(1 - z^n)^{M_n} \equiv 1 - \sum_{k=1} C_k z^k. \tag{7.83}$$

If the above expansion contains a finite number of terms, it is called a topological polynomial. The topological entropy is determined by the roots of the polynomial. For example, there exists a complete symbolic dynamics for any set of N letters: any string of length n is allowed just as in the surjective unimodal map when there are only two letters. Since

$$\Omega(z) = \sum_{n=1}^{\infty} z^n N^n = \frac{Nz}{1 - Nz}, \tag{7.84}$$

we get

$$\zeta^{-1}(z) = 1 - Nz. \tag{7.85}$$

Of course, it satisfies (7.80). Therefore, from the zero of $1/\zeta(z)$ we obtain the topological entropy $h = \ln N$.

Denoting $t_p = z^{n_p}$, it follows from (7.79) that

$$\frac{1}{\zeta} = \prod_p (1 - t_p) = 1 - \sum_{p_1, p_2, \cdots, p_k} t_{p_1 + p_2 + \cdots + p_k}, \tag{7.86}$$

where

$$t_{p_1+p_2+\cdots+p_k} \equiv (-1)^{k+1} t_{p_1} t_{p_2} \cdots t_{p_k}. \tag{7.87}$$

The summation in (7.86) runs over all combinations of primitive periods. For $k > 1$, $t_{p_1+p_2+\cdots+p_k}$ corresponds to "pseudo-orbit", obtained by concatenation of short p_i-letter orbits.

Take, for example, the symbolic dynamics of the two symbols R and L. We have

$$\begin{aligned}\zeta^{-1} =& (1 - t_R)(1 - t_L)(1 - t_{RL})(1 - t_{RLL}) \cdots \\ =& 1 - t_R - t_L - t_{RL} - t_{RLL} - \cdots + (t_{R+L} + t_{R+RL} + t_{L+RL} + \cdots) \\ & -(t_{R+RL+RLL} + \cdots) + \cdots \\ =& 1 - t_R - t_L - (t_{RL} - t_R t_L) - [(t_{RLL} - t_{RL} t_L) \\ & +(t_{RLR} - t_{RL} t_R)] - \cdots .\end{aligned} \tag{7.88}$$

The last step in the above derivation is regrouping of terms according to their length. If the symbolic dynamics is complete, then positive and negative contributions in each parentheses cancel out completely. Only the contribution from the basic period R^∞ and L^∞ remain, leading to (7.85).

For the kneading sequence $K = RLC$ we take R and RL to be new symbols. The symbolic dynamics is complete with respect to these new symbols. Therefore, we have

$$\zeta^{-1} = 1 - t_R - t_{RL} = 1 - z - z^2, \tag{7.89}$$

it leads to topological entropy $h = \ln \dfrac{1+\sqrt{5}}{2}$.

A more complicated example is a symbolic dynamics of three letters $\{a, b, c\}$ with the string ab forbidden. It corresponds to a complete symbolic dynamics with respect to the new letters $\{a, cb^k\}_{k=0}^{\infty}$ in addition to the single symbolic sequence b^∞. Therefore, its ζ function satisfies

$$\begin{aligned}\zeta^{-1} &= (1 - t_b)\left(1 - t_a - \sum_k t_{cb^k}\right) \\ &= (1 - z)\left(1 - z - \frac{z}{1 - z}\right) \\ &= 1 - 3z + z^2,\end{aligned} \tag{7.90}$$

which yields a topological entropy $h = \ln\dfrac{3 + \sqrt{5}}{2}$.

There exist similar expansions for other characteristics averaged over chaotic motion, e.g., escape rate from a strange repeller, dimensions and Lyapunov exponents of chaotic attractors, etc. However, they are associated with metric property of the system, the corresponding ζ functions do not look so simple as that for topological entropy due to corrections in contribution to a period from pseudo-orbits of the same length.

8
Symbolic Dynamics and Grammatical Complexity

Symbolic representation of orbits in dynamical systems fits well into the framework of formal languages. Therefore, symbolic sequences may be studied from the viewpoint of language and grammar complexity. This approach was initiated by S. Wolfram (Wolfram [1984]) when he applied formal language theory to compare the complexity of various cellular automata. Many authors have used this approach to study symbolic sequences occurring in one-dimensional maps, see, e.g., Grassberger [1986a and b, 1988a and b], Auerbach and Procaccia [1990], Crutchfield and Young [1990], Hao [1991], and recent papers of the group of H. M. Xie (Xie [1993a and b, 1995, 1996]; Wang and Xie [1994]; Wang [1997]). Although most of the results obtained so far are restricted to either unimodal or subcritical circle maps, they have indeed provided a deeper understanding and a finer classification of orbit types. For example, now we know:

1. Periodic and eventually periodic symbolic sequences are the only regular languages in unimodal maps. This essential result closes the problem of regular languages in unimodal maps and opens the question of how to go beyond regular languages.

2. An effective way to go beyond regular languages consists in looking at the infinite limits of block-concatenation schemes. The block structure of Stefan matrices helps us to comprehend these infinite limits.

3. The Feigenbaum limiting set and their generalized counterparts as well as so-called Fibonacci maps all belong to context-sensitive language.

4. There are two types of Fibonacci sequences, even and odd ones. Though the even Fibonacci sequences are simpler than the odd ones, they both belong

to the same class of context-sensitive languages.

5. A quite plausible conjecture: there is no proper context-free language in the languages generated by unimodal maps.

When we say symbolic sequences in the above list, we always have in mind the kneading sequences of unimodal maps.

In this chapter we shall not develop formal language theory, nor provide detailed derivation of the important new results. Our aim is to demonstrate the usefulness of the approach. For a systematic exposition of formal language and automaton theory one may consult, e.g., the book by Hopcroft and Ullman [B1979]. For most of the new results related to symbolic dynamics there is an excellent new book by Xie [B1996]. We will confine ourselves to a general explanation of the basic concepts and main results without giving any proofs or mathematical derivations.

This chapter is organized as follows.

Section 8.1 is a brief review of formal language theory. The Chomsky hierarchy of grammatical complexity based on sequential production rules and the Lindenmayer system based on parallel productions will be described.

Section 8.2 introduces the notion of finite automata on the example of the simplest, regular, language. As a continuation of Section 2.8.4 of Chapter 2, we also explore the connection of the Stefan matrices to the transfer functions in the automaton theory. This will help us to go beyond regular languages in Section 8.3.

Section 8.3, the main body of this chapter, discusses how to construct symbolic sequences which go beyond regular languages in their grammatical complexity. We shall first see that even in unimodal maps one may generate infinitely many Fibonacci sequences, i.e., periodic sequences with periods growing as Fibonacci numbers. The two known types of Fibonacci sequences provide examples of two distinctive classes of context-sensitive language. Then two different decompositions of a kneading sequence will be introduced in order to reveal further characteristics of non-regular languages.

In Section 8.4 we conclude this chapter with a summary of what is known on the grammatical complexity of different types of symbolic sequences in the unimodal maps.

8.1 Formal Languages and Their Complexity

Formal languages are defined in a fairly general manner. A formal language is just a set of finite symbolic sequences or strings. The specification of production rules, i.e., how the words in a language are generated from a small initial set, defines a grammar. In principle, there may be infinitely many classification schemes for languages. We touch only two of them: the Chomsky hierarchy based on sequential production and the Lindenmayer system (or L-system for short) based on parallel production. Time evolution in dynamical systems singles out a special class of languages—dynamical languages. All the nice properties we are going to describe have their roots in the dynamical nature of the language.

8.1.1 Formal Language

First, take an alphabet, i.e., a finite set S of letters, say, $S = \{R, L\}$ or $S = \{0, 1, 2, 3\}$. Next, form all possible finite strings (words) made of these letters, including the empty string, which is represented by ϵ hereafter. The collection of all possible strings is denoted by S^*. Any subset $\mathcal{L} \in S^*$ is called a *language*.

With such a general definition one cannot go any further. One has to indicate how the subset $\mathcal{L}$ is obtained. For example, one can specify a set S_0 of initial letters and a set P of production rules. By repeated application of rules from P to the letters in S_0 and to strings thus obtained, one generates all words in the language. In short, the collection of $\{S, P, S_0\}$ specifies a *grammar* G. Then the language may be denoted as $\mathcal{L}(G)$.

We need a few simple notions related to a string and its substrings. If a word $x = uv$, then u is said to be a prefix of x and v a suffix of x. If $u \neq \epsilon$ it is a *proper prefix* of x; if $v \neq \epsilon$ it is a *proper suffix* of x.

We mention in passing that a language defined in the above way turns out to be closely related to a well-known algebraic structure called a *free monoid.* A set of objects $\{a_i\}$ closed with respect to a binary operation ("multiplication") between any two objects makes a *semigroup.* If, in addition, there is an unit element ϵ among the objects such that $\epsilon a_i = a_i\epsilon = a_i$, then it is a *monoid.* A free monoid may be generated from objects that can be treated as independent symbols. A typical free monoid is the above set S^*. If the inverse further exists for any element then a semigroup becomes a group, a notion more familiar to physicists. In our case, the multiplication operation is the concatenation of two strings and the unit is given by the empty string. The free monoid point of view helps in providing some simple proofs in language theory (Xie [B1996]), but we will not go into details.

Dynamical languages

Any symbolic sequence producible by the dynamics with an appropriate partition of the phase space is an admissible word in the dynamical language. The time evolution imposes severe restrictions on these words. First, any substring in an admissible word is also produced by the same dynamics, hence it must be an admissible word. Second, with time going on an admissible word gets longer and longer. Therefore, there is at least one letter in the alphabet that one can append to an admissible word to get another admissible word. These two properties are sometimes called *factorizability* and *prolongability* and they define the *dynamical languages.* We will not formalize these properties as we do not go into any proofs. We notice only that these conditions were recognized in very early work on symbolic dynamics (Morse and Hedlund [1938]).

Forbidden words and distinct excluded blocks

A word x is forbidden, if it does not belong to the language $\mathcal{L}$; clearly, it belongs to the complement of $\mathcal{L}$, i.e., $x \in \mathcal{L}' \equiv S^* - \mathcal{L}$. If a string x belongs to $\mathcal{L}'$, but all its proper substrings belong to $\mathcal{L}$, it is called a *distinct excluded block* (abbreviated as DEB). The collection of all DEBs of a language $\mathcal{L}$ is denoted by $\mathcal{L}''$. Any string, containing a substring $x \in \mathcal{L}''$, is necessarily a forbidden word. A dynamical language $\mathcal{L}$, or rather its complement $\mathcal{L}'$, may be completely specified by $\mathcal{L}''$. In fact, the following holds :

$$\mathcal{L}' = S^*\mathcal{L}''S^* = \{xyz \,|\, x, z \in S^*,\ y \in \mathcal{L}''\}.$$

In general, this is not true for languages other than dynamical ones.

8.1.2 Chomsky Hierarchy of Grammatical Complexity

In the mid 1950s when the design of programming languages had just started, a question of principle arose: how to judge whether a programming language is too complex to guarantee that a computer would run all programs written in that language. This led to the problem of classification of languages and computers according to their complexity. The problem was solved by N. Chomsky in the framework of formal language theory. We describe briefly the Chomsky classification. For details one may consult the book by Hopcroft and Ullman [B1979] or the new book by H. M. Xie [B1996] and references therein.

In the Chomsky scheme symbols in the alphabet S are further subdivided into terminal and non-terminal ones. Non-terminal symbols are used as variables in the process of production; the final strings or "words" in the language consist of terminal symbols only.

Chomsky classified all languages into four classes. The simplest one, Class 3 according to Chomsky, is *Regular Language* (RGL), which is accepted by a finite automaton without memory. We will describe this class in some detail in the subsequent sections, using examples from unimodal maps.

The next, Class 2, is *Context-Free Language* (CFL), which needs a "push-down" automaton, i.e., an automaton with a memory which is organized as a push-down list, to recognize it. A push-down list (called also a stack or a first-in-last-out list) is a sequentially organized memory where a new content is always written into the top word in the memory while all existing contents are pushed down. When the memory is read, the topmost word is always fetched and all other words popped-up. Most of the present-day programming languages, e.g., FORTRAN or BASIC, belong to this class.

The Class 1 language is called *Context-Sensitive Language* (CSL). It requires a "linearly-bounded automaton" to accept it. This is an automaton with a memory, whose size is proportional to the input length.

The most complex Class 0 in the Chomsky hierarchy is *Recursively Enumerable Language* (REL), which must be accepted by a Turing machine with infinite memory.

For the time being, there has been accumulated a significant amount of knowledge on RGL and CFL, some knowledge on CSL, and a few results on REL. There exist, of course, languages even more complex than REL. However, by the very definition, it is impossible to construct any explicit examples of these languages, thus for the time being they are beyond the power of computers and algorithms.

If a dynamical language L belongs to Class 0 or 1 or 3, so does its set

L'' of DEBs. The question in case of Class 2 (CFL) is open, but there is a conjecture (Xie [B1996]):

Conjecture 1 A dynamical language L is context-free if and only if its L'' is context-free.

However, this might be an empty statement for unimodal maps, as at present we do not know any CFL generated by symbolic sequences of unimodal maps. In fact, there has been another conjecture (Xie [B1996]):

Conjecture 2 For unimodal maps if L is a context-free language, then it is a regular language. In other words, there is no proper CFL associated with unimodal maps.

8.1.3 The L-System

The Chomsky classification of languages is based on sequential production rules, which we did not elaborate. In 1968 the developmental biologist A. Lindenmayer introduced a symbolic system to describe the growth of multi-cell plants. His system was soon shown to be another classification scheme of formal languages based on parallel production rules. There are a few books on the L-systems (e.g., Rozenberg and Salomaa [B1980]; Prusinkiewicz and Hanan [B1989]; Prusinkiewicz and Lindenmayer [B1990]), so we are contented with touching a few terms only.

The simplest in the L-system is called a D0L, meaning Deterministic and no (0) interaction, i.e., all symbols in a word are replaced according to predefined rules in a context-free fashion without looking at their neighbors.

The generalized composition rules $R \to \rho$ and $L \to \lambda$ studied in Section 2.5.2, applied to all symbols in a word simultaneously, may be viewed as a set of parallel production rules.

There exist 0L and IL; "I" stands for Interaction, i.e., the substitutions performed depend on the neighboring letters. In other words, IL is context-sensitive. Both 0L and IL are more complex than D0L. If the production rules are chosen from a set of rules called a Table, the language becomes T0L or TIL. In the original L-systems, no distinction was made between terminal and non-terminal symbols as in the Chomsky system. Inclusion of this distinction has extended, say, T0L to ET0L.

There is no absolute scale of complexity in formal languages. A language located fairly high on the Chomsky ladder may happen to be quite simple in the L-system and *vice versa.* A scheme is drawn in Fig. 8.1 to show the relationship of various classes of languages. In this figure a class occupying a higher position along a connecting line contains the lower ones as proper subsets. For example, ET0L and E0L are beyond CFL and belong to CSL. In fact, one way to prove that the Feigenbaum limit of period-doubling cascade is a CSL consists in showing that it is an ET0L (Chen, Lu and Xie [1993]). In Fig. 8.1 a name not mentioned above is the indexed language (IND). It is a CSL, but not necessarily CFL. The Feigenbaum attractor has been shown to be an IND, hence a CSL (Crutchfield and Young [1990]), but not CFL.

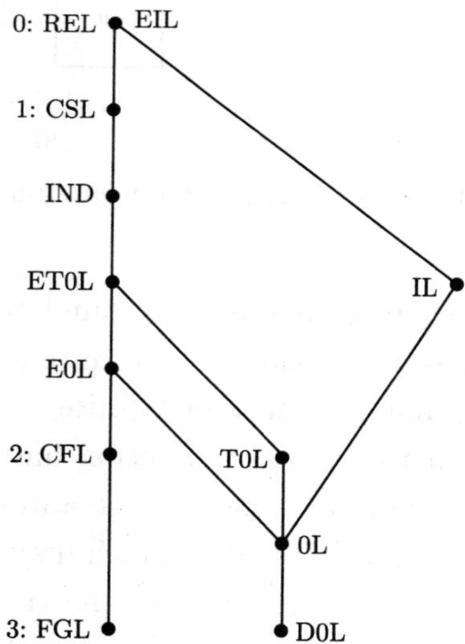

Figure 8.1 Relationship of various classes of languages

8.2 Regular Language and Finite Automaton

Regular languages may be defined with the help of finite automata and *vice versa.* Therefore, we start from the description of a finite automaton.

8.2.1 Finite Automaton

A finite automaton is shown schematically in Fig. 8.2. The box labeled by a or b represents an automaton in state a or b. The arrow on top of the box is a reading head which reads an input symbol from a tape. The string on the tape is a word from a certain language. The head reads one symbol at a time, then the tape moves one cell to the left, preparing for the next reading. The automaton changes its state according to a *transfer function* $\delta(q, s) = q'$, which says that the automaton in state "q" changes into state "q'" upon reading a symbol s from the tape. The situation shown in Fig. 8.2 is given by $\delta(a, R) = b$.

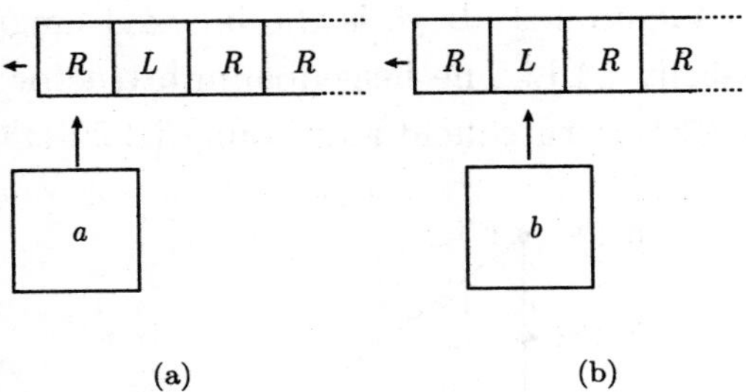

Figure 8.2 A finite automaton

When both the number of states and the number of different symbols on the tape are finite, the system is said to be a finite automaton. However, the length of the input string may be finite or infinite.

For a finite automaton the transfer function has discrete values and can be represented by a table. Suppose that for the automaton shown in Fig. 8.2 there are only two input symbols R and L which may appear on the tape, and there are four different states a, b, c, and d, the transfer function may look like the table shown in Section 8.2.3 below.

8.2.2 Regular Language

Among the symbols on the tape there may be one or more which tell the automaton to stop. If after reading a finite string of symbols from the tape the automaton stops, or, in the case of infinite input string, the automaton keeps changing its states without getting confused, the string is said to be accepted or recognized by the automaton. The automaton may stop when finishes reading a finite string. The collection of all finite strings acceptable by a given finite automaton makes a *regular language*. Conversely, a regular language may be accepted by one or another finite automaton, not necessarily unique.

For the sake of brevity, the language defined by the kneading sequence K in the symbolic dynamics of the quadratic map is called *Language* $\mathcal{L}(K)$.

It is well-known now that periodic symbolic sequences $(\Sigma C)^\infty$ and eventually periodic symbolic sequences $\rho\lambda^\infty$ correspond to the simplest regular language on the Chomsky ladder of complexity. We will show this by making use of an observation that the Stefan matrix of a periodic or eventually periodic symbolic sequence defines the transfer function of the corresponding automaton.

The converse statement has recently been proved by H. M. Xie [1993a]. Namely, in the language $\mathcal{L}(K)$ of the quadratic maps, periodic and eventually periodic symbolic sequences are the only two types of sequences which correspond to regular language. This strong result closes the problem of regular languages in the language $\mathcal{L}(K)$ and opens the question of how to construct more complex languages in $\mathcal{L}(K)$. Again Stefan matrices will give us some hints as how to go beyond regular languages.

Therefore, it is appropriate to recollect what has been said on Stefan matrices in Section 2.8.4 in order to make further connection with automaton theory.

8.2.3 Stefan Matrix as Transfer Function for Automaton

We pick up again the period 5 sequence $(RLRRC)^\infty$, which has been used in Section 2.8.4 as an example for the construction of its Stefan matrix. We reproduce Fig. 8.2 as Fig. 8.3 with the labels I_j of the subintervals changed to a, b, c, and d. We shall call each of these subintervals a *state*.

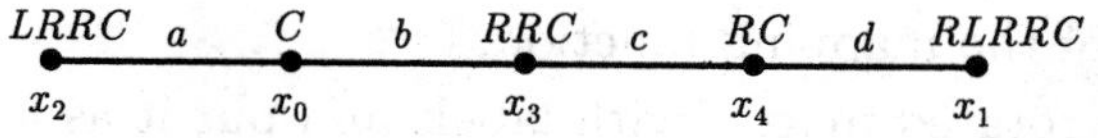

Figure 8.3 Subintervals, determined by a periodic kneading sequence $(RLRRC)^\infty$

If one starts from any of the orbital points, the orbit will be visited in the order given by the arrows in Fig. 8.4. As we have seen in Section 2.8.4, the four subintervals will transform into each other as follows:

$$\begin{aligned} a &\xrightarrow{L} c+d, \\ b &\xrightarrow{R} d, \\ c &\xrightarrow{R} b+c, \\ d &\xrightarrow{R} a. \end{aligned} \tag{8.1}$$

This is a simple consequence of the monotonicity of the mapping function on the corresponding subinterval. In fact, the symbols R and L on the arrows indicate which branch the state corresponds to. These symbols are the input letters in defining the transfer function of the automaton.

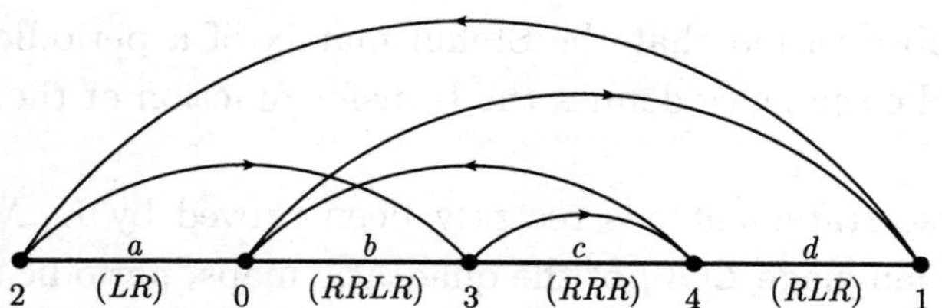

Figure 8.4 Order of visits in the periodic kneading sequence $(RLRRC)^\infty$

The transformation (8.1) is represented by a transfer matrix, i.e., the Stefan matrix

$$\mathbf{S} = \begin{bmatrix} 0 & 0 & 1 & 1 \\ 0 & 0 & 0 & 1 \\ 0 & 1 & 1 & 0 \\ 1 & 0 & 0 & 0 \end{bmatrix}. \tag{8.2}$$

Note that in the last column of the Stefan matrices there are always two adjacent 1's. Rotating the matrix counter-clockwise by 90 degrees, the non-zero elements in the matrix will form a one-hump pattern: from the lower left corner to the topmost row all 1's go in a monotone increasing fashion. Starting from the right 1 in the topmost row, all 1's are located in a monotone decreasing manner. It is easy to see that this pattern and the fact that in any column of the Stefan matrix there are at most two 1's follow from the unimodal shape of the mapping function.

We equip the rotated matrix with labels and put it as a table:

	R			
q	a	b	c	d
d	1	1	0	0
c	1	0	1	0
b	0	0	1	0
a	0	0	0	1

The last table may be put into the standard form of a discrete transfer function:

	R	L
a		c,d
b	d	
c	b,c	
d	a	

In fact, this transfer function may be written down directly from the transformation (8.1). Once we have a transfer function, it is straightforward to draw the finite automaton. We first get a *non-deterministic finite automaton* (NDFA). Then, using a standard procedure called *the subset construction*, we derive a *deterministic finite automaton* (DFA). NDFA and DFA are equivalent in their ability to accept a regular language. DFAs accepting the same regular language may have different number of nodes. Among them there is one with the minimal number of nodes. This number may be determined from the language and is used to compare complexity of regular languages (Wolfram [1984]).

Non-deterministic finite automaton

An automaton may be described by a graph. Fig. 8.5 shows the automaton defined by the transfer function given in the table above. A state is represented by a node in the graph. A symbol put on an arc indicates the transition between two states when reading in the symbol. Finite number of states leads to finite number of nodes, hence the term finite automaton. However, this is a *non-deterministic* finite automaton, because:

A. no starting node is indicated and one can begin tracing the graph from any state;

B. at some nodes, e.g., at node a or c, there are two choices upon reading in one and the same symbol.

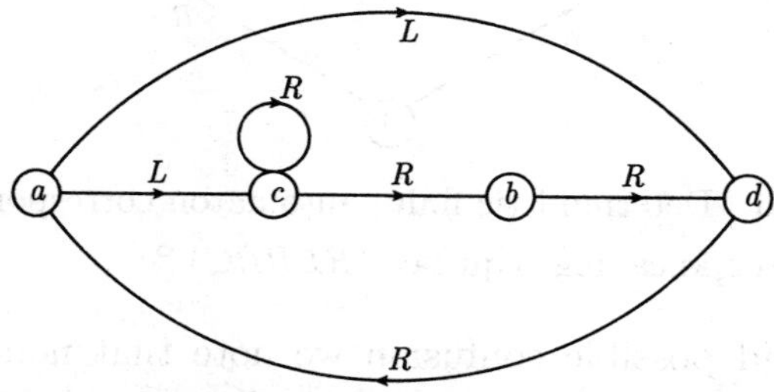

Figure 8.5 Non-deterministic finite automaton corresponding to the periodic kneading sequence $(RLRRC)^\infty$

Deterministic finite automaton

In the theory of language and automata there is a simple construction to obtain a *deterministic* finite automaton from a non-deterministic one. We show this *subset construction* on our example. Let us collect all states in the definition of the transfer function into a new state $1 = \{a, b, c, d\}$. From the original definition it follows that upon reading a symbol R this state goes into itself, while upon reading a symbol L it goes into a new state $2 = \{c, d\}$. Continue this construction, we get the transfer function:

	R	L
$1 = \{a, b, c, d\}$	$\{a, b, c, d\}$	$\{c, d\}$
$2 = \{c, d\}$	$\{a, b, c\}$	
$3 = \{a, b, c\}$	$\{b, c, d\}$	$\{c, d\}$
$4 = \{b, c, d\}$	$\{a, b, c, d\}$	

The finite automaton built according to this transfer function is drawn in Fig. 8.6. In this graph there is a unique starting state 1, which is represented by a double circle. Moreover, at each node there is only one definite choice upon reading a symbol. This shows that it is indeed a deterministic automaton. S. Wolfram [1984] suggested to use the number of nodes in a minimal deterministic automaton to characterize the complexity of the original dynamics. Of course, this works only when the number of nodes is finite, i.e., when one deals with regular language.

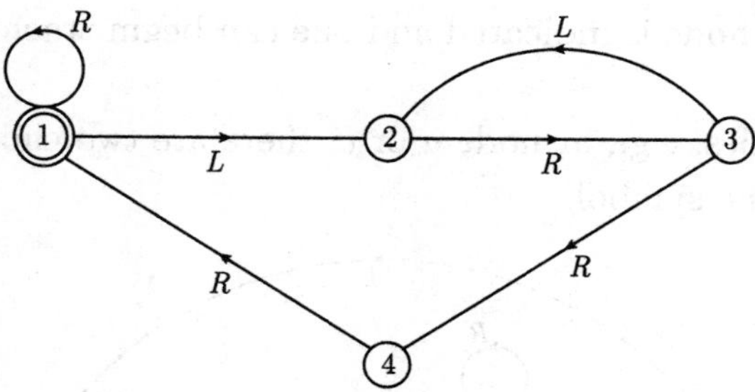

Figure 8.6 Deterministic finite automaton corresponding to the periodic kneading sequence $(RLRRC)^\infty$

In order to avoid possible confusion we note that non-deterministic and deterministic automata are entirely equivalent to each other. The usage of these terms in language theory is different from that in dynamics.

For an eventually periodic symbolic sequence, e.g., the sequence RLR^∞, which corresponds to the parameter in the bifurcation diagram where two chaotic bands merge into one band, the points on the interval will be visited as shown in Fig. 8.7. It is easy to write down the Stefan matrix and construct the finite automaton, both non-deterministic and deterministic. We leave this to the reader as an exercise.

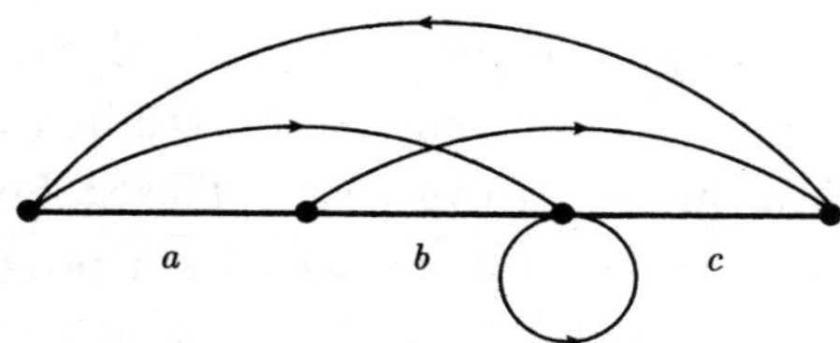

Figure 8.7 Order of visits in the eventually periodic kneading sequence RLR^∞

8.3 Beyond Regular Languages

One way to construct symbolic sequences whose complexity might go beyond regular language is to look at specially designed series of periodic sequences $(\Sigma_n C)^\infty$ or eventually periodic sequences $\rho_n \lambda_n^\infty$ obtained, for example, by applying the *doubling transformation* (2.70)

$$\begin{aligned} L &\to RR, \\ R &\to RL, \end{aligned} \tag{8.3}$$

repeatedly to R and L, and starting with R. For any finite n they are necessarily regular, but the $n \to \infty$ limits of such series may be more complex. From the result of Xie [1993a] mentioned at the end of Section 8.2.2, it follows that in order to prove that the limits are more complex than regular, it is enough to show that they are neither periodic nor eventually periodic.

This has been done (Xie [1993b]) for the limits of several "block concatenation" schemes, including the four cases given in Section 8.3.2 below. As the resulted automata are necessarily infinite ones, we need a way to comprehend the structure of these infinite objects. It turns out that the block structure of the Stefan matrices may be of great help to this end.

In order to reveal further distinctive features of non-regular languages, we need a few more new notions related to kneading sequences. We have in mind the odd and even *maximal primitive prefixes* of a given kneading sequence.

The *odd* maximal primitive prefixes lead to the definition of so-called *kneading map*. The *even* maximal primitive prefixes define the *distinct excluded blocks*, i.e., forbidden words, in the language $\mathcal{L}(K)$. We shall explain these notions by working out several examples.

8.3.1 Feigenbaum and Generalized Feigenbaum Limiting Sets

Repeatedly applying the doubling transformation (8.3) to what one obtains from a seed symbol R, we get all members R^{*n} of the Feigenbaum period-doubling cascade, where $*$ is the $*$-composition studied in Section 2.5.1. The length of these periodic sequences grows as 2^n. The Stefan matrices associated with these members may be generated easily by a program. Fig. 8.8 shows the Stefan matrix of the period-256 orbit. Although the size of the Stefan matrices goes to infinity with n, they all have a rather regular block structure.

It has been known for some time that the language $\mathcal{L}(R^{*\infty})$ belongs to context-sensitive language (Crutchfield and Young [1990]; Chen, Lu and Xie [1993]). In fact, one may take the $n \to \infty$ limit of W^{*n} with any primitive word W. This leads to the limit of the period-k-tupling sequence, where $k = |W|$ is the length of W. One may also look at the period-doubling sequence $W * R^{*n}$ in the one-band chaotic region or $R^m * W * R^{*n}$ in the 2^{m-1} band regions. These limiting sets are called *generalized Feigenbaum attractors*. They all belong to context-sensitive language (Lu [1994]).

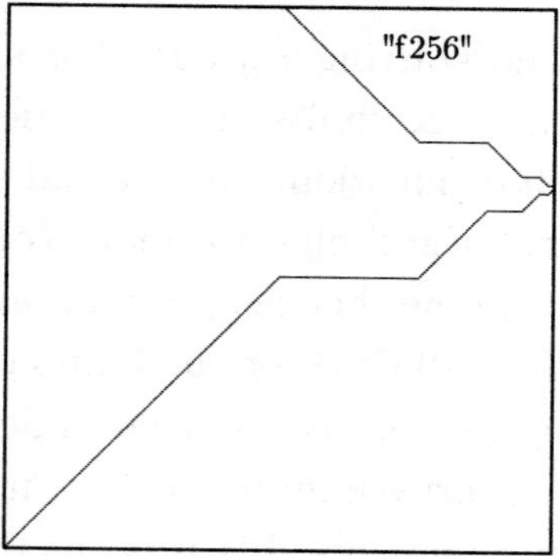

Figure 8.8 The Stefan matrix of the period-256 orbit in the Feigenbaum period-doubling cascade. A dot denotes a 1, while all other elements are zero

8.3.2 Even and Odd Fibonacci Sequences

There are many other ways to generate infinite kneading sequences which are neither periodic nor eventually periodic. For example, one may introduce the following "block concatenation" scheme (Hao [1991]):

$$\begin{aligned} b_{2n} &= b_{2(n-1)}b_{2n-1}, \\ b_{2n+1} &= b_{2n}b_{2n-1}, \end{aligned} \tag{8.4}$$

with given initial blocks b_0 and b_1. A word obtained in this way may not be shift-maximal. However, since we are interested in periodic sequences, any finite string may be cyclic-shifted to get the maximal word. The following four assignments all lead to sequences whose lengths grow as Fibonacci numbers. We call them *Fibonacci sequences.*

Case (a) $b_0 = L$, $b_1 = RR$. For the map (2.30) it converges to $\mu = 1.710\,398\,94\cdots$.

Case (b) $b_0 = R$, $b_1 = LR$ or $b_1 = RL$. It coincides with the Fibonacci sequence (2.92) that we have seen in Section 2.5.5 and converges to $\mu = 1.714\,744\,850\cdots$.

Case (c) $b_0 = L$, $b_1 = RL$ or $b_1 = LR$. It converges to $\mu = 1.858\,511\,400\,5\cdots$.

Case (d) $b_0 = R$, $b_1 = LL$. It converges to $\mu = 1.988\,787\,569\,39\cdots$.

It is easy to generate the Stefan matrices for periodic orbits in all four cases of the Fibonacci sequences. Examples of these matrices: for $F_{13} = 233$ are shown in Fig. 8.9. They are 232×232 matrices; an element 1 is represented by a dot and all zero elements are left blank. The structure of these Stefan matrices hints on different level of complexity of the corresponding language.

Case (a) always leads to matrices of a fixed number of blocks. If we examine the block structure only in the horizontal direction, there are in total 3 blocks, no matter how large n is. If we replace each block by an effective node when drawing an automaton for it, there are only 5 effective nodes in the $n \to \infty$ limit[1]. The other three cases look similar to each other. For instance, the number of blocks grows linearly with n, namely, for the n-th Fibonacci period F_n, this number is given by

$$2 \times \left[\frac{n+2}{3}\right], \quad 2 \times \left[\frac{n+1}{3}\right], \quad 2 \times \left[\frac{n}{3}\right],$$

[1]Taking this opportunity we correct a misprint in Hao [1991]: "Case 2" in the last sentence of §2.8 on p. 167 should read "Case 1".

for Cases (b), (c), and (d), respectively. It is not difficult to infer how new blocks come into play.

Intuitively, Case (a) looks simpler than other cases. The essential difference among them has been recognized recently (Wang [1997]). Wang and Xie have proved that, although all four cases belong to context-sensitive language, Case (a) is indeed simpler. In order to distinguish them, we will need a few notions to be explained in the next Section.

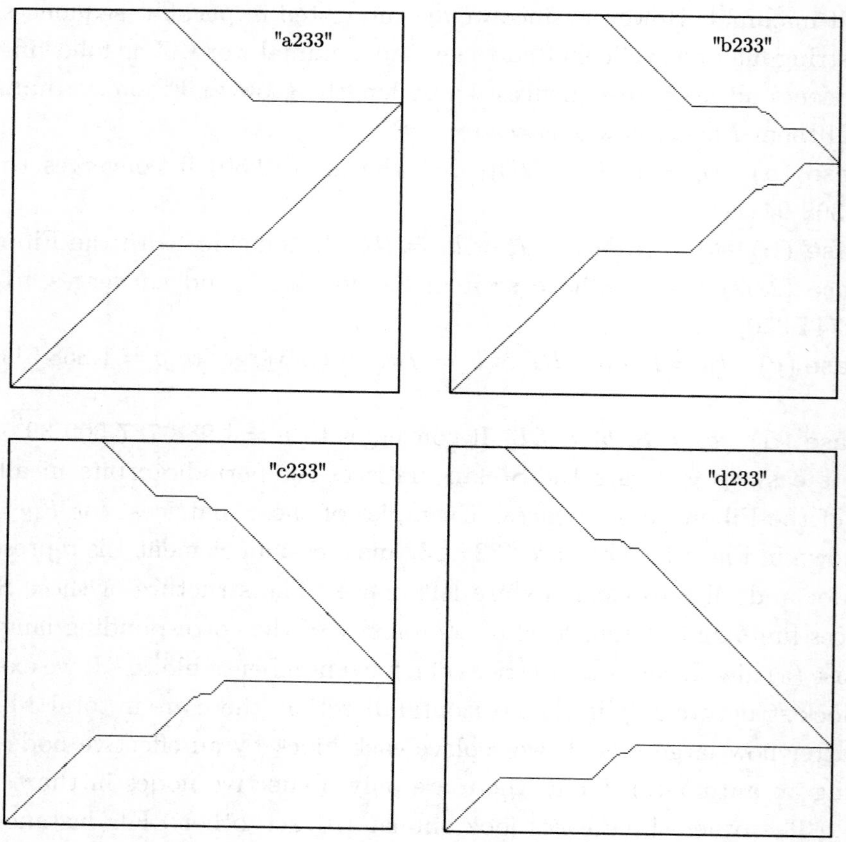

Figure 8.9 Stefan matrices for the Fibonacci sequences of period $F_{13} = 233$ at Cases (a) to (d), see text

The block-concatenation method described above consists of two operations: recursive generation of longer sequences and cyclic shifts to pick up the shift-maximal one. A more powerful and general method to construct infinite admissible words consists in using homomorphism on a free submonoid. Namely, one defines a homomorphism

$$h = (\alpha \to \rho, \beta \to \lambda)$$

and applies it to, say, α, then the limit

$$m_\infty = \lim_{n\to\infty} h^n(\alpha)$$

is a fixed point of h, i.e., $m_\infty = h(m_\infty)$. Here ρ and λ are strings consisting of α and β. Moreover, m_∞ is neither periodic nor eventually periodic, unless $\alpha\beta = \beta\alpha$. Consequently, the language thus generated is not regular, but Xie and coworkers have proved that it is at most context-sensitive. As examples, we list the homomorphisms that generate the four aforementioned Fibonacci sequences in Table 8.1 (Xie [1996]).

Table 8.1 Homomorphisms that generate the Fibonacci sequences

Case	α	ρ	β	λ
a	RLR	$RLRRLRRR$	RR	$RLRRR$
b	R	$RLRRL$	L	RRL
c	RL	$RLLRLRLL$	L	$RLRLL$
d	R	$RLLLL$	LL	$RLLRLLLL$

The generalized composition rules introduced in Section 2.5.5 are particular cases of homomorphisms when $\alpha = R$ and $\beta = L$. This tells us the close relation between the generalized composition rules, the parallel production rules in L-systems, and the homomorphisms on free submonoids.

8.3.3 Odd Maximal Primitive Prefixes and Kneading Map

A kneading sequence

$$K = \sigma_0\sigma_1\sigma_2\cdots\sigma_n\cdots$$

have infinitely many prefixes. Putting aside a few simple exceptions, we confine ourselves to those kneading sequences which start with RL, i.e., $\sigma_0 = R$ and $\sigma_1 = L$. We pick up longer and longer prefixes, which satisfy the conditions:

1. They contain an *odd* number of R.
2. They are *primitive*, i.e, not repetition of a shorter string.
3. They are the *maximal* ones under cyclic shifts of the primitive string.

Hence the name *odd maximal primitive prefixes* (OMPP). We denote these longer and longer OMPP by y_1, y_2, y_3, $\cdots$.

The Feigenbaum map

Take, for example, the kneading sequence that corresponds to the infinite limit of the Feigenbaum period-doubling cascade:

$$RLRRRLRLRLRRRLRRRLRRRLRLRLRRRLRLRLRRRLR\cdots.$$

We have

$$\begin{aligned} y_1 &= R, \\ y_2 &= RL, \\ y_3 &= RLRR, \\ y_4 &= RLRRRLRL, \\ y_5 &= RLRRRLRLRLRRRLRR, \\ &\cdots \end{aligned} \tag{8.5}$$

We note that the length of y_n grows as 2^{n-1}. Now, introduce a set $\{B_n\}$ of strings in the following recursive way:

$$\begin{aligned} y_1 &= B_0 = R, \\ y_2 &= y_1 B_1 = RL, \\ &\cdots \\ y_{n+1} &= y_n B_n, \\ &\cdots \end{aligned} \tag{8.6}$$

i.e., these $\{B_n\}$ lead to a decomposition of the kneading sequence

$$K = B_0 B_1 B_2 \cdots B_n \cdots. \tag{8.7}$$

These $\{B_n\}$ possess a nice property. Denote by $\overline{x}$ a string which differs from the string x only in the last symbol, i.e., by interchanging $R \leftrightarrow L$ at the last place one gets $\overline{x}$ from x and *vice versa*. For $n > 0$ each $\overline{B}_n$ is an OMPP, i.e. it coincides with one of the previous y_i. Consequently, $\overline{B}_n$ itself may be written as the concatenation of a number of consecutive B_i up to a certain B_m:

$$\overline{B}_n = B_0 B_1 \cdots B_m.$$

The dependence of the integer m on n is determined by the map. In other words, m is a function of n. Denoting this function by $Q(n)$, we have

$$\overline{B}_n = B_0 B_1 \cdots B_{Q(n)}, \tag{8.8}$$

which defines a map from integer n to $Q(n)$. It is called a *kneading map* (Hofbauer [1980]).

Using the same Feigenbaum map as an example, from (8.5) we have

$$\begin{aligned}
&B_0 = R, && \\
&B_1 = L, && \overline{B}_1 = y_1 = B_0, \\
&B_2 = RR, && \overline{B}_2 = y_2 = B_0B_1, \\
&B_3 = RLRL, && \overline{B}_3 = y_3 = B_0B_1B_2, \\
&B_4 = RLRRRLRR, && \overline{B}_4 = y_4 = B_0B_1B_2B_3, \\
&\quad\cdots && \quad\cdots
\end{aligned} \tag{8.9}$$

As suggested by these lines, the kneading map of the Feigenbaum limit $R^{*\infty}$ may be shown to be (de Melo and van Strien [B1993])

$$Q(n) = n - 1, \quad \forall n \geqslant 1. \tag{8.10}$$

If one looks at the numerical orbit of the Feigenbaum attractor, starting from the critical point $C = 0$, then there is a nice "recurrent" property: the orbital point x_i returns closer and closer to 0 when $i = 2^n, n > 1$.

The Fibonacci map

There are other maps with this kind of regular recurrent property. For example, one may require that x_i returns closer and closer to the vicinity of 0 when $i = F_n$ with n increasing, where F_n is the n-th Fibonacci number. This is realized in so-called *Fibonacci map.* The symbols in the kneading sequence of the Fibonacci map

$$K = \sigma_1\sigma_2\sigma_3\sigma_4\cdots$$

are defined in the following way.

1. At the closest returns to zero let

$$\sigma_{F_n} = \begin{cases} R, & \text{for } n = 0, 1 \mod 4, \\ L, & \text{for } n = 2, 3 \mod 4. \end{cases}$$

2. Between any two consecutive closest returns σ_{F_n} and $\sigma_{F_{n+1}}$ just insert the first $F_{n-1} - 1$ symbols $\sigma_1\sigma_2\cdots\sigma_{F_{n-1}-1}$ already obtained in the construction. Clearly, this is nothing but the symbolic way of saying that a close return has taken place at σ_{F_n}. Thus we obtain the kneading sequence

$$K = RLLRRRLRRLLRLRLLRRRLLRLLRRR\cdots.$$

In a quadratic map there is only one parameter $\mu = 1.870\ 528\ 632\ 1\cdots$ (Lyubich and Milnor [1993]), where the Fibonacci recurrence happens. We remind the reader that the form (2.30) or (2.31) is used for the map.

For this kneading sequence the length of y_n grows as F_n. The first $\{B_n\}$ are

$$\begin{array}{ll} B_0 = R, & \\ B_1 = L, & \overline{B}_1 = y_1 = B_0, \\ B_2 = L, & \overline{B}_2 = y_1 = B_0, \\ B_3 = RR, & \overline{B}_3 = y_2 = B_0B_1, \\ B_4 = RLR, & \overline{B}_4 = y_3 = B_0B_1B_2, \\ B_5 = RLLRL, & \overline{B}_5 = y_4 = B_0B_1B_2B_3, \\ \cdots & \cdots \end{array} \tag{8.11}$$

suggesting that the kneading map of the Fibonacci map is

$$Q(n) = n - 2, \quad \forall n \geqslant 2. \tag{8.12}$$

Indeed, it is (de Melo and van Strien [B1993], Example 3.5).

Just a reminder: please do not confuse Fibonacci map with maps which have periodic kneading sequences with the period growing as Fibonacci numbers, e.g., the Cases (a) to (d) studied in Section 8.3.2.

Note that both maps (8.10) and (8.12) are "unbounded", as compared to the kneading map of a periodic orbit, e.g., $K = (RLR)^\infty$. However, it can be shown that $Q(n) < n$ for any kneading map.

The unboundedness of $Q(n)$ is equivalent to the unboundedness of $\{|B_n|\}$. Here, as before, $|x|$ denotes the length of the string x. Recently, Wang and Yang (see, e.g., Wang [1997]) have proved that the language $\mathcal{L}(K)$ of all unimodal maps with unbounded kneading map $Q(n)$ belongs to context-sensitive or more complex languages, but not context-free. The Feigenbaum attractor and Fibonacci map are particular cases of this general result.

8.3.4 Even Maximal Primitive Prefixes and Distinct Excluded Blocks

Instead of odd maximal primitive prefixes of a kneading sequence K, one may look at *even maximal primitive prefixes* (EMPP) as well (Xie [B1996]). EMPP lead to another decomposition similar to (8.7) for kneading sequence K:

$$K = b_0b_1b_2\cdots b_n\cdots. \tag{8.13}$$

The importance of the $\{b_n\}$ consists in that the set $\{b_0 b_1 \cdots \overline{b}_n\}$ is nothing but the set of all *distinct excluded blocks* of the language $\mathcal{L}(K)$ (Xie [1995, B1996]). For example, for $K = (RLL)^\infty$, $b_0 = RLLR$, $b_1 = L$ and $b_2 = L$, $K = b_0 b_1 b_2 K$, and $L'' = \{RLLL, RLLRR, RLLRLR\}$. One again distinguishes the cases of $\{|b_n|\}$ being bounded or not. The odd Fibonacci sequence F_o has unbounded $\{|b_n|\}$, but the even F_e has bounded $\{|b_n|\}$.

8.4 Summary of Results

Grammatical complexity analysis of the language $\mathcal{L}(K)$ has enriched the classification of symbolic sequences in the unimodal map. Besides the periodic and eventually periodic sequences, we have explicitly some more complex orbit types, represented by even and odd Fibonacci sequences, Feigenbaum attractors, and Fibonacci maps. They are circumscribed by boundedness of $Q(n)$, $\{|B_n|\}$, $\{|b_n|\}$, and a few more characteristics not mentioned. Without going into technicalities of the mathematical proofs, we summarize the results in Table 8.2. For those readers who wish to learn more we recommend the book by Xie [B1996] and recent publications of Xie's group.

Table 8.2 A summary of grammatical complexity of symbolic sequences

Language	Symbolic Sequences	Remarks
REL		
CSL	Fibonacci map	$Q(n)$ and $\{\lvert B_n\rvert\}$ unbounded
	Feigenbaum attractor(ET0L)	
	Odd Fibonacci F_o	$\{\lvert b_n\rvert\}$ unbounded
	Even Fibonacci F_e (ET0L)	$\{\lvert b_n\rvert\}$ bounded
CFL	?	?
RGL	Σ^∞ and $\rho\lambda^\infty$	

The fact that no real chaotic orbits have been assigned a definite place on the ladder of grammatical complexity involves the very nature of chaos. So far the best we can do in uncovering the structure of chaotic orbits is to indicate the building blocks; any random concatenation of these blocks yields an admissible chaotic orbit. With the advance of symbolic dynamics and language theory we will be able to elucidate more types of chaotic orbits, but this can never be done in an exhaustive way. There is always more chaos to be appreciated.

9

Symbolic Dynamics and Knot Theory

Knot theory studies closed loops and their intertwining in three-dimensional space by looking for their classification scheme and various invariants. Ever since the discovery of the new linking polynomials and their relation with partition functions of some exactly solvable models in statistical mechanics, knot theory has drawn more and more attention of physicists.[1]

However, there has been another development which is more directly associated with the notion of knots and links but less known to physicists. This is the application of knot theory to the characterization of chaotic attractors in nonlinear systems. Both knot theory and symbolic dynamics are topological in nature. The relation between the two topological approaches has been addressed by mathematicians in a general but abstract setting many years ago (Franks [1980, 1981]), but operable applications of knot theory to dynamical systems started much later by physicists.

In a three-dimensional phase space periodic orbits, stable or unstable alike, appear as knots. In a chaotic attractor there are infinitely many unstable periodic orbits. Once periodic orbits in a chaotic attractor are viewed as knots, many questions come to one's mind. For example, given a nonlinear system in chaotic regime, are the unstable periodic orbits knotted? If so, what are their knot types? Taking two or more unstable periodic orbits, are they linked? If so, what are the link types?

In fact, these questions have been first studied in the context of the Lorenz model and periodic orbits in this model are shown to be knotted (Birman

[1]For a nice review along the line of integrable systems see F. Y. Wu, *Rev. Mod. Phys.*, 1992(**64**): 1099.

and Williams [1983a and b]). Later on relative rotation rate of periodic orbits were introduced and were shown to be associated with linking numbers of knots (Solari and Gilmore [1988a]; Tuffilaro, Solari and Gilmore [1990]). This approach has been applied to many systems described by differential equations, e.g., the NMR-laser model (Tuffilaro *et al.* [1991]), the Duffing equations (Solari and Gilmore [1988b]; McCallum and Gilmore [1993]). In many cases it is enough to study the knot property of a few short periods, say, period 1 and period 2, in order to construct the underlying "knot-holder" or "template". Most of the systems studied so far correspond to a horseshoe type template.

However, we will not touch these developments in this book except for giving the few references. A further remark consists in the following. The above application of knot theory utilizes the forward contraction of the stable manifold, thus is related to semi-infinite symbolic sequences of 1D map only. In particular, the calculation of linking numbers of closed orbits on a horseshoe essentially depends on the ordering of sequences in symbolic dynamics of unimodal maps (Le Sceller, Letellier and Gouesbet [1994]).

In this chapter we demonstrate that knot and link types of periodic orbits may be obtained from a simple construction, based on the ordering rule of one-dimensional maps. We summarize a few basic notions of knot theory in Section 9.1. In Section 9.2 we show how knots and links come out from an unimodal map. Section 9.3 summarizes the knot and link types of a few periodic sequences that have frequently appeared in the previous chapters. To keep up with the style of this book we will present the material in an intuitive and visual way, having in mind its possible application to physical systems.

9.1 Knots and Links

We need a few notions from knot theory which will not be explained in detail. The reader may consult, e.g., Kauffman's book (Kauffman [B1991]) for further information.

A close loop in a three-dimensional space makes a knot. If it can be transformed into a simple circle by continuous deformation, it is a trivial knot or an *unknot*. Otherwise, it is said to be knotted. Knot types have been classified and tabulated by mathematicians. For our purpose it is enough to refer to *torus link* of type $(2, n)$ or simply a K_n link in Kauffman's notation, see Fig. 9.1.

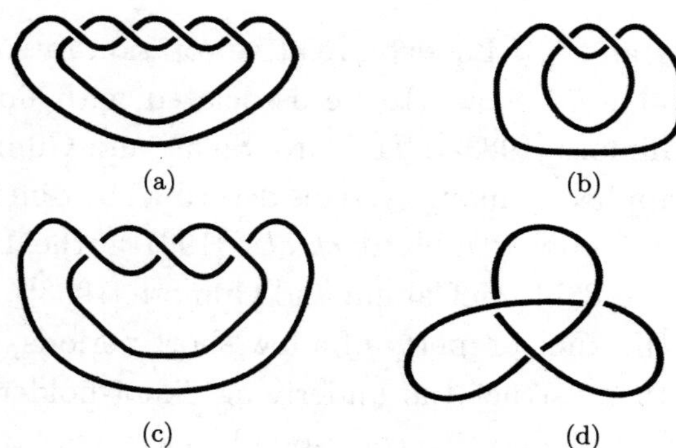

Figure 9.1 Examples of K_n torus links. (a) A K_4 link. (b) A K_2 (Hopf) link. (c) A K_3 link. (d) The trefoil which is equivalent to K_3

Although knots and links are three-dimensional objects, they are often studied in two-dimensional projections. In order to keep track of the original 3D loop, one has to indicate the way of crossings in a planar projection. A K_n torus link has n crossings. A K_4 link is shown in Fig. 9.1 (a). It is easy to see that when the number of crossings n is odd, the link is actually a single-component knot; when n is even K_n consists of two intertwined loops. A K_1 link is nothing but a trivial unknot. A K_2 link gives the simplest link of two loops. It has a special name *Hopf link* and is shown in Fig. 9.1 (b). The K_3 link shown in Fig. 9.1 (c) represents the simplest non-trivial knot. It is equivalent to a *trefoil* drawn in Fig. 9.1 (d).

Continuous deformation of a close curve in three-dimensional space may be realized in a 2D projection by applying some local moves to its parts in a number of consecutive steps. Three types of local moves are enough for this purpose. These are so-called Reidemeister moves of types I, II, and III, shown in Fig. 9.2. For example, it is easy to transform Figs. 9.1 (c) and (d) by applying these moves. It is the type III Reidemeister move that embodies the Yang-Baxter equations in exactly solvable models of statistical mechanics. No special meaning of type III Reidemeister move has been known in simple dynamics.

Any knot or link can be brought into a standard form of a *braid*. For instance, the K_4 link shown in Fig. 9.1 (a) may be transformed into the braid shown in Fig. 9.3. We put arrows on the two ascending strands in accordance

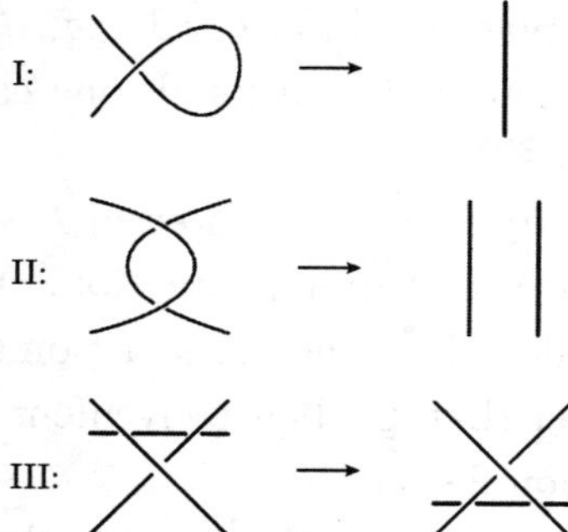

Figure 9.2 The three Reidemeister moves

with the convention that a braid is shown as a combination of descending strands. Having agreed on this convention, we can keep only the strands between the two horizontal dashed lines without drawing the ascending ones. There is an algebraic representation of braids that leads to the notion of braid groups. As braid groups are not used in what follows, we merely mention that the braid groups are generalization of symmetric groups. That is all what we need from knot theory.

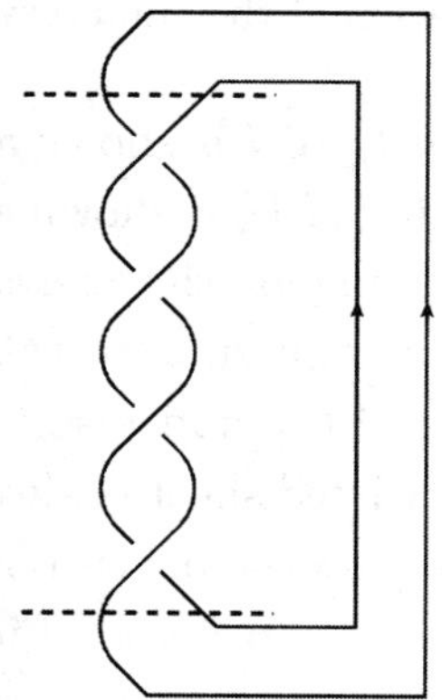

Figure 9.3 A K_4 link as a braid

9.2 Knots and Links from Unimodal Maps

For one-dimensional maps such as the quadratic map

$$x_{n+1} = 1 - \mu x_n^2,$$

the whole dynamics takes place on an interval, e.g., $I = [-1, 1]$. In spite of the fact that the phase space is one-dimensional, one can derive knots and links in the following way.

To construct symbolic dynamics the interval I is partitioned according to the monotone branches of the mapping function. We denote by I_L the left subinterval which is smaller than the critical point C and by I_R the right subinterval which is greater than C. It is convenient to consider the surjective case with $\mu = 2$ (see Section 2.6.1).

Under one iteration the subinterval I_L is stretched to the whole interval I with orientation preserved, while the subinterval I_R is stretched to the whole interval I with orientation reversed. One can visualize this transformation by connecting the corresponding intervals before and after one iteration by a sheet of canvas, as shown schematically in Fig. 9.4.

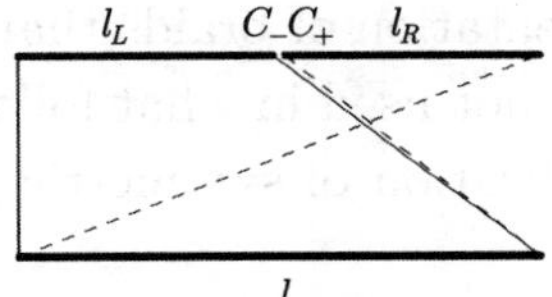

Figure 9.4 Transformation of the subintervals under one iteration

In Fig. 9.4 the stretching of I_L to I is shown as the two intervals connected by two solid lines. The stretching of I_R is shown as the two intervals connected by two dashed lines; the twisting of the canvas represents the orientation reversing. In order to show the two subintervals I_L and I_R clearer, we split the critical point C into C_- and C_+ and insert a gap in between. We agree from now on that the second, twisted sheet is always put behind the first sheet.

Now we are going to draw periodic orbits on the canvas. We recollect the ordering rule of symbolic sequences formulated in Section 2.3.2. Based on the natural order of real numbers on the interval

$$L < C < R,$$

one introduces an ordering for all symbolic sequences. Given two symbolic sequences

$$\Sigma_1 = \Sigma^* R \cdots ,$$

and

$$\Sigma_2 = \Sigma^* L \cdots ,$$

where Σ^* is their common leading string. If Σ^* contains an even number of R, then the order of $\Sigma_1 > \Sigma_2$. If Σ^* contains an odd number of R, then $\Sigma_1 < \Sigma_2$.

Let us take the superstable period 4 orbit in the main period-doubling cascade of the unimodal map. We will see that it gives rise to the simplest knotted loop—the trefoil.

To this end we determine the order of orbital points of the period 4 orbit $(RLRC)^\infty$ according to the ordering rule. We have

$$\begin{aligned} x_1 &= RLRC \cdots , \\ x_2 &= LRCR \cdots , \\ x_3 &= RCRL \cdots , \\ x_4 &= CRLR \cdots . \end{aligned} \tag{9.1}$$

These points are ordered as

$$x_2 < x_4 < x_3 < x_1$$

and are visited as shown in Fig. 9.5 (a). Let us imagine the map is created by intersection of a flow with a surface of section. To recover the flow from the map, it is convenient to represent two successive intersections of the flow with the section in two separate sections, and then identify the two sections. That is, in accordance with Fig. 9.4 we can split and separate the interval into two parallel lines. The points alternate as shown in Fig. 9.5 (b), where a number i labels the point x_i. One has to treat the crossings carefully. As the right, twisted sheet is always put behind the left sheet, the $2 \to 3$ and $4 \to 1$ lines go above all lines on the other sheet. We note that the point 4 is $x_4 = C$. Lines going from the upper right subinterval I_R down to I will cross each other due

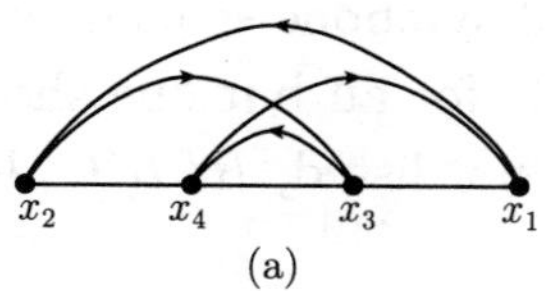

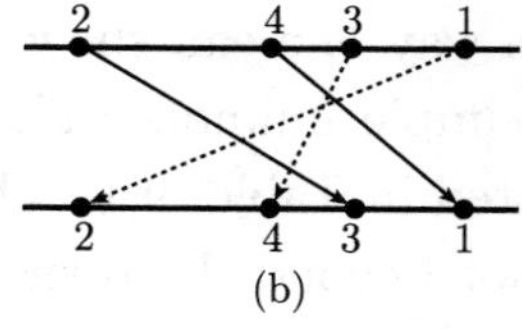

Figure 9.5 The alternation of orbital points for $(RLRC)^\infty$. (a) The order of visits. (b) The alternation of points in two successive intersections with the surface of section

to the twisting. Think the phase space as a strip. The dynamics of the flow is first stretching the strip and then folding it back. To comply with our first rule that the right sheet is put behind the left, the sense of the twist should be counterclockwise when viewing from the top. So, the correct crossing rule for I_R is: a line coming down from a point to the right of another point goes under the line coming down from the latter. For simplicity, we have ignored the possibility of any extra rotation of the whole flow. In this way we have resolved all five crossings in this example.

We further deform the lines in Fig. 9.5 (b) continuously to rearrange the crossings into what is shown between the two dashed horizontal lines in Fig. 9.6. Now we can connect points with the same number on the upper and lower lines by non-intersecting curves in the plane, thus transforming the orbit into a braid. This braid may be simplified using the Reidemeister moves. The result is a K_3 type torus link, i.e., a trefoil.

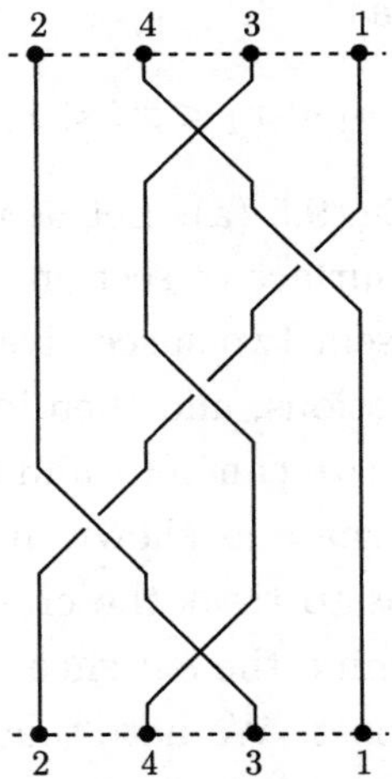

Figure 9.6 The alternation of points in $(RLRC)^\infty$ shown as a braid

In the same way one can show that all symbolic sequences of RL^nC type correspond to simple unknots. The results for all but one short words up to period 6 are listed in Table. 9.1. The one not listed, RLR^3C, leads to a more complicated knot beyond the torus links.

One can take more orbits and look at their link type. For example, we know that there is only one period-2 orbit $(RC)^\infty$ and one period 3 orbit $(RLC)^\infty$ in the unimodal maps. Both orbits belong to $(RL^nC)^\infty$ type, thus correspond to unknot. In a chaotic attractor, they become unstable periodic orbits $(RL)^\infty$ and $(RLR)^\infty$ or $(RLL)^\infty$. However, these two trivial knots are linked to each other. The five orbital points in $(RL)^\infty$ and $(RLL)^\infty$ are ordered as

$$LLR < LR < LRL < RL < RLL.$$

Table 9.1 Knot types of short periods

Period	Word	Knot Type	Period	Word	Knot type
1	C	unknot	5	RL^2RC	K_3
2	RC	unknot	5	RLR^2C	K_5
3	RLC	unknot	6	RL^4C	unknot
4	RL^2C	unknot	6	RL^3RC	K_3
4	$RLRC$	K_3	6	RL^2R^2C	K_5
5	RL^3C	unknot	6	RL^2RLC	K_5

With the convention to put the line going down from a greater point always below that from a smaller point, the same kind of construction leads to a braid, shown in Fig. 9.7. In the figure the two period-2 points are labeled by a and b, the three period 3 points by 1, 2, and 3. By using the Reidemeister moves this braid may be deformed to the K_4 type torus link, already shown in Fig. 9.1.

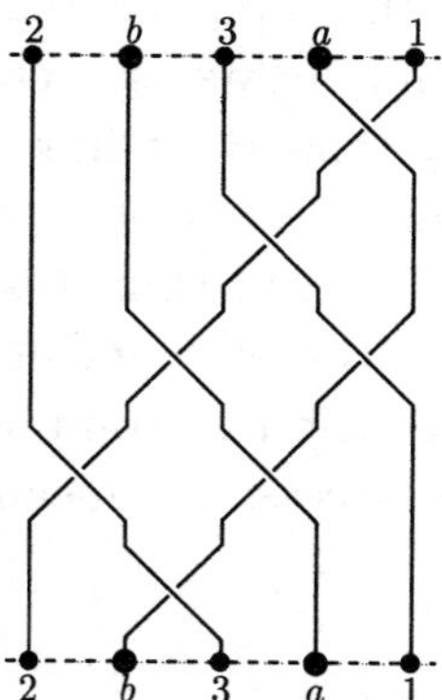

Figure 9.7 A braid formed by the $(RL)^\infty$ and $(RLL)^\infty$ orbits

9.3 Linking Numbers

The knot type of a single periodic orbit and link type of a set of periodic orbits are topological invariants. There are, in fact, many other invariants associated with a knot. Some are numbers, some are algebraic expressions or polynomials. Originally, these invariants are devised with the aim to distinguish different knots. Indeed, given two complicated knots, it is a non-trivial job to tell whether they are the same or not, e.g., by using repeatedly the Reidemeister moves. However, all invariants found so far are so-called "incomplete" invariants. The term means if two knots have different invariants, they are different for sure; however, having the same invariant does not imply that they are the same. Being interested primarily in dynamical aspects of knot theory, especially, its relation to symbolic dynamics, we will only touch a simple numerical invariant—the linking number.

Look at a two-component knot, for example, the Hopf link shown in Fig. 9.1 (b). First, assign an orientation to each of the loops by putting arrows on them. Second, define a sign $\epsilon(p)$ to each crossing p as shown in Fig. 9.8. The verbalized rule is: standing right at the crossing and turning the upper arrow toward the lower one in the shortest way, one has an $\epsilon = +1$ for the crossing if the turning is counterclockwise, and -1 otherwise. The linking number of the knot is defined as

$$Lk(\alpha, \beta) = \frac{1}{2} \sum_{p \in \alpha \cap \beta} \epsilon(p), \tag{9.2}$$

where α and β denote each of the components and the summation runs over crossings between different components; self-crossings of one and the same loop do not count. Therefore, the Hopf link has a linking number $Lk = \pm 1$, depending on the assignment of orientations. Clearly, if we change the orientation of one component, Lk will change sign. In general, the absolute value $|Lk|$ does not depend on the assignment of orientations. A non-zero linking number guarantees that the two components are linked. However, a zero linking number does not necessarily mean that the two loops are not linked.

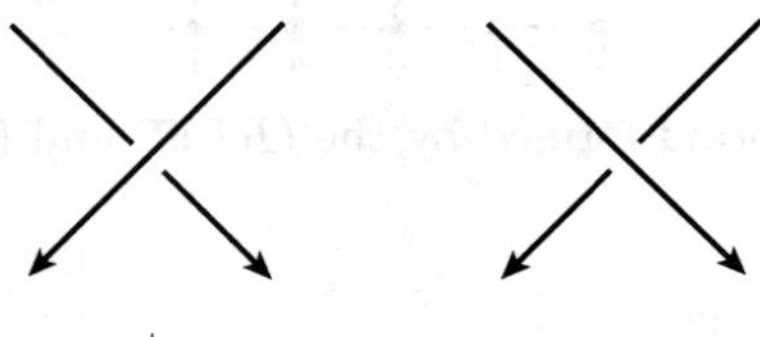

Figure 9.8 The sign of a crossing

Recollect our crossing convention for braids: a line starting from a point located to the right of another point always goes underneath a line coming out from the latter if they cross each other. Consequently, all crossings have the same sign -1. In order to calculate the linking number, it is enough to count the number of crossings between different periodic orbits. For example, from Fig. 9.7 it follows that the linking number between $(RL)^\infty$ and $(RLL)^\infty$ is -2.

The linking numbers of any pair of periodic orbits can be calculated by using similar braid construction. In Table 9.2 we collect linking numbers of the unstable period 2 orbit $(RL)^\infty$ with other periodic orbits up to length 5.

Table 9.2 Linking numbers of $(RL)^\infty$ with other low periods

RLR *RLL*	*RLRR*	*RLLR* *RLLL*	*RLLLR* *RLLLL*	*RLLRL* *RLLRR*	*RLRRR* *RLRRL*
−2	−3	−2	−2	−3	−4

Some remarks on Table 9.2 may be appropriate. The period 2 orbit $(RL)^\infty$ and its period-doubled partner $(RLRC)^\infty$ are linked with a linking number -3. Remember that symbolic names are borne by unstable periods. The linking of $(RL)^\infty$ to the two unstable period 3 orbits, $(RLR)^\infty$ and $(RLL)^\infty$, yields the same linking number -2. One can easily convince oneself that the two period 3 orbits are linked to each other with $Lk = -3$. Other pairs of unstable orbits born at tangent bifurcations are also linked with various linking numbers. A strong feeling is then that all periodic orbits in a chaotic attractor are linked to each other, although we do not know any mathematical theorem of this kind. What has been said deals with the surjective map when all unstable orbits are present. When the height of the map is lower than the surjective case, some periods are forbidden. However, this does not effect knot properties of orbits very much.

The method of calculating the linking numbers described here is simpler than what is suggested by Le Sceller, Letellier and Gouesbet [1994]. All calculated linking numbers agree with that of a horseshoe type template tabulated in Table V of Solari and Gilmore [1988a].

9.4 Discussion

The knot theory approach is restricted to periodic orbits only, while symbolic dynamics can deal with quasi-periodic and chaotic orbits as well. Moreover, symbolic dynamics provides a more refined characterization of the dynamics. For example, many different periodic sequences may belong to the same knot type, or, put in other words, it is possible to introduce a more crude classification of periodic orbits based on their knot types. In addition, knot theory has nothing to do with "period", but symbolic dynamics does. A properly constructed symbolic dynamics assigns a unique name to each unstable periodic orbit, determines their ordering and admissibility. In addition, we know the number of different periodic orbits in any continuous map with multiple critical points (see Chapter 7). These are topological characteristics of the dynamics.

If we confine ourselves to symbolic dynamics of two letters, there are only four basic prototypes, represented by the four piecewise linear maps considered in Section 2.8.6. They lead to four different knot holders or templates. The knot and link types of periodic orbits in each case may be calculated as we have just done with the unimodal (tent) map. In particular, the piecewise shift map gives rise to the Lorenz template of knots. Maps with multiple critical points would lead to more complicated types of templates. It would be interesting to learn what types of knots and links are allowed or forbidden in a given dynamics and to explore the dynamical implications of various knot invariants.

On the other hand, some deep questions of principle come to mind. Topology provides a skeleton, on which grows the "flesh" of metric and measure-theoretical property. There is a theorem which does not really touch on measure-theoretical property and which says that a flow with positive topological entropy must possess periodic closed orbits in infinitely many different knot type classes (Franks and Williams [1985]). Moreover, it seems plausible that unstable periodic orbits in a chaotic attractor must be all linked with each other in order to ensure ergodic property of the attractor. This kind of problems would lead us too far afield as physicists.

10
Appendix

In this Appendix we list three C programs. These programs may help to show how symbolic dynamics is implemented in practice, but no effort has been made to optimize them.

A.1 Program to Generate Admissible Sequences

The first program PERIOD.C generates all possible admissible symbolic sequences up to a given length for a map. We describe the map in a header file MAP.H.

As an example we list the header file for the Lorenz-Sparrow map, described in Section 3.8. As shown in Fig. 3.25, this map has four monotone branches denoted by the letters N, R, L, and M ("non-terminal symbols"). There are two critical points C, D and a "breaking" point B. In numerical study we split B into B_- and B_+. We also denote the two end points of the interval by B_1 and B_2. The natural order of these symbols

$$B_1 < N < C < R < B_- \leqslant B_+ < L < D < M < B_2$$

is embodied in using integers 1 to 10 to represent the letters. The 6 letters B_1, C, $B_\pm$, D, and B_2 are "terminal symbols". The symbols with odd parity, of total number `NO=2`, are given in the array `SO[NO]`. The program deals with integers all the time except for printing out a string.

When implementing admissibility conditions both C and B_2 are treated as minimal points, while D and B_1 as maximal points. This explains why `Nmax=Nmin=2`. The `0` in the arrays `SCmax` and `SCmin` are used to indicate that no more non-terminal symbols are involved in the corresponding admissibility condition.

The constant NC and array SC[NC] are used to deal with specific cases when more than two minima or maxima are bound to vary at the same time (see Section 7.1.2 and Fig. 7.1 for explanation). Since this does not happen in the Lorenz-Sparrow map, we skip any further details except for indicating that in spite of NC=0 we have to initialize the array SC[NC] with one zero element to avoid compilation error.

```
/* The header file describing the Lorenz-Sparrow map.  */
/* To be copied to map.h before compiling period.c */
#define NMAX 4 /* maximal period to calculate */
#define NN 4 /* Number of Non-terminal symbols */
#define NT 6 /* Number of Terminal symbols */
#define NTeff 4 /* Number of effective terminals */
#define NO 2 /* Number of Odd non-terminals */
#define Nmax 2 /* Number of max terminals */
#define Nmin 2 /* Number of min terminals */
#define Kmax 3
#define Kmin 3
#define NC 0
#define NTOTAL NN+NT
int SN[NN]={2, 4, 7, 9},
ST[NTeff]={1, 3, 8, 10},
SO[NO]={2, 9},
SC[1]={0},
SCmax[Nmax][Kmax]={{ 2, 0}, {7, 9, 0}},
SCmin[Nmin][Kmin]={{2, 4, 0}, {9, 0}},
STmax[Nmax]={1, 8},
STmin[Nmin]={3, 10};
int W[NTOTAL]={1,2,3,4,5,6,7,8, 9, 10};
char *CW[NTOTAL]={"B1", "N", "C", "R", "B-", "B+", "L",
                  "D", "M", "B2"};
```

Here is the program PERIOD.C:

```
/* period.c to generate admissible sequences up to a given */
/* length works for all continuous maps and some maps with */
```

```
/* discontinuities, e.g., the gap map, Lorenz-type map, */
/* 4-letter Lorenz-Sparrow map, etc.*/
#include <stdio.h>
#include <math.h>

/* the map is described in a separate .h file, which must be */
/* copied to map.h before compiling period.c */

#include "map.h"
#define MaxPeriod 30

/* function prototypes:  */
void innerloop(), printWord(), shift();
int compare(), parity(), pick();
int checkpmax(), checkpmin(), check(), checkother();

int n, count, s[MaxPeriod];

main()
 {
 int i, j, k;
 for(n=2; n<=NMAX; n++)
 { count=0;
    k=1;
    printf("Period = %d \n", n);
    innerloop(n, k);
    printf("Total number = %d \n", count);
    printf("\n");
 }
 }

void innerloop(n,k)
 int n, k;
 {
    int i, j;
    i=0;
    do
    { if(parity(s,k-1)==1) s[k-1]=SN[i];
    else s[k-1]=SN[NN-i-1];
    i++;
    if(n==k+1) for(j=0; j<NTeff; j++)
```

```
      { if(parity(s,k)==1) s[k]=ST[j];
      else s[k]=ST[NTeff-j-1];
      if(check(s, k+1)==0)
      { count++;
      if(n<=6) printWord(s, k+1);
      }
      }
      else innerloop(n, k+1);
      }
      while (i<NN);
  }

void shift(s, s1, n, k)
int n, k, s[ ], s1[ ];
{ int i;
    for(i=0; i<=n-k-1; i++) s1[i]=s[i+k];
}

void printWord(s, n)
 int n, s[ ];
 { int i;
     for(i=0; i<=n-1; i++) printf("%s", CW[s[i]-1]);
     printf("\n");
 }

compare(s1, s2, n1, n2)
int n1, n2, s1[ ], s2[ ];
{
    int i, j, key, nn;
    if(s1[0] != s2[0]) return((s1[0]>s2[0])?  1 :  -1);
    nn=(n1<n2)?  n1 :  n2;
    key=1;
    if(nn>1) for(i=1; i<=nn-1; i++)
    if(s1[i] != s2[i]) goto end;
    return 0;
end:
      if(((parity(s1, i)== 1) && (s1[i]<s2[i])) ||
    ((parity(s1, i)==-1) && (s1[i]>s2[i]))) key=-1;
    return key;
}
```

```
check(s, n)
int n, s[ ];
{
   if(NC != 0) if(checkother(s, n) != 0) return -1;
   if(Nmax != 0) if(checkpmax(s, n) != 0) return -1;
   if(Nmin != 0) if(checkpmin(s, n) != 0) return -1;
   return 0;
}
checkother(s, n)
int n, s[ ];
{
   int i,j,k,l, ni, nj, nni, nnj;
   int si[MaxPeriod], sj[MaxPeriod], ssi[MaxPeriod],
      ssj[MaxPeriod];
   if(n < 3) return 0;
   for(k=0; k < NC; k++)
   { i=SC[2*k];
   j=SC[2*k+1];
   ni=n;
   for(l=0;l<n;l++) si[l]=s[l];
repeat:
   if((nni=pick(si,ni,ssi,i))==0) goto endloop;
   nj=n;
   for(l=0;l<n;l++) sj[l]=s[l];
   if((nnj=pick(sj,nj,ssj,j))==0) goto endloop;
comp:
   if(compare(ssi,ssj,nni,nnj)>0) return -1;
   nj=nnj;
   for(l=0;l<nj;l++) sj[l]=ssj[l];
   if((nnj=pick(sj,nj,ssj,j))>0) goto comp;
   ni=nni;
   for(l=0;l<ni;l++) si[l]=ssi[l];
   goto repeat;
endloop:  ;
}
   return 0;
}

/* picking up a subsequence following the first encountered */
/* number i, returns the length of the subsequence */
```

```
pick(s, n, ss, i)
int n, i, s[ ], ss[ ];
{
    int j;
    for(j=0; j<n; j++)
    if(s[j]==i) { shift(s, ss, n, j+1);
    return (n-1-j);
    }
    return 0;
}
checkpmax(s, n)
int n, s[ ];
{
    int i, j, k, l, nk, n1, s1[MaxPeriod], s2[MaxPeriod];
    for(i=0; i<Nmax; i++) if(s[n-1]==STmax[i]) goto next;
    return 0;
next:
    for(j=0; j<Kmax; j++) /* Kmax was 3 */
    {
    n1=n;
    for(l=0; l<n; l++) s1[l]=s[l];
    if((k=SCmax[i][j]) != 0)
    { repeat:
    if((nk=pick(s1, n1, s2, k))==0) goto endloop;
    if(compare(s2, s, nk, n) > 0) return -1;
    n1=nk;
    for(l=0; l<n1;l++) s1[l]=s2[l];
    goto repeat;
    }
endloop:  ;
}
    return 0;
}

checkpmin(s, n)
int n, s[ ];
{
    int i, j, k, l, nk, n1, s1[MaxPeriod], s2[MaxPeriod];
    for(i=0; i<Nmin; i++) if(s[n-1]==STmin[i]) goto next;
    return 0;
```

```
next:
   for(j=0; j<Kmin; j++) /* Kmin was 3 */
   {
   n1=n;
   for(l=0; l<n; l++) s1[l]=s[l];
   if((k=SCmin[i][j]) != 0)
   { repeat:
   if((nk=pick(s1, n1, s2, k))==0) goto endloop;
   if(compare(s2, s, nk, n) < 0) return -1;
   n1=nk;
   for(l=0; l<n1;l++) s1[l]=s2[l];
   goto repeat;
   }
endloop:  ;
}
   return 0;
}


parity(s, n)
int n, s[ ];
{
   int i, j, key;
   if(n<=0) return 1;
   key=1;
   for(i=0; i<n; i++)
   if(NO != 0) for(j=0; j<NO; j++)
   if(s[i] == SO[j]) key=-key;
   return key;
}
```

A.2 Program to Draw Dynamical Foliations of a Two-Dimensional Map

The following program generates the dynamical foliations for the Hénon map. It is written in TURBO C and uses the TURBO C graphics. In order to watch its running step by step on the screen, a few pauses are inserted. One strikes any key to continue. (Instead of using `step()` and `curve()`, one can use any numerical integrator, say the Runge-Kutta method, to calculate foliations as the integral curves of the vector field found by `directn()`.)

```
#include <stdio.h>
 #include <math.h>
 #include <graphics.h> /* from Turbo C */

float aa, bb;
 double a,b;
 char id;
 int color;

void initgraphics(); /* use the one in Turbo C demo */
 void map(double *x, double *y, double u[ ]);
 double directn(double x, double y, char c);
 void step(double *x, double *y, double *slp, double *stp);
 void curve(double x, double y, double limit);

main()
 {
     double x,y,s1,s2,x3,y3;
     double k,h;
     char w[80];
     int i;
  loop:
      initgraphics();
     cleardevice();
     setactivepage(0);
     setvisualpage(0);
     rectangle(120,30,520,430);
     printf("input a and b:  (exit if a<0)\n");
     scanf("%f %f", &aa, &bb);
     a=aa; b=bb;
     if(a<0.0) {closegraph(); exit(0);}
     id='b';
     x=y=(b-1.+sqrt((b-1.)*(b-1.)+4.*a))/2./a; /* fixed point H+ */
     getch();exit();*/
```

```
  x3=y3=x;
  printf(" direction %f\n",k); getch(); */
  sprintf(w,"Henon map:  a=%5.3f b=%5.3f",a,b);
  outtextxy(120,20,w);
  color=15;
  circle(320+x*100, 230-y*105, 3);
  x=y=(b-1.-sqrt((b-1.)*(b-1.)+4.*a))/2./a; /* fixed point H- */
  circle(320+x*100, 230-y*105, 3);
  if(a-0.75*(1-b)*(1-b)>0.0)
  {
  x=(1-b-sqrt(-3+4*a+6*b-3*b*b))/2./a;
  y=(1-b+sqrt(-3+4*a+6*b-3*b*b))/2./a;
  circle(320+x*100, 230-y*105, 2);
  circle(320+y*100, 230-x*105, 2);
  }
  else printf("There is no 2P at a= %g b= %g\n", a, b); getch();
  id='b';
  x=x3; y=y3;
  curve(x,y,2.0); sound(300);delay(30);nosound(); getch();
 x=-2.; y=0.5; color=10;
 for(i=0; i<20; i++)
 {
  curve(x,y,2.0);
  x=-2.+0.2*(i+1);
  } sound(300);delay(30);nosound(); getch();
 id='f';
  x=x3; y=y3; color=14;
  curve(x,y,2.0); sound(300);delay(30);nosound(); getch();
 x=-2.; y=0.5; color=11;
 for(i=0; i<20; i++)
 {
 curve(x,y,2.0);
 x=-2.+0.2*(i+1);
 }
 sound(300);delay(30);nosound(); getch();
 if('p'==getch()) {
 }
 closegraph();
```

```
   goto loop;
}

void map(double *x, double *y, double u[])
{
   double aux,xx,yy;
    xx=*x; yy=*y;
    switch(id) {
    case 'f':  /* forward */
    aux=u[0]; u[0]=-2.0*a*xx*u[0]+b*u[2]; u[2]=aux;
    aux=u[1]; u[1]=-2.0*a*xx*u[1]+b*u[3]; u[3]=aux;
    aux=xx; *x=1.0-a*xx*xx+b*yy; *y=aux;
    break;
    case 'b':  /* backward */
    aux=u[0]; u[0]=u[2]*b; u[2]=(aux+2.*a*yy*u[2]);
    aux=u[1]; u[1]=u[3]*b; u[3]=(aux+2.*a*yy*u[3]);
    aux=xx; *x=yy; *y=(aux+a*yy*yy-1.)/b;}
    return;
}

double directn(double x, double y, char c)
 {
   int i,j;
    double xx,yy,v,v1,v2,tr,d;
    double u[]={1.,0.,0.,1.};
    xx=x; yy=y; id=c;
   for ( i=0; i<20; i++)
    {
    for (j=0; j<3; j++) map(&xx,&yy,u);
   v1=u[0]*u[0]+u[2]*u[2];
    v2=u[1]*u[1]+u[3]*u[3];
    v=u[0]*u[1]+u[2]*u[3];
    tr=v1+v2;
    v1 /=tr; v2=1.-v1; v /=tr;
    d=v1*v2-v*v; /* determinant */
    if (fabs(d)< 1.e-9) break;
    }
    if(i==20) {printf("fail!\n"); exit(1);}
```

```
   return((d-v1)/v);
}

void step(double *x, double *y, double *slp, double *stp)
 {
   int i;
    double x1,y1,dx,dy,h,k,k1,s;
    k=*slp ;h=*stp;
    for (i=0; i<5000; i++) {
    s=fabs(k)/(fabs(k)+1.);
    dx=h*s/k; dy=h*s;
    x1=(*x)+dx; y1=(*y)+dy;
    k1=directn(x1,y1,id);
    if(fabs(h)<.00001|| fabs((k-k1)/(1.+k*k1))<0.05) {
   if (k*k1<0.&&fabs(k)<1.0) h=-h;
    break;}
    h *=0.5; }
    h=(fabs(h)>0.01?h:4.*h);
    *stp =h; *slp =k1;
    *x =x1; *y =y1;
    if (i==5000) {
    printf("\n step() anomalistic!"); exit(1);}
    return;
 }

void curve(double x, double y, double limit)
 {
    int i,j;
    double x0,y0,k,h,slp,stp,sx,sy;
    x0=x; y0=y; h=stp=-.01;
    sx=200./limit; sy=200./limit;
    if(b<0.0 && x<sqrt(-b)/a) return;
    k=slp=directn(x,y,id);
    for(i=0;i<2;i++) /* draw in two directions */
   {
   while(fabs(x)<1.5*limit && fabs(y)<1.5*limit) {
   if(fabs(x)<limit && fabs(y)<limit)
         putpixel(320+x*sx,230-y*sy,color);
   step(&x,&y,&k,&h);}
```

```
    x=x0; y=y0; h=-stp; k=slp;
    }
    return;
}
```

A.3 A Greedy Program for Determining Partition Line

Below is a simplified version of a greedy algorithm to determine a partition line at a precision of up to a sequence length $n = 12$ for the Hénon map at $a = 1.4$ and $b = 0.3$, based on symbolic dynamics. The partition line is close to the y-axis, so it is $\bullet C$ instead of $C\bullet$. However, the difference between the methods to determine $\bullet C$ and $C\bullet$ is not so significant.

```
//determine partition line using long orbit only
#define a 1.4 #define b 0.3 #define K 12 //word length N=2^K
#define N 4096 //2^10=1024,2^20=1048576,2^15=32768 #define L
300000 //orbit length #include<stdio.h> #include<math.h> int
ix[L],ibl[N],iby[N],iyp[6000];
//ibl[],iby[]:lower-bound beta, xp[],yp[]:partition line
//iyp[]:beta of orbit point for updating yp[]
double xx[L],xp[6000],yp[6000]; int init(void)
//Generate pruning front, return total number of conflicts
//ibz:beta, ifz:alpha, ifp,ibp:parity,
{ int ibz,ifz,ibp,ifp,i,id,j,k,h;
  double q1,p,q;
  for(k=0;k<N;k++) ibl[k]=iby[k]=N-1;
  for(i=0;i<6000;i++) {
    p=.0005*i-1.5; q=xp[i];
    for(ifz=ifp=0,h=1,k=0;k<K;k++) {
      if(h) {
        q1=q; q=1.0-a*q*q+b*p; p=q1;
        j=2000*(p+1.5);
        if(j<0) j=0; else if(j>5999) j=5999;
        id=(q>xp[j]?1:0); }
      if(q<-5.) h=0;
      if(id) ifp=!ifp; ifz=2*ifz+ifp;}
   q=.0005*i-1.5; p=xp[i];
   for(ibz=0,ibp=h=1,k=0;k<K;k++) {
```

```
      if(h) {
        q1=q; q=(p+a*q*q-1.)/b; p=q1;
        j=2000*(q+1.5);
        if(j<0) j=0; else if(j>5999) j=5999;
        id=(p<xp[j]?1:0); }
      if(p>5.) h=0;
      if(id) ibp=!ibp; ibz=2*ibz+ibp;}
    if(ibz<ibl[ifz]) {
      ibl[ifz]=ibz; iby[ifz]=i; }}
  if(ibl[0]==N-1) ibl[0]=0;
  for(h=0,k=1;k<N;k++) {
    if(ibl[k]==N-1) ibl[k]=ibl[k-1];
    else if(ibl[k]<ibl[k-1]) h++; }
  return(h);
} void part(void)
//Greedy updating of partition line using a long orbit
{ int ibz,ifz,ibp,ifp,k,i;
  long nn;
  double q;
  for(i=0;i<6000;i++) iyp[i]=N-1;
  for(nn=1;nn<L;nn++) {
    k=2000*(xx[nn-1]+1.5); ix[nn-1]=(xx[nn]>xp[k]?1:0);}
  for(nn=K;nn<L-K-1;nn++) {
    for(ibz=ifz=ifp=0,ibp=1,k=1;k<=K;k++) {
      if(!ix[nn-k]) ibp=!ibp; ibz=2*ibz+ibp;
      if(ix[nn+k]) ifp=!ifp; ifz=2*ifz+ifp; }
    if(ibz<ibl[ifz]) {
      printf("%d %d %d\n",ifz,ibz,ibl[ifz]);
      q=xx[nn+1]; k=2000*(xx[nn]+1.5);
      if(fabs(q)<.08&&ibz<iyp[k]) {
        yp[k]=.1*xp[k]+0.9*q; iyp[k]=ibz;}}}
  for(k=0;k<6000;k++) xp[k]=yp[k];
} void main() { int i,k;
  long nn;
  double q;
  FILE *fq;
  if((fq=fopen("tmp","w"))==NULL) {
    printf("file error!\n"); exit(0); }
  for(i=0;i<6000;i++) xp[i]=yp[i]=0.;
  xx[0]=0.; q=1.; //initial for long orbit
```

```
  for(nn=1;nn<L;nn++) {
    xx[nn]=q; q=1.0-a*q*q+b*xx[nn-1]; }
  for(i=0;i<50;i++){
    printf("%d-th run: number of conflicts = %d\n",i,init());
    if('x'==getch())break;
    part(); }
  for(i=0;i<6000;i++) {
    if(xp[i]>1.e-9||xp[i]<-1.e-9) fprintf(fq,"%5d %6.4f\n",i,xp[i]);}
}
```

References

R.1 Books

Entries in this part are ordered by their years of publication and are referred to as, for example, Poincaré [B1899].

1. Bedford T, Keane M, Series C [B1991], eds.. *Ergodic Theory, Symbolic Dynamics, and Hyperbolic Space*, Oxford University Press.
2. Bowen R [B1979]. *Methods of Symbolic Dynamics.* a collection of Russian translations of Bowen's papers, ed. by V. M. Alekseev, Mir, Moscow.
3. Collet P, Eckmann J P [B1980]. *Iterated Maps on the Interval as Dynamical Systems.* Birkhäuser.
4. de Melo W, van Strien S [B1993]. *One-Dimensional Dynamics.* Springer-Verlag.
5. Devaney R L [B1986, B1989]. *An Introduction to Chaotic Dynamical Systems.* Addison-Wesley.
6. Glass L, Mackey M C [B1988]. *From Clocks to Chaos. The Rhythms of Life.* Princeton University Press.
7. Guckenheimer J, Moser J, Newhouse S E [B1980]. *Dynamical Systems.* C. I. M. E. Lectures, Progress in Mathematics **8**, Birkhäuser.
8. Guckenheimer J, Holmes P [B1983, B1990]. *Nonlinear Oscillations, Dynamical Systems, and Bifurcations of Vector Fields.* Springer-Verlag.
9. Hao B L [B1984], ed.. *Chaos.* An introduction and reprints volume, World Scientific.
10. Hao B L [B987], ed.. *Directions in Chaos.* vol. **1**, World Scientific.
11. Hao B L [B1988], ed.. *Directions in Chaos.* vol. **2**, World Scientific.
12. Hao B L [1989]. *Elementary Symbolic Dynamics and Chaos in Dissipative Systems.* World Scientific.
13. Hao B L [1990a], ed.. *Experimental Study and Characterization of Chaos.* vol. **3** of *Directions in Chaos.* World Scientific.

14. Hao B L [1990b], ed.. *Chaos II.* an update of Hao [B1984], World Scientific.
15. Hopcroft J E, Ullman J D [B1979]. *Introduction to Automata Theory, Languages, and Computation.* Addison-Wesley.
16. Kauffman L H [B1991]. *Knots and Physics.* World Scientific.
17. Rozenberg G, Salomaa A [B1980]. *The Mathematical Theory of L-systems.* Academic Press.
18. Sparrow C [B1982]. *The Lorenz Equations, Bifurcation, Chaos and Strange Attractors.* New York: Springer-Verlag.
19. Thompson J M T, Stewart H B [B1986]. *Nonlinear Dynamics and Chaos. Geometrical methods for engineers and scientists.* Wiley.
20. Poincaré H [B1899]. *Les Méthodes Nouvelles de la Mécaniqu Céleste.* tom **3**, Gauthier-Villars.
21. Kauffman L H [B1991]. *Knots and Physics.* World Scientific.
22. Prusinkiewicz P, Hanan J [B1989]. *Lindenmayer Systems, Fractals, and Plants, Lecture Notes in Biomathematics.* **79**, Springer-Verlag.
23. Prusinkiewicz P, Lindenmayer A [B1990]. *The Algorithmic Beauty of Plants.* Springer-Verlag.
24. Xie H M [B1996]. *Grammatical Complexity and One-Dimensional Dynamical Systems.* World Scientific.
25. Zhang S Y [B1991]. *Bibliography on Chaos.* vol. **5**. in *Directions in Chaos*, ed. by Hao B L, World Scientific.
26. Zheng W M, Hao B L [B1994]. *Applied Symbolic Dynamics* (in Chinese). Shanghai Scientific and Technological Education.

R.2 Papers

1. Aizawa Y [1983]. Symbolic dynamics approach to intermittent chaos. *Progr. Theor. Phys.*, **70**, 1249.
2. Alekseev V M, Yakobson M V [1981]. Symbolic dynamics and hyperbolic dynamic systems. *Phys. Reports*, **75**, 287.
3. Alseda L, Mañosas F [1990]. Kneading theory and rotation intervals for a class of circle maps of degree one. *Nonlinearity*, **3**, 413.
4. Arneodo A, Coullet P, Tresser C [1981]. A possible new mechanism for the onset of turbulence. *Phys. Lett.*, **A81**, 197.
5. Artuso R, Aurell E, Cvitanović P [1990]. Recycling of strange sets I. Cycle expansions and II. Applications. *Nonlinearity*, **3**, 325, 361.

6. Auerbach D, Procaccia [1990]. Grammatical complexity of strange sets, *Phys. Rev.* **A41**, 6602.

7. Bernhardt C [1982]. Rotation intervals of a class of endomorphisms of the circle. *Proc. Lond. Math. Soc.* III Ser., **45**, 258.

8. Biham O, Kvale M [1992]. Unstable periodic orbits in the stadium billiard. *Phys. Rev.*, **A46**, 6334.

9. Biham O, Wenzel W [1990]. Unstable periodic orbits and the symbolic dynamics of the complex Hénon map. *Phys. Rev.*, **A42**, 4639.

10. Birman J S, Williams R F [1983a]. Knotted periodic orbits in dynamical systems: I. Lorenz equations. *Topology*, **22**, 47.

11. Birman J S, Williams R F [1983b]. Knotted periodic orbits in dynamical systems: II. Knots holders for fibered knots. *Low Dimensional Topology, Contemp. Math. AMS*, **20**, 1.

12. Block L [1978]. Homoclinic points of mappings of the interval. *Proc. AMS*, **72**, 576.

13. Bowen R [1970]. Markov partitions and minimal sets for Axiom A diffeomorphisms. *Am. J. Math.*, **92**, 907.

14. Bowen R [1973a]. Symbolic dynamics for hyperbolic flows. *Am. J. Math.*, **95**, 429.

15. Bowen R [1973b]. Topological entropy for noncompact sets. *Trans. AMS*, **184**, 125.

16. Bowen R [1975a]. A horseshoe with positive measure. *Invent. Math.*, **29**, 203.

17. Bowen R [1975b]. Equilibrium states and the ergodic theory of Anosov diffeomorphisms. *Lect. Notes in Math.*, **470**, 108.

18. Bowen R, Ruelle D [1975]. The ergodic theory of Axiom A flows. *Invent. Math.*, **29**, 181.

19. Boyland P L [1986]. Bifurcations of circle maps: Arnol'd tongues, bistability and rotation intervals. *Commun. Math. Phys.*, **106**, 353.

20. Brucks K M [1987]. MSS sequences, colorings of necklaces, and periodic points of $f(z) = z^2 - 2$. *Adv. in Appl. Math.*, **8**, 434.

21. Bunimovich L A [1974]. On the ergodic properties of some billiards. *Funct. Anal. Appl.*, **8**, 254.

22. Bunimovich L A [1979]. On the ergodic properties of nonwhere dispersing billiards. *Comm. Math. Phys.*, **65**, 295.

23. Chang S J, McCown J [1985]. Universality behaviors and fractal dimensions associated with M-furcations. *Phys. Rev.*, **A31**, 3791.

24. Chavoya-Aceves O, Angulo-Brown F, Piña E [1985]. Symbolic dynamics of the cubic map. *Physica*, **D14**, 374.
25. Chen X, Lu Q H, Xie H M [1993]. Grammatical complexity of Feigenbaum attractor. *Advan. in Math.* (China), **22**, 185.
26. Chenciner A, Gambaudo J M, Tresser C [1984]. Une remarque sur la structure des endomorphisms degreé 1 du cercle. *C. R. Acad. Sci.*, **299**, 253.
27. Chossat P, Golubitsky M [1988]. Symmetry-increasing bifurcation of chaotic attractors. *Physica*, **D32**, 423.
28. Coffman K G, McCormick W D, Swinney H L [1986]. Multiplicity in a chemical reaction with one-dimensional dynamics, *Phys. Rev. Lett.*, **56**, 999.
29. Collet P, Crutchfield J P, Eckmann J P [1983]. Computing the topological entropy of maps. *Commun. Math. Phys.*, **88**, 257.
30. Crutchfield J P, Packard N H [1982]. Symbolic dynamics of one-dimensional maps: entropies, finite precision, and noise. *Int. J. Theor. Phys.*, **21**, 433.
31. Crutchfield J P, Young K [1990]. Computation at the onset of chaos. *Complexity, Entropy, and Physics of Information.* edited by W. Zurek, Addison-Wesley, 223.
32. Curry J H [1979]. On the Hénon transformation. *Commun. Math. Phys.*, **68**, 129.
33. Cvitanović P [1988]. Invariant measurement of strange sets in terms of cycles. *Phys. Rev. Lett.*, **61**, 2729.
34. Cvitanovic P, Gunaratne G H, Procaccia I [1988]. Topological and metric properties of Hénon-type strange attractors. *Phys. Rev.*, **38**, 1503.
35. D'Alessandro G, Isola S, Politi A [1991]. Geometric properties of the pruning front. *Prog. Theor. Phys.*, **86**, 1149.
36. D'Alessandro G, Grassberger P, Isola S, Politi A [1990]. On the topology of the Hénon map. *J. Phys.*, **A23**, 5285.
37. D'Humieres D, Beasley M R, Huberman B A, Libchaber A [1982]. Chaotic states and routes to chaos in the forced pendulum. *Phys. Rev.*, **26A**, 3483.
38. Doi S [1993]. On periodic orbits of trapezoid maps. *Adv. Appl. Math.*, **14**, 184.
39. Dawson S P, Grebogi C [1991]. Cubic maps as models of two-dimensional anti-monotonicity. *Chaos, Sol. & Frac.*, **2**, 137.
40. Dawson S P, Grebogi C, Yorke J A, Kan I, Koçak H [1992]. Anti-monotonicity: inevitable reversals of period-doubling cascades. *Phys. Lett.*, **A162**, 249.
41. De Sousa Vieira M C, Lazo E, Tsallis C [1987]. New road to chaos. *Phys. Rev.*, **A35**, 945.

42. Derrida B, Gervois A, Pomeau Y [1978]. Iteration of endomorphisms on the real axis and representation of numbers. *Ann. Inst. Henri Poincaré*, **29**, 305.

43. Dias de Deus J, Norouha da Costa A [1987]. Symbolic approach to intermittency. *Phys. Lett.*, **A120**, 19.

44. Ding M Z, Hao B L, Hao X [1985]. Power spectrum analysis and the nomenclature of periods in the Lorenz model. *Chinese Phys. Lett.*, 2, 1.

45. Ding M Z, Hao B L [1988]. Systematics of the periodic windows in the Lorenz model and its relation with the antisymmetric cubic map. *Commun. Theor. Phys.*, **9**, 375; reprinted in Hao [B1990].

46. Eckmann J P, Ruelle D [1985]. Ergodic theory of chaos and strange attractors. *Rev. Mod. Phys.*, **57**, 617.

47. El-Hamouly H, Mira C [1982a]. Lien entre les propriétés d'un endomorphisme et celles d'un difféomorphisme. *C. R. Acad. Sci. Paris*, **A293**, 525.

48. El-Hamouly H, Mira C [1982b]. Singularités dues au feuilletage du plan des bifurcations d'un difféomorphisme bi-dimensionnel. *C. R. Acad. Sci. Paris*, **A294**, 387.

49. Erber T, Johnson P, Everett P [1981]. Chebysev mixing and harmonic oscillator models, *Phys. Lett.*, **A85**, 61.

50. Fang H P [1994a]. Dynamics for strongly dissipative systems, *Phys. Rev. E*, **49**, 5025.

51. Fang H P [1994b]. Symbolic dynamics for a two-dimensional antisymmetric map. *J. Phys. A*, **27**, 5187.

52. Fang H P, Hao B L [1996]. Symbolic dynamics of the Lorenz equations. *Chaos, Solitons and Fractals*, **7**, 217.

53. Feigenbaum M J [1978]. Quantitative universality for a class of nonlinear transformations, *J. Stat. Phys.*, **19**, 25.

54. Feigenbaum M J [1979]. The universal metric properties of nonlinear transformations, *J. Stat. Phys.*, **21**, 669.

55. Feit S D [1978]. Characteristic exponents and strange attractors, *Commun. Math. Phys.*, **61**, 249.

56. Feng K [1985]. On Difference schemes and symplectic geometry. *Proceedings of the 1984 Beijing Symposium on Differential Geometry and Differential Equations—Computation of Partial Differential Equations*, ed. by Feng Kang, Science Press, Beijing, 42.

57. Feng K, Wu H M, Qin M Z, Wang D L [1989]. Construction of canonical difference schemes for Hamiltonian formalism via generating functions. *J Comp. Math.*, **7**, 1.

58. Feng K, Qin M Z [1991]. Hamiltonian algorithms for Hamiltonian dynamical systems, *Progress in Natural Science* (China), **1(2)**, 105.

59. Feng K, Wang D L [1994]. Dynamical systems and geometric construction of algorithms. *Computational Mathematics in China.* Contemporary Mathematics of AMS, **163**, ed. by Shi Z C and Yang C C, 1.

60. Finardi M, Flepp L, Parisi J, Holzner R, Badii R, Brun E [1992]. Topological and metric analysis of heteroclinic crisis in laser chaos. *Phys. Rev. Lett.*, **68**, 2989.

61. Fine N J [1958]. Classes of periodic sequences. *Illinois J. Math.*, **2**, 285.

62. Flepp L, Holzner R, Brun E, Finardi M, Badii R [1991]. Model identification by periodic-orbit analysis for NMR-laser chaos. *Phys. Rev. Lett.* **67**, 2244.

63. Franceschini V [1980]. A Feigenbaum sequence of bifurcations in the Lorenz model. *J. Stat. Phys.*, **22**, 397.

64. Franceschini V, Russo L [1981]. Stable and unstable manifolds of the Hénon mapping. *J. Stat. Phys.*, **25**, 757.

65. Franks J M [1980]. Symbolic dynamics, homology, and knots. *Lect. Notes in Math.*, **819**, 146.

66. Franks J M [1981]. Knots, links, and symbolic dynamics. *Ann. Math.*, **113**, 529.

67. Franks J M, Williams R F [1985]. Entropy and knots, *Trans. AMS*, **291**, 241.

68. Fukuda W, Katsura S [1986]. Exactly solvable models showing chaotic behavior II. *Physica*, **A136**, 588.

69. Gallas J A C [1993]. Structure of the parameter space of the Hénon map. *Phys. Rev. Lett.*, **70**, 2714.

70. Gambaudo J M, Glendinning P, Tresser C [1985]. Stable cycles with complicated structure, *J. de Phys. Lett.* **46**, L653.

71. Gaspard P, and Nicolis G [1983]. What can we learn from homoclinic orbits in chaotic dynamics?. *J. Stat. Phys.*, **31**, 499.

72. Gilbert E N, Riordan J [1961]. Symmetry types of periodic sequences. *Illinois J. of Math.*, **5**, 657.

73. Grassberger P [1986a]. How to measure self-generated complexity. *Physica*, **A140**, 319.

74. Grassberger P [1986b]. Toward a quantitative theory of self-generated complexity. *Int. J. Theor. Phys.*, **25**, 907.

75. Grassberger P [1988a]. Complexity and forecasting in dynamical systems. *Lecture Notes in Phys.*, **314**, 1.

76. Grassberger P [1988b]. On symbolic dynamics of one-humped maps of the interval. *Z. Naturforsch.*, **43a**, 671.

77. Grassberger P [1989]. Problems in quantifying self-generated complexity. *Helv. Phys. Acta*, **62**, 489.

78. Grassberger P, Kantz H [1985]. Generating partitions for the dissipative Hénon map. *Phys. Lett.*, **A113**, 235.

79. Grassberger P, Kantz H, Moenig U [1989]. On the symbolic dynamics of the Hénon map. *J. Phys.*, **A22**, 5217.

80. Grebogi C, Ott E, Yorke J A [1983a]. Crises, sudden changes in chaotic attractors, and transient chaos. *Physica*, **D7**, 181.

81. Grebogi C, Ott E, Yorke J A [1983b]. Fractal basin boundaries, long-lived chaotic transients, and unstable-unstable pair bifurcation. *Phys. Rev. Lett.*, **50**, 935; Erratum, **51**, 942.

82. Greene J M [1983]. Some order in the chaotic regimes of two-dimensional maps. *Long-Time Prediction in Dynamics*, ed. by W. Horton, L. Reichl and V. Szebehely, Wiley, 135.

83. Grossmann S, Thomae S [1977]. Invariant distributions and stationary correlation functions of the one-dimensional discrete processes, *Z. Naturforsch.*, **32a**, 1353.

84. Gu Y [1987]. Most stable manifolds and destruction of tori in dissipative dynamical systems. *Phys. Lett.*, **A124**, 340.

85. Gu Y [1988]. Most stable manifolds and transition to chaos in dissipative systems with competing frequencies. in Hao [B1988], 109.

86. Guckenheimer J [1977]. On the bifurcations of maps of the interval. *Invent. Math.*, **39**, 165.

87. Guckenheimer J [1980]. One-dimensional dynamics, *Ann. N. Y. Acad. Sci.*, **357**, 343.

88. Guckenheimer J, Williams R [1979]. Structural stability of the Lorenz attractor,. *Publ. Math. IHES*, **50**, 307.

89. Gumowski I, Mira C [1980]. Recurrences and discrete dynamic systems. *Lect. Notes in Math.*, **809**.

90. Gunaratne G H, Procaccia I [1987]. The organization of chaos. *Phys. Rev. Lett.*, **59**, 1377.

91. Hadamard J [1898]. Les surfaces á courbures opposées et leur lignes géodésic. *J. Math. Pures Appl.*, **4**, 27.

92. Hansen K T, Cvitanović P [1994]. Symbolic dynamics and Markov partitions for the stadium billiard (preprint).

93. Hao B L [1981]. Universal slowing-down exponent near period-doubling bifurcation points. *Phys. Lett.*, **A86**, 267.
94. Hao B L [1982]. Two kinds of entrainment-beating transitions in a driven limit cycle oscillator. *J. Theor. Biol.*, **98**, 9.
95. Hao B L [1983]. Bifurcation, chaos, strange attractor, turbulence and all that. *Progress in Phys.*, **3**, 329 (in Chinese).
96. Hao B L [1985]. Bifurcations and chaos in a periodically forced limit cycle oscillator. *Advances in Science of China: Physics*, vol. **1**, ed. by Zhu Hong-yuan, Zhou Guang-zhou, and Fang Li-zhi, Science Press, 113.
97. Hao B L [1986]. Symbolic dynamics and systematics of periodic windows. *Physica*, **A140**, 85.
98. Hao B L [1987]. Bifurcations and chaos in the periodically forced Brusselator. *Collected Papers Dedicated to Professor Kazuhisa Tomita on the Occasion of his Retirement from Kyoto University*, Kyoto University, 82.
99. Hao B L [1988]. Elementary symbolic dynamics. Chapter 14. *Order and Chaos in Nonlinear Physical Systems*, ed. by Lundqvist S, March N H, Tosi M P, Plenum, 387.
100. Hao B L [1991]. Symbolic dynamics and characterization of complexity. *Physica*, **D51**, 161.
101. Hao B L [1993]. Symbolic dynamics approach to chaos. *The First International Workshop on Nonlinear Dynamics and Chaos*, ed. by H. Lee, Pohang Institute of Science and Technology, 1.
102. Hao B L [1994]. Symbolic dynamics and description of complexity. *On Self-Organization*, ed. by R. K. Mishra, D. Maaß, E. Zwierlein, Springer Series in Synergetics, vol. **61**, Springer-Verlag, 197.
103. Hao B L [1997]. Symbolic dynamics and knot theory. *Look in the 21st Century*, Proceedings of the 1st Joint Meeting of Chinese Physical Societies, World Scientific, 458.
104. Hao B L [2005]. Crtical slowing down in one-dimensional maps and beyond. *J. Stat. Phys.*, **121(3–4)**, 749–757.
105. Hao B L, Wang G R, and Zhang S Y [1983]. U-sequences in the periodically forced Brusselator, *Commun. Theor. Phys.*, **2**, 1075.
106. Hao B L, Xie F G [1993]. Chaotic systems: counting the number of periods. *Physica*, **A194**, and in *Proceedings of STATPHYS 18*, North-Hollan, 77.
107. Hao B L, Zeng W Z [1986]. Information dimensions in unimodal mappings. *Proceedings of the Sino-Japan Bilateral Workshop on Statistical Physics and Condensed Matter Theory*, ed. by Xie Xi-de, World Scientific, 24.

108. Hao B L, Zeng W Z [1987]. Number of periodic windows in one-dimensional mappings. *The XV International Colloquium on Group Theoretical Methods in Physics*, ed. by R Gilmore, World Scientific, 199.

109. Hao B L, Zhang S Y [1982a]. Subharmonic stroboscopy as a method to study period-doubling bifurcations. *Phys. Lett.*, **A87**, 267.

110. Hao B L, Zhang S Y [1982b]. Hierarchy of chaotic bands and periodicities embedded in them in a forced nonlinear oscillator. *Commun. Theor. Phys.*, **1**, 111.

111. Hao B L, Zhang S Y [1982c]. Hierarchy of chaotic bands. *J. Stat. Phys.*, **28**, 769.

112. Hao B L, Zhang S Y [1983]. Subharmonic stroboscopic sampling method for study of period-doubling bifurcation and chaotic phenomena in forced nonlinear oscillators. *Acta Phys. Sinica*, **32**, 198 (in Chinese).

113. Hao B L, Zheng W M [1989]. Symbolic dynamics of unimodal maps revisited. *Int. J. Mod. Phys.*, **B3**, 235.

114. Hao B L, Liu J X, Zheng W M [1998]. Symbolic dynamics analysis of the Lorenz equations. *Phys. Rev.*, **E57**, 5378.

115. Hénon M [1976]. A two-dimensional mapping with a strange attractors. *Commun. Math. Phys.*, **50**, 69.

116. Hénon M [1982]. On the numerical computation of Poincaré maps. *Physica*, **D5**, 412.

117. Hénon M, Pomeau Y [1977]. Two strange attractors with a simple structure. *Lect. Notes in Math.*, Springer-Verlag, **565**, 29.

118. Hofbauer F [1980]. The topological entropy of the transformations $x \mapsto ax(1-x)$. *Monatsh. Math.*, **90**, 114.

119. Hsu C S, Kim M C [1984]. Method of constructing generating partitions for entropy evaluation *Phys. Rev.*, **A30**, 3351.

120. Hsu C S, Kim M C [1985]. Construction of maps with generating partitions for entropy evaluation. *Phys. Rev.*, **A31**, 3253.

121. Hu G, Hao B L [1990]. Two kinds of singularities in planar differential systems and the response to external forcing. *Phys. Rev.*, **A42**, 3335.

122. Huang Y N [1985]. Determination of the stable periodic orbits for the Hénon map by analytical method, *Chinese Phys. Lett.*, **2**, 98.

123. Huang Y N [1986]. An algebraic analytical method for exploring periodic orbits of the Hénon map, *Scientia Sinica* (Series A), **29**, 1302.

124. Huang Y N [1994]. The window analysis of one-dimensional m-th degree polynomial real maps. *Adv. in Math.*, **23**, 536 (in Chinese).

125. Ito R [1981]. Rotation sets are closed. *Math. Proc. Camb. Phil. Soc.*, **89**, 107.

126. Kai T, Tomita K [1979]. Stroboscopic phase portrait of a forced nonlinear oscillator. *Prog. Theor. Phys.*, **61**, 54.

127. Kan I, Yorke J A [1990]. Anti-monotonicity: concurrent creation and annihilation of periodic orbits. *Bull. AMS*, **23**, 469.

128. Kaplan H [1983]. New method for calculating stable and unstable periodic orbits. *Phys. Lett.*, **A97**, 365.

129. Kaplan H [1993]. Type-I intermittency for the Hénon-map family. *Phys. Rev.*, **E48**, 1655.

130. Kaplan J L, Yorke J A [1979]. Pre-turbulence: a regime observed in a fluid flow model of Lorenz. *Commun. Math. Phys.*, **67**, 93.

131. Katsura S, Fukuda W [1985]. Exactly solvable models showing chaotic behavior. *Physica*, **A130**, 597; Erratum, **A203** [1994], 159.

132. Keolian R, Putterman S J, Turkevich L A, Rudnick I, Rudnick J A [1981]. Subharmonic sequences in the Faraday experiment: departures from perioddoubling. *Phys. Rev. Lett.*, **47**, 1133.

133. Knobloch K, Moore D R, Toomre J, Weiss N O [1986]. Transitions to chaos in two-dimensional double-diffusive convection, *J. Fluid Mech.* **166**, 409.

134. Knobloch E, Weiss N O [1981]. Bifurcations in a model of double-diffusive convection. *Phys. Lett.*, **A85**, 127.

135. Kumar K, Agarwal A K, Bhattacharjee J K, Banerjee K [1987]. Precursor transition in dynamical systems undergoing period doubling. *Phys. Rev.*, **A35**, 2334.

136. Lazutkin V F [1973]. The existence of caustics for a billiard problem in a convex domain. *Math. USSR Izvestiya*, **7**, 185.

137. Le Sceller L, Letellier C, Gouesbet G [1994]. Algebraic evaluation of linking numbers of unstable periodic orbits in chaotic attractors. *Phys. Rev.*, **E49**, 4693.

138. Li J N, Hao B L [1989]. Bifurcation spectrum in a delay-differential system related to optical bistability. *Commun. Theor. Phys.*, **11**, 265.

139. Li T Y, Yorke J A [1975]. Period three implies chaos. *Am. Math. Monthly*, **82**, 985.

140. Liu J X, Zheng W M [1995a]. Symbolic analysis of NMR-laser chaos, *Phys. Rev.* **E51**, 3735.

141. Liu J X, Zheng W M [1995b]. Symbolic dynamics of the forced Brusselator. *Commun. Theor. Phys.*, **23**, 315.

142. Liu J X, Wu Z B, Zheng W M [1996]. Symbolic analysis of NMR-laser chaos. *Commun. Theor. Phys.*, **25**, 149.

143. Liu J X, Zheng W M, Hao B L [1996]. From annular to interval dynamics—symbolic analysis of the periodically forced Brusselator. *Chaos, Solitons and Fractals*, **7**, 1427.

144. Liu Z R, Cao Y L [1991]. Discussion on the geometric structure of strange attractor. *Chin. Phys. Lett.*, **8**, 503.

145. Liu Z R, Xie H M, Zhu Z X, Lu Q H [1992]. The strange attractor of Lozi mapping. *Int. J. Bif. & Chaos*, **2**, 831.

146. Lorenz E N [1963]. Deterministic nonperodic flow. *J. Atmos. Sci.*, **20**, 130.

147. Lorenz E N [1980]. Noisy periodicity and reverse bifurcation. *Ann. N. Y. Acad. Sci.*, **357**, 282.

148. Losson J, Mackey M C, Longtin A [1993]. Solution multi-stability in first-order nonlinear differential delay equations. *Chaos* (AIP), **3**, 167.

149. Lozi R [1978]. Un attracteur étrange (?) du type attracteur de Hénon. *J. Physique*, **39** (Coll. C5), 9.

150. Lu Q H [1994]. Grammatical complexity of generalized Feigenbaum attractors, *J. Nonl. Dyn.*, **1**, 182. (in Chinese)

151. Lutzky M [1988]. Counting stable cycles in unimodal iterations. *Phys. Lett.*, **A131**, 248.

152. Lyubich M, Milnor J [1993]. The Fibonacci unimodal map. *J. AMS*, **6**, 425.

153. MacKay R S, Tresser C [1987]. Some flesh on the skeleton: the bifurcation structure of bimodal maps. *Physica*, **D27**, 412.

154. MacKay R S, Tresser C [1988]. Boundary of topological chaos for bimodal maps of the interval. *J. London Math. Soc.*, **37**, 164.

155. Mackey M C, Glass L [1977]. Oscillation and chaos in physiological control systems. *Science*, **197**, 287.

156. Manneville P, Pomeau Y [1979]. Intermittency and the Lorenz model. *Phys. Lett.*, **A75**, 1.

157. Mao J M, Hu B [1988]. Multiple scaling and the fine structure of period doubling. *Int. J. Mod. Phys.*, **B2**, 65.

158. Marotto F R [1979]. Chaotic behavior in the Hénon mapping. *Commun. Math. Phys.*, **68**, 187.

159. May R M [1976]. Simple mathematical models with very complicated dynamics. *Nature*, **261**, 459.

160. May R M [1979]. Bifurcations and dynamic complexity in ecological systems. *Ann. N. Y. Acad. Sci.*, **316**, 517.

161. Meiss J D [1992]. Cantori for the stadium billiard. *Chaos* (AIP), **2**, 267.

162. Metropolis N, Stein M L, Stein P R [1973]. On finite limit sets for transformations on the unit interval. *J. Comb. Theor.*, **A15**, 25.

163. McCallum J W L, Gilmore R [1993]. A geometric model for the Duffing oscillator. *Int. J. Bif. & Chaos*, **3**, 685.

164. Milnor J, Thurston W [1977, 1988]. On iterated maps of the interval. Preprint in 1977. *Lect. Notes in Math.*, **1342**, 465.

165. Misiurewicz M [1979]. Horseshoes for mappings of the interval. *Bull. Acad. Pol. Ser. Sci. Math.*, **27**, 167.

166. Misiurewicz M [1980]. Strange attractors for the Lozi mapping. *Ann. N. Y. Acad. Sci.*, **375**, 348.

167. Moore D R, Weiss N O [1990]. Dynamics of double convection. *Phil. Trans. R. Soc. Lond.*, **A332**, 121.

168. Morse M [1921]. Recurrent geodesics on a surface of negative curvature. *Trans. AMS*, **22**, 84.

169. Morse M, Hedlund G A [1938]. Symbolic dynamics. *Am. J. Math.*, **60**, 815; reprinted in *Collected Papers of M. Morse*, vol. **2**, World Scientific, 1986.

170. Morse M, Hedlund G A [1940]. Symbolic dynamics II. Sturmian trajectories. *Am. J. Math.*, **61**, 1; reprinted in *Collected Papers of M. Morse*, vol. **2**, World Scientific, 1986.

171. Newhouse S E, Palis J, Takens F [1983]. Bifurcations and stability of families of diffeomorphisms. *Publ. Math. IHES*, **57**, 5.

172. Ni W S, Tong P Q, Hao B L [1989]. Homoclinic and heteroclinic intersections in the periodically forced Brusselator. *Int. J. Mod. Phys.*, **B3**, 643.

173. Ostlund S, Kim S H [1985]. Renormalization of quasiperiodic mapping, *Phys. Scripta*, **T9**, 193.

174. Peng S L, Zhang X S [1996]. The generalized Milnor-Thurston conjecture and dual star products of symbolic dynamics in order topological space of three symbols. Yunnan University preprint CNCS-YNU-96-04.

175. Percival I, Vivaldi F [1987a]. Arithmetical properties of strongly chaotic motions. *Physica*, **D25**, 105.

176. Percival I, Vivaldi F [1987b]. A linear code for the sawtooth and cat maps. *Physica*, **D27**, 373.

177. Piña E [1983]. Order in the chaotic region. *Lect. Notes in Phys.*, **189**, Springer-Verlag, 402.

178. Piña E [1984]. Comment on 'Study of a one-dimensional map with multiple basins' (Testa and Held [1983]). *Phys. Rev.*, **A30**, 2132.

179. Piña E [1986]. Kneading theory of the circle map. *Phys. Rev.*, **A34**, 574.

180. Procaccia I, Thomae S, Tresser C [1987]. First return maps as a unified renormalization scheme for dynamical systems. *Phys. Rev.*, **A35**, 1884.

181. Proctor M R E, Weiss N O [1990]. Normal forms and chaos in thermosolutal convection. *Nonlinearity*, **3**, 619.

182. Robbins K A [1979]. Periodic solutions and bifurcation structure at high R in the Lorenz model. *SIAM J. Appl. Math.*, **36**, 457.

183. Rössler O E [1976]. An equations for continuous chaos. *Phys. Lett.*, **A57**, 397.

184. Saltzman B [1962]. Finite amplitude convection as an initial value problem. *J. Atoms. Sci.*, **19**, 329.

185. Sarkovskiǐ A N [1964]. Coexistence of cycles of a continuous map of a line into itself. *Ukranian Math. J.*, **16**, 61 (in Russian).

186. Sato S, Sano M, Sawada Y [1983]. Universal scaling property in bifurcation structure of Duffing's and of generalized Duffing's equations. *Phys. Rev.*, **A28**, 1654.

187. Shimada T, Nagashima T [1978]. The iterative transition phenomenon between periodic and turbulent states in a dissipative dynamical system. *Prog. Theor. Phys.*, **59**, 1033.

188. Shimada T, Nagashima T [1979]. A numerical approach to ergodic problem of dissipative dynamical systems. *Prog. Theor. Phys.*, **61**, 1605.

189. Shimizu T, Morioka N [1978]. Chaos and limit cycles in the Lorenz model. *Phys. Lett.*, **A66**, 182.

190. Simó C [1979]. On the Hénon-Pomeau attractor. *J. Stat. Phys.*, **21**, 465.

191. Simoyi R H, Wolf A, Swinney H L [1982]. One-dimensional dynamics in a multi-component chemical reaction. *Phys. Rev. Lett.*, **49**, 245.

192. Singer D [1978]. Stable orbits and bifurcations of maps of the interval. *SIAM J. Appl. Math.*, **35**, 260.

193. Smale S [1991]. Dynamics retrospective: great problems. attempt that failed, *Physica.*, **D51**, 267.

194. Smale S, Williams R F [1976]. The qualitative analysis of a difference equation of population growth. *J. Math. Biol.*, **3**, 1.

195. Solari H G, Gilmore R [1988a]. Relative rotation rates for driven dynamical systems. *Phys. Rev.*, **A37**, 3096.

196. Solari H G, Gilmore R [1988b]. Organization of periodic orbits in the driven Duffing oscillator. *Phys. Rev.*, **A38**, 1566.

197. Stefan P [1977]. A theorem of Sarkovskiĭ on the existence of periodic orbits of continuous endomorphisms of the real line. *Commun. Math. Phys.*, **54**, 237.

198. Swift J W, Wiesenfeld K [1984] Suppression of period doubling in symmetric systems, *Phys. Rev. Lett.* **52**, 705.

199. Tél T [1983]. Invariant curves, attractors, and phase diagram of a piecewise linear map with chaos. *J. Stat. Phys.*, **33**, 195.

200. Tomita K, Kai T [1978]. Stroboscopic phase portrait and strange attractors. *Phys. Lett.*, **A66**, 91.

201. Tomita K, Tsuda I [1980]. Towards the interpretation of the global bifurcation structure of the Lorenz system: a simple one-dimensional model. *Prog. Theor. Phys. Supp.*, **69**, 185.

202. Tufillaro N B, Holzner R, Flepp L, Brun E, Finardi M, Badii R [1991]. Template analysis for a chaotic NMR laser. *Phys. Rev.*, **A44**, R4786.

203. Tufillaro N B, Solari H G, Gilmore R [1990]. Relative rotation rates: fingerprints for strange attractors. *Phys. Rev.*, **A41**, 5717.

204. Tyson J J [1973]. Some further studies of nonlinear oscillations in chemical systems. *J. Chem. Phys.*, **58**, 3919.

205. Ueda Y [1991]. Survey of regular and chaotic phenomena in the forced Duffing oscillator. *Chaos, Solitons & Fractals*, **1**, 199.

206. Ulam S M, von Neumann J [1947]. On combinations of stochastic and deterministic processes. *Bull. AMS*, **53**, 1120.

207. Veerman P [1986]. Symbolic dynamics and rotation numbers. *Physica*, **A134**, 543.

208. Wang G R, Chen S G, Hao B L [1983]. Intermittent chaos in the forced Brusselator, *Chinese Phys.* (AIP), **4**, 284; Chinese Orig. *Acta Phys. Sinica*, **32**, 1139.

209. Wang G R, Chen S G, Hao B L [1984a]. Kolmogorov capacity and Lyapunov dimension of strange attractors in the forced Brusselator. *Acta Phys. Sinica*, **33**, 1246 (in Chinese).

210. Wang G R, Chen S G, Hao B L [1984b]. On the nonconvergence problem in computing the capacity of strange attractors. *Chinese Phys. Lett.*, **1**, 11.

211. Wang G R, Chen S G [1986]. Universal constants and universal functions of period-n-tupling sequences in one-dimensional unimodal mappings. *Acta Phys. Sinica*, **35**, 58 (in Chinese).

212. Wang G R, Chen S G [1990]. Chaotic measures and scaling laws for supercritical circle map. *Acta Phys. Sinica*, **39**, 1705 (in Chinese).

213. Wang G R, and Hao B L [1984]. Transition from quasiperiodic regime to chaos in the forced Brusselator. *Acta Phys. Sinica*, **33**, 1321 (in Chinese).

214. Wang Y [1997]. A study of grammatical complexity of non-regular unimodal languages. *PhD Thesis*, Suzhou University.

215. Wang Y, Xie H M [1994]. Grammatical complexity of unimodal maps with eventually periodic kneading sequences. *Nonlinearity*, **7**, 1419.

216. Wiesenfeld K A, Knobloch E, Miracky R F, Clarke J [1984]. Calculation of period doubling in a Josephson circuit. *Phys. Rev.*, **A29**, 2102.

217. Williams R F [1979a]. The structure of Lorenz attractors. *Publ. Math. IHES*, **50**, 101.

218. Williams R F [1979b]. The bifurcation space of the Lorenz attractor. *Ann. N.Y. Acad. Sci.*, **316**, 393.

219. Wolfram S [1984]. Computation theory of cellular automata. *Commun. Math. Phys.*, **96**, 15.

220. Xie F G [1994]. Symbolic dynamics for the general quartic map. *Commun. Theor. Phys.*, **22**, 43.

221. Xie F G, Hao B L [1993]. Symbolic dynamics of the sine-square map. *Chaos, Solitons & Fractals*, **3**, 47.

222. Xie F G, Hao B L [1994]. Counting the number of periods in one-dimensional maps with multiple critical points. *Physica*, **A202**, 237.

223. Xie F G, Hao B L [1995]. The number of periods in one-dimensional gap map and Lorenz-like map. *Commun. Theor. Phys.*, **23**, 175.

224. Xie F G, Zheng W M [1994]. Analysis of dynamics in a parametrically damped pendulum. *Phys. Rev.*, **E49**, 1888.

225. Xie F G, Zheng W M, Hao B L [1995]. Symbolic dynamics of the two-well Duffing equation. *Commun. Theor. Phys.*, **24**, 43.

226. Xie H M [1993a]. On formal languages of one-dimensional dynamical systems. *Nonlinearity*, **6**, 997.

227. Xie H M [1993b]. The finite automata of eventually periodic unimodal maps on the interval. *J. Suzhou University*, **9**, 112.

228. Xie H M [1995]. Distinct excluded blocks and grammatical complexity of dynamical systems. *Complex Systems*, **9**, 73.

229. Xie H M [1996]. Fibonacci sequences and homomorphisms of free submonoid for unimodal maps. *Nonlinearity*, **9**, 1469.

230. Yang W M, Hao B L [1987]. How the Arnold tongues become sausages in a piecewise linear circle map. *Commun. Theor. Phys.*, **8**, 1.

231. Yorke J A, Alligood K T [1985]. Period doubling cascades of attractors: a prerequisite for horseshoes. *Commun. Math. Phys.*, **101**, 305.

232. Zeng W Z [1985]. A recursion formula for the number of stable orbits in the cubic map. *Chinese Phys. Lett.*, **2**, 429.

233. Zeng W Z [1987]. On the number of stable cycles in the cubic map. *Commun. Theor. Phys.*, **8**, 273.

234. Zeng W Z, Ding M Z, Li J N [1985]. Symbolic description of periodic windows in the antisymmetric cubic map, *Chinese Phys. Lett.*, **2**, 293.

235. Zeng W Z, Ding M Z, Li J N [1988]. Symbolic dynamics for one-dimensional mappings with multiple critical points. *Commun. Theor. Phys.*, **9**, 141.

236. Zeng W Z, Hao B L, Wang G R, Chen S G [1984]. Scaling property of period-n-tupling sequences in one-dimensional mappings. *Commun. Theor. Phys.*, **3**, 283.

237. Zeng W Z, Glass L [1989]. Symbolic dynamics and skeleton of circle maps. *Physica*, **D40**, 218.

238. Zeng W Z, Hao B L [1986]. Dimensions of the limiting sets of period-n-tupling sequences. *Chinese Phys. Lett.*, **3**, 285.

239. Zeng W Z, Hao B L [1987]. The derivation of a sum rule determining the q-th order information dimension. *Commun. Theor. Phys.* **8**, 295.

240. Zhang H J, Dai J H, Wang P Y, Jin C D [1983]. Chaos in liquid crystal hybrid optical bistable devices. *Laser Spectroscopy IV*, Springer Series Opt. Sci., **40**, 322.

241. Zhang H J, Dai J H, Wang P Y, Jin C D [1984]. Bifurcations and chaotic behavior in liquid crystal hybrid optical bistable devices. *Acta Phys. Sinica*, **33**, 1024 (in Chinese).

242. Zhang H J, Wang P Y, Dai J H, Jin C D, Hao B L [1985]. Analytical study of a bimodal mapping related to a hybrid optical bistable device using liquid crystal. *Chinese Phys. Lett.*, **2**, 5.

243. Zhang H J, Dai J H, Wang P Y, Jin C D [1986]. Bifurcation and chaos in a liquid crystal optical bistable device. *J. Opt. Soc. Am.*, **B3**, 231.

244. Zhang H J, Dai J H, Wang P Y, Jin C D, Hao B L [1987]. Analytical study of a bimodal map related to optical bistability. *Commun. Theor. Phys.*, **8**, 281.

245. Zhang H J, Dai J H, Wang P Y, Zhang F L, Xu G, Yang S P [1988]. Chaos in liquid crystal optical bistability. in Hao [B1988], 46.

246. Zhao H, Zheng W M, Gu Y [1992]. Determination of partition lines from dynamical foliations for the Hénon map. *Commun. Theor. Phys.*, **17**, 263.

247. Zhao H, Zheng W M [1993]. Symbolic analysis of the Hénon map. *Commun. Theor. Phys.*, **19**, 21.

248. Zheng W M [1986a]. Derivation of the spectrum for the standard mapping: a simple renormalization group procedure. *Phys. Rev.*, **A33**, 2850.

249. Zheng W M [1986b]. Simple renormalization procedure for quasiperiodic maps. *Phys. Rev.*, **A34**, 2336.

250. Zheng W M [1989a]. Generalized composition law for symbolic itineraries. *J. Phys.*, **A22**, 3307.

251. Zheng W M [1989b]. The W-sequence for circle maps and misbehaved itineraries. *J. Phys.*, **A22**, 3647.

252. Zheng W M [1989c]. Symbolic dynamics of the gap map. *Phys. Rev.*, **A39**, 6608.

253. Zheng W M [1989d]. Construction of median itineraries without using the anti-harmonic. *Int. J. Mod. Phys.*, **B3**, 1703.

254. Zheng W M [1990a]. Applied symbolic dynamics for the Lorenz-like map. *Phys. Rev.*, **A42**, 2076.

255. Zheng W M [1990b]. Retrieval of the dimension for Feigenbaum's limiting set from low periods. *Phys. Lett.*, **A143**, 362.

256. Zheng W M [1991a]. Symbolic dynamics for the circle map. *Int. J. Mod. Phys.*, **B5**, 481.

257. Zheng W M [1991b]. Symbolic dynamics for the Lozi map. *Chaos, Solitons & Fractals*, **1**, 243.

258. Zheng W M [1992a]. Admissibility conditions for symbolic sequences of the Lozi map. *Chaos, Solitons & Fractals*, **2**, 461.

259. Zheng W M [1992b]. Symbolic dynamics for the Tél map. *Commun. Theor. Phys.*, **17**, 167.

260. Zheng W M [1994]. Kneading plane of the circle map. *Chaos, Solitons and Fractals*, **4**, 1221.

261. Zheng W M [1996]. Pairing of legs in the parameter plane of the Hénon map. *Commun. Theor. Phys.*, **25**, 55.

262. Zheng W M [1997a]. Symbolic dynamics of the stadium billiard. *Phys. Rev.*, **E56**, 1556.

263. Zheng W M [1997b]. Symbolic dynamics of the hyperbolic potential. *Phys. Rev.*, **E56**, 6317.

264. Zheng W M [1997c]. Predicting orbits of the Lorenz equations from symbolic dynamics. *Physica*, **D109**, 191.

265. Zheng W M, Hao B L [1989]. Symbolic dynamics analysis of symmetry breaking and restoration. *Int. J. Mod. Phys.*, **B3**, 1183.

266. Zheng W M, Hao B L [1990]. Applied symbolic dynamics. in Hao [B1990a], 363–459; Chinese Version: *Progr. in Phys.*, **10**, 316.

267. Zheng W M, Liu J X [1994]. Symbolic dynamics of attractor geometry for the Lozi map. *Phys. Rev.*, **E50**, 3241.

268. Zheng W M, Liu J X [1997]. Numerical study of the kneading theory of the Lorenz model. *Commun. Theor. Phys.*, **27**, 423.

269. Zheng W M, Lu L S [1991]. Boundary of chaos of the gap map. *Commun. Theor. Phys.*, **15**, 161.

Index